IMAGE TECHNOLOGY DESIGN
A Perceptual Approach

THE KLUWER INTERNATIONAL SERIES IN ENGINEERING AND COMPUTER SCIENCE

IMAGE TECHNOLOGY DESIGN

A Perceptual Approach

Jean-Bernard Martens
Eindhoven University of Technology
Eindhoven, The Netherlands

KLUWER ACADEMIC PUBLISHERS
Boston / Dordrecht / London

Distributors for North, Central and South America:
Kluwer Academic Publishers
101 Philip Drive
Assinippi Park
Norwell, Massachusetts 02061 USA
Telephone (781) 871-6600
Fax (781) 681-9045
E-Mail < kluwer@wkap.com>

Distributors for all other countries:
Kluwer Academic Publishers Group
Post Office Box 322
3300 AH Dordrecht, THE NETHERLANDS
Telephone 31 78 6576 000
Fax 31 78 6576 254
E-Mail < services@wkap.nl>

Electronic Services < http://www.wkap.nl>

Library of Congress Cataloging-in-Publication Data

Martens, Jean-Bernard
Image Technology Design: A Perceptual Approach / Jean-Bernard Martens
p.cm—(The Kluwer international series in engineering and computer science; SECS 735)
Includes bibliographical references and index.
ISBN 1-4020-7461-1 (alk. paper)
1. Image Processing. I. Title. II. Series.

Cover design is by Bernard Champoux.

Printed on acid-free paper.

Printed in the United States of America.

This book is dedicated
to Marleen

Contents

Designing Image Technology

FOREWORD

A common aspect of the systems that are discussed in this book is that they produce images at their output, and that these images are intended for human viewing. Optimal design of such imaging systems requires that a compromise is reached between technological complexity (and cost) and user appreciation. On the one hand, the image quality at the output, as perceived by the user, must be sufficient for the intended purpose. On the other hand, limitations of the visual system may be exploited to help reduce the complexity and cost of the imaging systems involved. Consequently, optimal design of image technology involves knowledge of such diverse disciplines as visual perception, physics, signal processing, mathematics, hardware and software design, etc. Since all this design-relevant knowledge is usually not present within a single person[1], designing inevitably involves extensive communication between all parties involved (physicists, engineers, vision scientists, marketing people, management, etc.). Useful communication is however only possible if there exists a common terminology and a realistic expectation about the information that the different partners can contribute to the process. This latter aspect usually requires some basic understanding of the limitations that govern the different disciplines involved. Within the technical perspective that is presented here on the design of image technology, the focus will be on the possible exchange of ideas from the fields of mathematics, signal (image) processing, visual psychophysics and psychometrics (i.e., the statistical modelling of experimental data).

[1] Obviously, the author is no exception to this rule. Therefore, those aspects of the design of image technology that are most closely related to his background, expertise and interest will inevitably receive more attention than others.

This book has been written from the point of view of an electrical engineer, interested in the display, processing and coding of images, and frequently involved in applying knowledge from visual psychophysics, experimental psychology, statistics, etc, to the design of imaging systems. More specifically, this book plans to describe some relevant insights and models that the author has acquired and has helped to develop over recent years. The main points of view of the author that have guided the development of this book can be summarized as follows:

1 The theory of linear systems (i.e., vector spaces) provides a powerful mathematical framework that can be used for communicating information between many of the above-mentioned scientific fields. Therefore, a systematic use of linear system theory is pursued throughout this book. This leads to both new perspectives on known phenomena and new applications of existing mathematical insights.

2 The bottom-up description of the visual system, available from visual psychophysics, electrophysiology and anatomy, has inspired mathematical descriptions of visual phenomena such as the detection and discrimination of simple patterns. Several implications of such models for image technology design will be discussed. It should however be realized that our understanding of the visual perception of complex patterns, such as images, is still very limited. This is partly due to the fact that many known phenomena within low-level vision cannot be modelled as yet, and partly due to the fact that sensations such as perceived image quality are not only determined by low-level vision but also by (top-down) cognitive processes such as attention, expertise, etc. This cognitive point of view will for instance lead to the insight that a one-to-one reproduction of reality does not necessarily lead to optimal quality from the point of view of the human observer.

3 Because of the limitations in our a priori knowledge, there is (and will be) a need for a posteriori testing of the overall performance of imaging systems, not only from a technical perspective, but also from the point of view of the quality impression by the end user. Part of this book is therefore devoted to discussing how psychological attributes such as image quality can be measured and described in statistical models.

4 Many applications would benefit from instrumental (or objective) measures for image quality. The measures that have been proposed up to now are however not sufficiently accurate and/or general to replace human judgement.

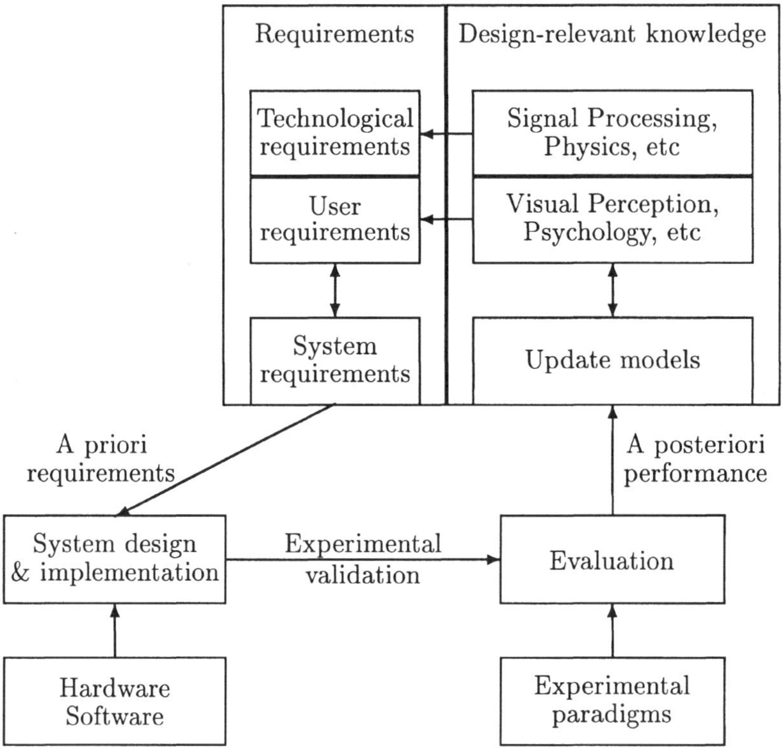

Figure 0.1. Principal components in a typical design cycle for image technology.

IMAGE TECHNOLOGY DESIGN CYCLE

Some of the aspects that can be distinguished in a typical design cycle in image technology are identified in Figure 0.1.

Probably the easiest place to start discussing the design cycle of Figure 0.1 is with the design-relevant knowledge. Often, this design-relevant knowledge will be formulated in terms of models. The complexity of these models can range from a mathematical function fitted through experimental data, as is often the case in psychophysics, to a full-fledged theory derived from basic principles, as is customary in physics and mathematics. A distinction can be made between technology-relevant and user-relevant knowledge:

1 Several enabling technologies and theories have a direct impact on how imaging systems can be characterized in technical terms. For example, physical descriptions of radiation (light), image sensors and displays, often in terms of electromagnetism and quantummechanics,

are especially useful for describing how imaging devices interact with the physical world. Signal and image processing theories and algorithms, which are in turn based on mathematical disciplines such as the theory of vector spaces, describe how signals, such as images, can be transformed, coded, etc.

2 Psychophysics studies the relationship between physical causes and psychological effects, and is relevant for understanding the impact of physical signals, such as images, on a human observer. Psychophysics can supply data and/or models on color and brightness perception, the detection and discrimination of spatially and/or temporally varying patterns, etc. Techniques from experimental psychology can also be applied to describe overall subject behavior when judging image quality.

Based on existing knowledge, a set of a priori requirements for an imaging device can be formulated. Within the limitations of available hardware and software, a system can then be designed and implemented. Since the design-relevant knowledge is typically incomplete or only partly reliable, most systems will contain parameters that need to be optimized. These parameters reflect design choices that are very difficult, or even impossible, to make, based on the a priori requirements. Experimental verification and optimization of design parameters is therefore usually required. This may take the form of an informal and interactive tuning of the parameters involved, but can also consist of a more formal psychophysical experiment. In this latter case, a priori predictions for the system performance by instrumental (or objective) image quality measures may also be compared against a posteriori determined performance. In case of serious discrepancies, an update of some of the models underlying the initial requirement analysis may be necessary and a new design cycle may have to be undertaken.

A SHORT TOUR OF THE BOOK

In chapter 1, we start by discussing how electromagnetic radiation passes from the outside world to the human visual system. This discussion also offers the opportunity to illustrate how Linear System Theory (LST) has contributed to the understanding of many aspects of electromagnetism, optics and visual perception. Subsequently, the definitions and properties from LST that are judged to be most relevant for image technology design are summarized in chapter 2. As much as possible, these theoretical concepts within LST are related to physical equivalents.

The impact of LST on visual perception and image technology is most easily illustrated for the case of color perception, as is shown in chapter 3. Important physical measures that correlate well with psychological sensations such as brightness and hue have been standardized within photometry and colorimetry and are also discussed in this chapter. Next, some basics of managing color information within technical systems are discussed in chapter 4. This chapter also introduces experimental evidence that supports the conclusion that human observers often prefer images that differ (slightly) from what is physically the most accurate reproduction. This illustrates that a top-down study of image quality is required to complement the bottom-up approach that is inspired on low-level models of the human visual system.

Most engineers involved in designing image technology are unfamiliar with the psychological techniques that are available for measuring and modelling sensations such as image quality. Therefore, these techniques are treated in detail in chapter 5.

The first four chapters of this book concentrate on the wavelength description of images, including the relationship with human color perception. Starting from chapter 6, the focus is shifted towards describing the spatial and temporal variations in images. Chapter 6 introduces some mathematical concepts such as lattices, Fourier transformations and convolutions that are required for this purpose. These concepts are derived as special cases of the more general theory of linear systems that is presented in chapter 2. Next, in chapter 7, a theory of image sampling and interpolation is presented. In order to more closely study the impact of sampling and interpolation on human perception, some aspects of the human perception of spatiotemporal patterns are treated in chapter 8. More specifically, mathematical models of the optical transfer function (OTF) of the human eye and of the contrast sensitivity function (CSF) of the overall visual system are introduced. They are subsequently applied to the optimization of image sampling structures in chapter 9.

After considering the variations in wavelength, space and time, chapter 10 concentrates on the variation in amplitude of color image components. More specifically, image quantization discusses how amplitude resolution can be reduced while preserving image quality as much as possible. Although not discussed explicitly in this book, much of the treatment on image quantization is also relevant for image coding.

Throughout the different chapters, many examples of experimental evaluations of image quality are presented. While these experiments are often required to guide parameter optimizations, they are time consuming and by necessity limited to a small number of subjects and images. Therefore, there has been a growing interest into obtaining a better understanding of the quality discrimination process that underlies the human judgement of images. It is however fair to state that the models available today are not (yet) sufficiently reliable and general to replace the human observer within the design process. Nevertheless, it is worthwhile to discuss a representative selection of the modelling approaches that have been pursued by different researchers, in order to identify the main problems and challenges faced by these models. In chapter 11, we focus on models that aim at predicting the image quality differences between pairs of images. Alternative models that perform predictions in the absence of a reference image are discussed in chapter 12.

Over the past ten years, two different undergraduate courses have been organized based on the material contained in this book. A first course has concentrated on teaching students how to set up and conduct psychophysical experiments within the context of image technology, and on how to process and interpret the experimental data. For example the comparison of bit-rate-to-image-quality characteristics of alterative image coding algorithms has been undertaken within this context. This first course has been based almost entirely on chapter 5. For this reason, this chapter is fairly self-containing. A second, image-processing related course has focused on how knowledge about visual perception can be applied in image processing and display design. It has been based mostly on the material in chapters 1, 2, 3, 6, 7, 9 and 10. The exercises in these chapters have been included to test the student's understanding of the theory in the main text.

ACKNOWLEDGEMENT

I have been very fortunate over the past fifteen years to be able to pursue my interests in image technology design in the multidisciplinary environment of the Institute for Perception Research (IPO) at the Eindhoven University of Technology, the Netherlands. I would like to thank my colleagues from the former Vision Group, especially the group leader Jacques Roufs, who have introduced me to new and unfamiliar disciplines, whose discussions have always inspired me, and whose work is at the basis of much of the material that is covered in this book. Running the risk of unduly forgetting someone, I want to quote their names, in alphabetical order: Najoua Belaïd, Frans Blommaert, Martin Boschman, Huib de Ridder, Boris Escalante Ramírez, Elena Fedorovskaya, Roelof Hamberg, Wijnand Ijsselstein, Ruud Janssen, Vishwakumara Kayargadde, Victor Koselka, Betty Majoor, Lydia Meesters, Marco Nijenhuis, Jan Pellegrino, Jacques Roufs, Gerard Schouten, Peep Stalmeier, Kees Teunissen, Paul Theelen, Bart van de Braak, Toon van Dijk, Ineke van Overveld, Arthur van Tongeren, Johan Vereijken, Joyce Westerink and Sergeij Yendrikhovskij.

I am also greatly indebted to the former director of the IPO, Herman Bouma, for providing the inspiring work environment in which I have been able to spent so many years of my research career, and to all my colleagues at the IPO who helped to create the fabulous social atmosphere that we used to call the "IPO feeling". The original motivation for writing this book is the fact that the IPO has recently ceased to exist as a research institute and that, consequently, I am increasingly involved in new activities. I nevertheless wanted to preserve some of the legacy of our former research group, and I hope that this book is able to convey some of the ideas and approaches that originated in this Vision Group of the IPO.

Last, but certainly not least, I want to thank Marleen and Linde for constantly reminding me of the truly important things in live. They provide the love and support that make my life worthwhile and without which I might never have found the necessary motivation for writing this book.

Chapter 1

VISUAL PERCEPTION AND LINEAR SYSTEM THEORY

1. INTRODUCTION

One of the goals of this book is to illustrate how many important characteristics of both the human visual system and technical systems for the acquisition and processing of images can be described using **linear system theory** (LST), i.e., the theory of linear vector spaces. Such a common mathematical framework has obvious practical advantages, since it simplifies the transfer of existing insights into the visual system to the optimization of technical imaging systems. A general mathematical perspective can also lead to new insights, since known mathematical theorems may have unforeseen implications. They may for instance lead to new algorithms for image processing or image coding and/or to new models for experimentally observed mechanisms in the visual system.

In this and following chapters, it is shown how LST has contributed to a better understanding of many aspects of visual perception. Experimental knowledge about the visual system has been gathered in such diverse disciplines as anatomy, electrophysiology and behavioral psychology (psychophysics). A true understanding of these findings implies that mathematical models can be constructed that agree with all experimental evidence. It turns out that most existing models in vision research are either linear systems or combine linear subsystems with simple nonlinearities.

Motivating the importance of LST for image technology in general, and vision research in particular, can be done in several ways. One possible approach is to start with an in-depth treatment of the mathematical theory of linear vector spaces, and to ask the reader to postpone his/her interest in the practical relevance of the different definitions and properties to a later stage. An alternative approach, that will be adopted here, is to start with a loose definition of LST and to illustrate its relevance by means of a specific example, before entering a more in-depth treatment of LST. The example that we will discuss in some detail in this chapter is color perception. This example was selected, partly because it is one of the most extensively studied and best understood aspects of visual perception, and partly because it allows to introduce some basic facts and terminology about electromagnetic radiation, light, radiometry, photometry, optics and the early visual system.

It should be mentioned that several topics that are treated in this chapter have been inspired by the excellent book "Foundations of Vision" by Brian Wandell [Wan95] that explains in much more depth the impact of LST on vision research.

2. LINEAR SYSTEMS

We loosely define a **linear system** (LS) as a system that linearly transforms a vector of input numbers, representing an input signal (or image), into a vector of output numbers, representing the corresponding output signal (or image). The number of input and output numbers may be different. A linear system can hence be characterized by a matrix multiplication, where the matrix need not be square.

We will see later that most signals of interest can indeed be represented by a discrete set of numbers. For example, images that are produced by a display device such as a computer monitor are usually stored as arrays of numbers in computer memory. These numbers encode the gray values or color values for the individual picture elements, or pixels. A transformation on this array of numbers corresponds to a transformation of the image produced on the display.

A **linear transformation** T between input and output vectors must satisfy *homogeneity*

$$T\left(a \cdot \vec{f}\right) = a \cdot T(\vec{f}), \qquad (1.1)$$

and *superposition*

$$T\left(\vec{f} + \vec{g}\right) = T(\vec{f}) + T(\vec{g}), \qquad (1.2)$$

for all vectors $\vec{f}$ and $\vec{g}$ containing possible inputs to the system, and all scalar values a. The system output $T(\vec{f})$ is a vector of output numbers. The scalar multiplication $a \cdot \vec{f}$ corresponds to a multiplication of all components of the vector $\vec{f}$ by the same factor a, while the vector addition $\vec{f} + \vec{g}$ is used to denote the addition of corresponding vector components. The vector components and scalars may be real or complex numbers.

A system that satisfies homogeneity and superposition for all possible input vectors and scalars is by definition a linear system. The linearity of an existing system can be tested experimentally, by verifying that all outputs do indeed vary linearly as a function of the inputs.

Many systems of interest will not behave as linear systems in the above strict sense. A nonlinear system that is characterized by a transformation of the form $h[T(\vec{f})]$, in which the outputs $T(\vec{f})$ of a linear front-end *uniquely* determine the nonlinear overall response of the system, can however also be studied by means of LST. More specifically, such systems can be studied by means of **matching** experiments. Indeed, if two inputs $\vec{f}$ and $\vec{g}$ give identical outputs, i.e., if $h[T(\vec{f})] = h[T(\vec{g})]$, then $T(\vec{f}) = T(\vec{g})$, provided that there exists a one-to-one relationship between the possible system outputs and the parameters controlling the outputs. Note that this implies that the dimension of the output vector $h[T(\vec{f})]$ must be greater than or equal to the dimension of the intermediate vector $T(\vec{f})$. Such systems will be referred to as **linear front-end systems** (LFS), and most of the example systems discussed in this text (amongst others, human color vision) will belong to this class.

3. ELECTROMAGNETIC RADIATION AND RADIOMETRY

This section provides a brief summary of some important physical characteristics of **electromagnetic radiation**, meanwhile defining **radiometric units**. The radiometric units radiant flux, radiant intensity, radiance, irradiance and radiant emittance that will be introduced comprise a standardized terminology for describing properties of radiation sources and reflecting surfaces. This standardization of radiometric units has been pursued by the CIE (Commission Internationale de l'Eclairage, http://www.cie.co.at/cie/) and ANSI (American National Standards Institute, http://www.ansi.org). It is intended to remedy the confusion about terminology that still exists in many scientific communities, such as the image processing community, that nevertheless rely heavily on the use of radiation and optics.

The optical transformation in a typical imaging system such as a camera will provide a suitable context for introducing the radiometric units. Amongst others, we will show that an object that is perceived by a lens system, such as a camera or the human eye, can be characterized by its radiance $L_e(\lambda, x, y, t)$ as a function of wavelength λ, position (x, y) and time t.

3.1 ELECTROMAGNETIC RADIATION

In order to enable visual sensations, a physical carrier is needed between the observed object or substance and the observer. This carrier is **visible light**, defined in Webster's dictionary [Pub91] as the electromagnetic radiation to which the organs of sight react, ranging in wavelength from about 400 nm to 770 nm (1 nm = 10^{-9} m). Next to visible light, we can also distinguish **infrared light** (near IR or IR-A: 780-1400 nm, IR-B: 1.4-3.3 μm, far IR or IR-C: 3-10 μm, with 1 μm = 10^{-6} m) and **ultraviolet light** (UV-A: 315-380 nm, UV-B: 280-315 nm, UV-C: 100-280 nm) [JHG99] that is invisible to a human observer, but that can be recorded by specialized sensors.

Newton wondered about the nature of light and performed a number of classical experiments. Using a prism, he was able to separate a single white light ray into a continuum of colored light rays. The colored rays could not be further decomposed by a similar method and were hence considered to be 'elemental' or 'basic' components of light. Newton also showed, by reconverging the colored light rays, that the above decomposition is reversible.

Later research in electromagnetism has revealed that the colored rays can be characterized by one number, their **temporal frequency** ν, i.e., the number of alternations per second in their electric and magnetic field. In free space where no charges and currents are present, a specific solution to Maxwell's equations of electromagnetism is a harmonic planar wave that propagates linearly in space and time [HS93, JHG99]. The electric and magnetic field of such a harmonic planar wave are orthogonal to each other and to the direction of propagation and vary sinusoidally with temporal frequency ν and **spatial frequency** ν/v, where $v = c/n$ is the **propagation velocity** in the medium, and c is the speed of light in vacuum[1]. The **refraction index** n of a medium is the ratio of the speed of light c in a vacuum to the speed of light v in the medium. The end points of the electric and magnetic field vectors of a harmonic planar wave trace out an ellips in the plane orthogonal to the propagation direction (i.e., the radiation is elliptically polarized[2]). More general electromagnetic planar waves can be considered as the superposition of harmonic planar waves with different frequencies. The effect of this superposition is that the electric and magnetic field vectors of these planar waves behave in a much more irregular way than in the case of

[1]The speed of light in vacuum is $c = 2.9979\ 10^8$ m/s.

[2]The special cases of a flat ellipse (or line) and a circle are referred to as linear and circular polarization, respectively. An elliptically polarized wave can be regarded as a sum of two linearly polarized waves with a constant phase difference [HS93].

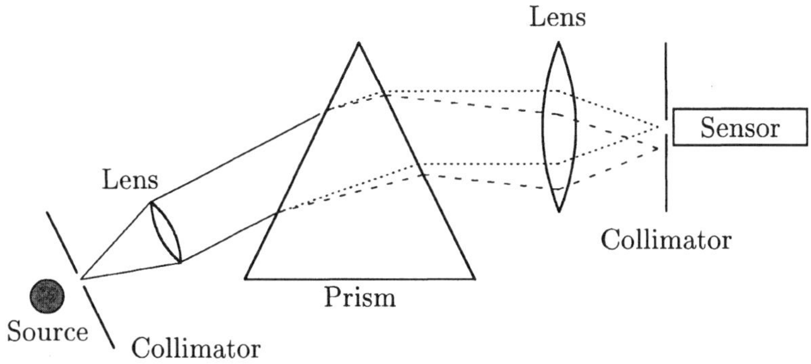

Figure 1.1. The glass prism in the spectroradiometer separates the incoming light into different wavelengths, because the refraction index of the glass is wavelength-dependent. By moving the recording device, i.e., the lens-collimator-sensor combination, the energy variation across wavelength can be measured.

the harmonic planar wave, i.e., the radiation can be only partially polarized or even completely unpolarized [HS93]. In **geometrical optics**, the electromagnetic waves are approximated locally by **light rays**, i.e., planar waves of zero (i.e., very small) cross section. This approximation neglects **diffraction** effects, i.e., the bending of light waves around objects in their path.

According to *Snell's law*, the angles of incidence θ_1 and refraction θ_2 of a light ray, measured relative to the surface normal, at a transition between two media, with respective refraction indices equal to n_1 and n_2, satisfy

$$\frac{\sin\theta_1}{\sin\theta_2} = \frac{n_2}{n_1} \quad \text{or} \quad n_1 \cdot \sin\theta_1 = n_2 \cdot \sin\theta_2. \tag{1.3}$$

The refraction angle within a glass prism, such as the one used in Newton's experiments, varies with the frequency ν because the refraction index of glass $n(\nu)$ is frequency-dependent [JHG99]. Consequently, light rays with different frequencies can be geometrically separated by the prism. A modern version of Newton's optical system is called a **spectroradiometer**, and is shown schematically in Figure 1.1. The optical part of the spectroradiometer behaves as a linear system, since Maxwell's equations of electromagnetism are linear in case of homogeneous media.

According to a more modern quantummechanical interpretation, electromagnetic radiation consists of a large numbers of **photons**, i.e., light quanta that have energy, that is the capacity to do work. It has been established that the energy E of a photon, expressed in J(oule), is proportional to its temporal frequency ν, i.e., $E = h \cdot \nu$, where h denotes Planck's constant[3]. The wavelength and particle interpretations of electromagnetic radiation can hence be related. The energy of a photon is often expressed in **electron volts**, where 1 eV (=1.602 10^{-19} J) is the energy of an electron after acceleration (from rest) in an electric field with a potential difference of one volt. This unit is especially useful for determining the potential interaction of electromagnetic radiation with a semiconductor sensor. Such a sensor can usually be characterized by its **bandgap energy** E_g in eV, which is the minimum photon energy required to lift electrons from the valence band to the conduction band. These electrons in the conduction band create a current through the sensor device that can be measured externally. As a rule of thumb, a semiconductor detector material is sensitive to radiation with energy $E = h \cdot \nu > E_g$. Frequently used semiconductor detector materials for imaging are indium antimonide (InSb, with $E_g = 0.18$ eV, i.e., sensitive to wavelengths $\lambda < \lambda_g = 6.9$ μm) and silicium (Si, $E_g = 1.1$ eV, i.e., sensitive to wavelengths $\lambda < \lambda_g = 1.1$ μm).

Although temporal frequency ν and number of photons are the natural units for expressing the properties of electromagnetic radiation, others units are often preferred, partly for historical reasons. Spatial **wavelength** $\lambda = c/\nu$ is often used instead of temporal frequency ν, while **radiation current** or **radiant flux** $\Phi_e(\lambda)$, expressed in W(atts) per unit wavelength, is usually preferred over number of photons

$$\Phi_p(\lambda) = \frac{\Phi_e(\lambda)}{h\,c/\lambda} = \frac{\lambda \cdot \Phi_e(\lambda)}{h\,c} \tag{1.4}$$

as a function of wavelength. The energy emitted, per unit interval of time, between wavelengths λ and $\lambda + d\lambda$ is equal to $\Phi_e(\lambda) \cdot d\lambda$. The total flux Φ_e is obtained by integrating $\Phi_e(\lambda)$ over all wavelengths.

3.2 OPTICS

The geometry of a typical viewing situation is depicted in Figure 1.2 [Hor86]. The camera or eye focuses on an object with the aid of a lens. The radiation that originates in a small area dS, at distance v, on the surface of the object is transferred by the lens to the area dS', at

[3]Planck's constant is $h = 6.626\ 10^{-34}$ J s.

distance b, on the camera target[4] (or eye retina[5]). The object and target distances from the lens are related by the familiar *lensmaker equation* from geometrical optics, i.e,

$$\frac{1}{f} = \frac{1}{v} + \frac{1}{b}, \tag{1.5}$$

where f denotes the **focal length**, or **focal distance**, of the lens[6]. An often-used alternative characterization of the focal length f is the **optical power**, defined as

$$d \text{ (in diopters)} = 1/f \text{ (in meters)}. \tag{1.6}$$

The focal length f of a combined lens system consisting of two lenses with respective focal lengths f_1 and f_2, at a distance a apart, is approximately equal to

$$\frac{1}{f} = \frac{1}{f_1} + \frac{1}{f_2} - \frac{a}{f_1 f_2}. \tag{1.7}$$

If the distance a is much smaller than the focal lengths f_1 and f_2, then optical powers can simply be added [JHG99]. A variable focus lens system can be created by varying the distance a between two lenses with fixed focal lengths.

A lower limit on the distance v between the object and the lens is imposed by the ability to focus properly. A noticeable difference exists between camera and eye in the way this focusing is accomplished. In the case of a camera, the focal distance f of the lens is often fixed, and an object at a distance v is brought into focus by changing the distance b between lens and target. The minimum obtainable object distance v is larger than the focal distance f, and limited by the maximum admissible value for b. In human vision, on the contrary, the distance b is fixed at approximately 17 mm, but the focal distance f can be varied by mechanically deforming the eye lens. This ability of the eye is

---

[4]In the case of a camera, the target is usually a planar surface, while in the case of the eye, the target is a spherical surface (i.e., the retina).

[5]The retina is the membrane which forms the inner lining of the back wall of the vertebrate eye, constituted of cells which respond to the stimulus of light, and sending nervous impulses to the brain through the optic nerve [Pub91].

[6]The lensmaker equation is derived under the assumption that all objects make small angles ρ with the optical axis (i.e. $\sin\rho \approx \rho$). It has to be generalized to $1/f = n_v/v + n_b/b$ if the refractive indices of the media at the object and detector side of the lens are n_v and n_b, respectively. The lensmaker equation cited in the text hence only holds if the lens is used in air, in which case $n_v = n_b = 1$. In case of the human eye, the index of refraction of the eye fluid is $n_b \approx 1.33$ and the physical distance between the cornea and the retina is 22.5 mm. The standard lens equation can still be used by adopting $b = 22.5/n_b = 17$ mm.

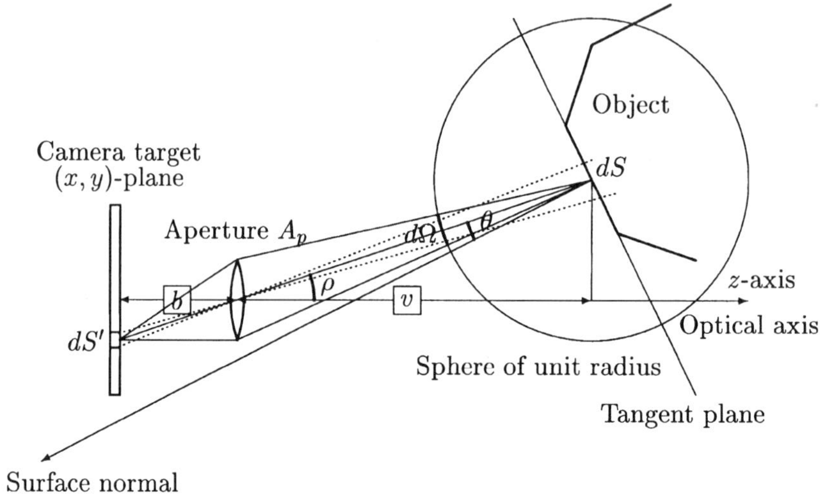

Figure 1.2. Geometry of a typical viewing situation. The area dS on the object surface is mapped by the lens with aperture A_p onto the area dS' on the camera target. The surface patch dS makes an angle ρ with the optical axis. The solid angle of the lens aperture A_p is $d\Omega$ when observed from the surface patch dS. The angle between the direction of the lens and the surface normal is θ.

called **accommodation**. For the human eye, the total optical power is obtained by combining the fixed optical power of the cornea[7] of approximately 43 diopters with the variable optical power of the deformable lens (between 16 and 26 diopters). The total optical power hence varies between $d = 58.8$ diopters (or $f = 17$ mm) for the unaccommodated eye to $d = 69$ diopters (or $f = 14.5$ mm) for the fully accommodated eye that focuses on nearby objects (up to a distance of 10 cm). The reader can easily verify experimentally that objects within a distance of 10 cm from the eye cannot be brought into focus. A simplified model of the unaccommodated eye is introduced in exercise 1.

A lens system can only focus at one object distance v at a time. Points at a distance $v' \neq v$ are focused at a distance $b' \neq b$ from the lens. Using the geometrical optics of Figure 1.2, we can easily derive that this implies that points at distance v' on the optical axis are mapped into spots with

[7]The cornea is the transparent portion of the external covering of the curved eyeball lying over the iris and pupil [Pub91].

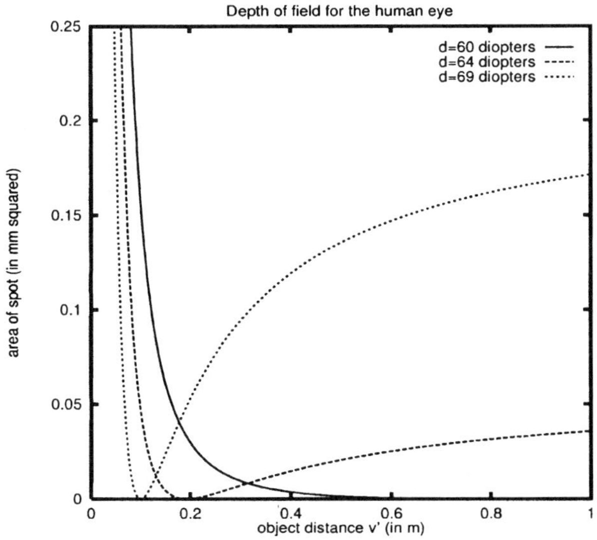

Figure 1.3. The area $A_s(v')$, in mm^2, of a spot on the retina that is the image of a point source at distance v' is plotted as a function of v', for three different values of the optical power (60, 64 and 69 diopters). The assumed pupil diameter is 3 mm.

an area equal to

$$A_s(v') = A_p \cdot \left(\frac{b' - b}{b'}\right)^2 = A_p\, f^2 \cdot \left[\frac{v' - v}{(v - f)\, v'}\right]^2 \tag{1.8}$$

on the target [Hor86], where A_p denotes the area of the lens aperture. The **depth of field** is the range of distances v' for which this area $A_s(v')$ of the spot is smaller than a preset threshold.

In Figure 1.3, we have plotted the area $A_s(v')$ of the spot on the human retina as a function of the object distance v' for three possible values of the optical power $d = 1/f$. The distance v at which the lens focuses is determined by the optical power, since the target distance is fixed at $b = 17$ mm in case of the human eye. We have assumed a pupil[8] diameter of $p = 3$ mm, and an circular aperture with area $A_p = (\pi/4)\,p^2$ (in mm^2). The curve for $d = 60$ diopters illustrates that accommodation of the eye is only needed for objects that are within 0.5 m of the observer. Focusing on such nearby objects results in a limited depth of field. The

[8]The pupil is the aperture in the iris of the eye, contracted or dilated by the muscles of the iris to control the amount of light entering the eye [Pub91]. In young people, the pupil diameter can vary between 2 and 8 mm. The range of pupil diameters decreases with age.

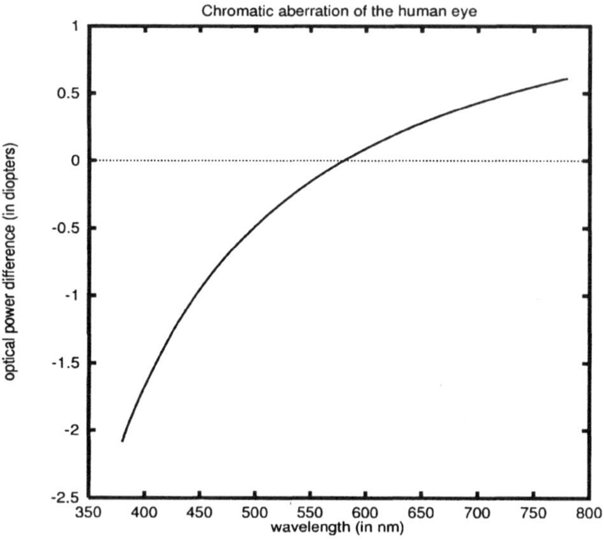

Figure 1.4. Chromatic aberration of the human eye. The curve shows the optical power that must be added to the human eye in order to bring the different wavelengths into a common focus with a light with wavelength $\lambda = 578$ nm.

depth of field furthermore decreases with decreasing focal distance, or increasing optical power.

In *far-field imaging*, which is the most frequently occurring situation in standard photography, an object at a distance $v \gg f$ is observed by a camera with a variable focal distance f (and variable target distance b). In this case, we obtain that

$$A_s(v') \approx A_p\, f^2 \cdot \left[\frac{v' - v}{v\, v'}\right]^2, \tag{1.9}$$

so that the depth of field decreases with increasing focal distance f. Tele-lenses with large focal distance f are indeed observed to have limited depth of field.

A similar blurring effect, called **chromatic aberration**, occurs across wavelengths. Since the focal length f of a lens is wavelength-dependent, not all wavelengths can be brought into focus simultaneously. The human eye has very strong chromatic aberration, as is demonstrated in Figure 1.4, where the dependence of the optical power on wavelength is shown [MW94, Wan95]. The optical power varies over 2 diopters for wavelengths in the range from 400 nm to 750 nm. The chromatic aberration is especially strong at short wavelengths. To demonstrate the

Figure 1.5. Simulated chromatic aberration showing an image at 440 nm (left) and at 580 nm (right) when the human eye focuses at 578 nm.

bad optical quality of the human eye, Figure 1.5 shows how an image that is focused at 578 nm would look at 440 nm. A linear model of the chromatic aberration in the human eye [MW94] was used to perform the simulation. This linear model is much more complex than the geometrical model discussed above, because it also takes into account wave properties of the radiation, such as the diffraction caused by the pupil. It will be described in more detail in chapter 7. The parameters of the simulation model were adjusted to correspond to the case that an image of size 6 cm $\times 6$ cm is viewed from a distance of 50 cm with an eye with a pupil diameter of $p = 3$ mm.

3.3 RADIOMETRY FOR A LENS SYSTEM

We now derive radiometric units for the above lens system by following the radiation on its path from the object to the camera target (or retina).

The radiant flux $\Phi_e(\lambda)$ at wavelength λ emitted by the patch dS in a given direction (in our case, the direction of the lens), per unit solid angle, is called the **radiation strength** or **radiant intensity**

$$I_e(\lambda) = \frac{\Phi_e(\lambda)}{d\Omega}, \tag{1.10}$$

expressed in W/sr (per unit wavelength), where sr denotes steradian[9]. We assume that the radiant flux is uniformly distributed across the solid

[9]The solid angle (in steradian, sr) is the angular spread of the vertex of a cone, expressed as the area of its intercept on the surface of the sphere of unit radius described about the vertex,

angle $d\Omega$. This assumption is likely to be valid if $d\Omega$ is small, i.e., if the object is viewed from a sufficiently large distance (i.e., if it is in the far field of the lens). Radiant intensity is mostly used to characterize point sources. A **point source** that radiates uniformly in all directions has intensity

$$I_{e,s}(\lambda) = \frac{\Phi_{e,s}(\lambda)}{4\pi\, s^2} \tag{1.11}$$

when observed from a distance s, where $\Phi_{e,s}(\lambda)$ denotes the total radiant flux emitted by the source at wavelength λ. This is called the **inverse square law of radiation.**

In case of the viewing situation of Figure 1.2, the solid angle subtended by the aperture of the lens, as observed from the object patch dS, is

$$d\Omega = \frac{A_p \cos\rho}{(v/\cos\rho)^2} = \frac{A_p}{v^2}\cos^3\rho, \tag{1.12}$$

because the apparent area of the lens is $A_p\ \cos\rho$, while the distance between the lens and the object patch is $v/\cos\rho$. The normal to the aperture makes an angle ρ with the viewing direction that connects the aperture to the surface patch. An object with **apparent area** $A_p \cdot \cos\rho$, positioned orthogonal to this viewing direction, will look equally large as the aperture with area A_p that is observed under an angle ρ.

The intensity of a radiating object in a specific direction, per unit of apparent area dS_{eff}, is called the **radiance**

$$L_e(\lambda) = \frac{I_e(\lambda)}{dS_{eff}} = \frac{I_e(\lambda)}{dS\ \cos\theta} = \frac{\Phi_e(\lambda)}{d\Omega\, dS\ \cos\theta}, \tag{1.13}$$

and is expressed in W/m^2·sr (per unit wavelength). Because the normal to the surface dS makes an angle θ with the camera direction, we again obtain that the apparent area dS_{eff}, as observed from the camera, is smaller than the actual area dS, i.e., $dS_{eff} = dS \cdot \cos\theta$.

Note that radiant intensity and radiance not only involve properties of the radiating source (i.e., the flux $\Phi_e(\lambda)$ and the surface area dS), but also properties of the measuring device (i.e., the solid angle $d\Omega$ spanned

---

the total solid angle subtended by such a sphere being 4π. Unit solid angle subtends an area of r^2 on a sphere of radius r [Pub91]. This definition of solid angle in three-dimensional (3-D) space is similar to the definition of planar angle in a two-dimensional (2-D) plane. Indeed, the planar angle (in radian, rad) is the angular spread of the vertex of a triangle, expressed as the length of its intercept on the circumference of the circle of unit radius described about the vertex, the total planar angle subtended by such a circle being 2π. Unit planar angle subtends a length of r on a circle of radius r [Pub91].

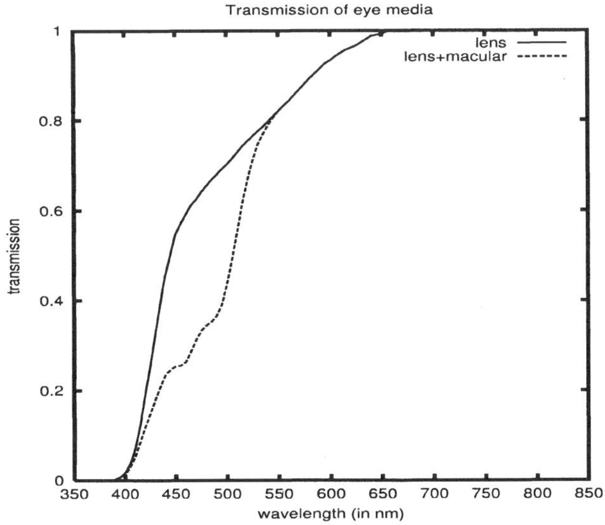

Figure 1.6. The upper curve shown the typical radiation transmission of the lens as a function of wavelength, while the lower curve combines the effects of absorption by the lens and absorption by the macular pigment in the retina.

by the lens aperture and the viewing direction θ with respect to the surface normal that accounts for the foreshortening of the surface dS into $dS_{eff} = dS \cdot \cos\theta$). We now derive why radiance is a suitable unit for characterizing extended radiation sources such as displays.

If we apply the above definitions to the viewing situation of Figure 1.2, we obtain that the radiant flux through the aperture of the lens can be expressed as follows

$$\Phi_e(\lambda) = L_e(\lambda) \cdot (dS \ \cos\theta) \cdot \frac{A_p \ \cos^3\rho}{v^2}. \tag{1.14}$$

in terms of the radiance $L_e(\lambda)$.

In the case of the eye, the radiant flux is attenuated by a factor of $\tau(\lambda)$ in the eye medium. The two major structures in the human eye that absorb radiation at visible wavelengths are the yellow pigment in the lens and the macular pigment[10] in the retina. In Figure 1.6 we plot a typical characteristic for lens absorption only, as well as for a combination of lens absorption and absorption by the macular pigment.

[10]The macular pigment protects the photoreceptors in the central part of the retina against possible harmful effects of intense radiation at short wavelengths.

Both characteristics are relevant for vision, since the macular pigment occurs only in the central part of the retina. The combined characteristic is hence mostly relevant for observations that are close to the optical axis, while the lens-only characteristic is relevant for peripheral vision. The densities of lens and macular pigment not only vary with eccentricity but also across individuals and with age[11] [Wea61, WS82], so that the transmission in the eye is usually modelled as

$$D(\lambda) = \log_{10} \tau(\lambda) = -\left[k_{lens} \cdot D_{lens}(\lambda) + k_{mac} \cdot D_{mac}(\lambda)\right], \qquad (1.15)$$

where $D_{lens}(\lambda)$ and $D_{mac}(\lambda)$ are optical densities for lens and macular pigments, respectively. The values k_{lens} and k_{mac} depend on eccentricity and can be adjusted to fit data from individual observers or groups of observers. Recent measurement data [SS99] (see http://www.cvrl.org) have been used to generate the plots of these optical density functions in Figure 1.7. The plots in Figure 1.6 and Figure 1.7 correspond to $k_{lens} = k_{mac} = 1$, which are typical values for an object on the optical axis of size 2 degrees of **visual angle**. Note that it is common practice in vision research to express the size of observed objects in visual angles. An object of size a, positioned orthogonal to the optical axis and viewed from a distance d, has a visual angle of $\theta = 2 \cdot \arctan(a/(2\,d)) \approx a/d$ radians or $\theta \cdot 180/\pi$ degrees. One **arcmin** is 1/60-th of a degree in visual angle[12].

For a monochrome camera, we can often assume that $\tau(\lambda)$ is constant and close to 1 for visual wavelengths. A filter with $\tau(\lambda)$ close to zero for wavelengths λ below 400 nm (ultraviolet filter) and/or above 750 nm (infrared filter) is sometimes used to avoid that the camera reacts to wavelengths outside the visible spectrum. In color cameras, the incoming light is split into three parts and directed towards three different camera targets. A different optical filter $\tau(\lambda)$ is positioned in each optical path in order to create three different images that are sensitive to long, medium and short wavelengths, respectively.

The (attenuated) radiant flux, per unit area of the radiated surface dS', is called the **irradiance**

$$E_e(\lambda) = \frac{\tau(\lambda)\,\Phi_e(\lambda)}{dS'}, \qquad (1.16)$$

expressed in W/m^2 (per unit wavelength). Irradiance is mostly used to characterize surfaces (such as sensors) that are not themselves radiation

[11] Absorbed short-wavelength radiation for instance makes the lens more yellow with age.

[12] For instance, the moon has a diameter of 3476 km, while the distance between earth and moon is 384400 km. This implies that a full moon spans 31 arcmin of visual angle for an observer on earth.

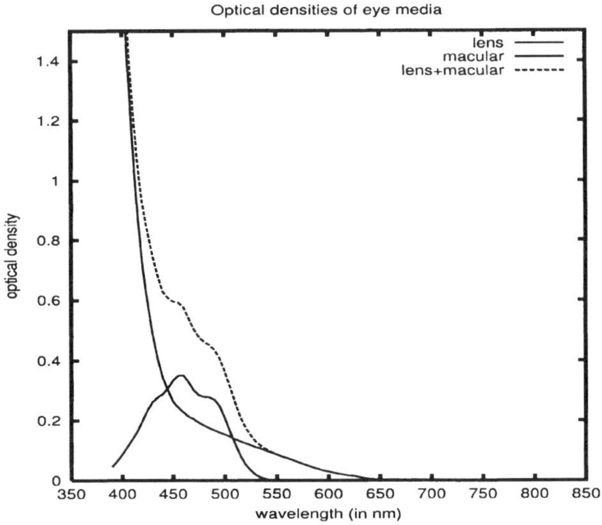

Figure 1.7. Optical densities of the lens pigment, macular pigment and combined lens+macular pigment as a function of wavelength. The curves are for a typical observer and light entering along the optical axis.

sources. For the small irradiated surface dS' in the camera target, we obtain that the irradiance

$$E_e(\lambda) = L_e(\lambda)\,\frac{dS}{dS'}\,\frac{\tau(\lambda)\,A_p}{v^2}\,\cos\theta\cos^3\rho \tag{1.17}$$

depends on the **area magnification factor** dS'/dS. A simple geometric argument can be used to determine this factor.

Rays passing through the center of the lens are not deflected (this is called **perspective projection** [Fau95]). As a result, the solid angle of the cone of rays leading to the patch on the object is equal to the solid angle of the cone of rays leading to the corresponding patch on the camera target. These cones are indicated by dotted lines in Figure 1.2. The apparent area of the object patch, as seen from the center of the lens, is $dS \cdot \cos\theta$, while the distance from the center of the lens is $v/\cos\rho$. Thus the solid angle subtended by this patch is $(dS \cdot \cos\theta)/(v/\cos\rho)^2$. Similarly, the solid angle of the target patch is $(dS' \cdot \cos\rho)/(b/\cos\rho)^2$. Because the two solid angles are equal, we obtain that the area magnification factor is equal to

$$\frac{dS'}{dS} = \left(\frac{b}{v}\right)^2 \frac{\cos\theta}{\cos\rho} = M_T^2\,\frac{\cos\theta}{\cos\rho}, \tag{1.18}$$

where $M_T = b/v$ is the transversal magnification factor and $M_L = M_T^2$ is the longitudinal magnification factor. Typical values for M_T are: $M_T < 0.1$ for far-field photography, between 0.1 and 1 for close-up photography, and between 1 and 50 for macrophotography.

By substituting the above expression for the area magnification factor in the expression for the irradiance, we obtain the following relationship

$$\boxed{E_e(\lambda) = L_e(\lambda)\,\frac{\tau(\lambda)\cdot A_p}{b^2}\,\cos^4\rho} \tag{1.19}$$

between object radiance and target irradiance.

We assume that the response of the photoreceptors in the image target (or eye retina) is determined by the irradiance, i.e.,

$$\begin{aligned} R_{sensor} &= h\left[\int_0^\infty R(\lambda)\,E_e(\lambda)\,d\lambda\right] \\ &= h\left[\cos^4\rho\int_0^\infty \frac{R(\lambda)\,\tau(\lambda)\,A_p}{b^2}\,L_e(\lambda)\,d\lambda\right] \end{aligned} \tag{1.20}$$

where $R(\lambda)$ is the spectral sensitivity of the photoreceptor positioned at dS', and h models the (possibly nonlinear) response of the sensor. For example, for many semiconductor devices, we expect that $R(\lambda)$ increases linearly with wavelength λ for wavelengths $\lambda < \lambda_g$, where λ_g is determined by the bandgap energy E_g of the device. This linear increase is a consequence of the fact that a photon with energy $E = h\,c/\lambda > E_g = h\,c/\lambda_g$ will typically generate only one electron, with fixed charge q, in the device, so that the transducer efficiency is proportional to $\lambda \cdot q/(h\,c)$. In case of the human eye, we will see later that the retina contains different kinds of receptors with varying spectral sensitivity $R(\lambda)$.

The radiance $L_e(\lambda)$ only involves parameters in front of the camera lens (i.e., external parameters), and is therefore often preferred for specifying the viewing conditions over the irradiance $E_e(\lambda)$, which also includes internal camera parameters. The overall spectral sensitivity $R(\lambda)\,\tau(\lambda)A_p/b^2$ combines the relevant internal parameters of the camera (or eye of the observer) into one function.

The unwanted factor $\cos^4\rho$ is called **field darkening** and is determined by the geometry of the viewing situation. For surface patches that are close to the optical axis, this factor is close to one. It gives rise to a predictable variation of the irradiance as a function of the angle ρ with the optical axis. Humans are known to be insensitive to such slow variations in the irradiance of the retina, so that field darkening is usually only noticeable in case of wide-angle lenses (such as for instance used in endoscopic surgery). In case of a camera, the effect of the field-darkening factor $\cos^4 \rho$, as well as other unwanted variations, such as variations across space in the sensor efficiency, can be corrected for by proper calibration (see section 3.4).

Radiance is the most versatile radiometric quantity, as all other quantities can be derived by integrating the radiance over solid angles or surface areas. In the remainder, we will therefore characterize visual displays such as Cathode Ray Tubes, CRTs, and Liquid Crystal Displays, LCDs, by their radiance $L_e(\lambda)$. In most applications, this radiance $L_e(\lambda)$ will vary as a function of time t and the position (x_d, y_d) on the (approximately flat) display surface, so that $L_e(\lambda, x_d, y_d, t)$ is a more complete characterization of the radiance. This radiance will be mapped into an irradiance $E_e(\lambda, x_c, y_c, t)$, where the spatial position (x_c, y_c) is measured in a coordinate system that is attached to the camera target. A geometric transformation[13] will relate the display coordinates (x_d, y_d) to the camera coordinates (x_c, y_c).

3.4 RADIOMETRY FOR REFLECTING SURFACES

For the sake of completeness, the radiometry of reflecting surfaces is also described in this section. This material is however not essential for the remainder of the discussion. Therefore, the reader who is more interested in how the visual system processes the radiation that is captured by the eye can skip the rest of the current section and continue with section 4.

The radiance of an object that is not itself a radiation source is determined by the irradiation of the object, and the fraction of this incident radiation that is reflected towards the observer. The geometry of the viewing situation, as depicted Figure 1.2, has been extended in Figure 1.8 to also include a radiation source. Let us denote the directions of the radiation source(s) and the observer, as viewed from a small area dS on the surface of the object, by spherical coordinates (θ_i, ϕ_i) and (θ, ϕ) respectively. The azimuth angles θ and θ_i are measured relative to the surface normal, while the polar angles ϕ and ϕ_i can be measured relative to any axis in the object plane, i.e., the plane tangent to the object (see Figure 1.8).

Let the flux that arrives at the area dS from the source direction (θ_i, ϕ_i) be denoted by $\Phi_{e,s}(\lambda; \theta_i, \phi_i)$. This implies that the irradiance

[13]This geometric transformation by the camera system is often approximated by a perspective projection [HS93].

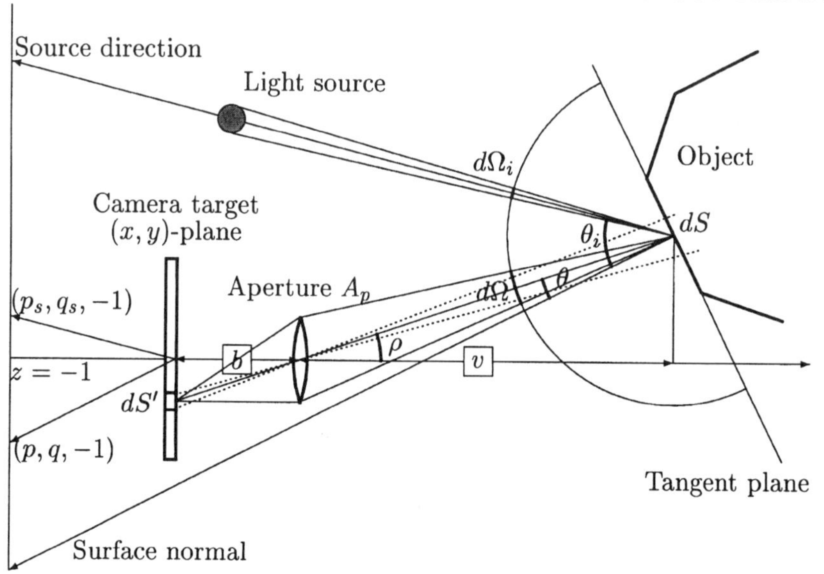

Figure 1.8. Geometry of a typical viewing situation. The area dS on the object surface is mapped by the lens with aperture A_p onto the area dS' on the camera target. The surface patch dS makes an angle ρ with the optical axis. The solid angles of the lens aperture and light source are $d\Omega = d\phi\, d\theta\, \sin\theta$ and $d\Omega_i = d\phi_i\, d\theta_i\, \sin\theta_i$, respectively, when observed from the surface patch. The azimuth angles θ and θ_i are measured relative to the surface normal, and the polar angles ϕ and ϕ_i are measured relative to any axis in the tangent plane. The surface normal and light source directions are parallel to the vectors $(p, q, -1)$ and $(p_s, q_s, -1)$, respectively.

from the source direction (θ_i, ϕ_i) is equal to

$$E_e(\lambda; \theta_i, \phi_i) = \frac{\Phi_{e,s}(\lambda; \theta_i, \phi_i)}{dS}. \tag{1.21}$$

From the point of view of the area dS, this flux can be considered to originate from a small patch, of area $d\Omega_i = d\phi_i\, d\theta_i\, \sin\theta_i$, in the direction (θ_i, ϕ_i) on the surface of the hemisphere at unit distance. In turn, viewed from this surface patch on the hemisphere, the area dS on the object surface occupies a solid angle of $dS \cdot \cos\theta_i$, so that the source intensity in the direction of the surface patch is

$$I_{e,s}(\lambda; \theta_i, \phi_i) = \frac{\Phi_{e,s}(\lambda; \theta_i, \phi_i)}{dS\, \cos\theta_i} = \frac{E_e(\lambda; \theta_i, \phi_i)}{\cos\theta_i}, \tag{1.22}$$

which can be rewritten as the well-known cosine relationship

$$E_e(\lambda; \theta_i, \phi_i) = I_{e,s}(\lambda; \theta_i, \phi_i) \cdot \cos\theta_i \tag{1.23}$$

between surface irradiance and source intensity. The radiance of the surface patch with (apparent) area $d\Omega_i$ on the hemisphere, as observed from the surface patch dS, is

$$L_{e,s}(\lambda;\theta_i,\phi_i) = \frac{I_{e,s}(\lambda;\theta_i,\phi_i)}{d\Omega_i} = \frac{E_e(\lambda;\theta_i,\phi_i)}{d\phi_i\, d\theta_i\, \sin\theta_i\, \cos\theta_i}, \tag{1.24}$$

so that the total irradiance of the surface

$$E_e(\lambda) = \int_{-\pi}^{+\pi} d\phi_i \int_0^{\pi/2} d\theta_i \cdot L_{e,s}(\lambda;\theta_i,\phi_i)\, \cos\theta_i\, \sin\theta_i, \tag{1.25}$$

which is obtained by integrating the irradiance $E_e(\lambda;\theta_i,\phi_i)$ over all source directions (θ_i,ϕ_i) in the hemisphere, can be related to the input radiance distribution. For instance, for a completely uniform radiation with a radiance $L_{e,s}(\lambda;\theta_i,\phi_i) = L_{e,u}(\lambda)$ that is independent of the input direction, we obtain that the total irradiance $E_e(\lambda) = \pi\, L_{e,u}(\lambda)$ is proportional to the input radiance. This **uniform source** model is sometimes used to approximate daylight conditions with a completely overcast sky. In computer graphics, the incoming radiation is often modelled by a linear sum of a uniform (or ambient) source with a limited number of point sources at known locations.

In the previous section, we have derived that the output of a measurement device (with a lens) can be made proportional to the radiance. This radiance of the surface area dS as observed from a surface patch of area $d\Omega = d\phi\, d\theta\, \sin\theta$ in the viewing direction (θ,ϕ) on the hemisphere, is usually expressed as

$$L_e(\lambda;\theta,\phi) = \int_{-\pi}^{+\pi} d\phi_i \int_0^{\pi/2} d\theta_i \cdot R(\lambda;\theta,\phi;\theta_i,\phi_i) \cdot L_{e,s}(\lambda;\theta_i,\phi_i)\, \cos\theta_i\, \sin\theta_i, \tag{1.26}$$

where $R(\lambda;\theta,\phi;\theta_i,\phi_i)$ denotes the **bidirectional reflectance distribution function** [Hor86, JHG99]. This function indicates how the irradiance $E_e(\lambda;\theta_i,\phi_i)$ from the source direction (θ_i,ϕ_i) contributes to the object radiance $L_e(\lambda;\theta,\phi)$ in the direction (θ,ϕ). It must satisfy the **Helmholtz reciprocity condition** $R(\lambda;\theta,\phi;\theta_i,\phi_i) = R(\lambda;\theta_i,\phi_i;\theta,\phi)$. For the uniform radiation source, the output radiance is proportional to the input radiance, i.e.,

$$L_e(\lambda;\theta,\phi) = L_{e,u}(\lambda) \int_{-\pi}^{+\pi} d\phi_i \int_0^{\pi/2} d\theta_i \cdot R(\lambda;\theta,\phi;\theta_i,\phi_i) \cdot \cos\theta_i\, \sin\theta_i, \tag{1.27}$$

where the integrated term on the right hand side is called the hemispherical reflectance. It determines the angular dependence of the output radiance $L_e(\lambda;\theta,\phi)$. If, on the other hand, the surface is irradiated with a point source with intensity $I_{e,s}(\lambda)$ in the direction (θ_i,ϕ_i), then the resulting surface radiance is

$$L_e(\lambda;\theta,\phi) = I_{e,s}(\lambda)\ \cos\theta_i \cdot R(\lambda;\theta,\phi;\theta_i,\phi_i). \tag{1.28}$$

This latter expression can be used to measure $R(\lambda;\theta,\phi;\theta_i,\phi_i)$.

The total radiant flux emitted by an object, per unit of surface area dS, is called the **radiant exitance** or **radiant emittance**, and is obtained by integrating the radiant flux in all directions, i.e.,

$$M_e(\lambda) = \frac{\Phi_e(\lambda)}{dS} = \int_{-\pi}^{+\pi} d\phi \int_0^{\pi/2} d\theta\ \cos\theta\ \sin\theta \cdot L_e(\lambda;\theta,\phi), \tag{1.29}$$

in W/m^2 (per unit wavelength).

According to the Torrance-Sparrow model for reflectance distributions [HS93, Hei94], at least two distinct processes are involved in the interaction of radiation with materials.

3.4.1 DIFFUSE REFLECTOR

The first process is scattering of photons inside the material. The scattered photons that are not absorbed leave the surface at a random orientation. For a **Lambertian** or **diffuse surface** this is the only active process. The radiance

$$L_e(\lambda;\theta,\phi) = L_e(\lambda) = \frac{1}{\pi}\, T_d(\lambda) \cdot E_e(\lambda) \tag{1.30}$$

of such a surface is by definition independent of the viewing direction, and proportional to the total irradiance $E_e(\lambda)$, so that the bidirectional reflectance distribution function

$$R_d(\lambda;\theta,\phi;\theta_i,\phi_i) = \frac{1}{\pi}\, T_d(\lambda) \tag{1.31}$$

is independent of the source and viewing directions, and only depends on the diffuse reflectance $T_d(\lambda)$. For a Lambertian surface, the radiance, radiant emittance and irradiance are all proportional, i.e.,

$$M_e(\lambda) = \pi \cdot L_e(\lambda) = T_d(\lambda) \cdot E_e(\lambda). \tag{1.32}$$

The normalization by the factor $1/\pi$ is performed to obtain that $T_d(\lambda) = 1$ for an ideal Lambertian surface that reflects all incident radiation.

Note that the radiance of a Lambertian surface depends on the source direction(s). For instance, if a diffuse reflector is irradiated by a single point source with intensity $I_{e,s}(\lambda)$ in the direction (θ_i, ϕ_i), then we obtain the well-known **Lambert cosine law**

$$L_e(\lambda) = I_{e,s}(\lambda)\ \cos\theta_i \cdot \frac{1}{\pi}\, T_d(\lambda) \tag{1.33}$$

for the radiance of the object as a function of the source direction. Hence, the radiance of a diffuse reflector is maximum if the surface normal points in the direction of the radiation source. It should be noted that the Lambertian surface model is usually only a good approximation for small angles of incidence θ_i.

An interesting special case of a Lambertian surface is the **black-body radiator**. It has been demonstrated by Planck that the radiance of this perfect absorber and emitter is completely determined by its temperature T (in degrees Kelvin), i.e.,

$$L_{e,b}(\lambda; T) = \frac{2\pi hc^2}{\lambda^5 \left[\exp\left(\frac{hc}{\lambda kT}\right) - 1\right]}, \tag{1.34}$$

where $k = 1.3805\ 10^{-23}\ J/K$ denotes Boltzmann's constant. The maximum radiance occurs at a wavelength $\lambda_{m,e}$ that is inversely proportional to the temperature, i.e.,

$$\lambda_{m,e}\, T = 2.891\ 10^{-3}\ \text{mK}. \tag{1.35}$$

(Wien's displacement law). The total emittance of such a black-body radiator is

$$M_e = \sigma\, T^4, \tag{1.36}$$

where $\sigma = 5.668\ 10^{-8}\ W/(m^2K^4)$ is Stefan-Boltzmann's constant. This implies that the absolute temperature T of a black-body radiator (such as the sun) can be derived from its radiance $L_e = M_e/\pi$.

So-called **grey-body radiators** have a radiance that is proportional to that of a black-body radiator, i.e., $L_{e,g}(\lambda; T) = \epsilon \cdot L_{e,b}(\lambda; T)$, and are characterized by their **color temperature** T. For instance, diffuse daylight can be approximated, at least within the range of visual wavelengths, by a grey-body radiator with a color temperature around $T = 6500^\circ K$.

3.4.2 SPECULAR REFLECTOR

The second process to be considered in the interaction between radiation and materials is the reflection of radiation at the material surface.

For a perfect **specular surface**, the incoming flux $\Phi_{e,s}(\lambda;\theta_i,\phi_i)$ from the direction (θ_i,ϕ_i) is attenuated to $T_s(\lambda)\cdot\Phi_{e,s}(\lambda;\theta_i,\phi_i)$ and radiated in the specular direction $(\theta=\theta_i,\phi=\phi_i+\pi)$. In practice, reflections will also occur in directions close to this specular direction, as a consequence of the roughness of the surface at the micrometer scale. This process, which is called **off-specular reflection**, is modelled by Phong shading in computer graphics [JHG99]. The maximum reflectance usually occurs when the angle between the viewing direction (θ,ϕ) and the direction of the specular reflection $(\theta_i,\phi_i+\pi)$ is zero, and falls off sharply with increasing angle. The fall-off is often modelled by a bidirectional reflectance distribution function of the form

$$R_s(\lambda;\theta,\phi;\theta_i,\phi_i)=T_s(\lambda)\,\frac{[\cos\beta(\theta,\phi;\theta_i,\phi_i)]^{n_s}}{N_s(\theta_i,\phi_i)}, \tag{1.37}$$

with

$$N_s(\theta_i,\phi_i)=\int_{-\pi}^{+\pi}d\phi\int_0^{\pi/2}d\theta\,\cos\theta\,\sin\theta\cdot[\cos\beta(\theta,\phi;\theta_i,\phi_i)]^{n_s}, \tag{1.38}$$

where

$$\cos\beta(\theta,\phi;\theta_i,\phi_i)=\cos\theta\,\cos\theta_i+\sin\theta\,\sin\theta_i\,\cos(\phi-(\phi_i+\pi)) \tag{1.39}$$

is the cosine of the angle between the viewing direction (θ,ϕ) and the optimal specular direction $(\theta_i,\phi_i+\pi)$, and $n_s=10$ is a typical value for the **specular reflection exponent**. For a point source with intensity $I_{e,s}(\lambda)$ in direction (θ_i,ϕ_i), the total radiant emittance is related to the incoming irradiance $E_{e,s}(\lambda;\theta_i,\phi_i)=I_{e,s}(\lambda)\cdot\cos\theta_i$ by

$$M_e(\lambda)=T_s(\lambda)\cdot E_{e,s}(\lambda;\theta_i,\phi_i), \tag{1.40}$$

so that the normalization by the factor $N_s(\theta_i,\phi_i)$ guarantees that the specular reflection coefficient $T_s(\lambda)$ has a value between 0 and 1.

In computer graphics, the bidirectional reflectance distribution functions of most surfaces are modelled by a linear combination of a diffuse and a specular component.

3.4.3 REFLECTANCE MAP

In many viewing situations, the source(s) and observer are at a sufficiently large distance from the object, so that the angles θ,ϕ,θ_i and ϕ_i only depend on the surface orientation. Hence, in neighboring points of the object surface, variations in these angles are entirely due to the curvature of the surface. Under such circumstances, a flat Lambertian surface will have constant radiance $L_e(\lambda)$ across space. Such a surface can for instance be used to calibrate a camera, since the photoreceptor outputs of the camera should be constant across space when a camera is directed at this surface.

Taking a reference frame attached to the observer, with the camera target as the (x, y)-plane and the optical axis as the z-axis, the surface of an object can be characterized by its distance $z = v(x, y)$ from the camera target. The surface normal lies along the direction $(p, q, -1)$ in the three-dimensional reference frame, with $p(x, y) = \partial v(x, y)/\partial x, q(x, y) = \partial v(x, y)/\partial y$. The radiance of an object is often locally approximated by an expression of the form [Hor86]

$$L_e(\lambda, x, y) = I_{e,s}(\lambda) \cdot T(\lambda, x, y) \cdot R_s\left[p(x, y), q(x, y)\right], \tag{1.41}$$

where $I_{e,s}(\lambda)$ depends on the intensity of the radiation source(s). It is usually a slowly varying function across space, which can locally be assumed to be (approximately) constant. The local spatial variations in the radiance can then be attributed to changes in the surface reflectance $T(\lambda, x, y)$ across the object, or in changes in the **reflectance map** $R_s[p(x, y), q(x, y)]$ due to curvature of the object surface.

For the special case of a diffuse reflector irradiated by a point source in the direction $(p_s, q_s, -1)$[14], the following explicit expression for the reflectance map arises

$$R_s(p, q) = \frac{1}{\pi} \cos\theta_i = \frac{1 + p_s \cdot p + q_s \cdot q}{\pi \sqrt{1 + p^2 + q^2}\, \sqrt{1 + p_s^2 + q_s^2}}. \tag{1.42}$$

In shape-from-shading algorithms [Hor86], this reflectance map is used to derive the shape of objects that can be assumed to be perfect diffuse reflectors. More specifically, if $R_s(p, q)$ can be measured for at least three known source directions (p_s, q_s), then the surface orientation (p, q) can be derived from these measurements, provided that the (ratios of the) source intensities $I_{e,s}(\lambda)$ are known.

4. COLOR PERCEPTION

Image technology includes the design of systems for capturing and displaying color images. Since color is a percept, i.e., a subjective sensation, it cannot be studied apart from human observers. We therefore present some relevant notions from **color perception**[15] and **colorimetry**[16].

---

[14] A radiation source can be either 'in front' of the viewer, in the direction $(p_s, q_s, 1)$, or 'behind' the viewer, in the direction $(p_s, q_s, -1)$.

[15] Color perception refers to the subjective sensation of electromagnetic radiation.

[16] Colorimetry describes how to objectively measure electromagnetic radiation in a way that correlates with human perception.

There exists a well-defined terminology for specifying both the *subjective* sensations of colors that are studied by color perception and the corresponding *objective* (or *instrumental*) measures that are adopted in colorimetry. This terminology, although standardized by the international committee of the CIE (Commission Internationale de l'Eclairage), seems to be fairly unknown, or at least heavily misused, by the engineering community. Such a standardized terminology is however the only way of guaranteeing a clear distinction between subjective and objective specifications of colors. It is also judged essential in communicating information between those that study the properties of the visual system i.e., the vision scientists, and those that apply the acquired insights in technical designs, i.e., the engineers.

According to CIE terminology [Hun78, Hun88, WP92], **(perceived) color** is 'the attribute of visual perception that can be described by color names: white, grey, black, yellow, orange, brown, red, green, blue, purple, etc., or by combinations of such names'. It will be noted that white, grey, and black are colors. Similarly, **hue** is defined as 'the attribute of a visual sensation according to which an area appears to be similar to one of the perceived colors red, yellow, green, blue, and purple (primary hues), or to a combination of the two (intermediate hues)'. Note that white, grey and black are excluded from the definition of hue. Hence, we can distinguish two classes of perceived colors, i.e., **achromatic (perceived) colors**, colors devoid of hue, and **chromatic (perceived) colors**, colors possessing hue.

Two distinct names for the second color dimension are needed, in order to distinguish between the two distinct modes of color perception. Color can be perceived as a property of light, such as the red of a traffic light, but also as a property of surface matter, such as the red of a fire engine. The visual system is usually able to tell the difference between *emitted* and *reflected* or *transmitted* light (also denoted by **aperture** or **unrelated** colors, and **object** or **related** colors). We can define **brightness** for both aperture and object colors as 'the attribute of a visual sensation according to which an area appears to be emitting more or less light (either by itself or from incident light)'. For object colors, the brightness depends on both the incident light and the surface reflectance. Since the observer can usually distinguish between both factors, object colors can also be characterized by their **lightness**, which is defined as 'the attribute of a visual sensation according to which a body seems to reflect (diffusely) or transmit a greater or smaller fraction of the incident light'.

Hue and brightness/lightness are only two of the three subjective color dimensions. In a chromatic perceived color, the hue may be exhibited weakly or strongly (as in pink and red, for instance). We can define **saturation** for both related and unrelated colors as 'the attribute of a visual sensation according to which the perceived color of an area appears to be more or less chromatic, judged in proportion to its brightness' (the saturation being low at relatively low and high brightness). Greys are called unsaturated and are assigned zero saturation, while monochromatic spectra, i.e., radiation sources containing energy at only one wavelength, have the highest saturation (for a given hue). Hunt [Hun78] argues that the alternative term **colorfulness** is more intuitive than the term saturation. Up to now, the term colorfulness has however not been accepted as part of the official CIE color terminology. **Perceived chroma** is only defined for nonluminous related colors as 'the attribute of a visual sensation according to which an area appears more colorful, judged in proportion to the average brightness of its surroundings'. Materials that reflect over a broad range of wavelengths are typically perceived as having little chroma. Fortunately, the triad of terms, hue, brightness/lightness and saturation/chroma (colorfulness) are all used in everyday speech, which contributes much to a better understanding of these terms.

A crucial observation in the above definitions of color percepts is that the human visual system measures and interprets the incoming light. It is interesting to know in more detail how this measurement is performed, amongst others because this will allow us to exploit the limitations of human color perception in technical imaging systems. In section 5, we will see that **brightness matching** experiments indicate that brightness is perceived by a LFS with one degree of freedom. Consequently, a single number, that can be derived by weighting the radiance of a light source with the **spectral sensitivity function of the eye**, determines the brightness impression induced by this light source. The resulting **luminance** is an example of a **photometric** unit, i.e., an objective number that correlates with the subjective sensation of brightness. By applying the same weighting by the spectral sensitivity function of the eye to the other radiometric units, they are also converted into photometric (light) units. In section 6, we describe how the experimental results from **color matching** indicate that color is perceived by a LFS with three degrees of freedom.

5. BRIGHTNESS MATCHING

5.1 AXIOMS OF ABNEY

The visual system is not equally sensitive to all wavelengths. It was internationally agreed to select the brightness impression as the measure for determining this sensitivity. By matching monochromatic spectra in brightness, the relative spectral sensitivity $V(\lambda)$ can be measured as a function of the wavelength.

Brightness matching is often performed by means of flicker photometry [WS82]. Two light sources are alternated at a sufficiently high temporal frequency (usually above 10 Hz). Because the mechanisms responsible for the perception of hue and saturation are fairly slow, brightness is the main attribute of color perception at these temporal frequencies. Alternating between light sources with unequal brightness is perceived as **flicker**. In a sensitivity measurement, the subject is instructed to adjust the energy of one light source until flicker becomes minimal, or disappears completely; the second (reference) light source usually has fixed energy. Suppose for example that the reference light source has a monochromatic spectrum with unit energy at wavelength λ_o nm, and assumed sensitivity $V(\lambda_o)$, and that brightness matching with a monochromatic spectrum at wavelength λ requires energy $E(\lambda)$, then the sensitivity at wavelength λ is equal to $V(\lambda) = V(\lambda_o)/E(\lambda)$. The experiments show that the maximum sensitivity occurs for $\lambda_m = 555$ nm, so that $V(\lambda_m) = 1$ is an obvious normalization.

For non-monochromatic source radiances, it turns out that brightness matching can be described by the axioms of Abney [Pra91], i.e.,

1. If stimulus $L_{e,1}(\lambda)$ matches $L_{e,2}(\lambda)$ in brightness, then $E \cdot L_{e,1}(\lambda)$ and $E \cdot L_{e,2}(\lambda)$ will also match for a broad range of energy values E. The spectral sensitivity is hence insensitive to the energy level.

2. If stimulus $L_{e,1}(\lambda)$ matches $L_{e,2}(\lambda)$ in brightness, and stimulus $L_{e,3}(\lambda)$ matches $L_{e,4}(\lambda)$, then the additive mixtures $L_{e,1}(\lambda) + L_{e,3}(\lambda)$ and $L_{e,2}(\lambda) + L_{e,4}(\lambda)$ will also match.

These axioms have been verified experimentally for a wide range of source radiances. They indicate that brightness matching is performed by a LFS with only one degree of freedom, and that the parameter value controlling the brightness perception is a linear weighted sum of the radiance $L_e(\lambda)$ across wavelength λ. This number

$$L = K_m \int_{380\,nm}^{780\,nm} V(\lambda) \cdot L_e(\lambda)\, d\lambda \tag{1.43}$$

is called the **luminance** of the stimulus with radiance $L_e(\lambda)$. The brightness B of a light source is then a (non-linear) function $B(L)$ of this luminance L.

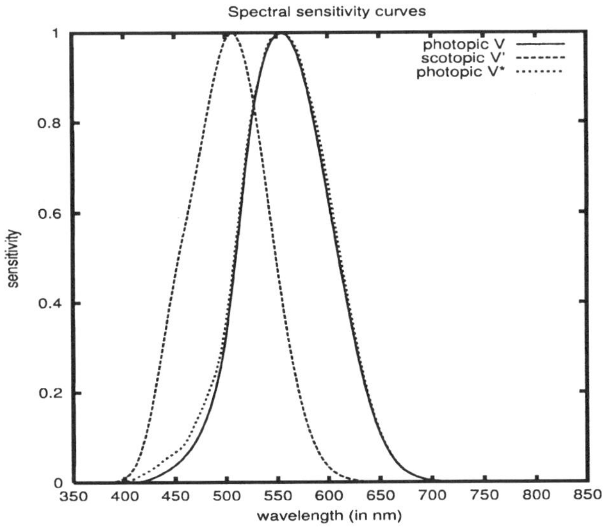

Figure 1.9. Standardized scotopic and photopic spectral sensitivity functions $V'(\lambda)$ and $V(\lambda)$ with peak sensitivities at wavelengths $\lambda'_m = 507$ and $\lambda_m = 555$ nm, respectively. Based on more recent measurements, the non-standardized curve $V^*(\lambda)$ has been proposed as an alternative for the photopic sensitivity function $V(\lambda)$.

5.2 SPECTRAL SENSITIVITY FUNCTION (SSF)

After pooling a large set of experimental data, the **spectral sensitivity function** (SSF) or **spectral luminous efficiency function** of the eye, most often denoted by $V(\lambda)$, was standardized by the CIE in 1924 [WS82]. This standardized curve is plotted as the continuous curve in Figure 1.9 and tabulated in Appendix 1.A. As the response of the human eye for radiation depends on a variety of physiological parameters, differing for individual human observers, the spectral sensitivity function corresponds only to an average normalized observer. The proportionality factor was standardized to $K_m = 683$ nit/W in 1980 [CIE83]. The **nit** is a light unit that indicates that the electromagnetic spectrum has been weighted by the eye sensitivity. Typical luminance levels for moonlight, indoor lighting and sunlight are 10^{-1}, 10^2 and 10^5 nit, respectively.

Based on more recent measurements [SS99], an alternative SSF, denoted by $V^*(\lambda)$, that deviates mostly in the short-wavelength region, has been proposed. It is also shown in Figure 1.9 and tabulated in

Appendix 1.A. Although this latter SSF provides a more accurate description of brightness perception, it has not (yet) been standardized.

All radiometric units can be weighted in a similar way as the radiance, resulting in the following corresponding light units:

1. Radiant flux $\Phi_e(\lambda)$, in W, maps to **light flux** or **luminous flux** Φ, in **lumen**, lm=cd·sr.

2. Radiant intensity $I_e(\lambda)$, in W/sr, maps to **light intensity** or **luminous intensity** I, in **candela**, cd.

3. Radiance $L_e(\lambda)$, in W/m^2·sr, maps to **luminance** L, in **nit**=cd/m^2.

4. Irradiance $E_e(\lambda)$, in W/m^2, maps to **illuminance** E, in **lux**=lm/m^2.

5. Radiant emittance $M_e(\lambda)$, in W/m^2, maps to **light emittance** or **luminous exitance** M, in lm/m^2.

The most versatile photometric quantity is luminance, since all other quantities can be derived by integrating luminance over solid angles and surface areas.

Partly due to the way in which the spectral sensitivity function was measured, i.e., by means of flicker photometry[17], the one-to-one mapping between luminance and brightness does not hold for all stimulus conditions. This is demonstrated by the *Helmholtz-Kohlrausch effect*: a colored light that contains predominantly energy at low or high wavelengths, and which energy level has been adjusted to match the luminance of a white light, will nevertheless appear to be much brighter. More specifically, a brightness match may require a difference of up to a factor of three in the luminance. For light sources that have energy at many different wavelengths, this effect is usually fairly small, so that luminance is most often a perceptually relevant attribute of the light source. The newly proposed photopic SSF $V^*(\lambda)$ with its higher sensitivity, especially at short wavelengths, should also reduce the Helmholtz-Kohlrausch effect.

A closer look at experimental data also reveals that the first axiom of Abney does not hold across all energy levels. The spectral sensitivity of the eye turns out to be different at low (< 0.01 cd/m^2), so-called **scotopic**, and high (> 10 cd/m^2), so-called **photopic**, light levels. Also, hue and saturation perception cease to exist at scotopic light levels. Therefore, next to the photopic spectral sensitivity function $V(\lambda)$, a

[17]Flicker photometry gives very reproducible results, but has the disadvantage that it suppresses the contribution of the fairly slow color channels to the brightness impression.

scotopic spectral sensitivity function $V'(\lambda)$ has also been measured and standardized by the CIE in 1951. This latter curve is shown as the dotted curve in Figure 1.9, and is also tabulated in Appendix 1.A. The peak sensitivity in scotopic viewing occurs at a shorter wavelength $\lambda'_m = 507$ nm than in the case of photopic viewing. Luminance under scotopic viewing conditions is standardized to

$$L = K'_m \int_{380\,nm}^{780\,nm} V'(\lambda) \cdot L_e(\lambda)\, d\lambda, \tag{1.44}$$

where $K'_m = 1754$ nit/W [CIE83]. Because the spectral sensitivity functions are normalized to 1 at different wavelengths λ_m and λ'_m for the photopic and scotopic case, respectively, different proportionality factors K_m and K'_m are needed to get a consistent definition of luminance across the entire energy range.

5.3 ANATOMICAL AND ELECTROPHYSIOLOGICAL EVIDENCE

The different behavior of the visual system at scotopic and photopic light levels is also supported by anatomical and electrophysiological evidence. Indeed, anatomical and electrophysiological studies of the human retina have revealed the following facts [Wan95]:

- The retina contains discrete cells, called photoreceptors, that are densely packed in a regular, approximately hexagonal, mosaic. Light falling on the retina has to cross the transparent cells in the retina before reaching the photo-sensitive substances, or photopigments, at the far ends of the photoreceptors. Henceforth, less than 10 percent of the light entering the eye is eventually absorbed by photopigments. Light that passes the photoreceptors is absorbed by the pigmented epithelium.
- In the central part of the retina, the fovea[18] centralis, all photoreceptors are of a similar type, called "cones". The distance between receptors in the fovea is approximately 0.5 arcmin. In the peripheral part of the retina, i.e., away from the optical axis, both the size and the spacing of the cone receptors grows with the distance from the optical axis (i.e., the eccentricity). The space between the cones is filled in with another type of small receptors, called "rods". The

[18]Fovea means small pit or depression. This depression in the central part of the retina is caused by the fact that the fovea, unlike other parts of the retina, contains only the layer of photoreceptors, and that the cells (neurons) responsible for transporting the responses of the photoreceptors are displaced outwards, out of the light path.

typical shape difference between rods and cones is reflected in their name.

- Several hundreds of peripheral rod cells can converge onto a single ganglion cell of the optic nerve, while cones in the fovea can have a one-to-one connection to the ganglion cells. Ganglion cells that are connected to rods can therefore react at much lower light levels than ganglion cells connected to cones. The optic nerve contains approximately $1.5 \cdot 10^6$ neurons that transport the visual information from the eye to the brain.

- Photoreceptor cells satisfy the principle of **univariance**, i.e., the same neural response can be obtained with photons of different wavelengths; the energy level required to obtain such an identical response does however vary with the wavelength. The photoreceptor cells are hence LFS with one degree of freedom.

- Light receptors can selectively absorb dyes, so that different types of receptors can potentially be visualized by chemical staining. Such staining has revealed that the fovea contains at least two types of cones. The ratio of the S-cones, i.e., the ones that respond to staining with a yellow dye, to the other cones is approximately 100:1. The average distance between S-cones is hence 10 times greater than the average distance between the other cones. The S-cones are mostly sensitive to short wavelengths, i.e, they absorb blue light with wavelengths around 450 nm. The photopigment in these S-cones was found to be different from the photopigment in the other cones. The latter L+M-cones respond to long and medium wavelengths only. The low spatial sampling density of S-cones is adequate because the radiation at short wavelengths is spatially blurred due to chromatic aberration, as was illustrated by the simulation in Figure 1.5.

- Historically, the evidence for three types of cones comes mostly from color matching experiments (see the next section). More recent experiments have provided direct evidence for the existence of two types of cones in the L+M-cone mosaic. Not surprisingly, the photopigments in the M- and L-cones have very similar spectral absorptions, with peak sensitivities at 540 and 565 nm, respectively. The ratio of the L-cones over the M-cones varies across species and individuals, with an average ratio L:M of 2 for humans [CN89, GC98].

Overall, there are many more rods in the retina than cones. Therefore, the photopigment rhodopsin, that is responsible for the light sensitivity of the rods, can be extracted much more easily than the photopigment(s)

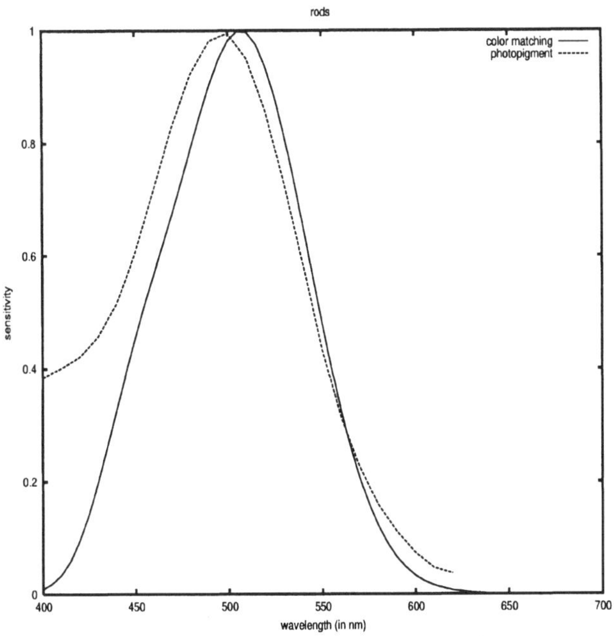

Figure 1.10. The spectral sensitivity function for rods. The drawn curve is the scotopic spectral sensitivity function $V'(\lambda)$ and is obtained through brightness matching. It contains the contribution of the spectral attenuation in the optical path. The dotted curve represents the photopigment (rhodopsin) absorption spectrum $R(\lambda)$, and differs mostly from $V'(\lambda)$ at short wavelengths. All curves are normalized such that the maximum response is one.

in the cones. The scotopic sensitivity $V'(\lambda)$ and the absorption spectrum $R(\lambda)$ of rhodopsin differ mostly at short wavelengths, as is illustrated in Figure 1.10. This difference can however be explained by the absorption $\tau(\lambda)$ of the optical media (lens and macular pigment)[Wan95]. Hence, a one-to-one correspondence exists between the psychophysical data from brightness matching at scotopic light levels and physiological data from analyzing rod receptors, i.e., $V'(\lambda) \propto R(\lambda) \cdot \tau(\lambda)$. We can hence conclude that the scotopic sensitivity function reflects the operation of the rods, while the photopic sensitivity function must reflect the operation of the other photoreceptors, i.e., the cones. At intermediate, so-called **mesopic**, light levels, both rods and cones are operational.

6. COLOR MATCHING

6.1 AXIOMS OF GRASSMAN

It is well-known that objects with different spectral radiance distributions can give rise to an identical color sensation (hue, saturation and brightness/lightness). Such radiometric functions are termed **metamere.** Metamerism is essential to many applications, including color photography, color television and color printing. For example, the color patches on a photographic slide can be made (approximately) metameric to the colors of the objects that were photographed. Almost surely, the slide colors and the object colors have very different spectra, i.e., energy as a function of wavelength, but they evoke (almost) the same color sensations. Without the metamerism of color mixture, systems for color photography, color television and color printing would be much more complex.

The classical approach to the study of metamerism is through color matching. Grassman has developed a set of axioms for additive color matching that were shown to hold over a wide range of conditions [Pra91]:

1. Any given color stimulus $L_e(\lambda)$ can be matched to an additive mixture of no more than three fixed color sources $L_{e,1}(\lambda)$, $L_{e,2}(\lambda)$ and $L_{e,3}(\lambda)$, called **primaries**, in one of the following ways

   (a) the stimulus $L_e(\lambda)$ matches $G_1 \cdot L_{e,1}(\lambda) + G_2 \cdot L_{e,2}(\lambda) + G_3 \cdot L_{e,3}(\lambda)$, or

   (b) the mixture of the stimulus and one of the primaries, for instance $L_e(\lambda) + G_1 \cdot L_{e,1}(\lambda)$, can be matched to a mixture of the remaining primaries, i.e., $G_2 \cdot L_{e,2}(\lambda) + G_3 \cdot L_{e,3}(\lambda)$, or

   (b) the mixture of the stimulus and two of the primaries, for instance $L_e(\lambda) + G_1 \cdot L_{e,1}(\lambda) + G_2 \cdot L_{e,2}(\lambda)$, can be matched to the remaining primary, i.e., $G_3 \cdot L_{e,3}(\lambda)$,

   for some values of the weights G_1, G_2 and G_3. These weights are called the **tristimulus values** of $L_e(\lambda)$ with respect to the primaries $L_{e,1}(\lambda)$, $L_{e,2}(\lambda)$ and $L_{e,3}(\lambda)$. If a primary is added to the stimulus in order to accomplish a match, then the corresponding weight is given a negative sign.

2. A color match at one energy level holds over a wide range of levels.

3. If color stimulus $L_{e,1}(\lambda)$ matches stimulus $L_{e,2}(\lambda)$, and color stimulus $L_{e,3}(\lambda)$ matches stimulus $L_{e,4}(\lambda)$, then the additive mixtures $L_{e,1}(\lambda) + L_{e,3}(\lambda)$ and $L_{e,2}(\lambda) + L_{e,4}(\lambda)$ will also match.

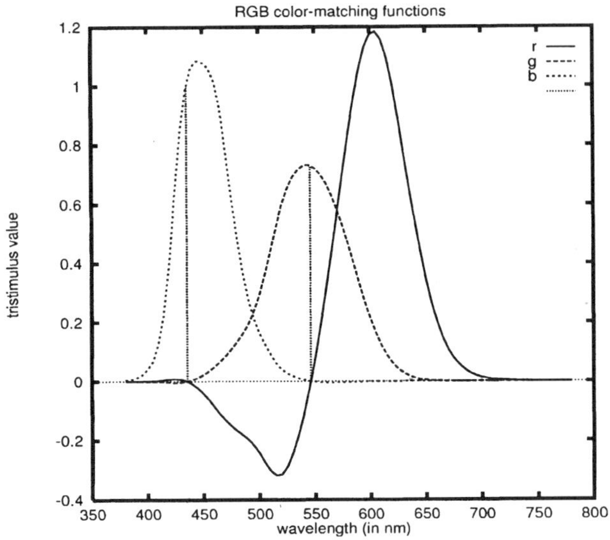

Figure 1.11. 1931 CIE color-matching functions $\bar{r}(\lambda)$, $\bar{g}(\lambda)$ and $\bar{b}(\lambda)$ for a 2 degree field and monochromatic primaries with wavelength $\lambda_r = 700$ nm, $\lambda_g = 546.1$ nm and $\lambda_b = 435.8$ nm.

4. If color stimulus $L_{e,1}(\lambda)$ matches stimulus $L_{e,2}(\lambda)$, and the color mixture $L_{e,1}(\lambda) + L_{e,3}(\lambda)$ matches the mixture $L_{e,2}(\lambda) + L_{e,4}(\lambda)$, then $L_{e,3}(\lambda)$ and $L_{e,4}(\lambda)$ will also match.

6.2 COLOR MATCHING FUNCTION (CMF)

The CIE has standardized a reference system [CIE71] with monochromatic primaries at wavelengths $\lambda_r = 700$ nm (red), $\lambda_g = 546.1$ nm (green) and $\lambda_b = 435.8$ nm (blue) in 1931. The relative energies of these primaries were chosen such that the tristimulus values for an **equal energy source**, i.e., a source with constant radiance for all wavelengths in the visual spectrum, are all equal. It was verified experimentally that this is realized if the radiances of the primaries are in the ratios $E_r = 72.0962 : E_g = 1.3791 : E_b = 1$. The tristimulus values $\bar{r}(\lambda)$, $\bar{g}(\lambda)$ and $\bar{b}(\lambda)$ for monochromatic spectra with unit radiance and variable wavelength λ were obtained by pooling the results from a large number of experiments, most noticeably those performed by Wright [Wri28] and Guild [Gui31], and standardized in the CIE 1931 observer. The results are plotted in Figure 1.11 [Hun88, WS82]. As expected, the tristimulus values for the monochromatic spectra at wavelengths λ_r, λ_g and λ_b

are equal to $(1/E_r, 0, 0)$, $(0, 1/E_g, 0)$ and $(0, 0, 1/E_b)$, respectively. The negative values for $\overline{r}(\lambda)$ for wavelengths λ close to 500 nm for instance indicate that the red primary must be added to the test light in order to obtain a color match with a mixture of the green and blue primaries.

If color matching is performed with two different sets of primaries, then different **color-matching functions** (CMFs) are obtained. It will be shown in chapter 3 that the CMFs for different sets of primaries are related by a linear transformation. Hence, color matching can be characterized by the tristimulus functions of Figure 1.11, but also by three arbitrary linear combinations of these curves, provided of course that the three linear combinations remain independent, i.e., one of the three cannot be written as a linear combination of the other two. The CIE has used this flexibility to also standardize a second set of CMFs

$$\begin{bmatrix} \overline{x}(\lambda) \\ \overline{y}(\lambda) \\ \overline{z}(\lambda) \end{bmatrix} = \begin{bmatrix} 0.49 & 0.31 & 0.2 \\ 0.17697 & 0.81240 & 0.01063 \\ 0.0 & 0.01 & 0.99 \end{bmatrix} \cdot \begin{bmatrix} \overline{r}(\lambda) \\ \overline{g}(\lambda) \\ \overline{b}(\lambda) \end{bmatrix}, \tag{1.45}$$

that are shown[19] in the upper part of Figure 1.12 and tabulated in Appendix 1.B. This second set was recommended for reasons of more convenient application in practical colorimetry. The most obvious advantage is that the second color-matching function $\overline{y}(\lambda)$ is equal to the photopic spectral sensitivity function $V(\lambda)$. In addition, the color-matching functions are positive for all possible wavelengths and normalized such that the tristimulus values of an equal energy source are all equal, i.e.,

$$\int \overline{x}(\lambda)\, d\lambda = \int \overline{y}(\lambda)\, d\lambda = \int \overline{z}(\lambda)\, d\lambda. \tag{1.46}$$

It will be proven in chapter 3 that the corresponding primaries are not physically realizable, since they contain negative energy at some wavelengths.

In 1951, Judd [Jud51] demonstrated that the CIE 1931 CMFs underestimate the human sensitivity at wavelengths below 450 nm and proposed the revised CMFs $\overline{x}_{51}(\lambda)$, $\overline{y}_{51}(\lambda)$ and $\overline{z}_{51}(\lambda)$ that are also tabulated in Appendix 1.B and plotted in the middle part of Figure 1.12. They will be referred to as the Judd-modified CMFs or the CIE 1951 2 degree standard observer. Additional corrections to these latter CMFs were suggested by Vos [Vos78] in 1978 and the CMFs $\overline{x}_{78}(\lambda)$, $\overline{y}_{78}(\lambda)$ and $\overline{z}_{78}(\lambda)$ that he proposed are plotted in the lower part of Figure 1.12.

[19] The photoreceptor sensitivities and color-matching functions that are plotted in the figures were obtained from the web site http://www.cvrl.org.

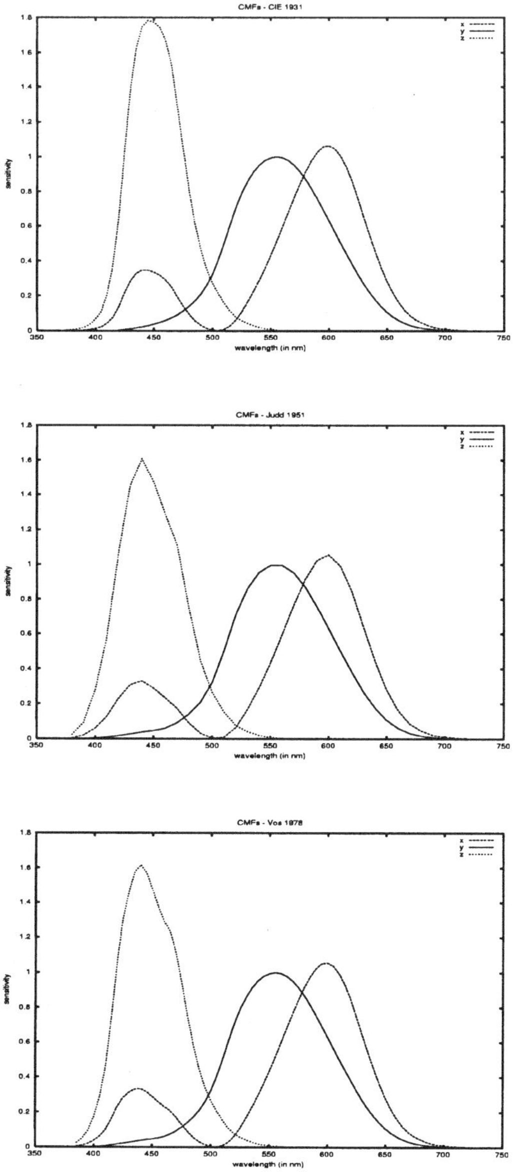

Figure 1.12. Color-matching functions $\overline{x}(\lambda)$, $\overline{y}(\lambda)$ and $\overline{z}(\lambda)$ for a 2 degree standard observer according to the 1931 CIE standard (top), as well as after modifications by Judd in 1951 (middle) and Vos in 1978 (bottom).

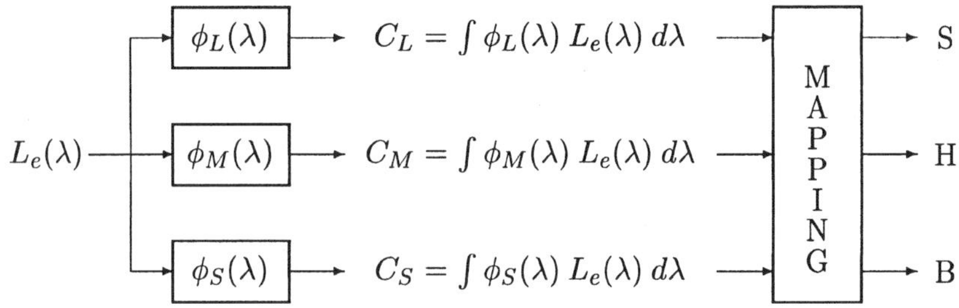

Figure 1.13. Model of human color vision. Three linear measurement functions, denoted by $\phi_L(\lambda)$, $\phi_M(\lambda)$ and $\phi_S(\lambda)$ (representing for instance the sensitivity functions of the long, medium and short-wavelength cones, and including the absorption by the lens and the macular pigment), are applied to the input radiometric function $L_e(\lambda)$ to obtain three coordinates (C_L, C_M, C_S) that encode the color information. These coordinates are mapped nonlinearly into the subjective color sensations brightness (B), hue (H) and saturation (S). Two radiometric functions are subjectively indistinguishable, or 'match', if they have the same coordinates.

They will be referred to as the Vos-modified CMFs or the CIE 1978 2 degree standard observer. Although it is broadly accepted that the modified CMFs of Judd and Vos provide a more adequate description of human color matching than the 1931 CIE standard observer, these modified CMFs have never been standardized by the CIE. Consequently, most available equipment, such as colorimeters, implement the 1931 CIE standard.

In 1964, the CIE supplemented the 1931 standard observer with a new set of CMFs $\bar{x}_{10}(\lambda)$, $\bar{y}_{10}(\lambda)$ and $\bar{z}_{10}(\lambda)$ that were recommended for use whenever a more accurate correlation with visual color matching of fields of large angular size (more than 4° at the eye of the observer) is desired. The experimental color-matching data were determined for a 10° field by Speranskaya [Spe59] and Stiles and Burch [SB59], and deviated from the 1931 observer mostly in the short-wavelength region [CIE71].

6.3 CONE SENSITIVITIES

The axioms of Grassman imply that color matching is performed by a LFS with three degrees of freedom. The tristimulus values must depend linearly on the radiance, as shown in Figure 1.13. This is obviously in agreement with physiological data that imply that there are three types of (cone) receptors, containing photopigments with different spectral sensitivity functions (i.e., weighting of the incoming spectral energy distribution), that cooperate to generate color vision. The spectral sensitivity functions of the cones must hence be linear combinations of the standardized color-matching functions.

Accurate descriptions of the spectral sensitivity functions of the different types of cones are largely based on color matching experiments with dichromats (i.e., subjects lacking one color receptor, see exercise 2 in chapter 3). Based on such experiments, Smith and Pokorny [SP75] proposed the following mapping between the Judd-Modified CMFs and the cone sensitivities

$$\begin{bmatrix} \phi_L(\lambda) \\ \phi_M(\lambda) \\ \phi_S(\lambda) \end{bmatrix} = \begin{bmatrix} 0.15514 & 0.54312 & -0.03286 \\ -0.15514 & 0.45684 & 0.03286 \\ 0 & 0 & 0.01608 \end{bmatrix} \cdot \begin{bmatrix} \bar{x}_{51}(\lambda) \\ \bar{y}_{51}(\lambda) \\ \bar{z}_{51}(\lambda) \end{bmatrix}. \tag{1.47}$$

To obtain cone sensitivities that are peak-normalized to one, this mapping can be rescaled to

$$\begin{bmatrix} \tilde{\phi}_L(\lambda) \\ \tilde{\phi}_M(\lambda) \\ \tilde{\phi}_S(\lambda) \end{bmatrix} = \begin{bmatrix} 0.2434 & 0.8522 & -0.0516 \\ -0.3956 & 1.1650 & 0.0838 \\ 0 & 0 & 0.6225 \end{bmatrix} \cdot \begin{bmatrix} \bar{x}_{51}(\lambda) \\ \bar{y}_{51}(\lambda) \\ \bar{z}_{51}(\lambda) \end{bmatrix}. \tag{1.48}$$

These normalized cone fundamentals are tabulated in Appendix 1.C and plotted in Figure 1.14. Because of the modifications introduced by Judd in the 1951 CMFs, the photometric spectral sensitivity function $V(\lambda)$, which is part of the 1931 CIE standard observer, cannot be written as a linear combination of the aforementioned cone sensitivities. Smith and Pokorny assumed a spectral sensitivity that was proportional to $\phi_L(\lambda) + \phi_M(\lambda)$ in their derivation[20].

Stockman and Sharpe [SS99] more recently analyzed a large body of experimental data, comprising spectral sensitivity measurements for the different types of cones and color matching data from normal observers and dichromats. They argued the need for a modified set of cone fundamentals. Unlike in earlier studies, they were able to use genetic testing to classify subjects according to their cone pigment genes[21]. Amongst others, this allowed to test which normal and dichromatic subjects shared

[20] Vos and Walraven [VW71] performed a similar analysis as Smith and Pokorny, but assumed a spectral sensitivity that was proportional to $\phi_L(\lambda) + \phi_M(\lambda) + \phi_S(\lambda)$ in their derivation.

[21] For example, slightly different genes $L(ser^{180})$ and $L(ala^{180})$ contain the genetic code for the L-cones in different subjects. They occur in 56 % and 44 % of the Caucasian population, respectively. Therefore, the L-cone spectral sensitivity has to be measured separately for each population, and the proposed L-cone characteristic is a weighted sum of both individual characteristics.

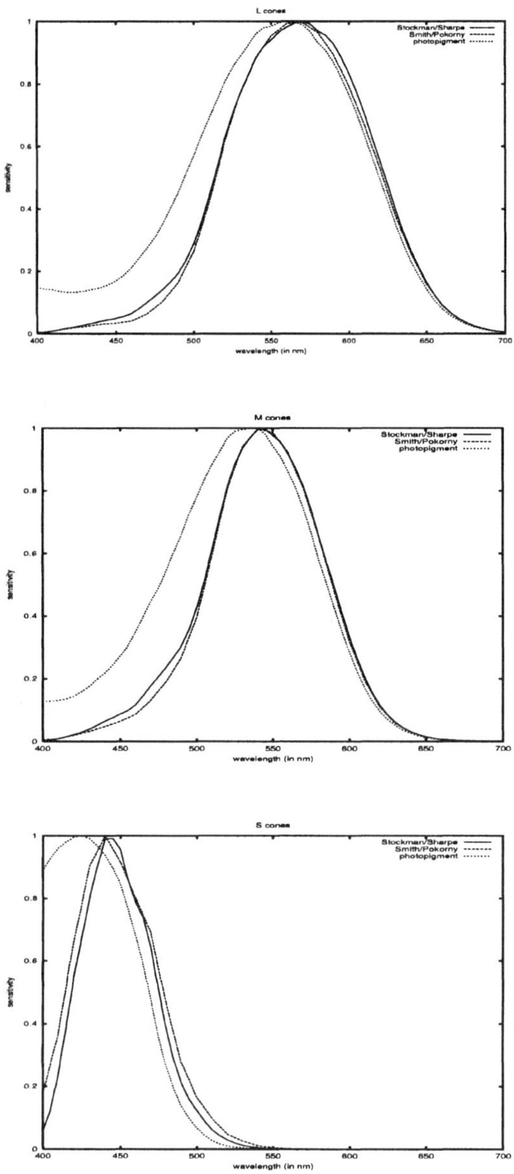

Figure 1.14. The spectral sensitivity functions of the three types of cones in the human retina, sensitive to long ($\tilde{\phi}_L(\lambda)$), medium ($\tilde{\phi}_M(\lambda)$) and short ($\tilde{\phi}_S(\lambda)$) wavelengths. The drawn [SS99] and dashed [SP75] curves are obtained through color matching. The dotted curves represent the photopigment sensitivities. All curves are normalized such that the maximum response is one.

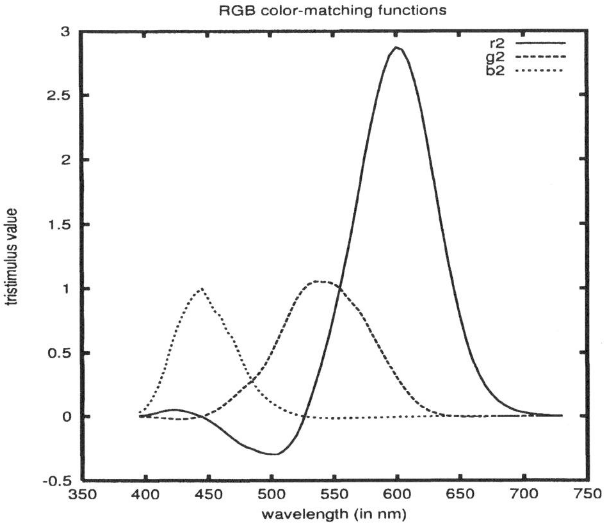

Figure 1.15. Color-matching functions $\bar{r}_2(\lambda)$, $\bar{g}_2(\lambda)$ and $\bar{b}_2(\lambda)$ for a 2 degree field and monochromatic primaries with wavelength $\lambda_r = 645$ nm, $\lambda_g = 526$ nm and $\lambda_b = 444$ nm.

the same cone receptors. The cone fundamentals were also required to be consistent with the tritanopic color matches performed by Stiles and Burch for 2 degree [SB55] and 10 degree [SB59] fields. More specifically, it was derived that

$$\begin{bmatrix} \phi_L(\lambda) \\ \phi_M(\lambda) \\ \phi_S(\lambda) \end{bmatrix} \propto \begin{bmatrix} 4.787127 & 16.782165 & 1 \\ 0.29089 & 12.24415 & 1 \\ 0 & 0.0163 & 1 \end{bmatrix} \cdot \begin{bmatrix} \bar{r}_2(\lambda) \\ \bar{g}_2(\lambda) \\ \bar{b}_2(\lambda) \end{bmatrix} \tag{1.49}$$

expresses the relationship between the cone sensitivities and the tristimulus values measured by Stiles and Burch. These tristimulus values $\bar{r}_2(\lambda)$, $\bar{g}_2(\lambda)$ and $\bar{b}_2(\lambda)$, which are plotted in Figure 1.15, were obtained for three monochromatic primaries with wavelength $\lambda_r = 645$ nm (red), $\lambda_g = 526$ nm (green) and $\lambda_b = 444$ nm (blue), respectively. The normalized cone sensitivities $(\tilde{\phi}_L(\lambda), \tilde{\phi}_M(\lambda), \tilde{\phi}_S(\lambda))$ of Stockman and Sharpe are also tabulated in Appendix 1.C and plotted in Figure 1.14. Stockman and Sharpe also proposed the following alternative expression

$$V^*(\lambda) = 0.4041 \cdot \left(1.572\, \tilde{\phi}_L(\lambda) + \tilde{\phi}_M(\lambda)\right), \tag{1.50}$$

for the photometric spectral sensitivity. This proposed characteristic, which is plotted in Figure 1.9, has however not (yet) been standardized.

The absorption spectra of the photopigments in the different cones have also been measured [SP72, KB96]. The differences between these absorption spectra and the above cone sensitivities can be attributed to the absorption $\tau(\lambda)$ in the optical pathway [BNS87, Wan95, SS99], i.e., $C(\lambda) \propto C_p(\lambda) \cdot \tau(\lambda)$, where $C(\lambda)$ is a psychophysically measured cone sensitivity (L, M or S), and $C_p(\lambda)$ is the absorption spectrum of the corresponding cone photopigment. This implies that color vision, as measured by color matching experiments, does indeed reflect the operation of the cones in the retina. The photopigment absorption spectra are shown as the dotted curves in Figure 1.14.

In order to perform a detailed analysis of the color matching experiments, more in-depth knowledge on linear vector spaces is required. We will therefore first provide an introduction to the theory of linear systems in the next chapter, and postpone the detailed analysis of the color matching data until chapter 3.

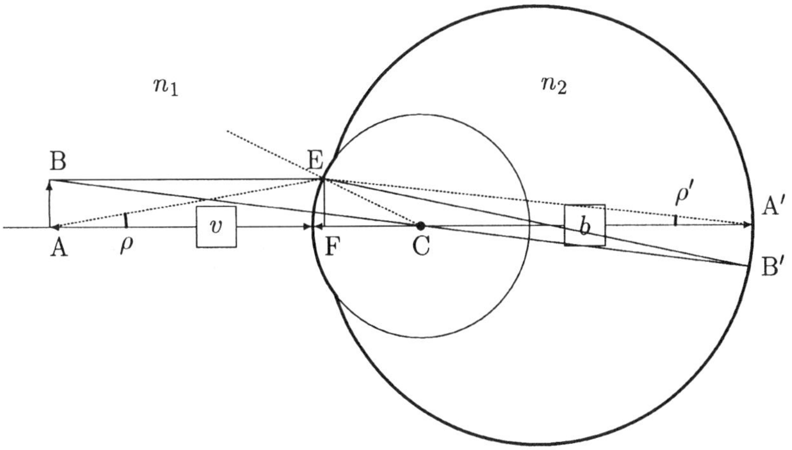

Figure 1.16. Simplified model of the unaccommodated eye. The eye is approximated by a uniform medium with an index of refraction $n_2 = 1.33$, immersed in a medium with a different index of refraction $n_1 = 1$. The light is deflected at the surface of the cornea, which is part of a sphere with center point C and radius $R = 5.6$ mm. The light is focused on the retina, which is part of a second, larger, sphere. The distance b between the cornea and the retina is approximately 22.6 mm.

EXERCISES

1 A simplified model of the unaccommodated eye is shown in Figure 1.16. Using the sine rule[22] on the triangles ACE and A′CE, and Snell's law, show that

$$\frac{\sin \rho}{\sin \rho'} = \frac{n_2\,(b - R)}{n_1\,(v + R)}.$$

For small angles ρ and ρ' with the optical axis, we can assume that the length ratio A′E/AE is approximately equal to b/v. Show that the above equation can be rewritten as a lensmaker equation using this small angle approximation. Derive the expression for the focal distance f in terms of the radius R and the indices of refraction n_1 and n_2. Verify that $f = 17$ mm for the unaccommodated eye.

[22]The sine rule states that the ratio of the length of one side in a triangle over the sine of the opposite angle is the same for all sides in a triangle.

2 Derive equation (1.8) on page 9 for the area of the blurring spot of an ill-focused object in a lens system (assuming geometrical optics).

3 Suppose that a surface patch dS is simultaneously irradiated by a uniform light source with radiance $L_{e,u}(\lambda)$ and a point source with radiant flux $\Phi_{e,s}$ at a distance D_s from dS that radiates equally in all directions. This source direction has coordinates (θ_s, ϕ_s) in the coordinate system spanned by the surface normal and tangent plane. The bidirectional reflectance distribution function is a linear combination of a diffuse component, see equation (1.31), and a specular component, see equation (1.37). Derive the radiance $L_e(\lambda; \theta, \phi)$ of the surface patch dS in the direction (θ, ϕ) of the camera, as well as the overall radiant emittance $M_e(\lambda)$. Refer to Figure 1.8 for the relevant geometry.

4 The *luminous efficacy* of a lighting system is defined as

$$K_s = \frac{\Phi}{P} = \frac{\Phi}{\Phi_e} \cdot \frac{\Phi_e}{P} = K_r \cdot \eta,$$

where P is the supply power, Φ_e is the radiant flux, and Φ is the light flux. The ratio

$$K_r = \frac{K_m \int_0^\infty \Phi_e(\lambda)\, V(\lambda)\, d\lambda}{\int_0^\infty \Phi_e(\lambda)\, d\lambda}$$

is called the *radiation luminous efficacy* and expresses the effectiveness of incident radiation to stimulate the perception of light. Which light source has the highest possible radiation luminous efficacy? How many lm/W can it produce? Just for the sake of reference, we cite typical efficacies for some popular light sources: a tungsten incandescent lamp ($K_r = 21.2$ lm/W, $K_s = 17.4$ lm/W), a fluorescent lamp ($K_r = 120$ lm/W, $K_s = 71$ lm/W), and a sodium discharge vapor lamp ($K_r = 524.6$ lm/W, $K_s = 197$ lm/W) [JHG99].

5 A good quality printer can generate output with 600 dots per inch (1 inch = 2.54 cm). Determine the visual angle (in arcmin) between successive dots for a viewing distance of 30 cm. In broadcast TV, the recommended viewing distance is 6 times the image height. The number of visible horizontal lines is 575. Determine the visual angle (in arcmin) between successive lines on the TV.

6 Photopic and scotopic metamerism:

- A yellow daisy and blue lilac may appear equally bright under scotopic conditions. Purkinje noticed that this does not occur under photopic conditions: the yellow flower is perceived to be much brighter. Explain this phenomenon which is known as the Purkinje shift [Wan95].

- Suppose a pair of lights are adjusted such that they are metamere under photopic vision. These lights will usually not be metamere under scotopic vision. Why not? Which condition should the scotopic spectral sensitivity function satisfy in order for photopic metamerism to imply scotopic metamerism?

APPENDIX 1.A: SPECTRAL SENSITIVITY FUNCTIONS

Table 1.A.1. CIE photopic, $V(\lambda)$, and scotopic, $V'(\lambda)$, spectral sensitivity functions at 10 nm intervals. The alternative photopic sensitivity function $V^*(\lambda)$ is based on recent measurements. It agrees better than $V(\lambda)$ with human brightness perception but has not been standardized.

λ	$V(\lambda)$	$V^*(\lambda)$	$V'(\lambda)$
380	0.0000	0.0000	0.0006
390	0.0001	0.0004	0.0022
400	0.0004	0.0024	0.0093
410	0.0012	0.0091	0.0348
420	0.0040	0.0205	0.0966
430	0.0116	0.0338	0.1998
440	0.0230	0.0517	0.3281
450	0.0380	0.0668	0.4550
460	0.0600	0.0881	0.5670
470	0.0910	0.1343	0.6760
480	0.1390	0.1843	0.7930
490	0.2080	0.2444	0.9040
500	0.3230	0.3564	0.9820
510	0.5030	0.5305	0.9970
520	0.7100	0.7292	0.9350
530	0.8620	0.8675	0.8110
540	0.9540	0.9617	0.6500
550	0.9950	0.9920	0.4810
560	0.9950	0.9942	0.3288
570	0.9520	0.9639	0.2076

λ	$V(\lambda)$	$V^*(\lambda)$	$V'(\lambda)$
580	0.8700	0.8797	0.1212
590	0.7570	0.7883	0.0655
600	0.6310	0.6648	0.0332
610	0.5030	0.5312	0.0159
620	0.3810	0.3991	0.0074
630	0.2650	0.2796	0.0033
640	0.1750	0.1815	0.0015
650	0.1070	0.1111	0.0007
660	0.0610	0.0620	0.0003
670	0.0320	0.0331	0.0001
680	0.0170	0.0168	0.0001
690	0.0082	0.0080	0.0000
700	0.0041	0.0039	0.0000
710	0.0021	0.0019	0.0000
720	0.0010	0.0009	0.0000
730	0.0005	0.0004	0.0000
740	0.0003	0.0002	0.0000
750	0.0001	0.0001	0.0000
760	0.0001	0.0001	0.0000
770	0.0000	0.0000	0.0000

APPENDIX 1.B: COLOR-MATCHING FUNCTIONS

Table 1.B.1. CIE 1931 standard CMFs (top) and Judd CMFs (bottom).

λ	$\bar{x}(\lambda)$	$\bar{y}(\lambda)$	$\bar{z}(\lambda)$	λ	$\bar{x}(\lambda)$	$\bar{y}(\lambda)$	$\bar{z}(\lambda)$
380	0.0014	0.0000	0.0065	580	0.9163	0.8700	0.0017
390	0.0042	0.0001	0.0201	590	1.0263	0.7570	0.0011
400	0.0143	0.0004	0.0679	600	1.0622	0.6310	0.0008
410	0.0435	0.0012	0.2074	610	1.0026	0.5030	0.0003
420	0.1344	0.0040	0.6456	620	0.8544	0.3810	0.0002
430	0.2839	0.0116	1.3856	630	0.6424	0.2650	0.0000
440	0.3483	0.0230	1.7471	640	0.4479	0.1750	0.0000
450	0.3362	0.0380	1.7721	650	0.2835	0.1070	0.0000
460	0.2908	0.0600	1.6692	660	0.1649	0.0610	0.0000
470	0.1954	0.0910	1.2876	670	0.0874	0.0320	0.0000
480	0.0956	0.1390	0.8130	680	0.0468	0.0170	0.0000
490	0.0320	0.2080	0.4652	690	0.0227	0.0082	0.0000
500	0.0049	0.3230	0.2720	700	0.0114	0.0041	0.0000
510	0.0093	0.5030	0.1582	710	0.0058	0.0021	0.0000
520	0.0633	0.7100	0.0782	720	0.0029	0.0010	0.0000
530	0.1655	0.8620	0.0422	730	0.0014	0.0005	0.0000
540	0.2904	0.9540	0.0203	740	0.0007	0.0003	0.0000
550	0.4334	0.9950	0.0087	750	0.0003	0.0001	0.0000
560	0.5945	0.9950	0.0039	760	0.0002	0.0001	0.0000
570	0.7621	0.9520	0.0021	770	0.0001	0.0000	0.0000

λ	$\bar{x}_{51}(\lambda)$	$\bar{y}_{51}(\lambda)$	$\bar{z}_{51}(\lambda)$	λ	$\bar{x}_{51}(\lambda)$	$\bar{y}_{51}(\lambda)$	$\bar{z}_{51}(\lambda)$
380	0.0045	0.0004	0.0224	580	0.9159	0.8700	0.0016
390	0.0201	0.0015	0.0925	590	1.0225	0.7570	0.0011
400	0.0611	0.0045	0.2799	600	1.0544	0.6310	0.0007
410	0.1267	0.0093	0.5835	610	0.9922	0.5030	0.0003
420	0.2285	0.0175	1.0622	620	0.8432	0.3810	0.0002
430	0.3081	0.0273	1.4526	630	0.6327	0.2650	0.0001
440	0.3312	0.0379	1.6064	640	0.4404	0.1750	0.0000
450	0.2888	0.0468	1.4717	650	0.2787	0.1070	0.0000
460	0.2323	0.0600	1.2880	660	0.1619	0.0610	0.0000
470	0.1745	0.0910	1.1133	670	0.0858	0.0320	0.0000
480	0.0920	0.1390	0.7552	680	0.0459	0.0170	0.0000
490	0.0318	0.2080	0.4461	690	0.0222	0.0082	0.0000
500	0.0048	0.3230	0.2644	700	0.0113	0.0041	0.0000
510	0.0093	0.5030	0.1541	710	0.0057	0.0021	0.0000
520	0.0636	0.7100	0.0763	720	0.0028	0.0010	0.0000
530	0.1668	0.8620	0.0412	730	0.0015	0.0005	0.0000
540	0.2926	0.9540	0.0200	740	0.0005	0.0003	0.0000
550	0.4364	0.9950	0.0088	750	0.0003	0.0001	0.0000
560	0.5970	0.9950	0.0039	760	0.0002	0.0001	0.0000
570	0.7642	0.9520	0.0020	770	0.0001	0.0000	0.0000

APPENDIX 1.C: CONE SENSITIVITIES

Table 1.C.1. Cone SSFs: Smith&Pokorny (top) and Stockman&Sharpe (bottom).

λ	$\tilde{\phi}_L(\lambda)$	$\tilde{\phi}_M(\lambda)$	$\tilde{\phi}_S(\lambda)$	λ	$\tilde{\phi}_L(\lambda)$	$\tilde{\phi}_M(\lambda)$	$\tilde{\phi}_S(\lambda)$
380	0.0003	0.0006	0.0139	580	0.9643	0.6514	0.0010
390	0.0014	0.0015	0.0576	590	0.8939	0.4775	0.0007
400	0.0043	0.0045	0.1742	600	0.7943	0.3181	0.0004
410	0.0087	0.0096	0.3632	610	0.6701	0.1935	0.0002
420	0.0157	0.0190	0.6612	620	0.5299	0.1103	0.0001
430	0.0233	0.0316	0.9042	630	0.3798	0.0584	0.0000
440	0.0300	0.0477	1.0000	640	0.2563	0.0297	0.0000
450	0.0342	0.0636	0.9161	650	0.1590	0.0144	0.0000
460	0.0412	0.0859	0.8018	660	0.0914	0.0070	0.0000
470	0.0626	0.1303	0.6930	670	0.0482	0.0033	0.0000
480	0.1019	0.1888	0.4701	680	0.0257	0.0016	0.0000
490	0.1620	0.2671	0.2777	690	0.0124	0.0008	0.0000
500	0.2628	0.3966	0.1646	700	0.0062	0.0003	0.0000
510	0.4230	0.5952	0.0959	710	0.0032	0.0002	0.0000
520	0.6166	0.8084	0.0475	720	0.0016	0.0002	0.0000
530	0.7731	0.9417	0.0256	730	0.0008	0.0000	0.0000
540	0.8832	0.9973	0.0125	740	0.0003	0.0000	0.0000
550	0.9537	0.9873	0.0055	750	0.0002	0.0000	0.0000
560	0.9930	0.9233	0.0024	760	0.0001	0.0000	0.0000
570	0.9972	0.8069	0.0012	770	0.0000	0.0000	0.0000

λ	$\tilde{\phi}_L(\lambda)$	$\tilde{\phi}_M(\lambda)$	$\tilde{\phi}_S(\lambda)$	λ	$\tilde{\phi}_L(\lambda)$	$\tilde{\phi}_M(\lambda)$	$\tilde{\phi}_S(\lambda)$
380	——	——	——	580	0.9694	0.6533	0.0001
390	0.0004	0.0004	0.0095	590	0.9277	0.4926	0.0000
400	0.0024	0.0023	0.0566	600	0.8340	0.3344	0.0000
410	0.0087	0.0088	0.2330	610	0.7057	0.2053	0.0000
420	0.0184	0.0217	0.5436	620	0.5542	0.1166	0.0000
430	0.0282	0.0395	0.8026	630	0.4007	0.0621	0.0000
440	0.0403	0.0648	0.9910	640	0.2658	0.0314	0.0000
450	0.0499	0.0871	0.9554	650	0.1651	0.0154	0.0000
460	0.0647	0.1163	0.7867	660	0.0930	0.0073	0.0000
470	0.0995	0.1759	0.6464	670	0.0499	0.0034	0.0000
480	0.1401	0.2358	0.3903	680	0.0254	0.0016	0.0000
490	0.1916	0.3036	0.2119	690	0.0121	0.0008	0.0000
500	0.2890	0.4278	0.1228	700	0.0059	0.0004	0.0000
510	0.4437	0.6155	0.0608	710	0.0028	0.0002	0.0000
520	0.6286	0.8166	0.0292	720	0.0013	0.0000	0.0000
530	0.7706	0.9357	0.0126	730	0.0006	0.0000	0.0000
540	0.8810	0.9952	0.0051	740	0.0003	0.0000	0.0000
550	0.9402	0.9772	0.0020	750	0.0002	0.0000	0.0000
560	0.9814	0.9178	0.0007	760	0.0001	0.0000	0.0000
570	1.0000	0.8135	0.0003	770	0.0000	0.0000	0.0000

Chapter 2

LINEAR SYSTEM THEORY AND VECTOR SPACES

1. INTRODUCTION

In this chapter, we describe a mathematical theory that is suited for describing both signals (such as images) and the operations on signals. We only discuss the case of linear operations on signals. This obviously simplifies the mathematical treatment. Many systems of interest are however linear systems (LS) or linear front-end systems (LFS) by themselves, or can be approximated by LS or LFS when operated under restricted conditions.

The processing of complex signals such as speech, music and images is nowadays mostly done by means of digital computers. Because of their inherent flexibility, computers allow processing of signals that is very hard, or even impossible, by analog means. A fundamental assumption in digital signal processing is that 'real-world' signals, which are mostly continuous, can be approximately specified by a discrete and finite set of (integer) numbers. These numbers, which are obtained by performing measurements on the input signal, are said to *represent* this signal if they can be used to generate an output signal that does not differ from the input signal by more that an *acceptable error*. A given signal may have many different **signal representations** with variable amounts of error. These representations may differ

- in the number of degrees of freedom (DOF), i.e., how many numbers are needed to describe the signal,

- in how the measurements on the signal are performed, i.e., how are these numbers derived from the input signal, and/or

- in the way the signal is approximated from these measurements, i.e., how are the numbers used to generate the output signal.

The theory of linear vector spaces that we discuss in this chapter allows to make this intuitive concept of signal representations mathematically precise for the simplest, but nevertheless very important, special case where signals are approximated by a linear sum of a priori selected functions. This theory will be applied to specific instances in the chapters to come.

What is considered an acceptable error between an original signal and its approximated output signal is in general dependent on the signal itself. Deciding *a priori* which errors are acceptable is difficult since we lack widely accepted fidelity criteria for complex signals such as images. This is especially true if the fidelity criterium is supposed to reflect the judgement of a human observer. Most existing signal representations are therefore based on simple error measures, such as average quadratic error, since designing and applying more advanced fidelity criteria, that correlate better with human perception, has turned out to be very difficult[1]. Some of the most widely used error measures will be discussed in this chapter.

A large number of alternative signal representations usually exist that satisfy a given fidelity criterium, so that additional objectives, such as ease of interpretation and manipulation, efficiency, robustness against errors, computational complexity, ..., are also involved in making a selection between alternative signal representations.

The design and evaluation of signal processing systems can often also be simplified by using suitable signal representations. A frequently used approach to describing and understanding the restrictions that are imposed by a signal processing system is to express the signals of interest as the sum of a discrete number of a priori selected **basis signals**, and to subsequently determine how these basis signals are transformed by the processing system. We will see later that linear systems can be completely characterized by how they transform a suitable set of basis signals. For instance, linear time- or space-invariant systems can be specified by how they transform harmonic signals (sinusoids). This explains the popularity of Fourier descriptions, which will be discussed in more detail in chapter 6. Proper selection of the basis signals sometimes requires intuition, and is also influenced by new insights in signal processing and human perception. Much ongoing research in signal pro-

[1]This issue or error criteria that correlate well will human judgement will be addressed in chapters 11 and 12.

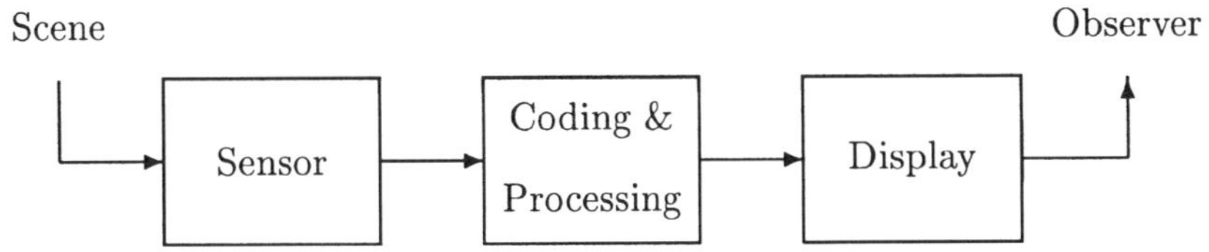

Figure 2.1. Image processing chain consisting of subsystems for image acquisition, processing and display.

cessing is aimed at finding basis signals with interesting (mathematical) properties and/or at designing techniques for adapting the basis signals to the (sub)system under study.

In many applications, the signal processing system consists of several subsystems, whose design and specification may benefit from using distinct signal representations. For instance, in image technology, a typical signal path comprises several stages, i.e., scenes are captured by a sensory device such as a camera, processed and/or coded for storage or transmission, displayed and eventually viewed by a human observer (see Figure 2.1). Some of the characteristic problems in the design of these subsystems that involve using signal representations are the following. Sensory devices perform a limited number of measurements on the incoming signals; these measurements have to be adequate for deriving useful signal approximations. Deciding which measurements to perform and how to use them most profitably for approximating the input signals are issues that need to be settled. Communication links between sensors and display devices limit the number and the nature of the signal components that can be transmitted. Typically, an interpretation of the signal components is needed to select adequate processing and/or coding for them. Display devices synthesize an image using a finite number of basic display responses, such as the phosphor dots on a cathode ray tube (CRT). Selecting the size and shape of these display responses, their configuration (number and relative position), and how they can be optimally controlled to obtain images with the required quality/fidelity are all aspects that need to be addressed.

In section 2, we discuss why the mathematical theory of vector spaces provides the proper framework for describing linear operations on signals. In section 3 we argue the need for a metric that expresses how closely related two different signals are. We also introduce the important concept of the basis of a vector space, and show how it can be used to establish a relationship between signals and signal representations. In section 4, we describe the theory of linear operators on vector spaces, and indicate that they provide a description for how signals are transformed by LS. In section 5, we concentrate on a specific kind of vector spaces, called Hilbert spaces. They are vector spaces with a very specific metric, i.e., one that is based on quadratic errors. Their main advantage is that they are more tractable than other vector spaces, and that they allow for unique solutions to problems that cannot be (uniquely) solved in more general vector spaces. In section 6, we discuss advanced properties of operators between Hilbert spaces and introduce concepts such as eigenvectors and eigenvalues.

Many of the definitions and properties that are discussed in the following sections were taken, without the accompanying proofs, from the books "Finite-Dimensional Vector Spaces", by P. Halmos [Hal87], and "An Introduction to Hilbert Space", by N. Young [You88]. We have tried to select those definitions and properties that are most relevant for signal processing, and as much as possible have illustrated the theoretical concepts with practical implications.

2. SIGNALS AND VECTORS

2.1 VECTOR SPACES

Consider a subset $\mathbb{F}$ of the real axis $\mathbb{R}$. A one-dimensional **signal** on $\mathbb{F}$ is a mapping that assigns to each point $x \in \mathbb{F}$ a value $f(x)$. Usually, we restrict ourselves to the case where $f(x) \in \mathbb{C}$ is a complex number. There are several possible physical interpretations of the argument x and the signal value $f(x)$. For instance, x may be wavelength and $f(x)$ some radiometric property of a light source, such as the radiant flux, that varies as a function of the wavelength. The signal is called **discrete** when the set $\mathbb{F}$ contains only a discrete number of argument values. If $\mathbb{F}$ consists of one or more continuous intervals, then the signal is called **continuous.** The set $\mathbb{F}$ is called **compact** if there exists a finite interval $[-a, a]$ that contains all of $\mathbb{F}$, i.e., if $x \in \mathbb{F}$ then $x \in [-a, a]$, for a finite value of a.

Let us denote the set of all signals over $\mathbb{F}$ that we want to consider by $V(\mathbb{F})$. In order to enable manipulation of these signals, two basic operations are defined, i.e., the **addition** of signals,

$$(f + g)(x) = f(x) + g(x), \tag{2.1}$$

and the **scalar product** of a complex number and a signal,

$$(c \cdot f)(x) = c \cdot f(x). \tag{2.2}$$

The addition and multiplication sign on the right-hand sides of the above equations are the familiar operations on complex numbers. The corresponding symbols on the left-hand sides define operations on signals. For instance, simultaneously projecting two light sources on a screen corresponds to a linear addition of the radiant flux coming from the individual light sources. Scalar multiplication can for instance be accomplished by putting a neutral density filter, i.e., a filter with a transmission coefficient that is independent of the wavelength, in between the light source and the screen, or by varying the supply voltage to the light source.

Note that the above definition of signals can be extended to signals of dimension $D > 1$ in a trivial way. A D-dimensional signal is a mapping that assigns to each point $\mathbf{x} = (x_1, \ldots, x_D)$ in a subset $\mathbb{F}$ of the D-dimensional real space $\mathbb{R}^D$ a complex value $f(\mathbf{x})$. For instance, in the previous chapter, we showed that a static image (or picture) can be physically specified as a radiance function in three dimensions, i.e., one wavelength dimension and two spatial dimensions. Similarly, a movie sequence can be physically specified as a radiance function in four dimensions, i.e., one wavelength dimension, two spatial dimensions and one time dimension. The definitions for discrete and continuous signals can be extended in a straightforward way. A subset $\mathbb{F}$ of $\mathbb{R}^D$ is compact if there exists a sphere $\{(x_1, \ldots, x_D) : x_1^2 + \ldots + x_D^2 \leq a^2\}$ of finite diameter a that contains $\mathbb{F}$.

Since the argument $\mathbf{x}$ of the signal can usually be derived from the context, we will often use the short-hand **vector** notation $\vec{f}$ to denote a signal $f(\mathbf{x})$. Similarly, if the set $\mathbb{F}$ can be inferred from the context, we will denote the set of signals $V(\mathbb{F})$ by V.

It is assumed in the remainder that the set V of signals under consideration is a **linear vector space**, i.e., satisfies the following properties

1 $V, +$ is a commutative group

- If $\vec{f}, \vec{g} \in V$, then $\vec{f} + \vec{g} \in V$ (closed)
- If $\vec{f}, \vec{g} \in V$, then $\vec{f} + \vec{g} = \vec{g} + \vec{f}$ (commutative)
- If $\vec{f}, \vec{g}, \vec{h} \in V$, then $(\vec{f} + \vec{g}) + \vec{h} = \vec{f} + (\vec{g} + \vec{h})$ (associative)
- There exists $\vec{0} \in V$ such that $\vec{f} + \vec{0} = \vec{f}, \forall \vec{f} \in V$ (zero vector)
- If $\vec{f} \in V$, then $\exists(-\vec{f}) \in V$ such that $\vec{f} + (-\vec{f}) = \vec{0}$ (complement)

2 Scalar product satisfies

- If $a \in \mathbb{C}$; $\vec{f} \in V$, then $a \cdot \vec{f} \in V$ (closed)
- If $a \in \mathbb{C}$; $\vec{f}, \vec{g} \in V$, then $a \cdot (\vec{f} + \vec{g}) = a \cdot \vec{f} + a \cdot \vec{g}$ (linearity in V)
- If $a, b \in \mathbb{C}$; $\vec{f} \in V$, then $(a + b) \cdot \vec{f} = a \cdot \vec{f} + b \cdot \vec{f}$ (linearity in $\mathbb{C}$)
- If $a, b \in \mathbb{C}$; $\vec{f} \in V$, then $(a \cdot b) \cdot \vec{f} = a \cdot (b \cdot \vec{f})$ (associative)

- If $\vec{f} \in V$, then $1 \cdot \vec{f} = \vec{f}$ (identity in $\mathbb{C}$)

These axioms are not claimed to be logically independent; they are merely a convenient characterization of the objects we wish to study. The condition that V is a vector space is not very restrictive, since the original set of signals can usually be extended to a vector space. For instance, by including theoretical light sources that can have negative flux at some wavelengths, we can include the physical light sources, which obviously have positive flux at all wavelengths, into a vector space.

On the other hand, sometimes one is specifically interested in a subset of all possible signals. A subset U of V is a **subspace** if $\vec{f} + \vec{g}$ belongs to U for all $\vec{f}, \vec{g} \in U$. For instance, all spectra that can be produced by a single light source, such as a projector, can usually be considered to belong to a subspace of the space of all possible spectra. It can easily be verified that a subspace is indeed also a vector space in its own right.

The rules of vector spaces only formalize the usual way in which signals are handle. We will therefore often refer to signals as vectors. If the signal values and scalars are restricted to real numbers, as will often be the case, then the term **real vector space** is used.

2.2 LINEAR FUNCTIONALS AND VECTOR PRODUCTS

Up to now, two basic operations have defined on signals: the addition of signals, which maps two input signals into an output signal, and the scalar product between a signal and a number, which maps an input signal and an input number into an output signal. We now define a third, and final, operation which maps a signal into a (complex or real) number. Formally, a **linear functional** L on a vector space V is a mapping that assigns a (complex or real) number $L\langle \vec{f} \rangle$ to each vector $\vec{f}$ of V, in such a way that

$$L\langle a \cdot \vec{f} + b \cdot \vec{g} \rangle = a \cdot L\langle \vec{f} \rangle + b \cdot L\langle \vec{g} \rangle, \tag{2.3}$$

for all $a, b \in \mathbb{C}$ and $\vec{f}, \vec{g} \in V$. A linear functional on V is **bounded** if the absolute value $|L\langle \vec{f} \rangle|$ is finite for all $\vec{f} \in V$. The main application of bounded linear functionals in signal processing is that they model how physical measurements on signals are performed, i.e., they specify how a finite number can be assigned to any signal of interest. The condition of linearity can be tested for most real measurement systems. For instance, in the previous chapter it was concluded that brightness matching is performed by a LFS with one degree of freedom, based on the experimental evidence that is summarized in the axioms of Abney.

This linear functional that fully determines the brightness was defined as the luminance.

In the remainder, we will be mainly interested in one specific kind of linear functional. The **vector product**, or **inner product**, is defined as a mapping that assigns a complex number[2], denoted by $\langle \vec{f}, \vec{g} \rangle$, to each combination of vectors $\vec{f}$ and $\vec{g}$. This mapping is assumed to satisfy the following properties

- If $\vec{f} \in V$, then $\langle \vec{f}, \vec{f} \rangle$ is real and ≥ 0; $\langle \vec{f}, \vec{f} \rangle = 0$ implies $\vec{f} = \vec{0}$
- If $\vec{f}, \vec{g} \in V$, then $\langle \vec{f}, \vec{g} \rangle \; = \; \langle \vec{g}, \vec{f} \rangle^\star$
- If $a, b \in \mathbb{C}$; $\vec{f}, \vec{g}, \vec{h} \in V$, then

$$\begin{aligned} \langle \vec{f}, a \cdot \vec{g} + b \cdot \vec{h} \rangle &= a \cdot \langle \vec{f}, \vec{g} \rangle + b \cdot \langle \vec{f}, \vec{h} \rangle \\ \langle a \cdot \vec{f} + b \cdot \vec{g}, \vec{h} \rangle &= a^\star \cdot \langle \vec{f}, \vec{h} \rangle + b^\star \cdot \langle \vec{g}, \vec{h} \rangle \end{aligned}$$

where $\star$ denotes complex conjugation. Mappings of the form $L\langle \vec{f} \rangle = \langle \vec{\phi}_L, \vec{f} \rangle$, with $\vec{\phi}_L$ a fixed vector, clearly satisfy the definition of a linear functional. The vector $\vec{\phi}_L$ is called the **sampling signal**, **measurement signal** or **test signal** of the linear functional L. Vector spaces in which a vector product is defined are called **inner-product spaces**, or **pre-Hilbert spaces**.

An important property of vector products is the *inequality of Cauchy-Schwartz*,

$$| \langle \vec{f}, \vec{g} \rangle |^2 \leq \langle \vec{f}, \vec{f} \rangle \cdot \langle \vec{g}, \vec{g} \rangle. \tag{2.4}$$

It guarantees that the following definition of the **relative angle** $\vec{\vartheta}(\vec{f}, \vec{g})$ between two vectors $\vec{f}$ and $\vec{g}$, i.e.,

$$\cos \vec{\vartheta}(\vec{f}, \vec{g}) = \frac{\langle \vec{f}, \vec{g} \rangle}{\sqrt{\langle \vec{f}, \vec{f} \rangle \cdot \langle \vec{g}, \vec{g} \rangle}}, \tag{2.5}$$

is meaningful in a real vector space. The sign of $\vec{\vartheta}(\vec{f}, \vec{g})$ is unknown, since only $\cos \vec{\vartheta}(\vec{f}, \vec{g})$ is uniquely determined, so that $\vec{\vartheta}(\vec{f}, \vec{g})$ is indeed a relative angle. Two vectors $\vec{f}$ and $\vec{g}$ that satisfy $\langle \vec{f}, \vec{g} \rangle = 0$ are said to be **orthogonal**, since the relative angle is then 90 degrees.

[2]In computer science terms, an inproduct can be interpreted as a function with two arguments, i.e., the vectors (signals) $\vec{f}$ and $\vec{g}$. The return value is a number that is denoted by $\langle \vec{f}, \vec{g} \rangle$. The operations of vector addition, scalar product and linear functional can be interpreted in a similar way.

Vector products that satisfy all the above properties are

$$\langle f(x), g(x) \rangle = \int_{\mathbb{F}} w^2(x)\, f^\star(x)\, g(x)\, dx, \tag{2.6}$$

for continuous signals, and

$$\langle f(x), g(x) \rangle = \sum_{x \in \mathbb{F}} w^2(x)\, f^\star(x)\, g(x), \tag{2.7}$$

for discrete signals, provided that $w(x)$ is a real function that differs from zero for all $x \in \mathbb{F}$ (for example, $w(x) = 1$)[3].

For instance, light measurements are typically performed by combining an optical filter with a photosensitive device, such as a photodiode, and can be modelled by a LFS $R(L)$ whose output is a nonlinear function of

$$L = \int_{\lambda_{min}}^{\lambda_{max}} \phi_L(\lambda)\, E_e(\lambda)\, d\lambda, \tag{2.8}$$

where $\phi_L(\lambda)$ and $E_e(\lambda)$ are the linear sensor response and the source irradiance falling onto the sensor, as a function of the wavelength λ, respectively. The photodiode converts electromagnetic radiation into an electrical signal (i.e., a number), while the combined spectral responses of the photodiode and the optical filter determine the measurement signal $\phi_L(\lambda)$ as a function of the wavelength λ.

3. VECTOR NORMS AND BANACH SPACES

3.1 NORMS AND DISTANCES

The basic operations needed to manipulate signals are defined in the previous section. In the current section, we concentrate on how signals can be compared. This capability is essential for constructing signal representations. Indeed, a typical problem will be to approximate, as closely as possible, a given signal by a linear combination of a priori given signals. Such an optimization requires a metric, i.e., a measure for how closely related two signals are. This is the motivation for introducing the norm of a vector and the distance between two vectors.

The **vector norm** is a *nonlinear* mapping that assigns a real number, denoted by $\|\vec{f}\,\|$, to each vector $\vec{f}$. In order for the norm to be useful, some restrictions need to be imposed on it. More specifically, it is required that $V, \|\cdot\|$ is a **normed space**, i.e., that V is a vector space with a norm $\|\cdot\|$ that satisfies the following properties

[3]This condition is needed to guarantee that $\langle \vec{f}, \vec{f}\,\rangle = 0$ implies $\vec{f} = \vec{0}$.

- If $\vec{f} \in V$, then $\| \vec{f} \| \geq 0$; $\| \vec{f} \| = 0$ implies $\vec{f} = \vec{0}$
- If $a \in \mathbb{C}$; $\vec{f} \in V$, then $\| a \cdot \vec{f} \| = |a| \cdot \| \vec{f} \|$
- If $\vec{f}, \vec{g} \in V$, then $\| \vec{f} + \vec{g} \| \leq \| \vec{f} \| + \| \vec{g} \|$ (*Minkowski's inequality*)

A vector for which $\| \vec{f} \| = 1$ is called **normalized**. Only normed spaces V in which all vectors $\vec{f} \in V$ have a finite norm $\| \vec{f} \|$ will be considered.

For instance, let $\Phi_e(\lambda)$ denote the flux of a light source as a function of the wavelength λ. The maximum absolute value of the flux over all wavelengths is a vector norm, called the supremum norm, with all the above properties.

In a normed space, the following **distance measure** (or **metric**)

$$d(\vec{f}, \vec{g}) = \| \vec{f} - \vec{g} \| \tag{2.9}$$

can be defined. This distance has the following properties

- If $\vec{f}, \vec{g} \in V$, then $d(\vec{f}, \vec{g}) \geq 0$; $d(\vec{f}, \vec{g}) = 0$ implies $\vec{f} = \vec{g}$
- If $\vec{f}, \vec{g} \in V$, then $d(\vec{f}, \vec{g}) = d(\vec{g}, \vec{f})$ (symmetry)
- If $\vec{f}, \vec{g}, \vec{h} \in V$, then $d(\vec{f}, \vec{h}) \leq d(\vec{f}, \vec{g}) + d(\vec{g}, \vec{h})$ (*triangle inequality*)

Hence, every normed space is also a **metric space**. For example, for the above supremum norm, the distance between two spectra is equal to the maximum absolute difference between the flux across wavelength.

An example of a norm is

$$\| f(x) \|_2 = \left[\sum_{x \in \mathbb{F}} w^2(x) \cdot |f(x)|^2 \right]^{1/2} . \tag{2.10}$$

The vector space on which this norm operates contains localized signals $f(x)$ that are defined for arguments $x \in \mathbb{F}$ that satisfy $w^2(x) \neq 0$. The corresponding distance measure

$$d_2[f(x), \hat{f}(x)] = \| f(x) - \hat{f}(x) \|_2, \tag{2.11}$$

allows to express the weighted quadratic error between a localized input signal $f(x)$ and its approximation $\hat{f}(x)$.

3.2 CAUCHY SEQUENCES AND BANACH SPACES

In a typical signal approximation situation, sequences of signals $\{\vec{f}_k\}$ will be constructed such that the distance $d(\vec{f}_k, \vec{f})$ to a given signal

$\vec{f}$ decreases with an increasing index k. For example, each successive signal $\vec{f_k}$ in the sequence may have one additional term that is selected to decrease the approximation error, up to the point where the error is judged acceptable. Such a sequence of vectors $\{\vec{f_k}\}$ in a normed space V is called a **Cauchy sequence** if, for every $\varepsilon > 0$, there exists an integer k_ε such that $k, l > k_\varepsilon$ implies that $d(\vec{f_k}, \vec{f_l}) < \varepsilon$. We will be uniquely interested here in the case of **complete normed spaces**, or **Banach spaces**, for which every Cauchy sequence converges to a well-defined limit in the space. Informally speaking, this implies that any sequence which *looks* to be converging *is* actually converging to a vector in the space V. A given vector space can always be extended to a Banach space by including all limit vectors.

The distinction between a normed space and a Banach space can be illustrated best with a simple example. The rational numbers $\mathbb{Q}$, with the familiar addition and multiplication operations, and with the absolute value as norm, constitute a normed space. However, some Cauchy sequences of rational numbers (such as $\{f_k = \sum_{n=0}^{k} 1/n!\}$) do not converge to a rational number but to a real number (in the above example, $f_k \to e = 2.71828...$ for $k \to \infty$). Such mathematical difficulties can be avoided in the Banach space of real numbers, where limits of Cauchy sequences always exist.

Well-known examples of Banach spaces are the vector spaces $\mathcal{L}^p(\mathbb{F})$ of all signals, defined on the set $\mathbb{F}$, for which the **Minkowski norm** or **p-norm**

$$\|f(x)\|_p = \left[\int_{\mathbb{F}} |f(x)|^p \, dx\right]^{1/p}, \tag{2.12}$$

is finite, where p is a positive real number. For $p = \infty$, one obtains as a special case the vector space $\mathcal{L}^\infty(\mathbb{F})$ of bounded signals $f(x)$ on the interval $\mathbb{F}$, i.e., $|f(x)|$ is finite for all $x \in \mathbb{F}$, together with the supremum[4] norm

$$\|f(x)\|_\infty = \sup\{|f(x)|; x \in \mathbb{F}\}, \tag{2.13}$$

i.e., the maximum signal amplitude in the interval $\mathbb{F}$.

---

[4]The supremum (sup) of a set is the smallest element that is larger than or equal to all elements in the set.

3.3 DUAL VECTOR SPACES

There are two important theorems concerning linear functionals on the Banach space $\mathcal{L}^p(\mathbb{F})$. First, *Hölder's inequality*

$$| \langle g(x), f(x) \rangle | \leq \|g(x)\|_q \cdot \|f(x)\|_p, \text{ if } \frac{1}{p} + \frac{1}{q} = 1, \tag{2.14}$$

for $f(x) \in \mathcal{L}^p(\mathbb{F})$ and $g(x) \in \mathcal{L}^q(\mathbb{F})$, guarantees that

$$L\langle f(x) \rangle = \langle \phi_L(x), f(x) \rangle \tag{2.15}$$

is a bounded linear functional on $\mathcal{L}^p(\mathbb{F})$ if the measurement function $\phi_L(x)$ belongs to $\mathcal{L}^q(\mathbb{F})$, with $\frac{1}{p} + \frac{1}{q} = 1$. More important however, this theorem also has an inverse, called *Riesz' representation theorem* which states that *all bounded linear functionals* on $\mathcal{L}^p(\mathbb{F})$, for p finite, are of this form[5]. Hence, for all practical purposes, **the inner product formalizes how we perform (linear) measurements on signals**, i.e., by letting the signal $f(x)$ 'interact' with a measurement signal $\phi_L(x)$.

The **dual vector space** V' of a given vector space V is defined as the space of all bounded linear functionals on V. It can be shown that V' is indeed a vector space with basic operations

$$\begin{aligned} (L_1 + L_2)\langle \vec{f} \rangle &= L_1\langle \vec{f} \rangle + L_2\langle \vec{f} \rangle \\ (a \cdot L)\langle \vec{f} \rangle &= a \cdot L\langle \vec{f} \rangle. \end{aligned} \tag{2.16}$$

Linear functionals are therefore sometimes referred to a **covectors**, a term which is especially popular in tensor calculus [Flo97]. The above theorems imply that $\mathcal{L}^p(\mathbb{F})$ and $\mathcal{L}^q(\mathbb{F})$ are dual spaces if $\frac{1}{p} + \frac{1}{q} = 1$. It can also be shown that $\mathcal{L}^q(\mathbb{F}) \subset \mathcal{L}^p(\mathbb{F})$ if $p \leq q$.

It was argued by Florack [Flo97] that the characterization of physical systems should start by specifying the relevant **device space** V. This is the space of all feasible (linear) test functions. Information on the outside world can only be obtained by observing with these test functions. A functional definition of the outside world, which he called **state space**, is hence that it consists of all linear functionals that can interact with the test signals to produce measurements of finite amplitude. For all practical purposes, the outside world is equal to the dual space V', since there is no point in defining signals to be different in state space if this difference cannot be observed. Such signals that give identical responses for all test functions in V are called metameric. The metamerism that

[5]This statement is still true for $p = \infty$ if $\mathbb{F}$ is compact, or if we restrict ourselves to signals that converge to zero at infinity for non-compact sets $\mathbb{F}$.

was discussed in section 6 of chapter 1, and that will be explored in more depth in chapter 3, is an illustration of these general ideas for the case of human color perception. In the example of color vision, the device space V is the vector space consisting of all possible linear combinations of the cone receptor curves. Many physically different light spectra give rise to identical color sensations.

3.4 SCHWARTZ FUNCTIONS AND TEMPERED DISTRIBUTIONS

System designers have little control over the signals that enter their systems. They can however specify the test functions that are used to measure incoming signals. Imposing more restrictions on the test functions is the way to guarantee that the outputs stay well-behaved, i.e., are finite and continuously related to the inputs, for a large class of possible input signals. Two intuitive restrictions are worth mentioning. First, one can require test functions to be *localized*, in the sense that they only use information about the input signal in a limited region around the position where the test function is centered. This avoids that a measurement at a specific position is influenced by signal aspects far removed from this position. Second, one can require that test functions are *smooth*, so that the measurement will not change abruptly when the test function is moved over a small distance. It was argued by Florack [Flo97] that these conditions impose that test functions should belong to the vector space $\mathcal{S}(\mathbb{R}^D)$ of Schwartz functions.

The function $\phi(\mathbf{x})$ is a **Schwartz function**, or **smooth test function**, if it is *infinitely differentiable* and

$$\sup_{\mathbf{x}\in\mathbb{R}^D} |\mathbf{x}^{\mathbf{n}}\nabla_{\mathbf{m}}\, \phi(\mathbf{x})| = \sup_{\mathbf{x}\in\mathbb{R}^D} \left| \prod_{i=1}^{D} x_i^{n_i} \frac{\partial^{m_i}}{\partial x_i^{m_i}}\, \phi(\mathbf{x}) \right| < \infty \tag{2.17}$$

is finite for all *multi-indices* $\mathbf{n} = (n_1, \ldots, n_D), \mathbf{m} = (m_1, \ldots, m_D) \in \mathbb{N}^D$ of finite rank $|\mathbf{n}| = n_1 + \ldots + n_D$ and $|\mathbf{m}| = m_1 + \ldots + m_D$ (i.e., having positive integer values with finite sum)[6]. The case $\mathbf{n} = \mathbf{0}$ corresponds to the demand that there can be no discontinuities or infinite values in either the smooth test function itself, or in any of its partial derivatives. The more general condition $\mathbf{n} \neq \mathbf{0}$ guarantees the localization of the test function, because the test function and all its derivatives must decrease faster than can be counteracted by any polynomial.

[6] We use $\partial^{m_i}\phi(\mathbf{x})/\partial x_i^{m_i}$ to denote the partial derivative of order m_i along the i-th dimension x_i of the multi-dimensional function $\phi(\mathbf{x})$.

Since the interest is mostly in one- and two-dimensional signals, we provide explicit definitions for one-dimensional Schwartz functions $\phi(x)$, i.e.,

$$\sup_{x\in\mathbb{R}} \left| x^n \frac{d^m\,\phi(x)}{d\,x^m} \right| < \infty \tag{2.18}$$

and two-dimensional Schwartz functions $\phi(x, y)$, i.e.,

$$\sup_{(x,y)\in\mathbb{R}^2} \left| x^{n_x}\, y^{n-n_x} \frac{\partial^m\,\phi(x,y)}{\partial x^{m_x}\,\partial y^{m-m_x}} \right| < \infty \tag{2.19}$$

for $n_x = 0, \ldots, n$ and $m_x = 0, \ldots, m$, where n and m are restricted to be finite integers. Gaussian functions, as well as their derivatives (see section 5.2), are smooth functions that have exponential decrease and are examples of smooth test functions.

It can be shown that $\mathcal{S}(\mathbb{R}^D)$, equipped with a norm such as

$$\nu_{N,M}[\phi(\mathbf{x})] = \sup_{\mathbf{x}\in\mathbb{R}^D, |\mathbf{n}|\leq N, |\mathbf{m}|\leq M} |\mathbf{x}^{\mathbf{n}} \nabla_{\mathbf{m}}\,\phi(\mathbf{x})|\,, \tag{2.20}$$

for positive polynomial rank N and derivative rank M, is a Banach space.

The space of bounded linear functionals T on $\mathcal{S}(\mathbb{R}^D)$ is the space of **tempered distributions**, denoted by $\mathcal{S}'(\mathbb{R}^D)$, and defined by the condition that $T \in \mathcal{S}'(\mathbb{R}^D)$ if there exists a positive constant $C > 0$ and multi-indices $\mathbf{n}$ and $\mathbf{m}$ such that

$$|\,T\langle\,\phi(\mathbf{x})\,\rangle\,| \leq C \cdot \sup_{\mathbf{x}\in\mathbb{R}^D} |\,\mathbf{x}^{\mathbf{n}} \nabla_{\mathbf{m}}\,\phi(\mathbf{x})\,| \tag{2.21}$$

for all $\phi(\mathbf{x}) \in \mathcal{S}(\mathbb{R}^D)$. The number $T\langle\,\phi(\mathbf{x})\,\rangle$ is used to denote the result of letting the linear functional T interact with the test function $\phi(\mathbf{x})$.

A sub-class of special interest are the **regular tempered distributions**

$$T_p\langle\,\phi(\mathbf{x})\,\rangle = \int_{\mathbb{R}^D} p(\mathbf{x}) \cdot \phi(\mathbf{x})\,d\mathbf{x} \tag{2.22}$$

that can be defined by means of functions $p(\mathbf{x}) \in \mathcal{P}(\mathbb{R}^D)$ of **polynomial growth**, i.e., for which there exist constants $F > 0$ and $Q \geq 0$ such that

$$|p(\mathbf{x})| \leq F\,\left(1 + \|\mathbf{x}\,\|^2\right)^Q, \tag{2.23}$$

for all $\mathbf{x} \in \mathbb{R}^D$.

Not all tempered distributions are regular tempered distributions. The mapping

$$\delta_{\mathbf{y}} : \mathcal{S}(\mathbb{R}^D) \mapsto \mathbb{R} : \phi(\mathbf{x}) \mapsto \phi(\mathbf{y}) \tag{2.24}$$

for $\mathbf{y} \in \mathbb{R}^D$, for instance defines a tempered distribution, called the **Dirac point distribution**, that is not a regular tempered distribution. It is common practice to abuse the notation and to write any tempered distribution T in the form T_f, even if no function $f(\mathbf{x})$ of polynomial growth exists. For instance, the Dirac point distribution $\delta_{\mathbf{y}}$ is usually associated with a function under the integral $\delta_{\mathbf{y}}(\mathbf{x}) = \delta(\mathbf{x} - \mathbf{y})$. Hence, for the case of Schwartz functions, one can again state that all linear functionals can be written as inproducts, i.e.,

$$T_f\langle\, \phi(\mathbf{x})\,\rangle = \langle \phi(\mathbf{x}), f(\mathbf{x})\rangle = \int_{\mathbb{R}^D} \phi^\star(\mathbf{x}) \cdot f(\mathbf{x})\, d\mathbf{x}, \tag{2.25}$$

provided that the inproduct definition is interpreted in the same loose sense as is common for tempered distributions.

3.5 BASES

An important concept in Banach spaces is the concept of bases. It is a very central concept in signal representations, since it describes how signals, that are in themselves untractable objects, can be mapped into numbers that are much easier to handle and study. There can be different types of bases in a Banach space. The set $B_\varphi = \{\vec{\varphi}_i; i = 1, 2, \dots\}$ is called a **Schauder basis** of the Banach space $V, \|\cdot\|$ if, for every vector $\vec{f}$ of V, there exist *unique* numbers $\hat{f}_\varphi(i), i = 1, 2, \dots,$ such that

$$\vec{f} = \lim_{N\to\infty} \sum_{i=1}^{N} \hat{f}_\varphi(i) \cdot \vec{\varphi}_i, \tag{2.26}$$

or, equivalently,

$$\left\| \vec{f} - \sum_{i=1}^{N} \hat{f}_\varphi(i) \cdot \vec{\varphi}_i \right\| \to 0 \text{ if } N \to \infty. \tag{2.27}$$

For instance, applied to the quadratic norm of equation (2.10), this translates into

$$\left[\sum_{x\in\mathbb{F}} w^2(x) \cdot \left\{ f(x) - \sum_{i=1}^{N} \hat{f}_\varphi(i) \cdot \varphi_i(x) \right\}^2 \right]^{1/2} \to 0 \text{ if } N \to \infty, \tag{2.28}$$

which is equivalent to making the weighted mean square error between a given signal $f(x)$ and its approximation equal to zero.

Vector spaces that have a **countable** basis, i.e., one that can be indexed by (a subset of) the integers, are termed **separable vector spaces**[7]. In a countable vector space, the vector $\vec{f}$ is described by its equivalent column vector $[\vec{f}]_{\varphi} = [\hat{f}_{\varphi}(1)\ \hat{f}_{\varphi}(2)\ \dots]^T$, which is referred to as the **representation** (or **coordinates**) of the vector $\vec{f}$ in the basis B_{φ}. Note that such a representation can be considered as a discrete signal on the set $I = \{1, 2, \dots\}$. The introduction of a basis for signals has similar consequences as the introduction of coordinates in geometry; it allows for a quantitative treatment of signals, i.e., geometric theorems can be replaced by algebraic theorems, and vice versa. Changing the set of basis vectors is equivalent to changing the coordinate system in geometry.

The uniqueness requirement of the coordinates $\hat{f}_{\varphi}(i)$ forces the vectors $\vec{\varphi}_i$ to be **linearly independent** in the sense that none of the vectors can be expressed as a linear combination of the other vectors. The linear independence of the basis vectors $\{\vec{\varphi}_i; i = 1, 2, \dots\}$ also implies that the zero vector $\vec{f} = \vec{0}$ has representation $[\hat{f}_{\varphi}(i) = 0;\ i = 1, 2, \dots]^T$.

The ordering of the basis vectors may be important in a Schauder basis. Reordering the basis vectors may for instance result in different coordinates, which is clearly an unwanted property. Therefore, in order for the intuitive notion of basis to apply, an additional requirement is needed. The Schauder basis $B_{\varphi} = \{\vec{\varphi}_i; i = 1, 2, \dots\}$ is called an **unconditional basis** of the Banach space $V, \|\cdot\|$ if it satisfies the following property

$$\sum_i \hat{f}_{\varphi}(i) \cdot \vec{\varphi}_i \in V \Rightarrow \sum_i |\hat{f}_{\varphi}(i)| \cdot \vec{\varphi}_i \in V. \tag{2.29}$$

For an unconditional basis, the order in which the basis vectors are taken does not matter. The most interesting Banach spaces, such as $\mathcal{L}^p(\mathbb{F})$, for $1 < p < \infty$, have unconditional bases[8].

An important distinction can be made between finite-dimensional and infinite-dimensional vector spaces V, depending on whether the basis of V has a finite number of basis vectors or not. The number of basis vectors is called the **dimension** of the vector space. Since all possible

[7] Most spaces of interest to us, such as $\mathcal{L}^p(\mathbb{F})$, are separable.

[8] It has been proven that $\mathcal{L}^p(\mathbb{F})$, with $\mathbb{F}$ continuous, does not contain an unconditional basis for $p = 1$ and $p = \infty$; even in these cases, we can however restrict our interest to a suitable subset of $\mathcal{L}^p(\mathbb{F})$, such as the set of infinitely differentiable functions, that has an unconditional basis.

bases of a finite-dimensional vector space contain the same number of basis vectors, this dimension is indeed a unique property of the space. Practical systems are restricted to a finite number of test functions, so that their device space is a finite-dimensional vector space.

4. LINEAR OPERATORS BETWEEN BANACH SPACES

4.1 LINEAR OPERATORS

Since all operations on vectors that we consider are composed of the elementary operations of vector addition, scalar multiplication and linear functionals, and since all these operations are linear, the overall operations of interest to us are also linear. The fact that there exists an extensive mathematical theory that deals exactly with such linear operations on vector spaces, is an important motivation for studying vector spaces. In this section, the main definitions and properties of this linear system theory are summarized for the case of Banach spaces.

A **linear operator** or **linear transformation** T from a Banach space V to a Banach space W is a mapping that assigns a *unique* vector $T\vec{f}$ of W to each vector $\vec{f}$ of V, in such a way that

$$T(a \cdot \vec{f} + b \cdot \vec{g}) = a \cdot T\vec{f} + b \cdot T\vec{g}, \tag{2.30}$$

for all $a, b \in \mathbb{C}$ and $\vec{f}, \vec{g} \in V$. A linear operator on the Banach space V is a linear operator from V to V. Linear functionals on a Banach space V can be viewed as vector operators from V to the vector space $\mathbb{C}$ of complex numbers.

The set $\operatorname{Ran} T = \{T\vec{f}; \vec{f} \in V\}$ of all images $T\vec{f}$ of vectors $\vec{f}$ of V is called the **range** of T. This set is a subspace of W. If this subspace is finite-dimensional with dimension n, then the operator T has **rank** n. The rank of the operator T is infinite when this subspace is not finite-dimensional. The set $\operatorname{Ker} T = \{\vec{f} \in V : T\vec{f} = \vec{0}\}$ of all vectors of V that are mapped onto the zero vector of W is called the **kernel** of T; $\operatorname{Ker} T$ is also a subspace of V.

An operator T is **continuous** if $\|T\vec{f} - T\vec{g}\,\|_W$ can be made arbitrarily small by making $\|\vec{f} - \vec{g}\,\|_V$ arbitrarily small, where $\|\cdot\|_V$ and $\|\cdot\|_W$ denote the vector norms in V and W respectively. From an engineering point of view, continuity of the signal processing operations is often a reasonable demand, since a small variation in the input signal, which may for instance be due to noise, should not result in a large change in output signal. An operator T is **bounded** if there exists a real bound $B_T \geq 0$ such that

$$\|T\vec{f}\,\|_W \leq B_T \cdot \|\vec{f}\,\|_V, \tag{2.31}$$

$\forall \vec{f} \neq 0 \in V$. The **norm of a bounded operator** is defined as the non-negative real number

$$\|T\| = \sup\{\|T\vec{f}\|_W; \vec{f} \in V \text{ and } \|\vec{f}\|_V \leq 1\}, \tag{2.32}$$

and can be interpreted as the largest factor by which T stretches any vector of V. It can be shown that all bounded operators are continuous, and vice versa.

The space $\mathcal{L}(V, W)$ of all *continuous* linear operators from the normed space V to the normed space W is itself a normed space with vector operations

$$\begin{aligned} (T_1 + T_2)\vec{f} &= T_1\vec{f} + T_2\vec{f}, \\ (a \cdot T)\vec{f} &= a \cdot T\vec{f}, \end{aligned} \tag{2.33}$$

for all $a \in \mathbb{C}$; $T_1, T_2, T \in \mathcal{L}(V, W)$ and $\vec{f} \in V$, and operator norm $\|T\|$. The left-hand sides in the above equations define the addition and scalar product of operators, while the right-hand sides are the addition and scalar product of vectors in W. It can be shown that $\mathcal{L}(V, W)$ is a Banach space if W is a Banach space. Note that $\mathcal{L}(V, W)$ is usually not an inner-product space.

The **composition** of two operators $T_1 \in \mathcal{L}(V, W)$ and $T_2 \in \mathcal{L}(W, U)$ is an operator of $\mathcal{L}(V, U)$, defined by $(T_2 \cdot T_1)\vec{f} = T_2(T_1\vec{f}), \forall \vec{f} \in V$. Even if both operators T_1 and T_2 are in $\mathcal{L}(V, V)$, then the composition of operators is in general non-commutative, i.e., $T_2 \cdot T_1 \neq T_1 \cdot T_2$. Note that the composition of operators is a generalization of matrix multiplication, and many of the familiar properties of matrices will have a counterpart in operator theory. For instance, an operator $T \in \mathcal{L}(V, W)$ is **invertible** if there exists a *continuous* operator, denoted by $T^{-1} \in \mathcal{L}(W, V)$, such that $T^{-1} \cdot T = I_V$ and $T \cdot T^{-1} = I_W$, where I_V and I_W are the identity operators of V and W, respectively (i.e., $I_V\vec{f} = \vec{f}, \forall \vec{f} \in V$).

If V is a Banach space, then there exists a simple useful way of producing inverses in $\mathcal{L}(V, V)$. More specifically, if $\|I_V - T\| < 1$, then the operator T is invertible with inverse

$$T^{-1} = \sum_{n=0}^{\infty} (I_V - T)^n. \tag{2.34}$$

4.2 DUAL BASES

One important application of inverse operators in $\mathcal{L}(V, V)$ is the concept of **dual basis**, which is now discussed in more detail. This concept will play a very important role in the remainder of the book.

Suppose that V is a *finite-dimensional* Banach space of dimension N, with an unconditional basis $B_\psi = \{\vec{\psi}_i; i = 1, \ldots, N\}$. This means that every vector in V can be expressed uniquely in terms of its coordinates, i.e.,

$$\vec{f} = \sum_{j=1}^{N} \hat{f}_\psi(j) \cdot \vec{\psi}_j \tag{2.35}$$

and that the inproduct can be rewritten as

$$\begin{aligned} \langle \vec{f}, \vec{g} \rangle &= \langle \sum_{i=1}^{N} \hat{f}_\psi(i) \cdot \vec{\psi}_i, \sum_{j=1}^{N} \hat{g}_\psi(j) \cdot \vec{\psi}_j \rangle \\ &= \sum_{i,j=1}^{N} \hat{f}^*_\psi(i) \cdot g_{\psi\psi}(i,j) \cdot \hat{g}_\psi(j) = [\vec{f}]^\dagger_\psi \, G_{\psi\psi} \, [\vec{g}]_\psi, \end{aligned} \tag{2.36}$$

for all $\vec{f}, \vec{g} \in V$, where $[\vec{f}]^\dagger_\psi$ denotes the Hermitian conjugate[9] of the coordinate vector $[\vec{f}]_\psi = [\hat{f}_\psi(1) \ldots \hat{f}_\psi(N)]^T$. The $N \times N$ **Gramian matrix** $G_{\psi\psi} = [g_{\psi\psi}(i,j)]_{i,j=1,\ldots,N}$ that occurs in the above expression has elements $g_{\psi\psi}(i,j) = \langle \vec{\psi}_i, \vec{\psi}_j \rangle$ for $i,j = 1, \ldots, N$. The definition of the inproduct implies that this Gramian matrix is Hermitian, i.e., $G^\dagger_{\psi\psi} = G_{\psi\psi}$. The basis B_ψ is called **orthogonal** if the Gramian matrix is diagonal, i.e., $g_{\psi\psi}(i,j) = g_i \cdot \delta_{ij}$, where δ_{ij} denotes Kronecker's delta function[10]. The basis B_ψ is called **orthonormal** if the Gramian matrix is equal to the identity matrix.

The operator defined by

$$G_{\psi\psi} \, \vec{f} = \sum_{i=1}^{N} \langle \vec{\psi}_i, \vec{f} \rangle \, \vec{\psi}_i, \tag{2.37}$$

for all vectors $\vec{f} \in V$, is a linear operator of $\mathcal{L}(V, V)$. It is called the **Gramian operator** or **normal operator** of the basis B_ψ. In terms of coordinates, it can be expressed as

$$G_{\psi\psi} : \vec{f} \mapsto G_{\psi\psi} \, \vec{f} = \sum_{i=1}^{N} \left[\sum_{j=1}^{N} g_{\psi\psi}(i,j) \, \hat{f}_\psi(j) \right] \vec{\psi}_i. \tag{2.38}$$

[9]The Hermitian conjugate of a matrix is obtained by element-wise complex conjugation, followed by matrix transposition.

[10]Kronecker's delta function: $\delta_{ij} = 1$ if $i = j$, and $\delta_{ij} = 0$ if $i \neq j$. The special case δ_{i0} is sometimes denoted by δ_i.

The coordinate vector $[G_{\psi\psi}\,\vec{f}]_{\psi}$ is hence obtained by a matrix multiplication of the Gramian matrix $G_{\psi\psi} = [g_{\psi\psi}(i,j)]_{i,j=1,\dots,N}$ with the coordinate vector $[\vec{f}]_{\psi}$. Because the vector operator and the associated matrix are so closely related, we do not introduce a separate notation to distinguish between them. The distinction is usually obvious from the context.

It can be shown that B_{ψ} is an unconditional basis if and only if the Gramian matrix $G_{\psi\psi}$ has an inverse $G_{\chi\chi} = G_{\psi\psi}^{-1} = [g_{\chi\chi}(i,j)]_{i,j=1,\dots,N}$. This inverse Gramian matrix can be used to define the **dual Gramian operator**

$$G_{\chi\chi} : \vec{f} \mapsto G_{\chi\chi}\,\vec{f} = \sum_{i=1}^{N} \left[\sum_{j=1}^{N} g_{\chi\chi}(i,j)\,\hat{f}_{\psi}(j) \right] \vec{\psi}_i. \tag{2.39}$$

The coordinate vector $[G_{\chi\chi}\,\vec{f}]_{\psi}$ is hence obtained by a multiplication of the matrix $G_{\chi\chi}$ with the coordinate vector $[\vec{f}]_{\psi}$.

Since $\vec{f} = (G_{\chi\chi} \cdot G_{\psi\psi})\,\vec{f}$, for all $\vec{f} \in V$, we obtain that

$$\vec{f} = G_{\chi\chi} \left(\sum_{i=1}^{N} \langle\, \vec{\psi}_i, \vec{f}\,\rangle\, \vec{\psi}_i \right) = \sum_{i=1}^{N} \langle\, \vec{\psi}_i, \vec{f}\,\rangle\, \vec{\chi}_i, \tag{2.40}$$

where $\vec{\chi}_i = G_{\chi\chi}\,\vec{\psi}_i$ is called the **dual basis vector** corresponding to the basis vector $\vec{\psi}_i$, for $i = 1,\dots,N$. Applying the above expression to the dual basis vector $\vec{\chi}_j$ shows that the following **biorthogonality** condition

$$\langle\, \vec{\psi}_i, \vec{\chi}_j\,\rangle = \delta_{ij} \tag{2.41}$$

must be satisfied by the basis vectors and the dual basis vectors. This in turn implies that $\vec{\psi}_i = G_{\psi\psi}\,\vec{\chi}_i$, for $i = 1,\dots,N$.

Any vector $\vec{f} \in V$ can be written in a unique way as a linear combination of the N dual basis vectors $\vec{\chi}_i$. This means that $B_{\chi} = \{\vec{\chi}_i; i = 1,\dots,N\}$ must also be a basis of V. It is called the **dual basis** of B_{ψ}. A vector $\vec{f}$ is represented by the coordinates

$$[\vec{f}]_{\chi} = [\hat{f}_{\chi}(i) = \langle\, \vec{\psi}_i, \vec{f}\,\rangle; i = 1,\dots,N]^T \tag{2.42}$$

in the basis B_{χ}.

Repeating the above reasoning, but now starting from the basis B_{χ}, one obtains that

$$\vec{f} = \sum_{i=1}^{N} \langle\, \vec{\chi}_i, \vec{f}\,\rangle\, \vec{\psi}_i, \tag{2.43}$$

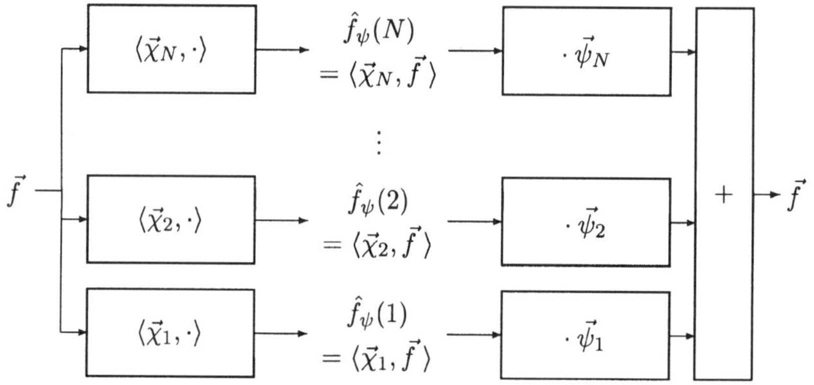

Figure 2.2. Signal analysis is performed by taking inner products between the incoming signal $\vec{f}$ and the dual basis vectors in B_χ. Signal synthesis is performed by applying the resulting representation $[\vec{f}]_\psi = [\hat{f}_\psi(i) = \langle \vec{\chi}_i, \vec{f} \rangle, i = 1, \ldots, N]^T$ to the basis vectors in B_ψ.

and

$$G_{\chi\chi}\, \vec{f} = \sum_{i=1}^{N} \langle \vec{\chi}_i, \vec{f} \rangle\, \vec{\chi}_i, \tag{2.44}$$

for all $\vec{f} \in V$, for $i = 1, \ldots, N$. Hence, the vectors in the dual basis B_χ specify which linear measurements have to be applied to a signal $\vec{f}$ to obtain its representation

$$[\vec{f}]_\psi = [\hat{f}_\psi(i) = \langle \vec{\chi}_i, \vec{f} \rangle; i = 1, \ldots, N]^T \tag{2.45}$$

in the basis B_ψ. The resulting signal analysis and synthesis is depicted in Figure 2.2.

Based on the above expressions, the following alternative formulae for the inner product

$$\langle \vec{f}, \vec{g} \rangle = \sum_{i=1}^{N} \langle \vec{f}, \vec{\psi}_i \rangle \langle \vec{\chi}_i, \vec{g} \rangle = \sum_{i=1}^{N} \langle \vec{f}, \vec{\chi}_i \rangle \langle \vec{\psi}_i, \vec{g} \rangle, \tag{2.46}$$

and

$$\langle \vec{f}, \vec{g} \rangle = [\vec{f}]_\psi^\dagger\, G_{\psi\psi}\, [\vec{g}]_\psi = [\vec{f}]_\chi^\dagger\, G_{\chi\chi}\, [\vec{g}]_\chi, \tag{2.47}$$

for all $\vec{f}, \vec{g} \in V$, can be derived. Amongst others, these expressions imply that the elements in the dual Gramian matrix are given by $g_{\chi\chi}(i,j) = \langle \vec{\chi}_i, \vec{\chi}_j \rangle$ for $i, j = 1, \ldots, N$.

In the special case of a basis of orthogonal vectors, i.e., $\langle \vec{\psi}_i, \vec{\psi}_j \rangle = g_i\, \delta_{ij}$, the vector product simplifies to

$$\langle \vec{f}, \vec{g} \rangle = \sum_i g_i\, f^\star_\psi(i)\, g_\psi(i), \tag{2.48}$$

and the dual vectors are given by $\vec{\chi}_i = \vec{\psi}_i / g_i$.

In case of a basis of non-orthogonal vectors, the following mechanical procedure arises for finding the dual basis. First, the Gramian matrix $G_{\psi\psi} = [\langle \vec{\psi}_i, \vec{\psi}_j \rangle]_{i,j,=1,\dots,N}$ is constructed from all pairwise combinations of the basis vectors in B_ψ. Second, this matrix is inverted to obtain the dual Gramian matrix $G_{\chi\chi} = [g_{\chi\chi}(i,j)]]_{i,j,=1,\dots,N}$. Third, the dual vectors are obtained through

$$\vec{\chi}_j = \sum_{i=1}^{N} g_{\chi\chi}(i,j)\, \vec{\psi}_i, \tag{2.49}$$

for $j = 1, \dots, N$. The representation $[\vec{\chi}_j]_\psi$ of the j-th dual basis vector is hence equal to the j-th column of the inverse Gramian matrix $G_{\chi\chi}$. This procedure is illustrated by means of an example in Appendix 2.A.

In conclusion, an N-dimensional vector space can be specified by either a set B_ψ of N basis vectors or a set B_χ of N dual basis vectors. It depends on the application which alternative is preferred. For instance, in the case of an image display, where images arise by combining basic display responses, a specification by means of basis vectors is straightforward. In the case of an image sensor, where the linear measurements performed on an image are known, a specification by means of dual basis vectors is obvious. Several problems in image technology, some of which are discussed in subsequent chapters, can be reduced to finding the dual basis that corresponds to a given basis.

The concept of dual basis can be extended in a straightforward way to infinite-dimensional vector spaces, provided that the inverse operator $G_{\chi\chi}$ of the Gramian operator $G_{\psi\psi}$ can be found. The existence of such an inverse is not guaranteed in the most general case. An example of the construction of a dual basis in case of an infinite-dimensional vector space will be presented in section 5.2.

4.3 OPERATORS AND MATRICES

A vector space can be described by many different bases. It is useful to know how the representations of a given signal in two different bases can be related. For example, in image technology, the different subsystems in an imaging chain can have very different functionality (for example, display versus coding), so that it is often advisable to adopt distinct signal representations to study them. To describe how a signal passes through the entire imaging chain hence requires that the output representation from one subsystem can be mapped into the input representation for the next subsystem.

Suppose that $B_\vartheta = \{\vec{\vartheta}_i; i = 1, \ldots, N\}$ is an alternative basis of V. This basis also contains N linearly independent vectors, and the matrix[11]

$$G_{\psi\vartheta} = [\langle \vec{\psi}_i, \vec{\vartheta}_j \rangle]_{i,j=1,\ldots,N} \tag{2.50}$$

is nonsingular. The representations $[\vec{f}]_\psi$ and $[\vec{f}]_\vartheta$ of a given vector $\vec{f}$ in the two bases are related by

$$[\vec{f}]_\psi = G_{\psi\psi}^{-1} \cdot G_{\psi\vartheta} \cdot [\vec{f}]_\vartheta = G_{\chi\vartheta} \cdot [\vec{f}]_\vartheta, \tag{2.51}$$

with

$$G_{\chi\vartheta} = [\langle \vec{\chi}_i, \vec{\vartheta}_j \rangle]_{i,j=1,\ldots,N}, \tag{2.52}$$

where $B_\chi = \{\vec{\chi}_i; i = 1, \ldots, N\}$ is the dual basis of B_ψ. In the specific case that the alternative basis B_ϑ is equal to the dual basis B_χ, the transformation rules $[\vec{f}]_\psi = G_{\chi\chi} \cdot [\vec{f}]_\chi$ and $[\vec{f}]_\chi = G_{\psi\psi} \cdot [\vec{f}]_\psi$ are obtained.

The theory of operators in $\mathcal{L}(V, W)$ is closely linked to the theory of (possibly infinite-dimensional) matrices. Indeed, suppose that the vector spaces V and W have respective bases $B_\psi = \{\vec{\psi}_n, n \in N\}$, and $B_\varphi = \{\vec{\varphi}_m, m \in M\}$. If the basis B_φ of W has a dual basis $B_\phi = \{\vec{\phi}_m, m \in M\}$, then

$$T_{\phi\psi} = [t_{\phi\psi}(m,n)]_{m\in M, n\in N} = [\langle \vec{\phi}_m, T\vec{\psi}_n \rangle_W]_{m\in M, n\in N} \tag{2.53}$$

is called the **matrix of the operator** T with respect to the above bases. An arbitrary input vector $\vec{f}$ of V is mapped to the vector

$$T\vec{f} = \sum_{m\in M} \left[\sum_{n\in N} t_{\phi\psi}(m,n) \cdot \hat{f}_\psi(n) \right] \vec{\varphi}_m \tag{2.54}$$

of W, or equivalently, in terms of vector representations

$$[Tf]_\varphi = T_{\phi\psi} \cdot [\vec{f}]_\psi. \tag{2.55}$$

Note that the matrix notation depends on the specific choice of the bases, while the operator notation does not have this dependency. Therefore,

[11]The matrix notation $G_{\alpha\beta}$ has a nice tensor property, which makes it easy to work with. This property is elaborated on in exercise 8.

the operator notation is often preferable, partly because there exists an extensive body of literature on linear vector operators.

Some well-known properties of matrices, such as eigenvalue and singular value decompositions, and inverses, cannot be easily generalized for all bounded operators, but only for a subset, called compact operators. An operator T of $\mathcal{L}(V, W)$ is said to be **compact** if, for every bounded sequence $\{\vec{f_k}\}$ in V (i.e., $\|\vec{f_k}\| < \infty, \forall k$), the mapped sequence $\{T\vec{f_k}\}$ contains a convergent subsequence in W. This is a fairly technical definition that is mostly of theoretical interest. Indeed, practical operators all have finite rank, since the output signals of a practical device are generated by controlling a finite number of variables, and hence $\operatorname{Ran} T$ is finite-dimensional. It can be shown that all bounded operators of finite rank are also compact. In section 6, we will be interested in specific operators of infinite rank, for which there exists a simplified test for compactness. At this time, it suffices to say that most operators of interest behave very similar to matrix multiplications.

5. VECTOR LENGTHS AND HILBERT SPACES

5.1 HILBERT SPACES

In section 2, we already restricted the discussion to pre-Hilbert spaces, i.e., vector spaces in which a vector product can be defined. It is now discussed how this vector product can also be used to define a specific norm.

Since $\langle \vec{f}, \vec{f} \rangle$ is real and ≥ 0, the following definition of the **length**

$$\|\vec{f}\| = \sqrt{\langle \vec{f}, \vec{f} \rangle} \tag{2.56}$$

of a vector $\vec{f}$ makes sense. It is easily verified that this length defines a norm with all the required properties, so that every inner-product space can be transformed into a normed space by adopting this norm as a valid signal approximation measure. A Banach space with a length norm is called a **Hilbert space**.

Note that most Banach spaces are not Hilbert spaces, since most norms cannot be written as inner products. For instance, the p-norm for $p \neq 2$ cannot be derived from an inner product, so that only $\mathcal{L}^2(\mathbb{F})$ is a Hilbert space. It can be shown that the *parallelogram law*

$$\|\vec{f} + \vec{g}\|^2 + \|\vec{f} - \vec{g}\|^2 = 2 \cdot \|\vec{f}\|^2 + 2 \cdot \|\vec{g}\|^2, \tag{2.57}$$

for all $\vec{f}, \vec{g} \in V$, which is clearly satisfied by an inner-product norm, characterizes Hilbert spaces in the class of Banach spaces. The parallel-

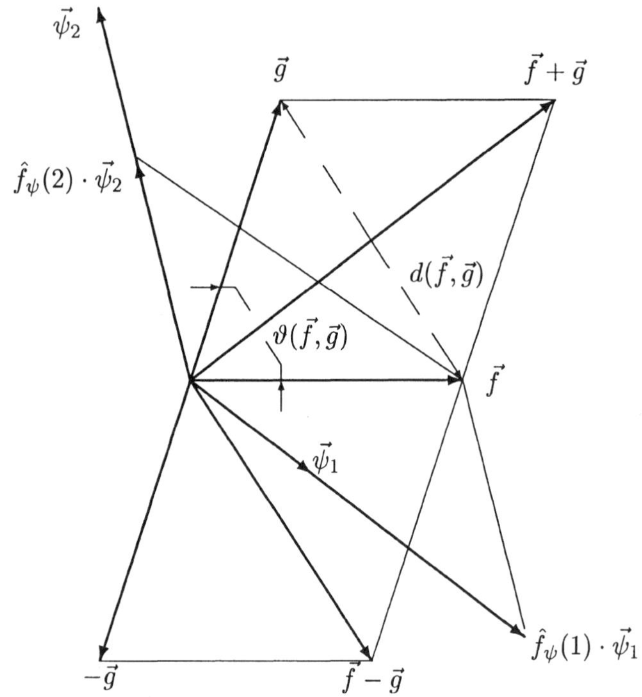

Figure 2.3. Two-dimensional Euclidean space with basis vectors $\{\vec{\psi}_1, \vec{\psi}_2\}$. Illustrated are: the addition, inversion and subtraction of vectors; the scalar product of a vector with a (real) number; the distance and relative angle between vectors.

ogram construction of vector addition in Figure 2.3 is hence appropriate for Hilbert spaces (see exercise 6).

A real Hilbert space is called a **Euclidean space**. The Euclidean spaces of dimensions 1, 2 and 3 are very familiar to us. We can use our geometrical insights, based on everyday experience, for studying signals that belong to a Euclidean space. For instance, in the case of a two-dimensional Euclidean space, most of the definitions and properties can be illustrated in a plane, as is demonstrated in Figure 2.3. In the next chapter, we will derive how the front end of human color vision can for instance be described in a three-dimensional Euclidean space.

Because of the following *polarization identity* (with $j^2 = -1$)

$$\langle \vec{f}, \vec{g} \rangle = \frac{1}{4} \sum_{n=0}^{3} (-j)^n \cdot \| \vec{f} + j^n \vec{g} \|^2, \tag{2.58}$$

for all $\vec{f}, \vec{g} \in V$, the definition of the norm and the vector product are interchangeable in a complex Hilbert space. For a real vector space, where $\langle \vec{f}, \vec{g} \rangle = \langle \vec{g}, \vec{f} \rangle$, the following relationship

$$\langle \vec{f}, \vec{g} \rangle = \frac{1}{4} \left(\| \vec{f} + \vec{g} \|^2 - \| \vec{f} - \vec{g} \|^2 \right) \tag{2.59}$$

can be used to derive the inner product from the norm.

Since all Hilbert spaces are Banach spaces, they also contain bases. An unconditional basis in a Hilbert space is called a **Riesz basis**. A Riesz basis $B_\psi = \{\vec{\psi}_1, \vec{\psi}_2, \dots\}$ has the property that there exist real numbers $a_\psi > 0$ and $b_\psi < \infty$ so that

$$a_\psi \| \vec{f} \|^2 \leq \sum_i |\langle \vec{\psi}_i, \vec{f} \rangle|^2 \leq b_\psi \| \vec{f} \|^2, \tag{2.60}$$

for all $\vec{f} \in V$. A specific property of separable Hilbert spaces is that they contain **orthonormal bases**, i.e., the basis $B_e = \{\vec{e}_1, \vec{e}_2, \dots\}$ can be chosen such that

$$\langle \vec{e}_i, \vec{e}_j \rangle = \delta_{ij}, \tag{2.61}$$

(δ_{ij} is the Kronecker delta) and

$$\| \vec{f} \|^2 = \sum_i |\langle \vec{e}_i, \vec{f} \rangle|^2. \tag{2.62}$$

It can be shown that any Riesz basis can be obtained by mapping an orthonormal basis through a bounded operator T.

A well-known example of an orthonormal basis is the Fourier basis that is discussed in detail in chapter 6. This basis is especially useful for describing position-invariant systems (see chapter 6 for a precise definition of position-invariant systems). Many systems that are of interest in image technology are however not position-invariant. This is one of the reasons why a more general theory for signal descriptions that only assumes linearity of the systems involved is discussed here. In the next section, a less-familiar example of a countable (infinite-dimensional) basis for signals in $\mathcal{L}^2(\mathbb{R})$, i.e., the basis of orthogonal polynomials, is presented.

5.2 BASES OF ORTHOGONAL POLYNOMIALS

GENERAL CASE

Consider the subset $\mathbb{F} = \{x \in \mathbb{R}; w(x) \neq 0\}$ of the real axis where a given real function $w(x)$ differs from zero. The set $\mathbb{F}$ can be either a discrete or a continuous subset of the real axis. The set of all real-valued signals $f(x)$ on the set $\mathbb{F}$ for which $f(x)\,w(x)$ is square-integrable, i.e., for which $\int_{\mathbb{F}} w^2(x)\, f^2(x)\, dx$ is finite, is denoted by $V(\mathbb{F})$. The vector product

$$\langle f(x), g(x) \rangle = \int_{\mathbb{F}} w^2(x)\, f(x)\, g(x)\, dx \tag{2.63}$$

can be shown to exist for all signals $f(x), g(x) \in V(\mathbb{F})$. This vector product can be used to define a length norm, i.e., $\|f(x)\|^2 = \langle f(x), f(x)\rangle$.

We show how to derive polynomials $\psi_n(x)$ of degree$[\psi_n(x)] = n, n \in \mathbb{N}$, that form a complete orthonormal basis for $V(\mathbb{F})$. The index set $\mathbb{N}$ contains the positive integers, including zero. There exists a generic, but not very efficient, algorithm for constructing these **orthonormal polynomials** [Sze59] for an arbitrary window function $w(x)$, i.e.,

$$\psi_n(x) = \frac{1}{\sqrt{M_{n-1} \cdot M_n}} \begin{vmatrix} c_0 & c_1 & \dots & c_n \\ c_1 & c_2 & \dots & c_{n+1} \\ \vdots & \vdots & & \vdots \\ c_{n-1} & c_n & \dots & c_{2n-1} \\ 1 & x & \dots & x^n \end{vmatrix}, \tag{2.64}$$

where the determinant M_n is defined by

$$M_n = |c_{i+j}|_{i,j=0,\dots,n}\ ,\ M_{-1} = 1 \tag{2.65}$$

and

$$c_n = \int_{\mathbb{F}} x^n\, w^2(x)\, dx \tag{2.66}$$

is the n-th order moment, for $n \in \mathbb{N}$. For example,

$$\psi_1(x) = \frac{1}{\sqrt{M_0 \cdot M_1}} \begin{vmatrix} c_0 & c_1 \\ 1 & x \end{vmatrix} = \frac{1}{\sqrt{c_0\,(c_0\, c_2 - c_1^2)}}\,(c_0 \cdot x - c_1). \tag{2.67}$$

This algorithm also applies to discrete signals if the expression for the n-th order moment is replaced by the discrete sum

$$c_n = \sum_{x \in \mathbb{F}} x^n\, w^2(x), \tag{2.68}$$

and provided that the polynomials $\psi_n(x)$ are only evaluated at the positions $x \in \mathbb{F}$. The number of orthonormal polynomials is finite in case the discrete window $w(x)$ is only different from zero at a finite number of positions.

Instead of using the above generic algorithm, it is usually more efficient to first determine the orthogonal polynomials $\tilde{\psi}_n(x)$ using recursion, and to normalize these polynomials afterwards, i.e., to map them to $\psi_n(x) = \tilde{\psi}_n(x)/\sqrt{\beta_n}$, where $\beta_n = \langle \tilde{\psi}_n(x), \tilde{\psi}_n(x) \rangle = \| \tilde{\psi}_n(x) \|^2$. By definition, $x \cdot \tilde{\psi}_n(x)$ is a polynomial of degree $n+1$, and can hence be written as the sum of orthogonal polynomials up to degree $n+1$, i.e.,

$$x \cdot \tilde{\psi}_n(x) = \tilde{\psi}_{n+1}(x) + \sum_{i=0}^{n} \gamma_i \cdot \tilde{\psi}_i(x), \tag{2.69}$$

for $n > 0$, with $\tilde{\psi}_0(x) = 1$ and

$$\gamma_i = \frac{\langle \tilde{\psi}_i(x), x \cdot \tilde{\psi}_n(x) \rangle}{\langle \tilde{\psi}_i(x), \tilde{\psi}_i(x) \rangle}, \tag{2.70}$$

for $i = 0, \ldots, n$. Because $\tilde{\psi}_n(x)$ is orthogonal to any polynomial of degree smaller than n,

$$\langle \tilde{\psi}_k(x), x \cdot \tilde{\psi}_n(x) \rangle = \langle \tilde{\psi}_n(x), x \cdot \tilde{\psi}_k(x) \rangle = 0, \tag{2.71}$$

for $k < n-1$. This leads to the following simplification

$$x \cdot \tilde{\psi}_n(x) = \tilde{\psi}_{n+1}(x) + \gamma_n \cdot \tilde{\psi}_n(x) + \gamma_{n-1} \cdot \tilde{\psi}_{n-1}(x), \tag{2.72}$$

or equivalently

$$\tilde{\psi}_{n+1}(x) = (x - \gamma_n) \cdot \tilde{\psi}_n(x) - \gamma_{n-1} \cdot \tilde{\psi}_{n-1}(x), \tag{2.73}$$

for the recursion relationship. The coefficients γ_n and γ_{n-1} can be expressed in terms of β_n, β_{n-1} and the inproduct $\alpha_n = \langle x \cdot \tilde{\psi}_n(x), \tilde{\psi}_n(x) \rangle$, so that

$$\tilde{\psi}_{n+1}(x) = \left(x - \frac{\alpha_n}{\beta_n} \right) \cdot \tilde{\psi}_n(x) - \frac{\beta_n}{\beta_{n-1}} \cdot \tilde{\psi}_{n-1}(x) \tag{2.74}$$

is an alternative formulation of the recursion relationship.

If the window is symmetric, i.e., if $w(x) = w(-x)$, then $\alpha_n = 0$, so that $\tilde{\psi}_1(x) = x$ and $\tilde{\psi}_n(-x) = (-1)^n \tilde{\psi}_n(x)$. The orthogonal polynomials of increasing order are hence alternatively odd and even in case of an even window function.

The orthogonal polynomials form a basis for $V(\mathbb{F})$, i.e., any function $f(x) \in V(\mathbb{F})$ can be approximated by a weighted sum of these polynomials [Sze59]. Using the definition of a basis, this implies that the weighted-mean-square error, i.e.,

$$\int_{\mathbb{F}} w^2(x) \left[f(x) - \sum_{n=0}^{N} \hat{f}_\psi(n) \cdot \psi_n(x) \right]^2 dx, \tag{2.75}$$

must converge to zero if $N \to \infty$. This can alternatively be stated as

$$w(x) \left[f(x) - \sum_{n=0}^{N} \hat{f}_\psi(n) \cdot \psi_n(x) \right] \to 0 \tag{2.76}$$

converges to zero at all positions $x \in \mathbb{R}$ if $N \to \infty$, which implies that the difference between the signal $f(x)$ and its polynomial approximation can be made arbitrarily small at the positions where $w(x) \neq 0$. **The series expansion is said to approximate the signal $f(x)$ within the window** $w(x)$. Such local signal approximations play a very prominent role in image processing and human vision.

The representation of the signal $f(x)$ in the basis of orthonormal polynomials has coefficients

$$\hat{f}_\psi(n) = \langle\, \psi_n(x), f(x) \,\rangle = \int_{\mathbb{F}} w^2(x)\, \psi_n(x) \cdot f(x)\, dx, \tag{2.77}$$

since $\psi_n(x)$ is orthonormal and hence identical to its dual basis vector. In order to obtain the polynomial coefficients, the following test functions

$$\phi_n(x) = w^2(x)\, \psi_n(x) \tag{2.78}$$

for $n = 0, 1, \ldots,$ must be applied to the input signal $f(x)$.

If $w(x) \neq 0$ for all $x \in \mathbb{R}$, then $f(x)$ can be replaced by $g(x)/w(x)$ in the above expressions. The condition that $\int w^2(x)\, f^2(x)\, dx$ must be finite is then equivalent to the demand that $\int g^2(x)\, dx$ must be finite, i.e., $g(x)$ must belong to $\mathcal{L}^2(\mathbb{R})$. This shows that the set of weighted polynomials, i.e.,

$$B_\varphi = \{\varphi_n(x) = w(x) \cdot \psi_n(x); n \in \mathbb{N}\}, \tag{2.79}$$

is a countable basis for $\mathcal{L}^2(\mathbb{R})$. More sophisticated countable bases for $\mathcal{L}^2(\mathbb{R})$ have recently been derived in the theory of wavelets [Dau92].

KRAWTCHOUK POLYNOMIALS

By way of example, we derive the orthogonal polynomials for the specific case of a binomial window function $w^2(x)$ =[1 2 1]/4 of size 3 (with

$x = -1, 0, 1$). These polynomials can be used to locally approximate a discrete signal. Since the moments of this binomial window satisfy

$$\begin{aligned} c_0 &= \sum_{x=-1}^{1} w^2(x) = 1 \\ c_{2k-1} &= \sum_{x=-1}^{1} x^{2k-1}\, w^2(x) = 0 \\ c_{2k} &= \sum_{x=-1}^{1} x^{2k}\, w^2(x) = \frac{1}{2} \end{aligned}$$

for $k = 1, 2, \ldots,$ we obtain, using either of the two methods described above, that the (non-zero) orthonormal polynomials are given by

$$\begin{aligned} \psi_0(x) = 1 &\quad\rightarrow\quad [1\ 1\ 1] \\ \psi_1(x) = \sqrt{2}\, x &\quad\rightarrow\quad [-\sqrt{2}\ 0\ \sqrt{2}] \\ \psi_2(x) = 2\, x^2 - 1 &\quad\rightarrow\quad [1\ -1\ 1] \end{aligned}$$

This result in turn implies that the test functions that have to be applied to the original signal $f(x)$ in order to obtain the polynomial coefficients are equal to

$$\begin{aligned} \phi_0(x) = \psi_0(x)\, w^2(x) &\quad\rightarrow\quad \frac{1}{4}\,[1\ 2\ 1] \\ \phi_1(x) = \psi_1(x)\, w^2(x) &\quad\rightarrow\quad \frac{1}{4}\,[-\sqrt{2}\ 0\ \sqrt{2}] \\ \phi_2(x) = \psi_2(x)\, w^2(x) &\quad\rightarrow\quad \frac{1}{4}\,[1\ -2\ 1] \end{aligned}$$

The orthonormal polynomials for a binomial window of arbitrary size $M + 1$, i.e.,

$$w^2(x) = \frac{1}{2^M} C_M^x = \frac{M!}{2^M\,(M-x)!\,x!}, \tag{2.80}$$

for $x = 0, \ldots, M$, are known as the **Krawtchouk polynomials**

$$\psi_n(x) = \frac{1}{\sqrt{C_M^n}} \sum_{k=0}^{n} (-1)^{n-k} C_{M-x}^{n-k}\, C_x^k, \tag{2.81}$$

for $x, n = 0, \ldots, M$ [Sze59]. It is left up to the reader to verify that this general expression leads to the same results as obtained above in the example case of $M = 2$ (provided the origin of the Krawtchouk polynomials is shifted by $M/2$).

HERMITE POLYNOMIALS

Another interesting special case of orthogonal polynomials arises when $\mathbb{F} = \mathbb{R}$ and

$$w^2(x) = \frac{1}{\sigma\sqrt{\pi}} \exp\left[-\frac{x^2}{\sigma^2}\right] \tag{2.82}$$

is a Gaussian window. The orthonormal basis vectors in this case are equal to

$$\psi_n(x) = \frac{1}{\sqrt{2^n n!}} H_n\left(\frac{x}{\sigma}\right), \tag{2.83}$$

where the standard notation

$$H_n(x) = (-1)^n e^{x^2} \frac{d^n}{dx^n}\left(e^{-x^2}\right) = \sum_{m=0}^{\lfloor n/2 \rfloor} (-1)^m \frac{n!}{(n-2m)!\, m!}(2x)^{n-2m} \tag{2.84}$$

is used for the **Hermite polynomial** of order n. These Hermite polynomials satisfy the recurrence relation

$$H_n(x) = 2\,x \cdot H_{n-1}(x) - 2\,(n-1) \cdot H_{n-2}(x), \tag{2.85}$$

with $H_0(x) = 1$, $H_1(x) = 2x$, $H_2(x) = 2(2x^2 - 1)$, etc., and the derivative relation

$$\frac{dH_n(x)}{dx} = 2\,n \cdot H_{n-1}(x), \tag{2.86}$$

for $n = 1, 2, 3, \ldots$ [Sze59, Mar90c].

The resulting signal expansion[12] is

$$w(x)\left[f(x) - \sum_{n=0}^{\infty} \hat{f}_\psi(n)\, \psi_n(x)\right] = 0, \tag{2.87}$$

for $x \in \mathbb{R}$, with

$$\hat{f}_\psi(n) = \int_{-\infty}^{+\infty} w^2(x)\, \psi_n(x) \cdot f(x)\, dx, \tag{2.88}$$

for $n \in \mathbb{N}$. The test functions

$$\phi_n(x) = w^2(x)\, \psi_n(x) = \frac{(-1)^n}{\sqrt{2^n n!}} \frac{d^n}{d(x/\sigma)^n}\left[\frac{1}{\sigma\sqrt{\pi}} \exp\left(-\frac{x^2}{\sigma^2}\right)\right] \tag{2.89}$$

[12] Equality in this expansion actually means convergence in the mean square sense, as discussed in the previous section.

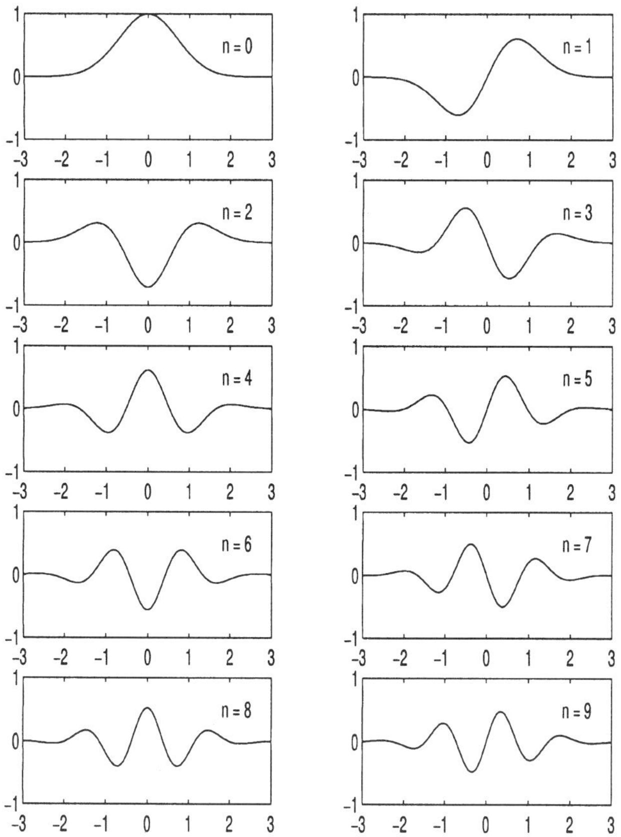

Figure 2.4. Derivative-of-Gaussian test functions for orders $n < 10$.

that are applied to the original signal $f(x)$ are hence proportional to **derivatives of Gaussians**. These smooth test functions are plotted in Figure 2.4 for orders $n < 10$ and have been shown to agree very well with the experimentally determined receptive fields in human spatial vision [You85]. This leads to the interpretation that the visual system may be locally approximating the incoming image by a polynomial expansion [Mar90c]. This polynomial description will be used in chapter 12 to construct image-quality measures.

5.3 ORTHOGONAL PROJECTION

An important property of Hilbert spaces is the *closest point property.* Consider a subset U of a complex vector space V. This subset is **convex** if $\lambda \cdot \vec{f}_1 + (1-\lambda) \cdot \vec{f}_2$, for $0 < \lambda < 1$, belongs to U for all $\vec{f}_1, \vec{f}_2 \in U$. The closest point property states that for any vector $\vec{g}$ of the Hilbert space V there exists a *unique* vector $P\vec{g}$ in the **closed convex set**[13] U which is closer to $\vec{g}$ than any other vector in U, i.e.,

$$\|\vec{g} - P\vec{g}\,\| = \inf_{\vec{f} \in U} \|\vec{g} - \vec{f}\,\|, \tag{2.90}$$

where inf denotes infimum[14]. In a Banach space, on the contrary, there may be infinitely many closest vectors. Note that P is an operator since it assigns a unique vector $P\vec{g}$ to any vector $\vec{g}$ of V. The closest point property does however not guarantee that P is a linear operator.

The closest point property formalizes a practical signal approximation situation in which the subset U contains all signals that a given system can produce (for example, all images that can be represented on a given display), and $\vec{g}$ is an input signal that needs to be approximated as closely as possible. Although the existence of a unique *best approximation* to the given input signal is guaranteed, one still needs an algorithm for finding this optimal approximation. If U is a *subspace* of the Hilbert space$V$, and hence also a closed convex set, then the problem is much simplified, since P becomes a linear operator for which an explicit expression exists.

Let the subspace U have basis $B_\varphi = \{\vec{\varphi}_i\}$ and let us assume that the dual basis $B_\phi = \{\vec{\phi}_i\}$ exists and has been derived. Any element of U, including the closest vector $P\vec{g}$, can be expressed as a linear combination of the basis vectors in U, i.e.,

$$P\vec{g} = \sum_i \widehat{Pg}_\varphi(i)\, \vec{\varphi}_i = \sum_i \widehat{Pg}_\phi(i)\, \vec{\phi}_i. \tag{2.91}$$

The closest vector can be found by minimizing the quadratic form

$$\|\vec{g} - P\vec{g}\,\|^2 = \langle \vec{g} - P\vec{g}, \vec{g} - P\vec{g}\,\rangle \tag{2.92}$$

in terms of the coordinates $\widehat{Pg}_\varphi(i), \forall i$ (or $\widehat{Pg}_\phi(i), \forall i$). This is done by equating the partial derivatives of the quadratic form with respect to

---

[13]The set has to be closed, because otherwise, not all 'border' elements are part of the set. In mathematical terms, it is required that all convergent series in U converge to an element in U.

[14]The infimum (inf) of a set is the largest element that is smaller than or equal to all elements in the set.

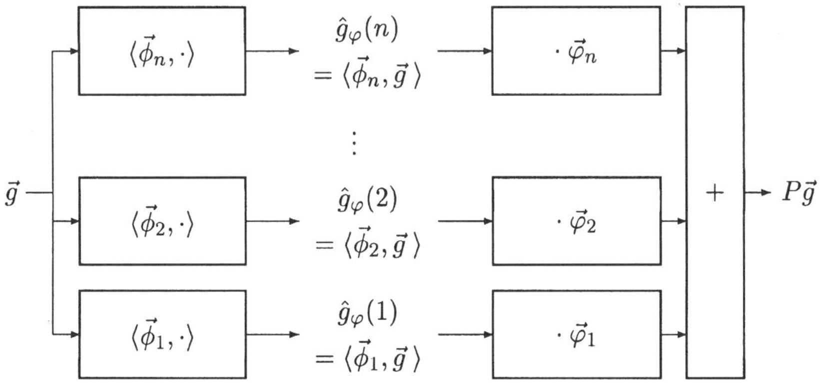

Figure 2.5. Signal analysis is performed by taking inner products between the incoming signal $\vec{g}$ and the dual basis vectors in B_ϕ. Signal synthesis is performed by applying the resulting representation $[\vec{g}]_\varphi = [\hat{g}_\varphi(i), i = 1, \ldots, n]^T$ to the basis vectors in B_φ. The output signal $P\vec{g}$ is the projection of the vector $\vec{g}$ onto the subspace U spanned by the basis B_φ.

these weights to zero, and solving the resulting set of linear equations. The solution is

$$\boxed{P\vec{g} = \sum_i \langle \vec{\phi}_i, \vec{g} \rangle \, \vec{\varphi}_i = \sum_i \langle \vec{\varphi}_i, \vec{g} \rangle \, \vec{\phi}_i,} \tag{2.93}$$

and is called the **orthogonal projection** $P\vec{g}$ of the vector $\vec{g}$ onto the subspace U. This projection operation is depicted in Figure 2.5.

It is easily verified that the linear operator P is rightfully called a projection operator, since it satisfies the basic requirement of **idempotency**, i.e., $P \cdot P = P$. Idempotency expresses the fact that all elements of U are mapped onto themselves. Note that the projection operator satisfies

$$\langle P\vec{f}, P\vec{g} \rangle = \langle P\vec{f}, \vec{g} \rangle = \langle \vec{f}, P\vec{g} \rangle, \tag{2.94}$$

for all vectors $\vec{f}, \vec{g} \in V$, and that the distance between a vector $\vec{g}$ and its projection $P\vec{g}$ can be derived from

$$d^2(\vec{g}, P\vec{g}) = \langle \vec{g}, \vec{g} \rangle - \langle P\vec{g}, P\vec{g} \rangle, \tag{2.95}$$

for all $\vec{g} \in V$.

If the vector space V has basis $B_\psi = \{\vec{\psi}_j\}$, with dual basis $B_\chi = \{\vec{\chi}_j\}$, then the following expressions for the original vector

$$\vec{g} = \sum_j \hat{g}_\psi(j) \, \vec{\psi}_j = \sum_j \hat{g}_\chi(j) \, \vec{\chi}_j \tag{2.96}$$

can be substituted into the expression for the orthogonal projection $P\vec{g}$. This allows to derive how the representations $[\vec{g}]_\psi$ and $[\vec{g}]_\chi$ of the original vector in the vector space V are mapped into the representations $[P\vec{g}]_\varphi$ and $[P\vec{g}]_\phi$ of the projected vector in the subspace U. A representation in a common basis B_ψ or B_χ of V is however required in order to compare the original vector $\vec{g}$ with the projected vector $P\vec{g}$. Based on $\langle \vec{\chi}_j, P\vec{g}\rangle = \langle P\vec{\chi}_j, \vec{g}\,\rangle$ and $\langle \vec{\psi}_j, P\vec{g}\,\rangle = \langle P\vec{\psi}_j, \vec{g}\,\rangle$, it is immediately obtained that

$$\boxed{P\vec{g} = \sum_j \langle P\vec{\chi}_j, \vec{g}\,\rangle\, \vec{\psi}_j = \sum_j \langle P\vec{\psi}_j, \vec{g}\,\rangle\, \vec{\chi}_j} \tag{2.97}$$

Evaluating these expressions for an arbitrary input signal $\vec{g}$ requires knowledge of the projected basis vectors

$$\begin{aligned} P\vec{\chi}_j &= \sum_i \langle \vec{\phi}_i, \vec{\chi}_j\,\rangle \vec{\varphi}_i = \sum_i \langle \vec{\varphi}_i, \vec{\chi}_j\,\rangle \vec{\phi}_i \\ &= \sum_k \left[\sum_i \langle \vec{\psi}_k, \vec{\varphi}_i\,\rangle\, \langle \vec{\phi}_i, \vec{\chi}_j\,\rangle \right] \vec{\chi}_k = \sum_k p_{\psi\chi}(k,j)\, \vec{\chi}_k, \end{aligned} \tag{2.98}$$

or

$$\begin{aligned} P\vec{\psi}_j &= \sum_i \langle \vec{\phi}_i, \vec{\psi}_j\,\rangle \vec{\varphi}_i = \sum_i \langle \vec{\varphi}_i, \vec{\psi}_j\,\rangle \vec{\phi}_i \\ &= \sum_k \left[\sum_i \langle \vec{\chi}_k, \vec{\varphi}_i\,\rangle\, \langle \vec{\phi}_i, \vec{\psi}_j\,\rangle \right] \vec{\psi}_k = \sum_k p_{\chi\psi}(k,j)\, \vec{\psi}_k, \end{aligned} \tag{2.99}$$

for all j. Once these projected vectors have been determined, the representations $[P\vec{g}]_\psi$ and $[P\vec{g}]_\chi$ of the projected vector $P\vec{g}$ can be expressed in terms of the corresponding representations $[\vec{g}]_\psi$ and $[\vec{g}]_\chi$ of the original vector, i.e.,

$$[P\vec{g}]_\psi = P_{\chi\psi} \cdot [\vec{g}]_\psi, \quad [P\vec{g}]_\chi = P_{\psi\chi} \cdot [\vec{g}]_\chi. \tag{2.100}$$

The matrices $P_{\chi\psi} = [p_{\chi\psi}(i,j) = \langle P\vec{\chi}_i, P\vec{\psi}_j\,\rangle]$, $P_{\psi\chi} = [p_{\psi\chi}(i,j) = \langle P\vec{\psi}_i, P\vec{\chi}_j\,\rangle]$ are called **projection matrices** and satisfy $P_{\chi\psi} = P^\dagger_{\psi\chi}$. Note that the j-th column of the projection matrix $P_{\chi\psi}$ is equal to the representation $[P\vec{\psi}_j]_\psi$ of $P\vec{\psi}_j$ in the basis B_ψ. Similarly, the j-th column of the projection matrix $P_{\psi\chi}$ is equal to the representation $[P\vec{\chi}_j]_\chi$ of $P\vec{\chi}_j$ in the basis B_χ. The projection matrices are characteristic for the subspace U, and independent of the choice of the basis B_φ of U.

The resulting mappings between the representations of the original and projected vectors are depicted in Figure 2.6, and are illustrated by an example in Appendix 2.A.

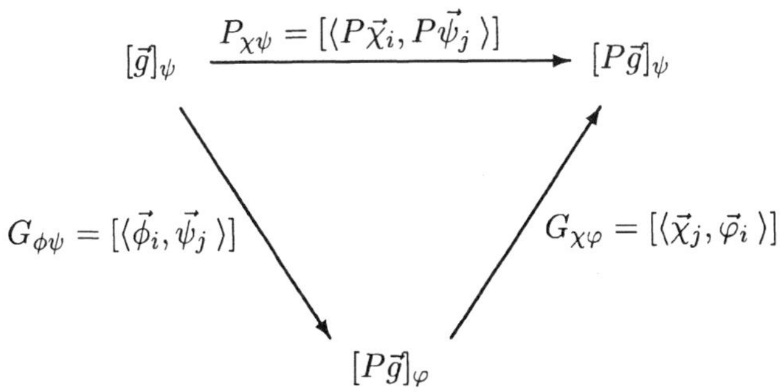

Figure 2.6. Mappings between representations $[\vec{g}]_\psi$ and $[P\vec{g}]_\psi$ of the original and projected vector, respectively.

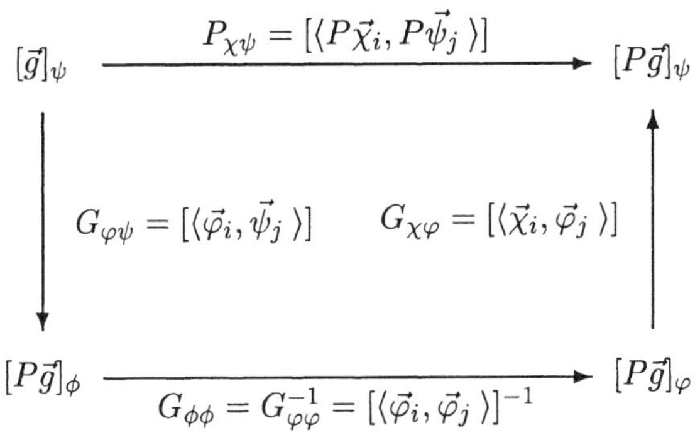

Figure 2.7. Mappings between the representations of the original and projected vectors when the basis vectors $\{\vec{\varphi}_i\}$ of the subspace are specified.

In Figure 2.7, the diagram of Figure 2.6 has been expanded to make the dependence on the basis B_φ more explicit. Similarly, in Figure 2.7, the diagram of Figure 2.6 has been expanded to make the dependence on the dual basis of B_ϕ more explicit.

Important applications of projection operators, some of which will be discussed in the following chapters, are: modelling of human color vision, conversions between analog and digital signals (sampling and interpolation), image descriptions with applications in image coding and image processing, etc.

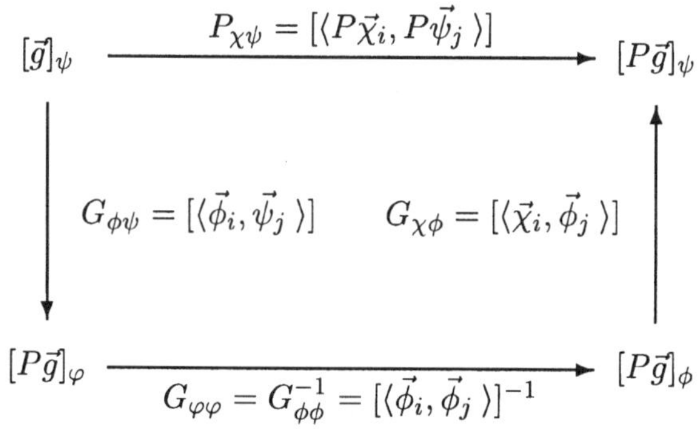

Figure 2.8. Mappings between the representations of the original and projected vectors when the test vectors $\{\vec{\phi}_i\}$ of the subspace are specified.

For some applications, amongst which the modelling of metamerism in human color vision, it is necessary to consider the elements of V that are not in the subspace U. The **orthogonal complement** of the subspace U is defined by $U^\perp = \{\vec{g} \in V; \langle \vec{f}, \vec{g} \rangle = 0 \text{ for all } \vec{f} \in U\}$. The following properties can be easily verified

1. $U^\perp$ is a subspace of V;
2. $U^\perp = \{\vec{g} \in V; \langle \vec{g}, \vec{\varphi}_i \rangle = 0, \forall i\}$ where $B_\varphi = \{\vec{\varphi}_i\}$ is a basis of U;
3. if $\hat{U}$ is a subspace of U $(\hat{U} \subset U)$, then $U^\perp \subset \hat{U}^\perp$;
4. if $\vec{g} \in V$ and $P\vec{g} \in U$, then $\vec{g} - P\vec{g} \in U^\perp$;
5. any vector $\vec{g} \in V$ can be written uniquely as $\vec{g} = P\vec{g} + (\vec{g} - P\vec{g})$, with $P\vec{g} \in U$ and $\vec{g} - P\vec{g} \in U^\perp$;
6. $(U^\perp)^\perp = U$.

It should be noted that the more advanced properties of linear operators between Hilbert spaces that are discussed in the remainder of this chapter are not needed for the discussion of human color vision in chapter 3. It may therefore be advisable to switch to chapter 3 at this time, because the specific example of human color vision may help to clarify some of the general results that have been derived up to now.

6. LINEAR OPERATORS BETWEEN HILBERT SPACES

Although operators between Hilbert spaces have all the properties that were mentioned before for operators between Banach spaces, they also have specific properties that make them much more useful and easy to handle than more general operators. Some of these properties are discussed next. It is implicitly assumed that all vector spaces that occur in this section are Hilbert spaces.

6.1 UNITARY OPERATORS

An operator $T \in \mathcal{L}(V, W)$ is an **isometry** if

$$\langle T\vec{f}, T\vec{g}\,\rangle_W = \langle \vec{f}, \vec{g}\,\rangle_V, \tag{2.101}$$

for all $\vec{f}, \vec{g} \in V$, where $\langle \cdot, \cdot \,\rangle_V$ and $\langle \cdot, \cdot \,\rangle_W$ denote the inner products in V and W respectively. This definition states that the inner product in W is essentially the same as the inner product in V, and that there exists a one-to-one correspondence between the elements of V and W. An isometry $T \in \mathcal{L}(V, V)$ is also called a **unitary** operator.

Since all Hilbert spaces contain orthonormal bases, it is very easy to classify all possible *separable* Hilbert spaces. Indeed, if V is a Hilbert space of finite dimension N, then the mapping

$$T : V \mapsto \mathbb{C}^N : f \mapsto \Big(\hat{f}_1 = \langle \vec{e}_1, f\,\rangle, \dots, \hat{f}_N = \langle \vec{e}_N, f\,\rangle\Big) \tag{2.102}$$

with $\{\vec{e}_i; i = 1, \dots, N\}$ an orthonormal basis of V, is an isometry. The vector product between two N-tuples $T\vec{f} = (\hat{f}_1, \dots, \hat{f}_N)$ and $T\vec{g} = (\hat{g}_1, \dots, \hat{g}_N)$ of $\mathbb{C}^N$ is equal to

$$\langle T\vec{f}, T\vec{g}\,\rangle_{\mathbb{C}^N} = \sum_{i=1}^{N} \hat{f}_i^\star \cdot \hat{g}_i. \tag{2.103}$$

The inverse mapping

$$T^{-1} : \mathbb{C}^N \mapsto V : \Big(\hat{f}_1, \dots, \hat{f}_N\Big) \mapsto \sum_{i=1}^{N} \hat{f}_i\, \vec{e}_i \tag{2.104}$$

associates a unique element of V with any N-tuple. If V is an infinite-dimensional Hilbert space, then the mapping

$$T : V \mapsto \mathbb{C}^{\mathbb{N}} : \vec{f} \mapsto \Big(\hat{f}_1 = \langle \vec{e}_1, \vec{f}\,\rangle, \hat{f}_2 = \langle \vec{e}_2, \vec{f}\,\rangle, \dots\Big) \tag{2.105}$$

with $\{\vec{e}_i; i \in \mathbb{N}\}$ an orthonormal basis of V, is an isometry. That implies that all elements of V can be represented by infinite sequences, which have to be square-summable. The inverse mapping

$$T^{-1} : \mathbb{C}^{\mathbb{N}} \mapsto V : \left(\hat{f}_1, \hat{f}_2, \dots\right) \mapsto \sum_{i \in \mathbb{N}} \hat{f}_i \, \vec{e}_i \tag{2.106}$$

associates a unique element of V with any such sequence.

6.2 ADJOINT OPERATORS

The **adjoint** of an operator T of $\mathcal{L}(V, W)$ is the operator $T^\dagger$ of $\mathcal{L}(W, V)$ that satisfies

$$\langle T\vec{f}, \vec{g} \rangle_W = \langle \vec{f}, T^\dagger \vec{g} \rangle_V, \tag{2.107}$$

for any $\vec{f} \in V$ and $\vec{g} \in W$. The adjoint operator $T^\dagger$ is uniquely defined if V and W are Hilbert spaces. The matrix of the operator $T^\dagger$ with respect to the bases B_ψ of V (with dual basis B_χ) and B_φ of W (with dual basis B_ϕ) is

$$T^\dagger_{\chi\varphi} = [\langle \vec{\chi}_n, T^\dagger \vec{\varphi}_m \rangle_V]_{n \in N, m \in M} = [\langle T\vec{\chi}_n, \vec{\varphi}_m \rangle_W]_{n \in N, m \in M} \tag{2.108}$$

so that this matrix is the Hermitian conjugate of the matrix $T_{\varphi\chi}$.

An operator T in $\mathcal{L}(V, V)$ is called **Hermitian** or **self-adjoint** if $T = T^\dagger$. The Gramian operators and the projection operator that were introduced earlier are examples of Hermitian operators. Note that the matrix of a Hermitian operator is Hermitian, i.e., it is mapped onto itself by element-wise conjugation followed by matrix transposition, as is to be expected from the terminology. Hermitian operators play a very prominent role in the theory of signal approximations, and they have unique properties, some of which will be discussed below.

If T is a Hermitian operator on a Hilbert space V, then an alternative way of determining the operator norm is using the property

$$\| T \| = \sup_{\|\vec{f}\| = 1} |\langle T\vec{f}, \vec{f} \rangle|. \tag{2.109}$$

It was stated in section 4 that compact operators and matrices have very similar properties, but that it is fairly difficult in general to decide if an operator is compact. However, for the case of separable Hilbert spaces, there is an easy test for identifying a large subclass of the compact operators. More specifically, if the operator T satisfies

$$\sum_n \| T\vec{e}_n \|_W^2 < \infty, \tag{2.110}$$

with $\{\vec{e}_n\}$ any orthonormal basis in V, then T is a **Hilbert-Schmidt** operator. It has been shown that all Hilbert-Schmidt operators are compact. Unfortunately, there are some operators, such as

$$T\vec{f} = \sum_{n=1}^{\infty} \frac{1}{\sqrt{n}} \langle \vec{e}_n, \vec{f} \rangle_V \, \vec{e}_n, \tag{2.111}$$

for all $\vec{f} \in V$, that are compact, but not Hilbert-Schmidt. In terms of the matrix notation that was introduced earlier, an operator T is Hilbert-Schmidt if and only if

$$\sum_{m,n} |t_{de}(m,n)|^2 = \sum_{m,n} |\langle \vec{d}_m, T\vec{e}_n \rangle_W|^2 \tag{2.112}$$

is finite, with $\{\vec{e}_n\}$ and $\{\vec{d}_m\}$ arbitrary orthonormal bases in V and W, respectively.

6.3 EIGENVALUE DECOMPOSITION

One of the first historical successes of operator theory was that the *eigenvalue decomposition*, familiar from the theory of matrices, was generalized to compact Hermitian operators on Hilbert spaces. If $T \in \mathcal{L}(V,V)$ is a compact Hermitian operator, then it can be expressed as

$$T\vec{f} = \sum_{n} \lambda_n(T) \, \langle \vec{e}_n, \vec{f} \rangle \, \vec{e}_n, \tag{2.113}$$

$\vec{f} \in V$, where $\lambda_n(T)$ are the *real* **eigenvalues** of T and $\vec{e}_n$ are corresponding **eigenvectors**, i.e.,

$$T\vec{e}_n = \lambda_n(T) \cdot \vec{e}_n. \tag{2.114}$$

The eigenvectors constitute an orthonormal basis for the subspace Ran$T$ of V, and are only uniquely determined if all eigenvalues are distinct. The practical interpretation of the eigenvalue decomposition is that the linear operator T is completely characterized by how it transforms the eigenvectors.

If the operator has rank N, then there are only N non-zero eigenvalues. If the rank of the operator is infinite, then $\lambda_n(T) \to 0$ for $n \to \infty$, since T is compact. The norm of the operator

$$\|T\| = \sup\{|\lambda_n(T)|; \forall\, n\} \tag{2.115}$$

is equal to the largest absolute value of the eigenvalues. The operator has inverse

$$T^{-1}\vec{f} = \sum_{n} \frac{1}{\lambda_n(T)} \, \langle \vec{e}_n, \vec{f} \rangle \, \vec{e}_n, \tag{2.116}$$

for $\vec{f} \in V$, provided that the smallest absolute value $\inf\{|\lambda_n(T)|; \forall\, n\}$ of the eigenvalues is nonzero and Ran$T = V$. Therefore, compact Hermitian operators of infinite rank cannot be invertible, which is an indirect way of realizing that the identity operator on an infinite-dimensional Hilbert space is for instance not compact.

If the Hermitian operator S commutes with the operator T, i.e., if $(S \cdot T)\,\vec{f} = (T \cdot S)\,\vec{f}$, for all $\vec{f} \in V$, then the operator S can be expressed as

$$S\vec{f} = \sum_n \lambda_n(S)\, \langle \vec{e}_n, \vec{f}\,\rangle\, \vec{e}_n, \tag{2.117}$$

$\vec{f} \in V$, where $\lambda_n(S)$ are the eigenvalues of S. Hence, both operators can be diagonalized using the same eigenvectors, and the composition of both operators can be expressed as

$$(S \cdot T)\vec{f} = (T \cdot S)\vec{f} = \sum_n \lambda_n(S)\, \lambda_n(T)\, \langle \vec{e}_n, \vec{f}\,\rangle\, \vec{e}_n, \tag{2.118}$$

$\vec{f} \in V$. We will for instance see in chapter 6 that Fourier signals are eigenvectors of all linear space-invariant operators in $\mathcal{L}^2(\mathbb{F})$, which is in agreement with the fact that these operators commute.

6.4 SINGULAR VALUE DECOMPOSITION

A more recent success of operator theory is that one of the most powerful constructs of matrix algebra, i.e., the *singular value decomposition (SVD)*, has also been generalized to compact operators on Hilbert spaces. For $T \in \mathcal{L}(V, W)$, with V and W Hilbert spaces, we can define

$$s_k(T) = \inf\{\|T - R\,\| : R \in \mathcal{L}(V, W) \text{ and rank } R \leq k\} \tag{2.119}$$

for any non-negative integer k. Note that $s_k(T)$ expresses how closely the operator T can be approximated by an operator R of rank at most k. The positive numbers

$$s_0(T) \geq s_1(T) \geq s_2(T) \geq \ldots \geq 0 \tag{2.120}$$

are called the **singular values** of the operator T. If T is a compact operator, then it can be expressed as

$$T\vec{f} = \sum_n s_n(T)\, \langle \vec{v}_n, \vec{f}\,\rangle\, \vec{w}_n, \tag{2.121}$$

for all $\vec{f} \in V$, where $s_n(T)$ are the singular values of T and $(\vec{v}_n, \vec{w}_n)$ is a corresponding **Schmidt pair of singular vectors**, i.e.,

$$T\vec{v}_n = s_n(T) \cdot \vec{w}_n, \quad T^\dagger \vec{w}_n = s_n(T) \cdot \vec{v}_n. \tag{2.122}$$

Note that the singular vectors satisfy

$$T^\dagger T\vec{v}_n = s_n^2(T)\cdot\vec{v}_n, \quad TT^\dagger\vec{w}_n = s_n^2(T)\cdot\vec{w}_n, \tag{2.123}$$

so that $\{\vec{v}_n\} \subset V$ and $\{\vec{w}_n\} \subset W$ are sets of orthonormal eigenvectors of the compact Hermitian operators $T^\dagger T$ and $TT^\dagger$, respectively. The singular vectors are only uniquely determined if all singular values are distinct. The practical interpretation of the singular value decomposition is that the linear operator T is completely characterized by how it transforms the singular vectors $\vec{v}_n$.

If T is a compact Hermitian operator on V, then the SVD reduces to an eigenvalue decomposition (i.e., $\vec{v}_n = \vec{w}_n$ and $s_n(T) = \lambda_n(T)$ for all n).

The SVD of the adjoint operator $T^\dagger \in \mathcal{L}(W,V)$ is very similar to the SVD of the operator $T \in \mathcal{L}(V,W)$, i.e.,

$$T^\dagger\vec{g} = \sum_n s_n(T)\,\langle\vec{w}_n,\vec{g}\,\rangle\,\vec{v}_n, \tag{2.124}$$

for all $\vec{g} \in W$. This adjoint operator $T^\dagger$ is completely characterized by how it transforms the singular vectors $\vec{w}_n$.

If the operator has rank N, then there are only N non-zero singular values. Truncating the SVD after K terms gives the best rank-K approximation to the operator T. If the rank of the operator is infinite, then $s_n(T) \to 0$ for $n \to \infty$, since T is compact. The norm of the operator is $\|T\,\| = s_0(T)$.

6.5 PSEUDO-INVERSE OPERATORS

A frequently-occurring problem in operator theory is the following: find $\vec{f} \in V$ such that $\|\vec{g} - T\vec{f}\,\|_W$, with $\vec{g} \in W$ known, is minimized. This problem is stereotypical for a measurement device, and is therefore very relevant in signal processing and human perception. The measurement device, specified by the operator T, performs a number of linear measurements $\vec{g}$ on the input signal $\vec{f}$, and the goal is to find the signals $\vec{f}$ that agree best with these measurements $\vec{g}$. The above problem has a solution in terms of the **pseudo-inverse operator** $T^\# \in \mathcal{L}(W,V)$ of the operator $T \in \mathcal{L}(V,W)$, i.e.,

$$T^\#\vec{g} = \sum_{s_n(T)\neq 0} \frac{1}{s_n(T)}\,\langle\vec{w}_n,\vec{g}\,\rangle\,\vec{v}_n, \tag{2.125}$$

for all $\vec{g} \in W$.

The pseudo-inverse satisfies the following four *Moore-Penrose conditions*

$$
\begin{aligned}
T^{\#}\, T\, \vec{f} = \sum_n \langle \vec{v}_n, \vec{f}\, \rangle\, \vec{v}_n, \forall \vec{f} \in V \quad &\rightarrow \quad T\, T^{\#}\, T = T \\
T\, T^{\#}\, \vec{g} = \sum_n \langle \vec{w}_n, \vec{g}\, \rangle\, \vec{w}_n, \forall \vec{g} \in W \quad &\rightarrow \quad T^{\#}\, T\, T^{\#} = T^{\#} \\
T\, T^{\#} &= (T\, T^{\#})^{\dagger} \\
T^{\#}\, T &= (T^{\#}\, T)^{\dagger}
\end{aligned}
\tag{2.126}
$$

These conditions imply that the Hermitian operator $T^{\#}\, T \in \mathcal{L}(V, V)$ is an orthogonal projection, i.e., $(T^{\#}\, T)^2 = T^{\#}\, T$, that maps any vector $\vec{f} \in V$ to the nearest vector $T^{\#}\, T\, \vec{f}$ in the subspace $\operatorname{Ran} T^{\dagger} = \operatorname{Ran} T^{\#} \subset V$. Likewise the Hermitian operator $T\, T^{\#} \in \mathcal{L}(W, W)$ is an orthogonal projection, i.e., $(T\, T^{\#})^2 = T\, T^{\#}$, that maps any vector $\vec{g} \in W$ to the nearest vector $T\, T^{\#}\, \vec{g}$ in the subspace $\operatorname{Ran} T \subset W$. If T is a compact Hermitian operator on V, then so is $T^{\#}$.

Let us now return to our measurement problem. All input signals

$$
\vec{f} = T^{\#} \vec{g} + (I_V - T^{\#}\, T) \vec{h} \tag{2.127}
$$

with $\vec{h}$ an arbitrary signal of V, have the same minimum error

$$
\| \vec{g} - T \vec{f}\, \| = \| (I_W - T\, T^{\#})\, \vec{g}\, \|_W, \tag{2.128}
$$

and are therefore equally suited candidates for the input signal. This is due to the fact that the operator T maps any vector of the form $(I_V - T^{\#}\, T)\vec{h}$ onto the zero vector of W, i.e., these signals cannot be observed by the measurement system. The unique input signal $\vec{f} = T^{\#} \vec{g}$ simultaneously minimizes $\| \vec{g} - T \vec{f}\, \|_W$ and $\| \vec{f}\, \|_V$.

The first possibility is that all measurements are independent, such that $TT^{\#} = I_W$, but that there are insufficient measurements to uniquely determine the input signal, such that $T^{\#} T \neq I_V$. Then there are several input signals that agree perfectly with the measurements, and $\vec{f} = T^{\#} \vec{g}$ is the unique input signal which also minimizes $\| \vec{f}\, \|_V$. This latter condition hence allows to find a unique solution for $\vec{f} \in V$ and is said to **regularize** the ill-posed problem of minimizing $\| \vec{g} - T \vec{f}\, \|$ over all possible input vectors in V.

The second possibility is that the measurements are dependent, such that $T T^{\#} \neq I_W$, but that there are sufficient measurements to uniquely determine the input signal, such that $T^{\#}\, T = I_V$. The pseudo-inverse can then be used to express the unique input signal $\vec{f} = T^{\#} \vec{g}$ that corresponds to the measurement $\vec{g}$, provided of course that $\vec{g}$ is a valid measurement. In practice, the measurement is often modelled by $\vec{g} = T\vec{f}+\vec{n}$, where $\vec{n}$ denotes the unwanted measurement error (or noise), with norm $\|\vec{n}\|_W$. The closest valid measurement vector $\vec{g}^{\#} = TT^{\#}\vec{g} = T\vec{f} + T\,T^{\#}\vec{n}$ corresponds to the input signal $\vec{f} + T^{\#}\vec{n}$. The estimation error $T^{\#}\vec{n}$ on the input signal is due to the measurement noise. By using the dependence between the measurements, the norm of the measurement error signal is reduced from $\|\vec{n}\|_W$ to $\|T\,T^{\#}\vec{n}\|_W$.

6.6 FRAMES

The pseudo-inverse can also be used to extend the concept of basis and dual basis to the more general concept of frames and dual frames. Suppose that V is a Hilbert space, and let $F_\varphi = \{\vec{\varphi}_i\}$ be a subset of V, which may contain *linearly-dependent* vectors. We will refer to such a set of vectors as a **frame**. The subset U of all vectors of the form

$$\vec{h} = \sum_i \tilde{h}_\varphi(i) \cdot \vec{\varphi}_i, \tag{2.129}$$

with $\tilde{h}_\varphi(i)$ arbitrary numbers, is obviously a subspace of V. Note that the frame F_φ is a subset of U. The numbers $[\tilde{h}_\varphi(i), \forall i]^T$ are called the **frame coordinates** of the vector $\vec{h} \in U$. Since the frame vectors are linearly dependent, the frame coordinates are *not unique*, i.e., the same vector $\vec{h} \in U$ can be described by (infinitely) many different frame coordinates.

In analogy with the definition of the Gramian operator of a basis, the **frame operator**

$$F_{\varphi\varphi}\,\vec{g} = \sum_i \langle \vec{\varphi}_i, \vec{g}\,\rangle\, \vec{\varphi}_i \tag{2.130}$$

can be defined for all vectors $\vec{g} \in V$. The operator $F_{\varphi\varphi}$ is a Hermitian operator from the space V to the subspace U. For the elements $\vec{h} \in U$, the frame operator can be written as

$$F_{\varphi\varphi}\,\vec{h} = \sum_i \left[\sum_j f_{\varphi\varphi}(i,j)\, \tilde{h}_\varphi(j) \right] \vec{\varphi}_i \tag{2.131}$$

in terms of the frame coordinates, with $f_{\varphi\varphi}(i,j) = \langle \vec{\varphi}_i, \vec{\varphi}_j\,\rangle, \forall i,j$.

Now suppose that the frame operator $F_{\varphi\varphi}$ has a pseudo-inverse $F_{\varphi\varphi}^{\#} \in \mathcal{L}(U,V)$ that satisfies the Moore-Penrose conditions. This implies that the operator $P = F_{\varphi\varphi}^{\#}\, F_{\varphi\varphi}$ is equal to the projection operator onto the subspace U, i.e.,

$$\boxed{P\vec{g} = \sum_i \langle \vec{\varphi}_i, \vec{g}\,\rangle\, \vec{\phi}_i,} \tag{2.132}$$

for all vectors $\vec{g} \in V$, where $\vec{\phi}_i = F^{\#}_{\varphi\varphi}\, \vec{\varphi}_i$ is called the **dual frame vector** corresponding to the **frame vector** $\vec{\varphi}_i$. Note that the dual frame vectors $\vec{\phi}_i$ also belong to the subspace U, since the pseudo-inverse satisfies $F^{\#}_{\varphi\varphi} = P\, F^{\#}_{\varphi\varphi}$, and hence $\vec{\phi}_i = P\vec{\phi}_i$. We refer to $F_\phi = \{\vec{\phi}_i\}$ as the **dual frame** of F_φ.

In the important special case that the frame F_φ contains a finite number N of frame vectors, standard matrix techniques can be used to determine the pseudo-inverse. More specifically, the matrix $F_{\varphi\varphi} = [f_{\varphi\varphi}(i,j)]_{i,j=1,\dots,N}$ is Hermitian and has an eigenvalue decomposition of the form $F_{\varphi\varphi} = E \cdot \Lambda \cdot E^{\dagger}$. The diagonal matrix Λ contains the eigenvalues, while the columns of the unitary matrix E are equal to the orthonormal eigenvectors. The matrix $F^{\#}_{\varphi\varphi} = [f_{\phi\phi}(i,j)]_{i,j=1,\dots,N} = E \cdot \Lambda^{\#} \cdot E^{\dagger}$, where $\Lambda^{\#}$ is obtained from Λ by inverting all nonzero eigenvalues, is the Moore-Penrose pseudo-inverse matrix of $F_{\varphi\varphi}$ [GvL83]. This pseudo-inverse can be used to express the **dual frame operator**

$$F^{\#}_{\varphi\varphi}\, \vec{h} = \sum_i \left[\sum_j f_{\phi\phi}(i,j)\, \tilde{h}_\varphi(j) \right] \vec{\varphi}_i, \tag{2.133}$$

for all $\vec{h} \in U$, in terms of the frame coordinates. The dual frame vectors are linear combinations of the frame vectors, i.e.,

$$\vec{\phi}_j = \sum_i f_{\phi\phi}(i,j)\, \vec{\varphi}_i \tag{2.134}$$

for $j = 1, \dots, N$.

The pseudo-inverse can be extended to an operator on V by expressing it as

$$F_{\phi\phi}\, \vec{g} = \sum_i \langle \vec{\phi}_i, \vec{g}\,\rangle\, \vec{\phi}_i = \sum_i \langle \vec{\phi}_i, P\vec{g}\,\rangle\, \vec{\phi}_i = F^{\#}_{\varphi\varphi}\, P\vec{g}, \tag{2.135}$$

for all vectors $\vec{g} \in V$. Obviously, $F_{\phi\phi}\, \vec{h} = F^{\#}_{\varphi\varphi}\, \vec{h}$ if $\vec{h} \in U$. The following alternative expression for the projection operator

$$\boxed{P\vec{g} = \sum_i \langle \vec{\phi}_i, \vec{g}\,\rangle\, \vec{\varphi}_i,} \tag{2.136}$$

for all $\vec{g} \in V$, with $\vec{\varphi}_i = F^{\#}_{\phi\phi}\, \vec{\phi}_i = F_{\varphi\varphi}\, \vec{\phi}_i$, is easily derived by analogy. The numbers $[\langle \vec{\phi}_i, \vec{g}\,\rangle, \forall\, i]^T$ are hence frame coordinates of the signal $P\vec{g}$ in the frame F_φ. A specific property of these frame coordinates is that, if $[\tilde{g}_\varphi(i) = \langle \vec{\phi}_i, \vec{g}\,\rangle + \varepsilon_i, \forall\, i]^T$ is any other set of frame coordinates of the signal $P\vec{g}$, then

$$\sum_i \mid \tilde{g}_\varphi(i) \mid^2 = \sum_i \mid \langle \vec{\phi}_i, \vec{g}\,\rangle \mid^2 + \sum_i \mid \varepsilon_i \mid^2 . \tag{2.137}$$

This implies that the frame coordinates obtained by using the dual frame vectors as test functions have minimum energy.

The following expression for the inner product

$$\langle P\vec{f}, P\vec{g}\,\rangle = \langle P\vec{f}, \vec{g}\,\rangle = \langle \vec{f}, P\vec{g}\,\rangle = \sum_i \langle \vec{f}, \vec{\varphi}_i\,\rangle\, \langle \vec{\phi}_i, \vec{g}\,\rangle, \tag{2.138}$$

for all $\vec{f}, \vec{g} \in V$, can be used to find the error $d^2(\vec{g}, P\vec{g}) = \langle \vec{g}, \vec{g}\rangle - \langle P\vec{g}, P\vec{g}\rangle$ due to projection onto the subspace U.

An important property of frames that was previously mentioned to hold for the specific case of Riesz bases is that there exists (unique) numbers a_φ and b_φ such that

$$a_\varphi\, \|\vec{h}\,\|^2 \le \sum_i |\langle \vec{\varphi}_i, \vec{h}\,\rangle|^2 \le b_\varphi\, \|\vec{h}\,\|^2, \tag{2.139}$$

for all $\vec{h} \in U$. The numbers $a_\varphi > 0$ and $b_\varphi < \infty$ are called **frame bounds**. Frames for which $a_\varphi = b_\varphi$ are called **tight frames**. They have the property that the dual frame vectors are equal to $\vec{\phi}_i = \vec{\varphi}_i / a_\varphi$, for $i = 1, 2, \ldots,$ so that a tight frame behaves very similar to an orthogonal basis.

The implications of frames for signal processing are that they make it possible to design measurement systems (or display systems) in which not all measurements (or display responses) are independent. The obvious disadvantage of signal representations with frames is that they require more numbers to specify a signal, and that these numbers are not unique. However, in practical systems with noise, the mutual dependence between the frame coordinates also has an advantage: for a fixed error between the input and output signal, the required accuracy on frame coordinates is less than the required accuracy on basis coordinates [Dau92].

EXERCISES

1 Consider a Euclidean vector space U of order 2 with basis $B_\varphi = \{\vec{\varphi}_1, \vec{\varphi}_2\}$, i.e., assume that all vectors in the space U can be written as unique linear combinations

$$f = \hat{f}_\varphi(1) \cdot \vec{\varphi}_1 + \hat{f}_\varphi(2) \cdot \vec{\varphi}_2,$$

where $\hat{f}_\varphi(1)$ and $\hat{f}_\varphi(2)$ are real numbers. Assume that the Gramian matrix is given by

$$G_{\varphi\varphi} = \begin{bmatrix} 1 & \frac{1}{2} \\ \frac{1}{2} & 1 \end{bmatrix}.$$

Solve the following questions:

- Express the vector product $\langle \vec{f}, \vec{g} \rangle$ in terms of the representations $[\vec{f}]_\varphi = [\hat{f}_\varphi(1), \hat{f}_\varphi(2)]^T$ and $[\vec{g}]_\varphi = [\hat{g}_\varphi(1), \hat{g}_\varphi(2)]^T$ of the vectors $\vec{f}$ and $\vec{g}$.
- Find the dual basis $B_\phi = \{\vec{\phi}_1, \vec{\phi}_2\}$ of B_φ.
- Find the lengths $\|\vec{\phi}_1\|$ and $\|\vec{\phi}_2\|$ of the dual basis vectors.
- Find the angles $\theta(\vec{\varphi}_1, \vec{\varphi}_2)$, $\theta(\vec{\phi}_1, \vec{\phi}_2)$, $\theta(\vec{\varphi}_1, \vec{\phi}_2)$ and $\theta(\vec{\varphi}_2, \vec{\phi}_1)$.
- Find the distances $d(\vec{\varphi}_1, \vec{\varphi}_2)$ and $d(\vec{\phi}_1, \vec{\phi}_2)$.
- Draw the basis vectors $\vec{\varphi}_1$, $\vec{\varphi}_2$ and the dual basis vectors $\vec{\phi}_1$, $\vec{\phi}_2$ in a Euclidean plane (similar to Figure 2.3). Verify the angles and distances found above.

2 Prove the inequality of Cauchy-Schwartz.

3 Prove that the supremum norm $\|f(x)\|_\infty$ and the vector length $\|f(x)\|_2$ do indeed satisfy the properties of a vector norm.

4 Derive equations (2.46) and (2.47) for the inner product of two vectors in terms of the inner products of the vectors with a set of basis and dual basis vectors.

5 Derive equation (2.51) for the transformation between coordinates in two finite-dimensional bases in a Banach space.

6 Explore the relationship between the parallelogram construction in geometry and the parallelogram law of the length norm in a Hilbert space, i.e., first: show that the sum of the squares of the lengths of the four sides in a parallelogram is equal to the sum of the squares of the diagonal lengths, and second: show that the length norm in a Hilbert space does indeed satisfy the parallelogram law.

7 Use the example of the supremum norm $\|f(x)\|_\infty$ to illustrate that the closest point property does not hold in a Banach space, i.e., show by means of an example that there may be infinitely many functions in a convex set that are equally distant from a given input function.

8 Consider a Banach space V with vector product $\langle \vec{f}, \vec{g} \rangle$, and let $B_\varphi = \{\vec{\varphi}_i; i = 1, \ldots, n\}$ be the basis of an n-dimensional subspace U of V. Denote the dual basis by $B_\phi = \{\vec{\phi}_i; i = 1, \ldots, n\}$. Let $S_\alpha = \{\vec{\alpha}_i; i = 1, \ldots, K\}$ and $S_\beta = \{\vec{\beta}_i; i = 1, \ldots, L\}$ denote arbitrary sets of vectors in V. We can define the $K \times n$ matrix

$$G_{\alpha\phi} = \left[\langle \vec{\alpha}_i, \vec{\phi}_j \rangle\right]_{i=1,\ldots,K; j=1,\ldots,n}$$

and the $n \times L$ matrix

$$G_{\varphi\beta} = \left[\langle \vec{\varphi}_i, \vec{\beta}_j \rangle\right]_{i=1,\ldots,n; j=1,\ldots,L}$$

in the usual way. Prove that

$$G_{\alpha\phi} \cdot G_{\varphi\beta} = G_{\alpha\beta} = \left[\langle \vec{\alpha}_i, \vec{\beta}_j \rangle\right]_{i=1,\ldots,K; j=1,\ldots,L}$$

if $S_\alpha \subset U$ and/or $S_\beta \subset U$. Is this relationship still valid if both sets S_α and S_β contain vectors that are not in the subspace U? Motivate the answer; think about the projection operator!

9 Derive the equations for the orthogonal projection $P\vec{g}$ of the vector $\vec{g} \in V$ on the subspace U of the Hilbert space V (i.e., those equations that are displayed in a box in section 6). Assume a Euclidean vector space, i.e., that all coefficients are real numbers and the vector product satisfies $\langle \vec{f}, \vec{g} \rangle = \langle \vec{g}, \vec{f} \rangle$.

10 Consider the Hilbert space of piecewise constant signals on the interval [0,6[, i.e.,

$$V = \left\{ f(x) = \sum_{i=0}^{5} \hat{f}_\psi(i) \cdot \psi_i(x); \hat{f}_\psi(i) \in \mathbb{R} \right\},$$

where the basis functions are $\psi_i(x) = 1$, for $x \in [i, i+1[$ (and zero elsewhere)[15], for $i = 0, \ldots, 5$. The inner product in the vector space V is defined by

$$\langle \sum_{i=0}^{5} \hat{f}_\psi(i) \cdot \psi_i(x), \sum_{i=0}^{5} \hat{g}_\psi(i) \cdot \psi_i(x) \rangle = \sum_{i=0}^{5} \hat{f}_\psi(i) \cdot \hat{g}_\psi(i).$$

[15][a,b[denotes the interval from a to b, including a and excluding b.

Consider the subspace U of V spanned by the basis vectors $B_\varphi = \{\varphi_i(x); i = 1, 2\}$, where $\varphi_1(x) = 1/2$, for $x \in [0, 4[$ (and zero elsewhere), $\varphi_2(x) = 1/2$, for $x \in [2, 6[$ (and zero elsewhere). The basis vector $\varphi_2(x)$ is drawn in the following figure.

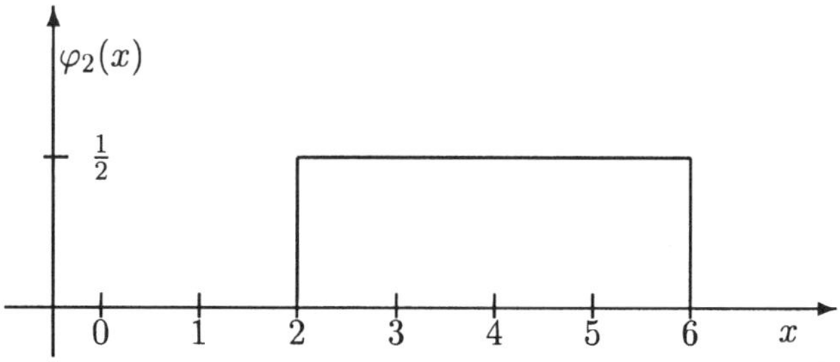

Solve the following questions:

- Find the 2×2 Gramian matrix $G_{\varphi\varphi}$.
- Find and draw the dual basis vectors $\{\phi_1(x), \phi_2(x)\}$ as a function of x; verify the biorthogonality between the basis and measurement vectors.
- Find the 2×6 matrix $G_{\phi\psi}$ that maps the representation $[\vec{f}]_\psi$ of a vector $\vec{f}$ in the basis B_ψ of V to the representation $[P\vec{f}]_\varphi$ of the orthogonal projection $P\vec{f}$ (onto the subspace U) in the basis B_φ of U. What is the interpretation of the columns of this matrix? Write $P\vec{\psi}_i$, for $i = 0, \ldots, 5$ in terms of $\vec{\varphi}_1$ and $\vec{\varphi}_2$.
- Find the 6×2 matrix $G_{\chi\varphi}$ that maps the vector representation $[P\vec{f}]_\varphi$ in the basis B_φ of U to the representation $[P\vec{f}]_\psi$ in the basis B_ψ of V. What is the interpretation of the columns of this matrix?
- Find the 6×6 projection matrix $P_{\chi\psi}$ that maps the representation $[\vec{f}]_\psi$ of the vector $\vec{f}$ in the basis B_ψ of V to the representation $[P\vec{f}]_\psi$ of the orthogonal projection $P\vec{f}$ (onto the subspace U). What is the rank of this matrix? What is the interpretation of the columns of this matrix?
- Find and draw (as a function of x) the projection $Pf(x)$ of the signal $f(x)$ with representation $[\vec{f}]_\psi = [\hat{f}_\psi(i) = i; i = 0, \ldots, 5]$. Determine the quadratic error between $f(x)$ and $Pf(x)$. Find the representation $[P\vec{f}]_\varphi$ (2 numbers) in the basis B_φ of U, as well as the representation $[P\vec{f}]_\psi$ (6 numbers) in the basis B_ψ of V.

11 Consider the Hilbert space of piecewise constant signals on the interval [0,8[, i.e.,

$$V = \left\{ f(x) = \sum_{i=0}^{7} \hat{f}_\psi(i) \cdot \psi_i(x); \hat{f}_\psi(i) \in \mathbb{R} \right\},$$

where the basis functions are $\psi_i(x) = 1$, for $x \in [i, i+1[$ (and zero elsewhere), for $i = 0, \ldots, 7$. The inner product in the vector space V is defined by

$$\langle \sum_{i=0}^{7} \hat{f}_\psi(i) \cdot \psi_i(x), \sum_{i=0}^{7} \hat{g}_\psi(i) \cdot \psi_i(x) \rangle = \sum_{i=0}^{7} \hat{f}_\psi(i) \cdot \hat{g}_\psi(i).$$

Consider the subspace U of V spanned by the basis vectors $B_\varphi = \{\varphi_i(x); i = 1, 2, 3\}$, where $\varphi_1(x) = 1/2$, for $x \in [0, 4[$ (and zero elsewhere), $\varphi_2(x) = 1/2$, for $x \in [2, 6[$ (and zero elsewhere), and $\varphi_3(x) = 1/2$, for $x \in [4, 8[$ (and zero elsewhere). Solve the following questions:

- Find the 3×3 Gramian matrix $G_{\varphi\varphi}$.
- Find and draw the dual basis vectors $\{\phi_1(x), \phi_2(x), \phi_3(x)\}$ as a function of x.
- Find the 8×8 projection matrix $P_{\chi\psi}$ that maps the representation $[\vec{f}]_\psi$ of the vector $\vec{f}$ in the basis B_ψ of V to the representation $[P\vec{f}]_\psi$ of the orthogonal projection $P\vec{f}$ (onto the subspace U). What is the rank of this matrix? What is the interpretation of the columns of this matrix?
- Find and draw (as a function of x) the projection $Pf(x)$ of the signal $f(x)$ with representation $[\vec{f}]_\psi = [\hat{f}_\psi(i) = i; i = 0, \ldots, 7]$. Find the representation $[P\vec{f}]_\varphi$ (3 numbers) in the basis B_φ of U, as well as the representation $[P\vec{f}]_\psi$ (8 numbers) in the basis B_ψ of V.

12 Consider the Hilbert space of piecewise constant signals on the interval [0,6[, i.e.,

$$V = \left\{ f(x) = \sum_{i=0}^{5} \hat{f}_\psi(i) \cdot \psi_i(x); \hat{f}_\psi(i) \in \mathbb{R} \right\},$$

where the basis functions are $\psi_i(x) = 1$, for $x \in [i, i+1[$ (and zero elsewhere), for $i = 0, \ldots, 5$. The inner product in the vector space V

is defined by

$$\langle \sum_{i=0}^{5} \hat{f}_\psi(i) \cdot \psi_i(x), \sum_{i=0}^{5} \hat{g}_\psi(i) \cdot \psi_i(x) \rangle = \sum_{i=0}^{5} \hat{f}_\psi(i) \cdot \hat{g}_\psi(i).$$

Consider the subspace U of V spanned by the frame $F_\varphi = \{\varphi_i(x); i = 1,2,3\}$, where $\varphi_1(x) = 1/2$, for $x \in [0,4[$ (and zero elsewhere), $\varphi_2(x) = -1/2$, for $x \in [2,6[$ (and zero elsewhere), $\varphi_3(x) = -1/2$, for $x \in [0,2[$, $\varphi_3(x) = 1/2$, for $x \in [4,6[$ (and zero elsewhere).

Solve the following questions:

- Find and draw the dual frame vectors $\{\phi_1(x), \phi_2(x), \phi_3(x)\}$ as a function of x. Use the eigenvalue decomposition of the matrix

$$\begin{bmatrix} 1 & -\frac{1}{2} & -\frac{1}{2} \\ -\frac{1}{2} & 1 & -\frac{1}{2} \\ -\frac{1}{2} & -\frac{1}{2} & 1 \end{bmatrix} = U^t \cdot \begin{bmatrix} 0 & 0 & 0 \\ 0 & \frac{3}{2} & 0 \\ 0 & 0 & \frac{3}{2} \end{bmatrix} \cdot U$$

  with

$$U = \begin{bmatrix} \frac{1}{\sqrt{3}} & \frac{1}{\sqrt{3}} & \frac{1}{\sqrt{3}} \\ -\frac{1}{\sqrt{2}} & 0 & \frac{1}{\sqrt{2}} \\ \frac{1}{\sqrt{6}} & -\frac{2}{\sqrt{6}} & \frac{1}{\sqrt{6}} \end{bmatrix}$$

- Find the 3×6 matrix $G_{\phi\psi}$ that maps the representation $[\vec{f}]_\psi$ of a vector $\vec{f}$ in the basis B_ψ of V to the representation $[P\vec{f}]_\varphi$ of the orthogonal projection $P\vec{f}$ (onto the subspace U) in the frame F_φ of U. What is the interpretation of the rows and columns of this matrix?
- Find the 6×3 matrix $G_{\chi\varphi}$ that maps the vector representation $[P\vec{f}]_\varphi$ in the frame F_φ of U to the representation $[P\vec{f}]_\psi$ in the basis B_ψ of V. What is the interpretation of the columns of this matrix?
- Find the 6×6 projection matrix $P_{\chi\psi} = G_{\chi\varphi} \cdot G_{\phi\psi}$ that maps the representation $[\vec{f}]_\psi$ of the vector $\vec{f}$ in the basis B_ψ of V to the representation $[P\vec{f}]_\psi$ of the orthogonal projection $P\vec{f}$ (onto the subspace U). Compare this projection matrix with the one in exercise 10.
- Draw the vectors $\vec{\varphi}_i$, $\vec{\phi}_i$, for $i = 1,2,3$, and $P\vec{\psi}_j$, for $j = 0, \dots, 5$, in a Euclidean plane (similar to Figure 2.3).

APPENDIX 2.A: EXAMPLE VECTOR SPACE

Consider the real vector space

$$V = \left\{ \vec{f} = (f_1, f_2, f_3)\,;\, f_i \in \mathbb{R} \right\}$$

of discrete signals of length 3. This vector space is a pre-Hilbert space if we define the vector product by

$$\langle \vec{f}, \vec{g} \rangle = f_1 \cdot g_1 + f_2 \cdot g_2 + f_3 \cdot g_3.$$

This vector product can be used to define a length norm, i.e.,

$$\|\vec{f}\|^2 = \langle \vec{f}, \vec{f} \rangle = f_1^2 + f_2^2 + f_3^2,$$

and hence to turn V into a Hilbert space.

A possible basis B_ψ for the vector space V consists of the signals

$$\begin{aligned} \vec{\psi}_1 &= (1,1,1) \\ \vec{\psi}_2 &= (-1,0,1) \\ \vec{\psi}_3 &= (1,-1,1) \end{aligned}$$

The Gramian matrix of this basis

$$G_{\psi\psi} = \begin{bmatrix} 3 & 0 & 1 \\ 0 & 2 & 0 \\ 1 & 0 & 3 \end{bmatrix}$$

is nonsingular with inverse

$$G_{\chi\chi} = \begin{bmatrix} 3/8 & 0 & -1/8 \\ 0 & 1/2 & 0 \\ -1/8 & 0 & 3/8 \end{bmatrix},$$

so that the dual basis B_χ consists of the signals

$$\begin{aligned} \vec{\chi}_1 &= \tfrac{3}{8}\,\vec{\psi}_1 - \tfrac{1}{8}\,\vec{\psi}_3 &&= \left(\tfrac{1}{4}, \tfrac{1}{2}, \tfrac{1}{4}\right) \\ \vec{\chi}_2 &= \tfrac{1}{2}\vec{\psi}_2 &&= \left(-\tfrac{1}{2}, 0, \tfrac{1}{2}\right) \\ \vec{\chi}_3 &= -\tfrac{1}{8}\,\vec{\psi}_1 + \tfrac{3}{8}\,\vec{\psi}_3 &&= \left(\tfrac{1}{4}, -\tfrac{1}{2}, \tfrac{1}{4}\right) \end{aligned}$$

The fact that B_χ is the dual basis of B_ψ can easily be checked by verifying that the signals $\vec{\chi}_i$, for $i = 1,2,3$, are linear combinations of the basis signals in B_ψ, and moreover satisfy the biorthogonality conditions $\langle \vec{\chi}_i, \vec{\psi}_j \rangle = \delta_{ij}$.

The representation of the signal $\vec{f} = (f_1, f_2, f_3)$ in the basis B_ψ is given by

$$[\vec{f}]_\psi = \begin{bmatrix} \hat{f}_\psi(1) = \langle \vec{\chi}_1, f \rangle = (f_1 + 2\,f_2 + f_3)/4 \\ \hat{f}_\psi(2) = \langle \vec{\chi}_2, f \rangle = (-f_1 + f_3)/2 \\ \hat{f}_\psi(3) = \langle \vec{\chi}_3, f \rangle = (f_1 - 2\,f_2 + f_3)/4 \end{bmatrix}.$$

Now consider the subspace U of V spanned by the basis B_φ of signals

$$\begin{aligned} \vec{\varphi}_1 &= (1,1,1) \\ \vec{\varphi}_2 &= (-1,2,-1) \end{aligned}$$

The Gramian matrix of this basis

$$G_{\varphi\varphi} = \begin{bmatrix} 3 & 0 \\ 0 & 6 \end{bmatrix}$$

is nonsingular with inverse

$$G_{\phi\phi} = \begin{bmatrix} 1/3 & 0 \\ 0 & 1/6 \end{bmatrix},$$

so that the dual basis B_ϕ consists of the signals

$$\begin{array}{lll} \vec{\phi}_1 & = \frac{1}{3}\,\vec{\varphi}_1 & = (\frac{1}{3}, \frac{1}{3}, \frac{1}{3}) \\ \vec{\phi}_2 & = \frac{1}{6}\,\vec{\varphi}_2 & = (-\frac{1}{6}, \frac{1}{3}, -\frac{1}{6}) \end{array}$$

We now derive how the orthogonal projection $P\vec{f}$ onto U of a signal $\vec{f} = (f_1, f_2, f_3)$ can be found. The representation of $P\vec{f}$ in the basis B_φ is given by

$$[P\vec{f}]_\varphi = \begin{bmatrix} \widehat{Pf}_\varphi(1) = \langle \vec{\phi}_1, \vec{f}\,\rangle = (f_1 + f_2 + f_3)/3 \\ \widehat{Pf}_\varphi(2) = \langle \vec{\phi}_2, \vec{f}\,\rangle = (-f_1 + 2\,f_2 - f_3)/6 \end{bmatrix}.$$

The representations $[\vec{\varphi}_i]_\psi$ of the signals $\vec{\varphi}_i$, for $i = 1, 2$, in the basis B_ψ are equal to the columns of the matrix

$$G_{\chi\varphi} = \begin{bmatrix} 1 & 1/2 \\ 0 & 0 \\ 0 & -3/2 \end{bmatrix},$$

while the representations $[\vec{\phi}_i]_\psi$ of the dual signals $\vec{\phi}_i$, for $i = 1, 2$, in the basis B_ψ are equal to the columns of the matrix

$$G_{\chi\phi} = \begin{bmatrix} 1/3 & 1/12 \\ 0 & 0 \\ 0 & -1/4 \end{bmatrix}.$$

Similarly, the representations $[P\vec{\psi}_i]_\varphi$ of the projected signals $P\vec{\psi}_i$, for $i = 1, 2, 3$, in the basis B_φ are equal to the columns of the matrix

$$G_{\phi\psi} = \begin{bmatrix} 1 & 0 & 1/3 \\ 0 & 0 & -2/3 \end{bmatrix},$$

while the representations $[P\vec{\chi}_i]_\varphi$ of the projected signals $P\vec{\chi}_i$, for $i = 1, 2, 3$, in the basis B_φ are equal to the columns of the matrix

$$G_{\phi\chi} = \begin{bmatrix} 1/3 & 0 & 0 \\ 1/12 & 0 & -1/4 \end{bmatrix}.$$

The overall projection matrix is given by

$$P_{\chi\psi} = G_{\chi\varphi} \cdot G_{\phi\psi} = \begin{bmatrix} 1 & 0 & 0 \\ 0 & 0 & 0 \\ 0 & 0 & 1 \end{bmatrix},$$

which is obviously an idempotent matrix of rank 2. From this matrix we can derive that $P\vec{\psi}_1 = \vec{\psi}_1, P\vec{\psi}_2 = 0, P\vec{\psi}_3 = \vec{\psi}_3$ and $P\vec{\chi}_1 = \vec{\chi}_1, P\vec{\chi}_2 = 0, P\vec{\chi}_3 = \vec{\chi}_3$.

An arbitrary vector $\vec{f}$ with representation $[\vec{f}]_\psi$ can be projected onto the subspace U. This projected signal $P\vec{f}$ has representations

$$[P\vec{f}]_\varphi = G_{\phi\psi} \cdot [\vec{f}]_\psi = \begin{bmatrix} \widehat{Pf}_\varphi(1) = \hat{f}_\psi(1) + \hat{f}_\psi(3)/3 \\ \widehat{Pf}_\varphi(2) = -2\,\hat{f}_\psi(3)/3 \end{bmatrix}$$

and

$$[P\vec{f}]_\psi = P_{\chi\psi} \cdot [\vec{f}]_\psi = \begin{bmatrix} \widehat{Pf}_\psi(1) = \hat{f}_\psi(1) \\ \widehat{Pf}_\psi(2) = 0 \\ \widehat{Pf}_\psi(3) = \hat{f}_\psi(3) \end{bmatrix}$$

in the bases B_φ and B_ψ respectively.

The squared distance between the original and projected signal is given by

$$d^2\left(\vec{f}, P\vec{f}\right) = 2\,\hat{f}_\psi(2)\,\hat{f}_\psi(2) = \frac{(f_1 - f_3)^2}{2},$$

which implies that U is the subspace of symmetric signals (for which $f_1 = f_3$).

Chapter 3

COLOR PERCEPTION AND COLORIMETRY

1. INTRODUCTION

In this chapter, the experimental results on color matching, as presented in chapter 1, are interpreted using the mathematical concepts introduced in chapter 2. It is demonstrated that the radiance distribution of a light source as a function of wavelength, also referred to as the **radiometric function**, the **spectrum** or the **spectral power distribution**, can be decomposed into a component that is relevant for color perception, called the fundamental metamere, and an irrelevant component, called the metameric black or residual. The fundamental metameres comprise a three-dimensional subspace of the vector space of all possible radiometric functions. This subspace is called the fundamental color space and is derived in section 2.

The color diagram that is described in section 3 is a two-dimensional color specification that is now mostly of historical interest, but nevertheless still frequently used. The color diagram can be derived from fundamental color space through a perspective transformation.

Different coordinate frames that are related by linear transformations can be used to describe fundamental color space. We will be especially interested in coordinate frames that can accomplish a correlation between coordinate values and subjective color sensations such as hue, saturation and brightness or lightness.

In chapter 1, it was derived that luminance, which can be derived by weighting the spectrum of the light source with the spectral sensitivity function (SSF) of human vision, is an important psychometric property of light. Coordinate frames for which luminance is one of the coordinates will therefore be preferred. In section 4 on photometry, the relationship between luminance and brightness (or lightness) is examined in more depth. The implications of this relationship for display characterization and standardization are also discussed.

In section 5, results from opponent-color perception are used to extend the luminance coordinate with two additional coordinates, in such a way that the coordinates in the resulting coordinate frame can be easily related to the color percepts hue and saturation. The relationship between this coordinate frame and the standardized uniform color space CIE(L^*,a^*,b^*) is also explained.

2. FUNDAMENTAL COLOR SPACE

2.1 VECTOR SPACE MODEL FOR COLOR MATCHING

In chapter 1, it was concluded that the color matching axioms imply that color vision is performed by a linear front-end system (LFS) with three degrees of freedom. The axioms do however not give detailed insight into how color matching is accomplished. In order to formulate this mechanism more accurately, Wyszecki hypothesized that each radiometric function $L_e(\lambda)$ is the sum of two parts, a **fundamental metamere** and a **metameric black** [WS82]. The fundamental metamere is the perceptually relevant part of the radiometric function. It carries all the information necessary for the unique color sensation evoked. A color match is equivalent to an equality of the fundamental metameres. The metameric black does not give rise to a visual sensation. Wyszecki showed that non-trivial metameric blacks cannot be physically realized, since they contain negative energy at some wavelengths. This does however not prohibit that they are contained in a physical stimulus.

Cohen and Kappauf [CK82, CK85, Coh88] interpreted the hypothesis of Wyszecki in terms of vector spaces and postulated that the fundamental metameres must be in a three-dimensional subspace of the vector space of all possible radiometric functions (including the ones with negative energy at some wavelengths). They called this subspace the **fundamental color space** and showed how it could be constructed, based on color matching data. Their derivation, in a somewhat adapted notation, is given below.

Let V be the Hilbert space of radiometric functions spanned by the basis $B_\psi = \{\psi_j(\lambda); j = 1,\ldots,N\}$. Any radiometric function $L_e(\lambda)$ of interest is hence assumed to be of the form

$$L_e(\lambda) = \sum_{j=1}^{N} \hat{L}_\psi(j) \cdot \psi_j(\lambda), \tag{3.1}$$

and has an equivalent representation $[L_e(\lambda)]_\psi = [\hat{L}_\psi(1) \ldots \hat{L}_\psi(N)]^T$. The vector product between two radiometric functions $L_1(\lambda)$ and $L_2(\lambda)$ is defined by

$$\langle L_1(\lambda), L_2(\lambda) \rangle = \frac{1}{\lambda_{max} - \lambda_{min}} \int_{\lambda_{min}}^{\lambda_{max}} L_1(\lambda)\, L_2(\lambda)\, d\lambda, \tag{3.2}$$

where the integration is over all wavelengths in the visual spectrum. The length norm

$$\| L_e(\lambda) \|^2 = \langle\, L_e(\lambda), L_e(\lambda)\, \rangle = \frac{1}{\lambda_{max} - \lambda_{min}} \int_{\lambda_{min}}^{\lambda_{max}} L_e^2(\lambda)\, d\lambda \tag{3.3}$$

is used to determine the distance between two radiometric functions. Note that all wavelengths in the range $[\lambda_{min}, \lambda_{max}]$ are a priori considered equally important when comparing two radiometric functions, since no weighting across wavelength is assumed in the definition of the inproduct and length norm.

In case of the so-called canonical basis (see chapter 6), the basis function $\psi_j(\lambda)$ approximates a monochromatic spectrum, i.e., it is equal to one in a small interval of wavelengths $\lambda_j \leq \lambda < \lambda_j + \Delta\lambda$, and zero elsewhere. We take $\Delta\lambda = (\lambda_{max} - \lambda_{min})/N$ and $\lambda_j = \lambda_{min} + (j-1) \cdot \Delta\lambda$, for $j = 1, \ldots, N$. By making $\Delta\lambda$ sufficiently small (typically, $\lambda_{min} =$ 380 nm, $\lambda_{max} =$ 780 nm, $\Delta\lambda =$ 5 nm and $N = 80$), we can make piecewise-constant approximations to radiometric functions that are sufficiently accurate for describing color vision. This assumption about the limited wavelength resolution of human vision has been verified experimentally by means of wavelength discrimination experiments and is also confirmed by the analysis that follows, the results of which do not change substantially if a slightly different value for $\Delta\lambda$ (such as $\Delta\lambda = 10$ nm) is used.

Using the canonical basis, the vector product can be expressed as

$$\langle\, L_1(\lambda), L_2(\lambda)\, \rangle = \frac{1}{N} \sum_{j=1}^{N} \hat{L}_{1\psi}(j) \cdot \hat{L}_{2\psi}(j). \tag{3.4}$$

The dual basis of the canonical basis B_ψ is $B_\chi = \{\chi_j(\lambda) = N \cdot \psi_j(\lambda); j = 1, \ldots, N\}$ since B_ψ is orthogonal.

Bases B_ψ of low dimension N have also been proposed as an alternative for the canonical basis; they are for instance based on a principal-components analysis of light sources and reflecting surfaces occurring in nature [MW92, Wan95]. They can be handled in a similar way as the canonical basis.

2.2 ANALYSIS OF COLOR MATCHING DATA

In a color matching experiment, three primary light sources $L_{e,1}(\lambda)$, $L_{e,2}(\lambda)$ and $L_{e,3}(\lambda)$ of V are used to match all the basis functions $\psi_j(\lambda)$, for $j = 1, \ldots, N$. A color match is assumed to correspond to an equality of the fundamental metameres, while fundamental color space U is assumed to be a three-dimensional subspace of V.

Since the fundamental color space U is a subspace of V, there exists a linear operator P that maps any radiometric function $L_e(\lambda)$ to its orthogonal projection $PL_e(\lambda)$ in U. This projection $PL_e(\lambda)$ is the fundamental metamere of the radiometric function $L_e(\lambda)$. The metameric black, or residual, is equal to $L_e(\lambda) - PL_e(\lambda)$, and belongs to $U^{\perp}$. The projected primaries $B_\varphi = \{\varphi_1(\lambda), \varphi_2(\lambda), \varphi_3(\lambda)\}$, with $\varphi_1(\lambda) = PL_{e,1}(\lambda)$, $\varphi_2(\lambda) = PL_{e,2}(\lambda)$, and $\varphi_3(\lambda) = PL_{e,3}(\lambda)$, form a basis of U, so that any projected radiometric function $PL_e(\lambda)$ can be expressed as a linear combination of the projected primaries. More specifically, the projected basis vectors of B_ψ are

$$P\psi_j(\lambda) = \sum_{i=1}^{3} g_{\phi\psi}(i,j) \cdot \varphi_i(\lambda), \tag{3.5}$$

for $j = 1, \ldots, N$. Since the fundamental metamere $P\psi_j(\lambda)$ fully determines the color impression of $\psi_j(\lambda)$, the numbers $g_{\phi\psi}(i,j)$, for $i = 1, 2, 3$, must be the tristimulus values of the radiometric function $\psi_j(\lambda)$ with respect to the primaries $L_{e,1}(\lambda)$, $L_{e,2}(\lambda)$ and $L_{e,3}(\lambda)$. These tristimulus values can be experimentally determined by color matching and subsequently assembled into an $3 \times N$ matrix

$$G_{\phi\psi} = [g_{\phi\psi}(i,j)]_{i=1,2,3;j=1,\ldots,N}\,. \tag{3.6}$$

The j-th column of $G_{\phi\psi}$ is the representation $[P\psi_j(\lambda)]_\varphi$ of the fundamental metamere $P\psi_j(\lambda)$ of the basis function $\psi_j(\lambda)$ in the basis B_φ of projected primaries.

This matrix of tristimulus values is sufficient information to construct the three-dimensional fundamental color space. From the previous chapter, see Figure 2.8, we can deduce that the representation of the fundamental metamere $PL_e(\lambda)$ of any radiometric function $L_e(\lambda)$ with (physical) representation $[L_e(\lambda)]_\psi$ can be written as

$$[PL_e(\lambda)]_\psi = P_{\chi\psi} \cdot [L_e(\lambda)]_\psi, \tag{3.7}$$

where

$$P_{\chi\psi} = G_{\chi\phi}\left(G_{\phi\psi} \cdot G_{\chi\phi}\right)^{-1} G_{\phi\psi} \tag{3.8}$$

denotes the idempotent $N \times N$ projection matrix of rank 3. In case of the orthogonal canonical basis, $G_{\chi\phi} = N \cdot G^T_{\phi\psi}$. The j-th column of the matrix $P_{\chi\psi}$ is the representation $[P\psi_j(\lambda)]_\psi$ of the fundamental metamere of basis function $\psi_j(\lambda)$, for $j = 1, \ldots, N$, while the three columns of the $N \times 3$ matrix

$$G_{\chi\varphi} = G_{\chi\phi} \left(G_{\phi\psi} \cdot G_{\chi\phi}\right)^{-1} \tag{3.9}$$

are the representations $[\varphi_i(\lambda) = PL_{e,i}(\lambda)]_\psi$ of the fundamental metameres of the primary light sources $L_{e,i}(\lambda)$, for $i = 1, 2, 3$. Note that $G_{\phi\psi} \cdot G_{\chi\varphi} = I_3$, where I_3 denotes the 3×3 identity matrix.

The inproduct $\langle PL_1(\lambda), PL_2(\lambda)\rangle$ between two fundamental metameres can be derived from the representations $[PL_1(\lambda)]_\psi$ and $[PL_2(\lambda)]_\psi$. This vector product completely determines the geometry of fundamental color space.

By considering alternative decompositions $P_{\chi\psi} = \hat{G}_{\chi\varphi} \cdot \hat{G}_{\phi\psi}$ of the projection matrix, where $\hat{G}_{\chi\varphi}$ and $\hat{G}_{\phi\psi}$ are $N \times 3$ and $3 \times N$ matrices that satisfy $\hat{G}_{\phi\psi} \cdot \hat{G}_{\chi\varphi} = I_3$, different coordinate frames for fundamental color space are obtained. Such a change of coordinate frame is accomplished by choosing $\hat{G}_{\chi\varphi} = G_{\chi\varphi} \cdot M^{-1}$ and $\hat{G}_{\phi\psi} = M \cdot G_{\phi\psi}$, where M is any invertible 3×3 matrix. The new decomposition still satisfies $P_{\chi\psi} = \hat{G}_{\chi\varphi} \cdot \hat{G}_{\phi\psi}$ and $\hat{G}_{\phi\psi} \cdot \hat{G}_{\chi\varphi} = I_3$. Hence, there are infinitely many possible sets of primaries, and each choice of primaries gives rise to different tristimulus values or color matching functions (CMFs). If $\hat{G}_{\phi\psi}$ is a matrix of tristimulus values for an alternative set of primaries $\hat{L}_{e,i}(\lambda)$, for $i = 1, 2, 3$, then the columns of the matrix $\hat{G}_{\chi\varphi}$ are the representations $[\hat{\varphi}_i(\lambda) = P\hat{L}_{e,i}(\lambda)]_\psi$ of the fundamental metameres of these primaries. The projection matrix, and therefore also the three-dimensional fundamental color space itself, are however independent of the choice of the primaries.

The symmetric projection matrix is displayed in Figure 3.1 for the case of a basis with $N = 80$ monochromatic spectra with $\Delta\lambda = 5$ nm, $\lambda_{min} = 380$ nm and $\lambda_{max} = 780$ nm. The color-matching functions $\overline{x}(\lambda)$, $\overline{y}(\lambda)$ and $\overline{z}(\lambda)$ of the 1931 CIE standard observer [CIE71], plotted in Figure 1.12, were used as input into the calculations, i.e., they constitute the rows of the input matrix $G_{\phi\psi}$. The entries in the resulting projection matrix $P_{\chi\psi}$ show three peaks along the main diagonal, corresponding to approximately 445 nm, 535 nm and 605 nm. These wavelengths are very efficient in stimulating the visual system [CK85]. Less efficient monochromatic spectra with adjacent wavelengths have fundamental metameres that are smaller, as is illustrated in the left part of Figure 3.2 by the length of the fundamental metameric vectors

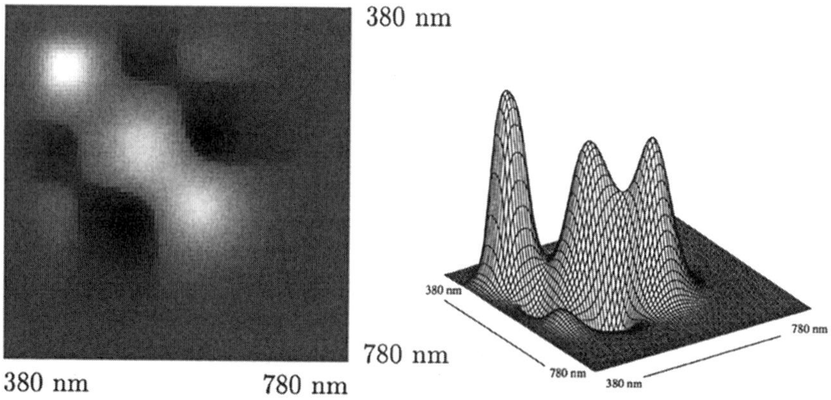

Figure 3.1. The projection matrix $P_{\chi\psi}$ onto fundamental color space for the canonical basis is rendered as a mesh plot on the right, and as an image on the left. The rows (and columns) of this matrix are the fundamental metameres of the monochromatic spectra. The values x in the projection matrix have been mapped to grey values $\text{sign}(x) \cdot \sqrt{x}$ (+offset) in order to improve the visual quality of the displayed image.

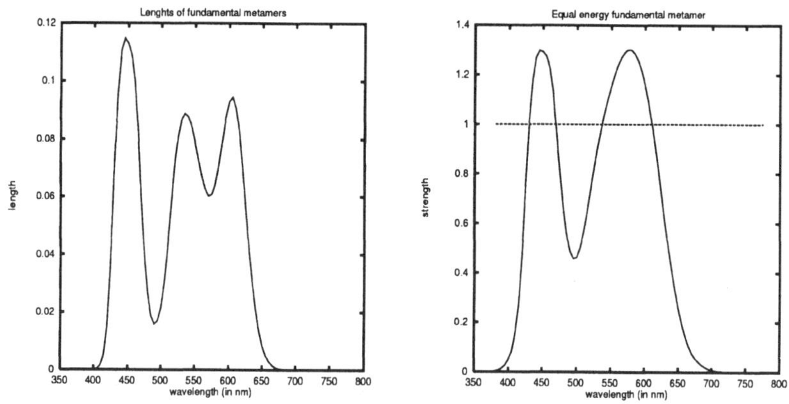

Figure 3.2. The square roots of the diagonal entries of the projection matrix are plotted in the left figure. They are equal to the lengths of the fundamental metameres of the monochromatic spectra. The right figure plots the sum of all rows in the projection matrix, which is the fundamental metamere $p_E(\lambda)$ of an equal energy source.

as a function of wavelength. These less-efficient fundamental metameres tend to peak at one of the summital wavelengths and do not maximize at their own wavelengths.

The sum of all columns (or rows) of the matrix $P_{\chi\psi}$ is the fundamental metamere of the equal-energy source, and is shown in the right part of Figure 3.2. The near-equal height of the two peaks in this fundamental metamere is typical for an unsaturated color source.

The (dual) basis $B_\varphi = \{p_x(\lambda), p_y(\lambda), p_z(\lambda)\}$ of primaries that corresponds to the basis $B_\phi = \{\overline{x}(\lambda), \overline{y}(\lambda), \overline{z}(\lambda)\}$ of the CIE color-matching functions can be determined in a straightforward way. The Gramian matrix

$$\begin{aligned} G_{\phi\phi} &= \begin{bmatrix} \langle\overline{x}(\lambda),\overline{x}(\lambda)\rangle & \langle\overline{x}(\lambda),\overline{y}(\lambda)\rangle & \langle\overline{x}(\lambda),\overline{z}(\lambda)\rangle \\ \langle\overline{y}(\lambda),\overline{x}(\lambda)\rangle & \langle\overline{y}(\lambda),\overline{y}(\lambda)\rangle & \langle\overline{y}(\lambda),\overline{z}(\lambda)\rangle \\ \langle\overline{z}(\lambda),\overline{x}(\lambda)\rangle & \langle\overline{z}(\lambda),\overline{y}(\lambda)\rangle & \langle\overline{z}(\lambda),\overline{z}(\lambda)\rangle \end{bmatrix} \\ &= \frac{1}{80}\begin{bmatrix} 14.394 & 11.333 & 5.125 \\ 11.333 & 15.441 & 1.701 \\ 5.125 & 1.701 & 28.078 \end{bmatrix} \end{aligned} \tag{3.10}$$

of the color-matching functions has to be inverted to obtain the Gramian matrix

$$\begin{aligned} G_{\varphi\varphi} &= \begin{bmatrix} \langle p_x(\lambda),p_x(\lambda)\rangle & \langle p_x(\lambda),p_y(\lambda)\rangle & \langle p_x(\lambda),p_z(\lambda)\rangle \\ \langle p_y(\lambda),p_x(\lambda)\rangle & \langle p_y(\lambda),p_y(\lambda)\rangle & \langle p_y(\lambda),p_z(\lambda)\rangle \\ \langle p_z(\lambda),p_x(\lambda)\rangle & \langle p_z(\lambda),p_y(\lambda)\rangle & \langle p_z(\lambda),p_z(\lambda)\rangle \end{bmatrix} \\ &= \begin{bmatrix} 14.447 & -10.382 & -2.008 \\ -10.382 & 12.677 & 1.127 \\ -2.008 & 1.127 & 3.147 \end{bmatrix} \end{aligned} \tag{3.11}$$

of the corresponding primaries $p_x(\lambda)$, $p_y(\lambda)$ and $p_z(\lambda)$. The primaries are then obtained by a weighted sum of the color-matching functions, with weights equal to the columns of the matrix $G_{\varphi\varphi}$. The resulting primaries are plotted in Figure 3.3. Note that they contain negative energies at some wavelengths, and can therefore not be physically realized.

The fundamental metameres of all physical light sources are linear combinations (with positive weights) of these basic primaries. More specifically, in the CIE standard observer reference system, the fundamental metamere of the radiometric function $L_e(\lambda)$ is specified by the tristimulus values $X = \langle\,\overline{x}(\lambda), L_e(\lambda)\,\rangle$, $Y = \langle\,\overline{y}(\lambda), L_e(\lambda)\,\rangle$ and $Z = \langle\,\overline{z}(\lambda), L_e(\lambda)\,\rangle$, and can be expressed as

$$PL_e(\lambda) = X \cdot p_x(\lambda) + Y \cdot p_y(\lambda) + Z \cdot p_z(\lambda). \tag{3.12}$$

The color perception evoked by any radiometric function can be derived from the position of its fundamental metamere in fundamental color space. Color-matching experiments do however not offer any information on how the mapping from fundamental color space to subjective

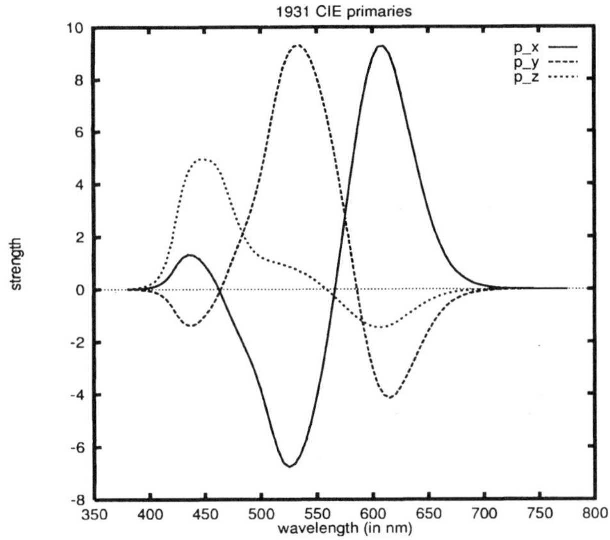

Figure 3.3. The 1931 CIE standard primaries $p_x(\lambda)$, $p_y(\lambda)$ and $p_z(\lambda)$ form a basis for fundamental color space. These primaries constitute the dual basis of the basis formed by the 1931 CIE standard color-matching functions.

color sensations such as hue, saturation and brightness/lightness is accomplished. We can imagine that these sensations can be described as functions operating on the coordinates in fundamental color space[1], for instance, brightness $B(X, Y, Z)$. In the following sections, it will be examined how the freedom in the choice of the coordinate frame can be used to select coordinate frames that simplify the description of these mappings between fundamental color space and visual sensations.

Two 3D renderings of fundamental color space are shown in Figure 3.4. The reference frame $\{p_E(\lambda), p_x(\lambda), p_z(\lambda)\}$ that is adopted in this figure will be discussed in more detail in the next sections. The first primary $p_E(\lambda) = p_x(\lambda) + p_y(\lambda) + p_z(\lambda)$ is the fundamental metamere of an equal energy source. The plane spanned by the red-green $(p_x(\lambda))$ and yellow-blue $(p_z(\lambda))$ primaries, is called the **chromaticity plane**. The axis orthogonal to the chromaticity plane corresponds to the spectral sensitivity function $\overline{y}(\lambda) = V(\lambda)$. The closed curves connect the fundamental metameric vectors $E \cdot P\psi_j(\lambda)$ of the scaled monochromatic spectra $E \cdot \psi_j(\lambda)$, for seven equally-spaced values of the energy factor

[1]This simplified view usually applies only to well-controlled experimental situations, where spatial and temporal color interactions between objects in a scene can be neglected (see later).

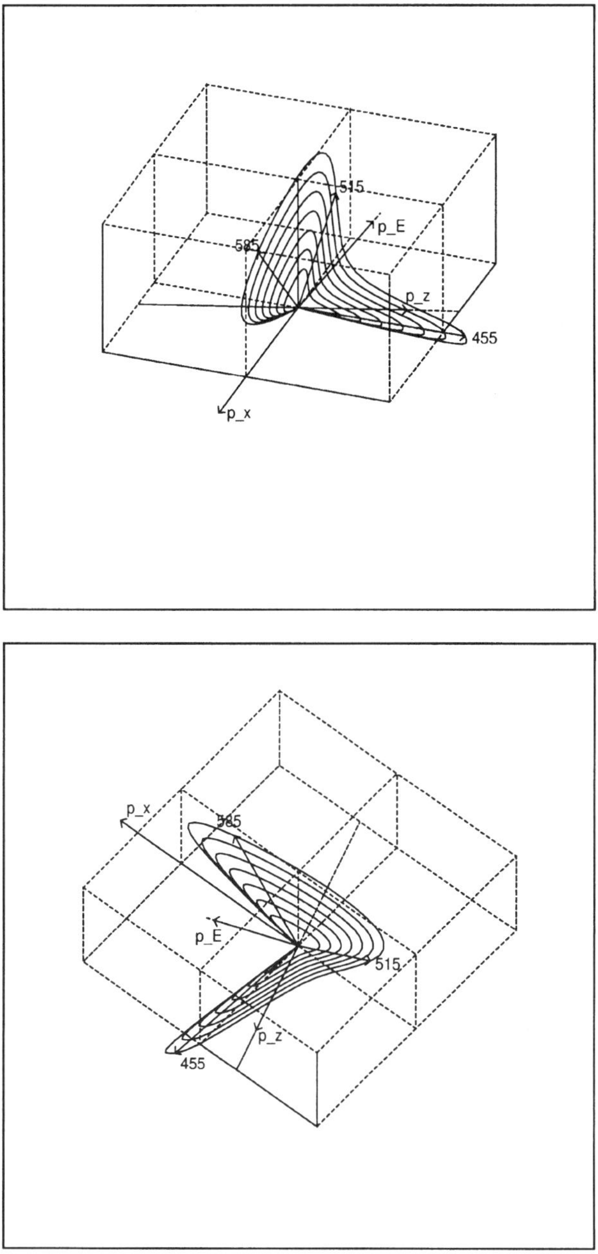

Figure 3.4. Two 3D renderings of fundamental color space for different observer positions. See text for explanation.

E. These curves lie on a butterfly-shaped surface. Since physical radiometric functions have a positive energy contribution at all wavelengths, only metameric vectors that lie in front of this surface can be physically realized. An interesting, approximately orthogonal, coordinate frame is formed by the fundamental metameres of the monochromatic spectra at wavelengths 455 nm, 515 nm and 585 nm [CK85].

3. COLOR DIAGRAM

Historically, i.e., before the introduction of fundamental color space and uniform color spaces (see later), colors were often specified by their **light efficiency** Y, or luminance $L = K_m \cdot Y$, and their two-dimensional **color coordinates**

$$x = \frac{X}{X+Y+Z}\ ,\ y = \frac{Y}{X+Y+Z}. \tag{3.13}$$

The reverse mapping back to tristimulus values is easily derived to be

$$X+Y+Z = \frac{Y}{y}\ ,\ X = \frac{x}{y}\,Y\ ,\ Z = \frac{z}{y}\,Y, \tag{3.14}$$

where $z = 1 - x - y$. These color coordinates are introduced here because they are still used frequently in practice.

Figure 3.5 shows the **color diagram**, which is the coordinate system spanned by the color coordinates (x, y). This color diagram allows to show simultaneously all colors that have the same luminance. Neutral (white) reference sources (C, D_{65} and E) have color coordinates close to $x = y = 1/3$. Note that the color coordinates are not linearly related to the tristimulus values[2].

If two colors with tristimulus values (X_1, Y_1, Z_1) and (X_2, Y_2, Z_2) are mixed additively, then the tristimulus values of the color mixture satisfy

$$X = k\,X_1 + (1-k)\,X_2\ ,\ Y = k\,Y_1 + (1-k)\,Y_2\ ,\ Z = k\,Z_1 + (1-k)\,Z_2, \tag{3.15}$$

where $0 \leq k \leq 1$ specifies the relative amount of both colors. The color coordinates of this color mixture are hence equal to

$$x = \alpha\,x_1 + (1-\alpha)\,x_2\ ,\ y = \alpha\,y_1 + (1-\alpha)\,y_2, \tag{3.16}$$

where

$$\alpha = \frac{k \cdot Y_1/y_1}{k \cdot Y_1/y_1 + (1-k) \cdot Y_2/y_2}. \tag{3.17}$$

---

[2] The color diagram is obtained by a perspective transformation of fundamental color space.

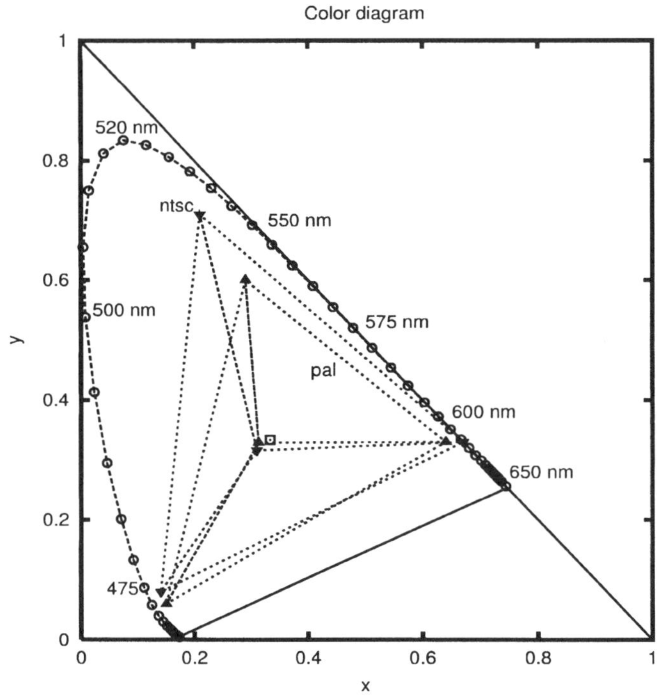

Figure 3.5. Color diagram with monochromatic spectra (in steps of 5 nm) and color triangles spanned by the standard PAL (triangles) and NTSC (inverted triangles) displays. The square indicates the color coordinates of an equal energy source E. The white reference sources in PAL and NTSC are D_{65} and C, respectively, and are indicated by the triangles in the center.

This implies that all additive mixtures of two colors with color coordinates (x_1, y_1) and (x_2, y_2) have color coordinates (x, y) that lie on the straight line connecting (x_1, y_1) and (x_2, y_2). These color coordinates can also be obtained by the graphical construction in Figure 3.6.

The rounded curve in Figure 3.5 shows how the monochromatic spectra map into the color diagram. Only colors that fall within this rounded shape can be physically realized, because all physical spectra are additive mixtures of monochromatic spectra. The color triangles spanned by the phosphor colors in two standard display systems (PAL and NTSC, see chapter 4) are also shown. Only colors that fall within the respective triangles can be realized by these displays.

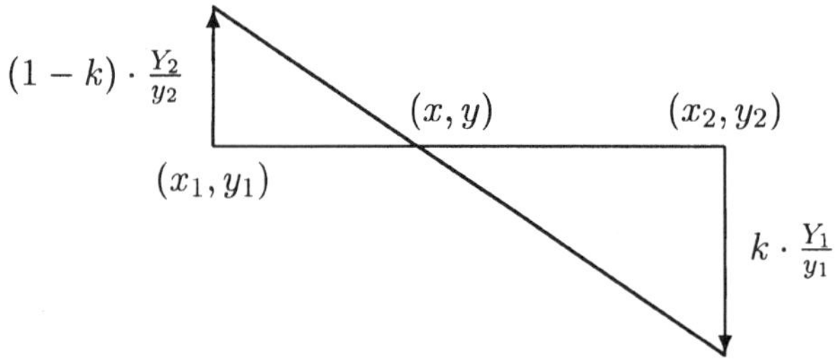

Figure 3.6. Additive mixtures of two colors have color coordinates that lie on the connecting line between the color coordinates of the component colors.

4. PHOTOMETRY

4.1 LUMINANCE AND LUMINANCE RATIO

The color sensation that is triggered by a radiometric function $L_e(\lambda)$ is determined by its fundamental metamere $PL_e(\lambda)$. In chapter 1, it was argued that the light efficiency

$$Y = \langle \overline{y}(\lambda), L_e(\lambda)\rangle = \langle P\overline{y}(\lambda), L_e(\lambda)\rangle = \langle \overline{y}(\lambda), PL_e(\lambda)\rangle, \tag{3.18}$$

where $\overline{y}(\lambda) = V(\lambda)$ denotes the spectral sensitivity of the human eye under photopic circumstances, is an important parameter of a light source. Because of the way in which the eye sensitivity curve was determined, it is known that the light efficiency Y, or luminance $L = K_m \cdot Y$, is the main determinant of subjective brightness.

Any coordinate system in fundamental color space for which $\overline{y}(\lambda)$ is the first test function will have Y as the first coordinate. Because of the bi-orthogonality condition, the second and third primaries of such a coordinate frame will have to be orthogonal to $\overline{y}(\lambda)$. In the 3D rendering of fundamental color space in Figure 3.4, we have therefore used $\overline{y}(\lambda)$ as the first axis. This axis points upwards in the figures. The chromaticity plane, which is spanned by the second and third primaries, and which will be discussed in greater detail in the next section, is orthogonal to the vector $\overline{y}(\lambda)$. The height above this chromaticity plane is proportional to the light efficiency Y and luminance L. Note for instance that the highest luminance is indeed obtained for a monochromatic stimulus of unit energy with $\lambda = 555$ nm, and that radiance components at short wavelengths (i.e., around $\lambda = 450$ nm) have a strong influence on the position in fundamental color space, and hence on the color perception, but little influence on the luminance, and hence on the brightness.

In color perception, an important distinction has to be made between primary light sources (or unrelated colors), i.e., surfaces or objects emitting light produced by a transformation of energy, and secondary light sources (or related colors), i.e., surfaces or objects that are not self-emitting, but receive light and redirect it, at least in part, by reflection or transmission. Section 6 of chapter 1 discussed the standardized terminology that is available for denoting the subjective color attributes that can occur in both cases [Hun78].

The case of related colors is complicated by the fact that the color of an object is judged in relation to its surround. For instance, the lightness of a nonluminous related color is defined as 'the attribute of a visual sensation according to which an area appears to reflect diffusely or transmit a greater or smaller fraction of incident light'. Hence, lightness is judged relative to the incident light. This assumes that the visual system is able to deduce the incident light, and can use it as a reference. The simplest objective measure that relates to the attribute of lightness is the **luminance factor** or **luminance ratio**, defined as 'the ratio of the luminance of the body to that of a perfectly reflecting or transmitting diffuser identically illuminated'. Hence, this luminance factor can be expressed as the ratio $Y/Y_o = L/L_o$ of the object luminance $L = K_m \cdot Y$ to the reference luminance $L_o = K_m \cdot Y_o$. In case of a display, such as a CRT or LCD, the maximum luminance is most often taken as the reference luminance L_o. We will however come back to this choice for L_o at a later stage.

Current display devices usually start from a digital image stored in computer memory. A monochrome image is characterized by its **grey value** V as a function of space (and time). These grey values are often integer numbers in the range [0,255]. The display converts the grey values of the digital image into luminances for the displayed image. The display may be a *soft-copy device* such as a CRT or an LCD (monitor or projector) that emits light, and hence can also be viewed in the dark, or it may be a *hard-copy device* such as a printer (output on paper) or film writer (output on transparency or film) that requires an external light source in order for a human observer to perceive the images. Such an external light source obviously also has an effect on the mapping from grey value to luminance of the displayed image.

The luminance factor can be used to compare display devices that have different luminance ranges, such as a CRT and a transparent film on a light box[3]. In Figure 3.7, we show $\log(L/L_o)$ (where log denotes the base-10 logarithm and L_o was taken equal to the maximum luminance

[3]A typical CRT for medical applications will for instance have a maximum luminance of 400 cd/m^2 or below, while the maximum luminance in case of a transparent film on a light box may easily exceed 2000 cd/m^2.

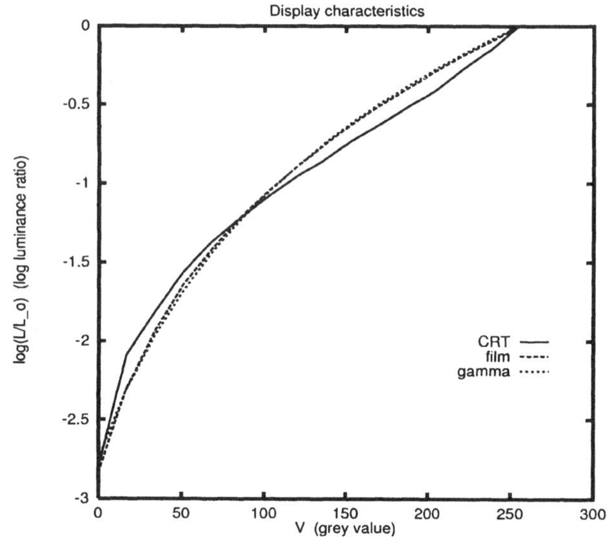

Figure 3.7. Display characteristics for CRT and film. The film characteristic can be closely approximated by the gamma characteristic of equation (3.19) with parameters $\gamma = 3.5$ and $V_i = 50$.

of the respective displays) as a function of input grey value in the digital input image for a commercially available CRT and film (+light box) used in radiological applications. Note that both display devices have similar **grey-value-to-luminance-ratio characteristics.**

Display characteristics can often be closely approximated by a power law (or **gamma characteristic**), i.e.

$$\frac{L}{L_o} = \left(\frac{V + V_i}{V_o + V_i}\right)^{\gamma} \quad \text{or} \quad \log\frac{L}{L_o} = \gamma \cdot [\log(V + V_i) - \log(V_o + V_i)]\,, \tag{3.19}$$

where V_o is the grey value that corresponds to the reference luminance L_o. If V_i is small, then the relationship between $\log L$ and $\log V$ is approximately linear with slope γ. A power-law approximation (with $\gamma = 3.5$ and $V_i = 50$) to the hard-copy display characteristic is also shown in Figure 3.7.

4.2 BRIGHTNESS MODELS AND DISPLAY STANDARDIZATION

Especially in professional applications, such as medical imaging or graphical arts, viewing an image on different display media (monitor, printer and/or film) is not supposed to affect the colors of the objects being displayed. The inconsistencies in luminance characteristics between different soft-copy displays as well as between hard-copy and soft-copy displays has for instance been a contributing factor why many radiologists still have reservations about performing diagnosis on soft copies. This problem can only be overcome if displays are standardized. If a luminance characteristic is to be proposed for standardization, it makes sense to try and optimize it in some way or other. We will show that this requires a better understanding of the relationship between measured luminance and perceived brightness.

The relationship between physical stimuli (e.g., the luminance on a display) and psychological reactions (e.g., the corresponding brightness response) has been primarily formulated by Weber and Fechner in the 19th century [Fal86, LK88]. Weber examined this relationship by focusing on the **just noticeable difference** (JND). Weber's law states that when L represents stimulus luminance then

$$\frac{\Delta L}{L} = k, \tag{3.20}$$

where ΔL is the just noticeable change in luminance required to see a difference between L and $L + \Delta L$, and k is a constant called the Weber fraction (a typical value is $k = 0.015$). Since this fraction is constant, it is implicitly understood that as L increases, ΔL must also increase. Weber's law is an approximation to experimental data that only holds for intermediate luminance values (which, fortunately, are the most relevant ones for practical applications). Experimental data can deviate significantly from Weber's law at very low and high luminances.

Fechner used Weber's law to derive a scale relating the strength of the physical stimulus (e.g., the luminance) to the strength of the observer's psychological sensation (e.g., the brightness). From Weber's law, it can be deduced that this relationship is not linear. Indeed, in order for k to be constant, the JNDs from the upper end of the scale require a larger increase in luminance compared to JNDs from the lower end of the scale. If it can be assumed that brightness $B(L)$ is a differentiable function of the luminance L, then approximately

$$\Delta B = \frac{dB(L)}{dL}\,\Delta L \tag{3.21}$$

for small ΔL. Fechner assumed that each JND in luminance (from L to $L + \Delta L$) resulted in an equal increase in brightness sensation ΔB, which he took as the unit of the brightness scale. This implies that

$$\frac{dB(L)}{dL} = \frac{1}{\Delta L}, \tag{3.22}$$

where ΔL is determined experimentally as a function of L. Substituting Weber's law for luminance JND's ΔL as a function of luminance L in the above equation results, after integration, in a logarithmic relationship

$$B \propto \log L, \tag{3.23}$$

between perceived brightness B and luminance L. This motivates why we have plotted the logarithm of the luminance as the relevant output in Figure 3.7

A related way of experimentally determining the relationship between luminance and brightness (or lightness) is my means of a supra-threshold technique called *bisection*. Two reference stimuli of luminance L_1 and L_2 are presented to the observer, who has to adjust the luminance L ($L_1 < L < L_2$) of the test stimulus such that the brightness difference induced by $L - L_1$ is equal to the brightness difference induced by $L_2 - L$. A similar integration technique as above can be used to derive the luminance-to-brightness relationship from such experimental data, since ΔB is again constant, although larger than 1 JND [Whi92].

A more direct way of determining the relationship between stimulus intensity and psychological response was adopted by Stevens in the 1950's [Ste57]. He used *magnitude estimation* to obtain judgements. In a magnitude estimation experiment, the observers are asked to give numbers that match their impression of psychological magnitude induced by the physical stimulus. Stevens's power law states that the magnitude of the brightness response R is proportional to the luminance L of the physical stimulus, which has been raised to some power n, i.e.,

$$R \propto L^n. \tag{3.24}$$

This relationship is supported by experimental evidence. Unlike Fechner's law, the power law of Stevens also includes the mapping from internal sensation (i.e., brightness) to external magnitude response (i.e., numbers), which explains why both descriptions differ [Wag75].

Brightness (response) models can be used to help standardize a display characteristic. Indeed, suppose that equal steps in grey scale evoke equal steps in brightness (response). Such **perceptual linearization** [Piz81, BM95, BOM97, Bel99] has the advantage that it maximizes the number of discernable levels in the display. Indeed, if the minimum and maximum luminance of a display are fixed, then the number of JNDs that fit into this range are also fixed. Suppose the input values are quantized to exactly this number of levels. If the display is perceptually linearized, then the difference between any two successive input levels will be exactly one JND across the entire range. If the display is not perceptually linearized, then some differences between successive input levels will be below 1 JND, while others will be above 1 JND. Hence, in order to guarantee that all differences are at most 1 JND, a finer quantization, i.e., a smaller step between successive input values, is required with respect to the case of perceptual linearization.

If a model such as $R \propto L^n$ is adequate to describe the luminance-to-brightness-response relationship, then using a display with the inverse power-law characteristic $L \propto V^{1/n}$ guarantees a linear relationship $R \propto V$ between input grey values V and output response R. As observed earlier, the characteristics of many commercially available displays can indeed be approximated by such a power law.

If an agreement could be reached on a single luminance-to-brightness (response) formula, then perceptual linearization holds the key to deriving a display standard from this relationship. Unfortunately, experimental results on brightness discrimination, such as in the case of Weber, and brightness estimation, such as in the case of Stevens, depend to some extent on the stimulus configuration and experimental conditions, so that the discussion concentrates around which experimental condition is most relevant for the perception of input values in complex scenes such as medical images. For instance, one might argue that the relevant task for diagnosis is to detect the brightness difference between a lesion area with luminance $L + \Delta L$ and a uniform surround with luminance L. The size of the lesion area and/or the surround luminance will for instance influence this detection task, in the sense that the JND discrimination data may be closely approximated by a power-law, i.e.,

$$\Delta L = a \cdot L^b \tag{3.25}$$

but both a and b will depend on the experimental conditions. Any derived brightness model will also share this dependency. A single display characteristic will hence be unable to realize perceptual linearization for all discrimination tasks [BM98a, Bel99], so that a compromise has to be sought.

There are two broadly adopted brightness/lightness models that can for instance be used to construct a linearizing display standard, i.e., the CIE psychometric lightness model [CIE71, CIE79] and the ACR/NEMA (American College of Radiology/ National Electrical Manufacturers Association) standard proposal [BM95, XI97]. These models were derived from experimental data on relatively large patches (2 to 4 degrees of visual angle), so that they are mostly relevant for describing the brightness/lightness judgement in sufficiently large and approximately uniform areas in the image. We only present the CIE lightness model, partly because it has the simplest mathematical description, and partly because it can be extended towards color vision, as we will see in the next section.

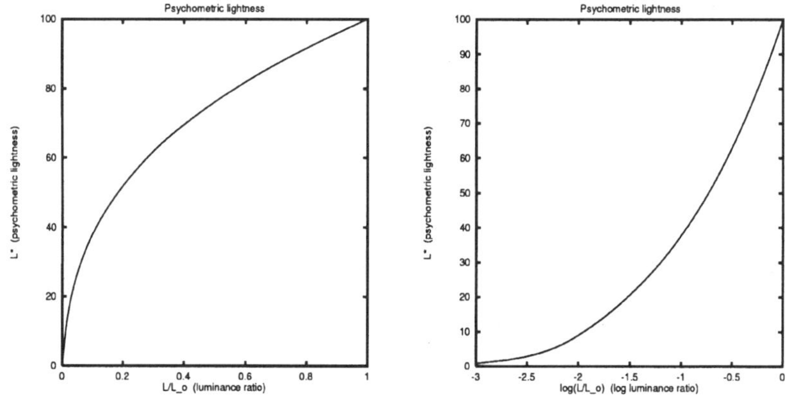

Figure 3.8. The CIE psychometric lightness as a function of the luminance factor (left) and as a function of the logarithm of the luminance factor (right).

The CIE has introduced the **psychometric lightness** measure

$$L^* = h_{lab}\left(\frac{L}{L_o}\right) = \begin{cases} 116\left(\frac{L}{L_o}\right)^{1/3} - 16 & \text{for} \quad \frac{L}{L_o} \geq 0.008856 \\ 903.3\left(\frac{L}{L_o}\right) & \text{for} \quad \frac{L}{L_o} \leq 0.008856 \end{cases} \tag{3.26}$$

as being approximately linearly related to perceived lightness. This function is depicted in Figure 3.8. The luminance L_o of the reference is assigned a psychometric lightness of 100. The psychometric lightness deviates from Weber's law at low luminances since the slope of the curve relating $\log(L/L_o)$ to psychometric lightness is not constant but decreases with $\log(L/L_o)$.

A display with a linear relationship between input grey values and output psychometric lightness values can be constructed as follows. It is assumed that the input grey values are in the range $[0, V_{max}]$, the corresponding output luminances are in the range $[L_{min}, L_{max}]$, and the reference luminance is L_o. The grey-value-to-luminance-characteristic for the linearized display is then equal to

$$L(V) = L_o \cdot h_{lab}^{-1}\left[L^*_{min} + (L^*_{max} - L^*_{min})\frac{V}{V_{max}}\right], \tag{3.27}$$

for $V \in [0, V_{max}]$, where $L^* = h_{lab}(L/L_o)$ is the psychometric lightness characteristic, $L^*_{min} = h_{lab}(L_{min}/L_o)$ and $L^*_{max} = h_{lab}(L_{max}/L_o)$.

In Figure 3.9, the original CRT characteristic of Figure 3.7 is displayed together with the linearizing display characteristics according to the two

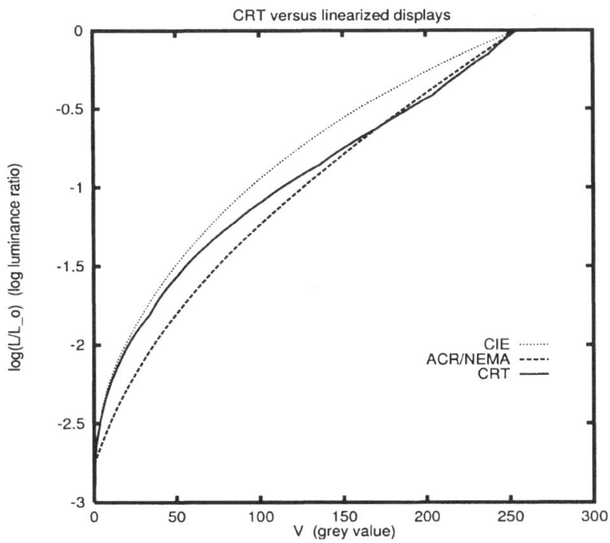

Figure 3.9. Comparison between an actual CRT characteristic and linearized display characteristics according to CIE and ACR/NEMA.

above-mentioned models (CIE and ACR/NEMA). The minimum and maximum luminances are determined by the original CRT at $L_{min} = 0.7$ and $L_{max} = 400$ cd/m^2, respectively, and $L_o = L_{max}$. Note that the actual CRT characteristic is in between the two standard curves. The display characteristic based on the CIE lightness standard will result in a larger average luminance than the display characteristic based on the ACR/NEMA proposal.

In Figure 3.10, the original film (+light box) characteristic of Figure 3.7 is displayed together with the linearizing display characteristics according to the two above-mentioned models. The minimum and maximum luminance are determined by the original film (+light box) at $L_{min} = 2$ and $L_{max} = 1262$ cd/m^2, respectively, and $L_o = L_{max}$. Note that the actual film characteristic is again in between the two standard curves, and is closer to the CIE characteristic than to the ACR/NEMA-characteristic.

4.3 BRIGHTNESS ADAPTATION

It was argued that display characteristics should be designed to approximate perceptual linearization, and that the grey-value-to-luminance-ratio characteristic of a display plays a major role in this linearization.

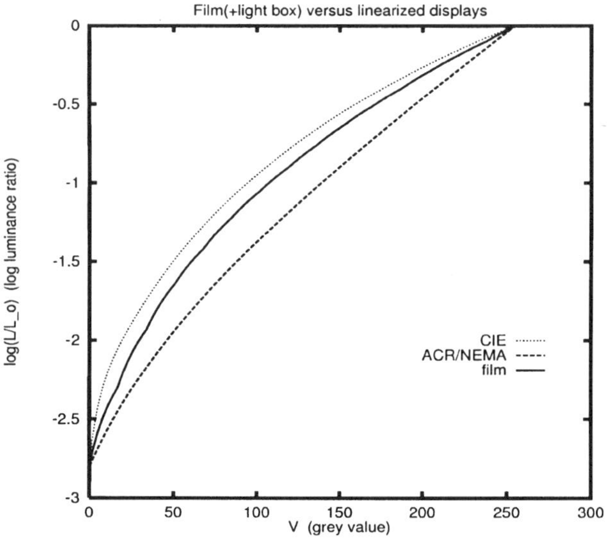

Figure 3.10. Comparison between an actual film characteristic and linearized display characteristics according to CIE and ACR/NEMA.

However, other factors also have to be considered in the design of display characteristics. Many of these factors are described in extensive detail in a recent book on **digital color management** [GM98] and are only briefly summarized here:

1 When an image is observed, the viewer should only see the light from the display itself. In most actual viewing conditions, the observer will however also see **flare light**, i.e., stray light containing no image information, that reflects from the surface of the film or the faceplate of the CRT. This light may originate from external light sources, or may be light from the display itself that is reflected back onto the display (for instance, by the walls of the room). Flare is mostly modelled as a uniform luminance offset that is added to the luminance coming directly from the display. For instance, in Figure 3.11, it is illustrated how the characteristic of the CRT is altered by a flare of 1 cd/m^2. Note that this correction tends to straighten the characteristic. Most displays have been designed with some expected flare in mind[4]. For instance, the glass plate in front of a CRT usually has

[4]The 'brightness' knob on a monitor allows to alter the grey-value-to-luminance characteristic for small grey values, and hence to compensate to some extent for flare.

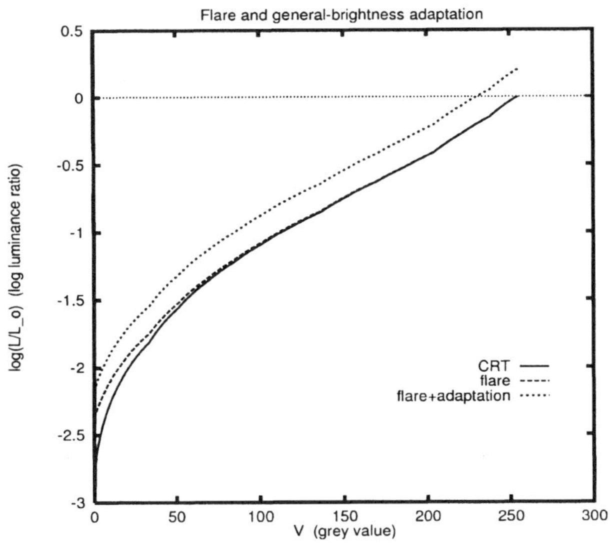

Figure 3.11. Effect of flare and global-brightness adaptation on a CRT display characteristic.

a transmission coefficient τ that is substantially smaller than 1. The flare light passes through the glass twice, and is hence reduced by a factor τ^2, while the light coming from the display is only reduced by a factor τ. Hence, although the luminance of the display is reduced, the effect of the flare light on the luminance ratio is also reduced [THM97].

2 In the definition of luminance ratio it was already indicated that there is some freedom in selecting the reference luminance L_o of the perfect reflector or transmitter. This choice of the reference luminance is especially relevant if different displays that operate in very different conditions (such as soft-copy and hard-copy displays) have to be compared. The observer has to judge from the image itself what the luminance of the reference white is. This judgement is not only influenced by the luminances in the displayed image, but also by the luminances in the surround. This effect is sometimes referred to as **general-brightness adaptation**. For instance, for a photographic print viewed in a light surround, the reference luminance will be mostly determined by the luminance of the incident light, which may for instance be judged very well from the light reflected from the table or wall on which the photograph is placed. For film and CRT, which

are typically viewed in a dark surround, the reference luminance is determined partly by the luminances in the scene[5], and partly by the dark surround luminance. As a result, the reference luminance adopted for film and CRT should be lower than the maximum luminance of the display. Values of $L_o = L_{max}/1.5$ and $L_o = L_{max}/1.6$ are proposed as typical for slide film and video display, respectively [GM98]. Similar values should most likely also hold for the case of film on a light box. In Figure 3.11, we have illustrated the combined effect of flare (of 1 cd/m^2) and general-brightness adaptation (by a factor of 1.6) on a CRT characteristic.

3 The fact that all graphs depicted up to now show (log) luminance ratio as the relevant output variable might indicate that the reference luminance L_o itself is not very relevant for the perception of a display. Although this may be true as a first approximation, there are some obvious limitations and second-order effects that should be mentioned. First, images with a larger average luminance are typically better appreciated, at least if the maximum luminance is not visually blinding. This is similar to the case of music, where appreciation increases with loudness over a large range. Second, if the average luminance in the scene is low (say, below 10 cd/m^2), then the perceived luminance contrast is reduced. This so-called **lateral-brightness adaptation** effect means that the same perceived contrast between two adjacent regions in an image requires a larger difference in log luminance ratio for low luminances than for high luminances [PYGR91]. This implies that equal values of $\log(L/L_o)$ do not necessarily correspond to equal perceived lightness at very distinct reference luminance levels L_o. Two display media that are intended to give similar subjective impressions in very distinct luminance ranges may hence require different $\log(L/L_o)$ characteristics. This is well-known in photography, where the characteristics of print and slides are different in order to compensate for this perceptual effect. The characteristic for slide film is steeper than for photographic prints at low values of $\log(L/L_o)$ in order to compensate for the fact that slides are typically displayed at lower luminance levels. The CRT and film (+light box) characteristics in Figure 3.7 are similar, because the luminance range under which they are operated is not too different in both cases.

[5]By adjusting the 'contrast' knob on a monitor, all luminances in the display are approximately multiplied by a constant factor, which will influence the choice of the reference luminance in case of a constant surround luminance.

5. COLORIMETRY

5.1 HERING PRIMARIES

The luminance $L = K_m \cdot Y$ of an unrelated color is an important objective measure, since it usually has a one-to-one relationship with the brightness of the color. Similarly, the luminance factor L/L_o is an important measure for a related color, since it usually has a one-to-one relationship with the lightness of the color. We will therefore be mainly interested in coordinate frames for fundamental color space for which the luminance (or luminance factor) is one of the coordinate values. Additional properties of the reference frame that can help in describing the relationship between coordinate values and the remaining subjective color percepts hue and saturation will now be discussed.

The hue of colored stimuli is usually expressed in the *complementary color* sensations red-green and yellow-blue. The term complementary refers to the fact that the hue pairs red/green and yellow/blue are opponent or antagonistic. Both members of a pair are not seen simultaneously in the same place.

In order to measure the amount of red/green and yellow/blue in monochromatic spectra, a null method can be used. This experimental method requires a **neutral** (or **white**) **reference** that contains (by definition) no red/green and yellow/blue. Suppose that we present a short-wavelength monochromatic spectrum, say at 420 nm, of unit energy. This stimulus has a certain amount of blueness. By adjusting the energy level of a long-wavelength monochromatic spectrum, say at 580 nm, that has a certain amount of yellowness, we can obtain that the additive color mixture contains no yellow or blue. For pure wavelengths, the relative spectral sensitivities for red/green and yellow/blue have been measured in this way, using an equal energy source as neutral reference. For moderate light levels, the resulting **color mixture functions** can be expressed analytically by

$$\begin{aligned} \overline{a}(\lambda) &= \overline{x}(\lambda) - \overline{y}(\lambda) \\ \overline{b}(\lambda) &= \overline{y}(\lambda) - \overline{z}(\lambda), \end{aligned} \tag{3.28}$$

and are illustrated in Figure 3.12 [Hur81]. Positive values for $\overline{a}(\lambda)$ indicate that the corresponding wavelengths are perceived as containing red, while negative values for $\overline{a}(\lambda)$ indicate that the corresponding wavelengths are perceived as containing green. Similarly, positive values for $\overline{b}(\lambda)$ indicate that the corresponding wavelengths are perceived as containing yellow, while negative values for $\overline{b}(\lambda)$ indicate that the corresponding wavelengths are perceived as containing blue.

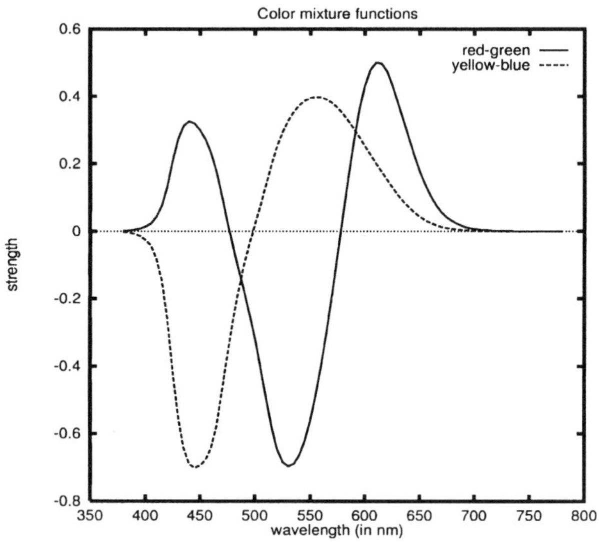

Figure 3.12. Color mixture functions $\overline{a}(\lambda)$ (red-green) and $0.4\,\overline{b}(\lambda)$ (yellow-blue).

The relative weight (of 0.4 on $\overline{b}(\lambda)$) between the color mixture functions in Figure 3.12 is adjusted in such a way that the curves cross at a wavelength of 590 nm that is subjectively judged to contain an equal amount of red and yellow. This relative weight needed to obtain equal amounts of red and yellow at 590 nm does however change with the energy level. It is typically observed that, for the same spectral distribution, red and green hues tend to dominate over yellows and blues at low energy levels, while the reverse is true at high energy levels [KB96]. This *Bezold-Brücke effect* can be understood by realizing that the red-green system has a lower threshold than the yellow-blue one, so that it reacts at lower energy levels, but that the response level of the latter system increases faster with energy level. One implication of the color mixture functions that has been verified experimentally for a wide range of energy levels, is that the wavelength $\lambda_g = 498$ nm for which $\overline{y}(\lambda) = \overline{z}(\lambda)$ is pure green (i.e., no yellow or blue) is independent of the energy level. A similar conclusion holds for the pure blue $\lambda_g = 477$ nm and pure yellow $\lambda_y = 578$ nm wavelengths, for which $\overline{x}(\lambda) = \overline{y}(\lambda)$.

For radiometric functions containing many different wavelengths, a reasonable agreement with subjective hue perception is obtained if the color mixture functions are used to weigh the radiometric function. The luminance sensitivity function $\overline{y}(\lambda)$, that is also referred to as the **white-**

black color mixture function in color theory, together with the two above color mixture functions, can hence be used as a basis of measurement functions for fundamental color space. The dual basis consists of a set of primaries, called the **Hering primaries** [Jud51]. The first primary, i.e., the one weighted by the white-black color mixture function, is equal to the fundamental metamere of the equal energy source $p_E(\lambda) = p_x(\lambda) + p_y(\lambda) + p_z(\lambda)$ (shown in Figure 3.2). This is easily verified, since $X = Y = Z$ for an equal energy source, so that only the first Hering primary has a nonzero weight. The second primary, i.e., the one weighted by the red-green color mixture function, has coordinates $Y = Z = 0$, and is hence proportional to the primary $p_x(\lambda)$ of Figure 3.3. The third primary, i.e., the one weighted by the yellow-blue color mixture function, has coordinates $X = Y = 0$, and is hence proportional to the primary $p_z(\lambda)$ of Figure 3.3. The relative angle between both primaries is $\theta[p_x(\lambda), p_z(\lambda)] = 107.3°$. The fundamental metamere of any radiometric function $L_e(\lambda)$ can be expressed as

$$PL_e(\lambda) = Y \cdot p_E(\lambda) + (X - Y) \cdot p_x(\lambda) - (Y - Z) \cdot p_z(\lambda) \qquad (3.29)$$

in terms of the Hering primaries.

5.2 CHROMATICITY PLANE

The hue perception of a colored light is determined by the projection of the fundamental metamere on the plane spanned by the two latter primaries. This plane is called the **chromaticity plane** and is specified by $Y = 0$, so that it must be orthogonal to $\overline{y}(\lambda)$. The projection of an arbitrary stimulus in fundamental color space on the chromaticity plane is performed parallel to the fundamental metamere of the equal energy source, since the origin must be the projection of the equal energy source, independent of the energy level. The orthogonal reference frame that is used in Figure 3.4 has $\overline{y}(\lambda)$ as first axis and $p_x(\lambda)$ as second axis. The third axis is oriented along the vector $p_z(\lambda) + 0.139\, p_x(\lambda)$. This vector lies in the chromaticity plane and is orthogonal to both $\overline{y}(\lambda)$ and $p_x(\lambda)$.

Similarly as in the previous section on brightness and lightness, a separate description of color perception is required for related colors. In the above discussion, it was implicitly assumed that the perceived color of a (uniform) patch is solely determined by its color coordinates. This is a reasonable assumption for a light source (i.e., an unrelated color), but it is easily demonstrated that this assumption is less accurate for related colors. For instance, the two small squares in Figure 3.13 have the same color coordinates, while their perceived brightness is obviously different. The square on the dark surround looks brighter than the square on the light surround. Similarly, the hue and the saturation of

Figure 3.13. Induction by the different surrounds results in a different brightness impression for the two small squares with identical color coordinates.

a related color can also be influenced by the hue and saturation of the surround color. This adaptation (or induction) effect of the surround color on the perceived color of a patch is difficult to predict in general, although some *color-appearance models* have recently been standardized by the CIE [LH98b, LH98a, LLH00, Hun98]. These models intend to summarize a large number of experimental findings [LH98c, HSC$^+$98] on how the color of a patch is influenced by the color of its surround, and they reflect our current understanding of the phenomenon. A detailed discussion of these models is beyond the scope of this introductory text. We limit ourselves to a simple adaptation of the above theory in order to make it applicable for related colors.

The most simple adaptation model is to assume that a change in neutral reference can account for the varying perception of related colors with changing surround color. Selecting adequate tristimulus values for this neutral reference color is then similar to selecting the reference luminance L_o for lightness perception (see the previous section). The mentioned color-appearance models can for instance assist in selecting this neutral reference in specific cases. For small color patches on a large uniform surround, the color of the surround will usually be selected as neutral reference.

According to this simple model, the perceived color of a patch with tristimulus values (X, Y, Z) and color coordinates (x, y) can be expressed in terms of the coordinates

$$\begin{aligned} \ell &= \frac{Y}{Y_n} \\ a &= \frac{X}{X_n} - \frac{Y}{Y_n} = \ell\left(\frac{x}{y}\frac{y_n}{x_n} - 1\right) \\ b &= \frac{Y}{Y_n} - \frac{Z}{Z_n} = \ell\left(1 - \frac{z}{y}\frac{y_n}{z_n}\right) \end{aligned} \tag{3.30}$$

in case the neutral reference has tristimulus values (X_n, Y_n, Z_n) and color coordinates (x_n, y_n). This choice guarantees that the reference color has coordinates $a = b = 0$, and is hence interpreted as neutral. Note that $a \geq -\ell$ and $b \leq \ell$, since X, Y and Z are always positive.

The proposed model uses the fundamental metamere of the neutral reference as one of the coordinate axes in fundamental color space. The coordinate ℓ for this primary $p_\ell(\lambda) = X_n\, p_x(\lambda) + Y_n\, p_y(\lambda) + Z_n\, p_z(\lambda)$ is equal to the luminance factor $\ell = Y/Y_n$. The remaining two primaries $p_a(\lambda) = X_n\, p_x(\lambda)$ and $p_b(\lambda) = -Z_n\, p_z(\lambda)$ are in the chromaticity plane $(Y = 0)$, and their orientation is not affected by the change in neutral reference. The relative angle between both primaries is $\theta[p_a(\lambda), p_b(\lambda)] = 72.7°$. The fundamental metamere of any radiometric function $L_e(\lambda)$ can be expressed as

$$PL_e(\lambda) = \ell \cdot p_\ell(\lambda) + a \cdot p_a(\lambda) + b \cdot p_b(\lambda) \tag{3.31}$$

in terms of these primaries. In the model, the coordinates a and b are proportional to the red-green and yellow-blue sensations, respectively.

The (a, b)-coordinate system of the chromaticity plane is depicted in Figure 3.14. The projections onto the chromaticity plane, parallel to the equal energy neutral reference $(X_n = Y_n = Z_n)$, of the monochromatic spectra with $\ell = 0.4$ are also shown[6]. Since vectors in fundamental color space can be (linearly) projected onto the chromaticity plane, we retain the simple interpretation of additive mixing of two primary colors that also holds in fundamental color space, i.e., that all colors with coordinates on the line connecting two primary colors can be realized by mixing these primaries. The projections of the primaries that are used in our current color-TV systems (PAL and NTSC, see section 4 of chapter 4) are also indicated in Figure 3.14. Only colors that fall within the (color) triangles can be physically realized by these systems.

[6]The curve connecting the projected monochromatic spectra is identical to the intersection of the butterfly-shaped surface in Figure 3.4 with a horizontal plane at height $\ell\, Y_n$. The intersection of this horizontal plane with the vector $p_E(\lambda)$, i.e., the fundamental metamere of the neutral reference, corresponds to the origin in the chromaticity plane.

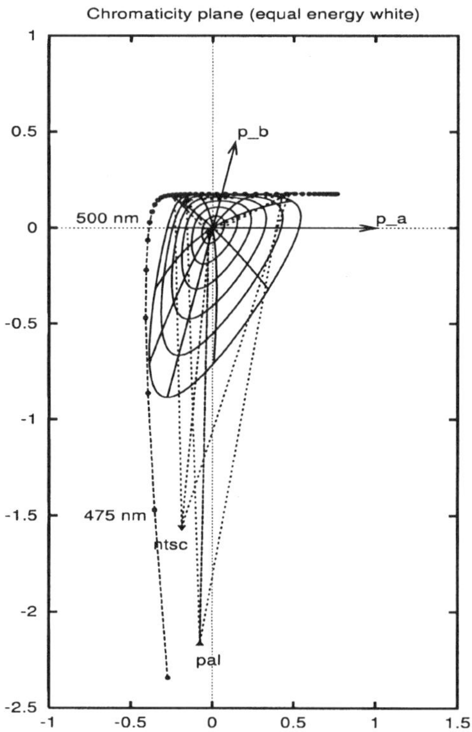

Figure 3.14. The chromaticity plane with $p_a(\lambda)$ and $p_b(\lambda)$ primaries. The open contour corresponds to the projections, parallel to the equal energy white reference (with $X_n/Y_n = Z_n/Y_n = 1$), of the monochromatic spectra with $\ell = 0.4$. The curves that radiate from the origin connect coordinates of constant psychometric hue (hue angle steps of 30°). The closed contours connect coordinates with constant psychometric chroma. Psychometric hue and chroma are according to the 1976 CIE(L^*,a^*,b^*) metric for $\ell = 0.4$. The triangles indicate the color coordinates that can be obtained in current color-TV systems (PAL and NTSC). The neutral references of PAL and NTSC differ slightly from the equal energy reference.

5.3 CIE STANDARD ILLUMINANTS

Because the color appearance of a reflecting body depends on the spectral distribution of the illuminant, it is important to be able to uniquely specify an illumination condition. The CIE has standardized a number of illuminants for this purpose in 1931. Standard illuminant A corresponds to a tungsten filament lamp at a color temperature of $T = 2856°$ K, while standards B and C, which are intended to approximate direct sunlight and diffuse daylight, are derived from illuminant A by applying specific

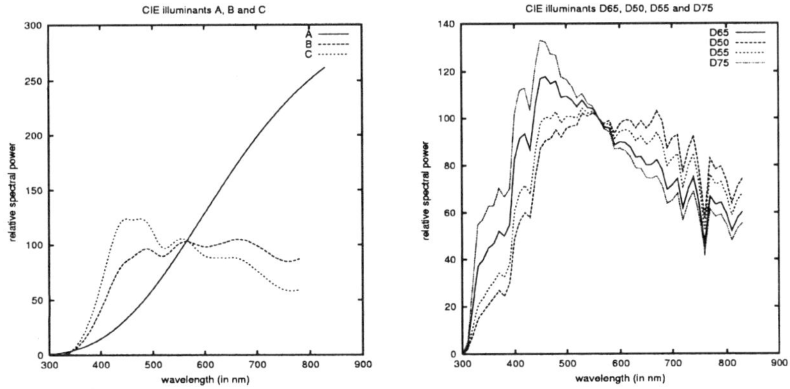

Figure 3.15. Relative spectral power distributions of CIE illuminants A, B, C (left image) and D_{50}, D_{55}, D_{65}, D_{75} (right image).

color filters [WS82]. In 1963, the CIE in addition introduced D illuminants that provide a better approximation to daylight conditions over a wider range of wavelengths (from 300 to 830 nm)[7]. These D illuminants are characterized by their **correlated color temperature**. This correlated color temperature is defined as the temperature of a grey-body radiator of equal brightness that matches the illuminant in color appearance [Hun88]. The color coordinates of illuminant D_n, where $T_c = 100 \cdot n$ is the correlated color temperature in degrees Kelvin, are

$$
\begin{aligned}
x_n &= \begin{cases} 0.244063 + 0.09911 \cdot d + 2.9678 \cdot d^2 - 4.6070 \cdot d^3, \\ \qquad\qquad\qquad\qquad 40 \leq n \leq 70 \\ 0.237040 + 0.24748 \cdot d + 1.9018 \cdot d^2 - 2.0064 \cdot d^3, \\ \qquad\qquad\qquad\qquad 70 \leq n \leq 250 \end{cases} \\
y_n &= -0.275 + 2.870 \cdot x_n - 3.000 \cdot x_n^2
\end{aligned}
\tag{3.32}
$$

where $d = 10/n = 1000/T_c$. The most frequently used D illuminants D_{50}, D_{55}, D_{65} ($T_c = 6504°$ K) and D_{75} are plotted in Figure 3.15, together with illuminants A, B ($T_c = 4874°$ K) and C ($T_c = 6774°$ K). Illuminants C and D_{65} are used as the white reference for the NTSC and PAL television system, respectively (see chapter 4), while D_{50} is

[7]The improved correspondence between daylight and illuminant D_{65} at short wavelengths, as compared to illuminants B and C, is especially important for modern colorants with fluorescing agents. In such agents, radiation that is absorbed at ultraviolet wavelengths is re-emitted as visible light.

mostly used in the printing industry. Illuminant A is the most appropriate choice of standard illuminant in case incandescent lighting is to be approximated.

5.4 THE MUNSELL BOOK OF COLOR

Both theoretical studies in color vision and practical applications involving color have benefited greatly from the availability of the Munsell system for ordering related colors. In this system, a large number of colors have been ordered based on three attributes: hue, value and chroma. They are the Munsell correlates of the standardized color attributes perceived hue, lightness and chroma.

The fact that there is a natural order of hues from red (R) over yellow (Y), green (G), blue (B), purple (P) back to red, has led A.H. Munsell to arrange the different hues into a circle. Based on experimental data, the different hues were arranged along a circle in a perceptual uniform way. The circle was subsequently divided into a number of sectors, which were given hue (H) names. The 10 original hue sectors of Munsell (with hue names R, YR, Y, GY, G, BG, B, PB, P and PR) have since been extended to 40 sectors (with hue names 2.5R, 5R, 7.5R, 10R, 2.5YR, 5YR, etc, where the current hue names 5R, 5YR, etc, correspond to the original hue names R, YR, etc).

The Munsell values V range from 0 for pure black to 10 for a perfect (white) reflector. Experimental bisection data were used to position colors on the value scale such that equal distances corresponded approximately to constant differences in lightness. Since the Munsell system has played and important role in developing the psychometric lightness scale, it is not surprising that there exists a close relationship between Munsell value and psychometric lightness, i.e., $V \approx L^*/10$.

Colors with the same hue and value can be ordered, based on how much they departure from the neutral color. The corresponding attribute is denoted by chroma (C). The scaling of chroma is again intended to be perceptually uniform. The scale starts at zero, for neutral colors, and the maximum obtainable chroma value depends on the hue and value. The chroma scale for normal reflecting materials extends to 20 in some cases, while some fluorescent materials have chroma as high as 30.

The complete Munsell notation for a chromatic color is H V/C, while neutral colors are denoted by N V/. For example 5R 6/14 is a saturated red with hue 5R, value 6 and chroma 14. The *Munsell Book of Color* contains a collection of reference samples or chips that are ordered according to the Munsell notation. Two versions of the book exist, containing matt and glossy samples, respectively. An unknown color can be compared against the chips in the Munsell book in order to derive the appropriate color specification.

A relationship between the Munsell color ordering system and standard color theory was established by determining the CIE tristimulus values (X, Y, Z) for all samples in the Munsell Book of Color under illuminant C. It was found that the transitions in the tristimulus values from one sample to the next in the original Munsell Book of Color were not always smooth, which led to the introduction of some modifications, known as the Munsell Renotation [KGN43, GNF43, NNJ43]. Current editions of the Munsell Book of Color agree by this renotation.

The Munsell data are available from the Munsell Color Science Laboratory[8] and can be used to illustrate the non-uniformity of fundamental color space. They have also assisted in deriving more uniform color spaces. Several alternative color ordering systems [Hun98] could also be used for this purpose. We have however selected the Munsell system because it is the best-known and most widely accepted.

5.5 CIELAB UNIFORM COLOR SPACE

Equal differences in (ℓ, a, b)-coordinates do usually not correspond to equal perceptual color differences, as is demonstrated in Figure 3.16 for the Munsell colors of value $V = 4$. If the chromaticity plane was uniform, then points representing colors with fixed hue H and constant increments in chroma, which lie on the lines that radiate from the origin, should lie on straight lines and be equidistant. Similarly, points representing colors with fixed chroma C and variable hue, which lie on the elliptic contours, should lie on circles and be separated by a constant angle. This is clearly not the case.

5.5.1 1976 CIELAB

A (non-Euclidean) metric is hence needed to more closely approximate subjective color differences. From the (a, b)-coordinates, two **chroma coordinates** were derived by the CIE by nonlinear transformations, i.e.,

$$\begin{aligned} a^* &= 500\left[f_{lab}\left(\frac{X}{X_n}\right) - f_{lab}\left(\frac{Y}{Y_n}\right)\right] &= 500\left[f_{lab}(\ell + a) - f_{lab}(\ell)\right] \\ b^* &= 200\left[f_{lab}\left(\frac{Y}{Y_n}\right) - f_{lab}\left(\frac{Z}{Z_n}\right)\right] &= 200\left[f_{lab}(\ell) - f_{lab}(\ell - b)\right], \end{aligned} \tag{3.33}$$

[8] http://www.cis.rit.edu/mcsl/online/munsell.shtml

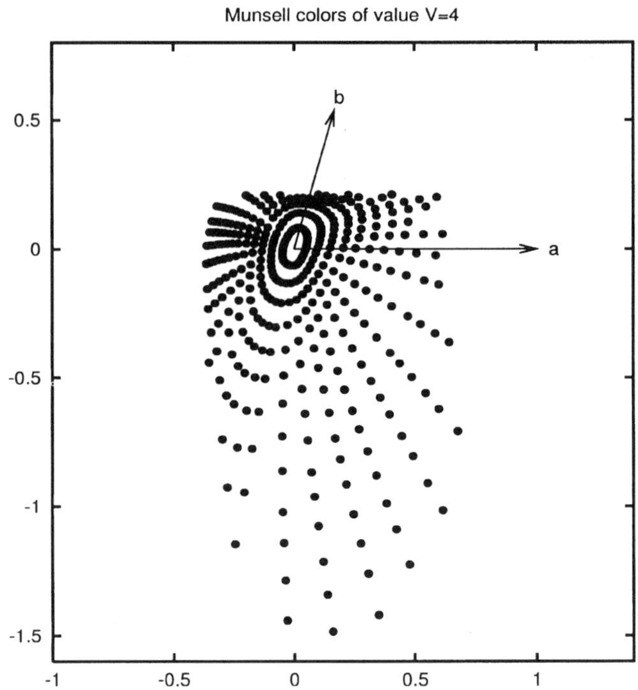

Figure 3.16. Munsell colors of value $V = 4$ are shown in the chromaticity plane with the proper primaries $p_a(\lambda)$ and $p_b(\lambda)$ for the C illuminant. The points of identical hue H lie one lines that radiate from the origin, while the points of identical chroma C lie on elliptic contours.

with

$$\begin{aligned} f_{lab}(x) &= 7.787 \cdot x + \tfrac{16}{116} && \text{for} \quad x \leq 0.008856 \\ &= x^{1/3} && \text{for} \quad x \geq 0.008856 \end{aligned}$$

Note that these transformations depend on the value of the luminance factor ℓ, as is illustrated for the mapping from a to a^* in Figure 3.17. Pure red/green stimuli with $b = 0$ have chroma coordinate $b^* = 0$, while pure yellow/blue stimuli with $a = 0$ have chroma coordinate $a^* = 0$. The neutral reference has chroma coordinates $a^* = b^* = 0$. For small a^* and b^* (i.e., for unsaturated colors), we obtain approximately $(a^*, b^*) \approx 100\, f'_{lab}(\ell) \cdot (5a, 2b)$. The derivative $f'_{lab}(\ell)$ decreases with increasing luminance factor ℓ.

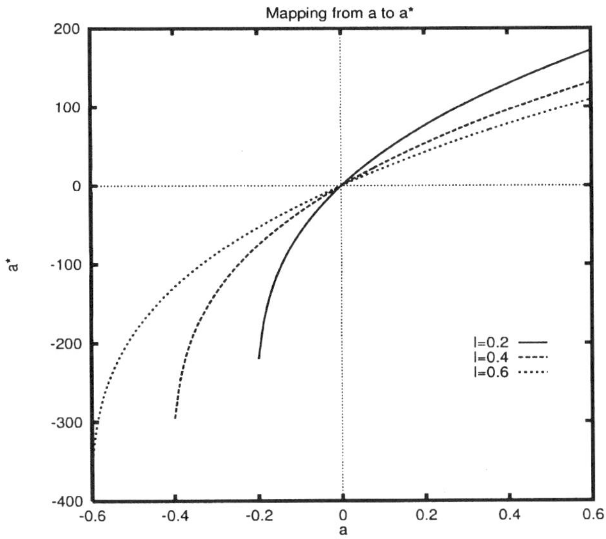

Figure 3.17. The mapping from the linear color coordinate a to the nonlinear CIE(L^*,a^*,b^*) chroma coordinate a^* depends on the luminance factor ℓ, as is illustrated for $\ell = 0.2$, $\ell = 0.4$ and $\ell = 0.6$.

The chroma coordinates (a^*, b^*), together with the previously defined psychometric lightness measure

$$L^* = h_{lab}\left(\frac{Y}{Y_n}\right) = 116 \cdot f_{lab}\left(\frac{Y}{Y_n}\right) - 16 = 116 \cdot f_{lab}(\ell) - 16, \quad (3.34)$$

define the 1976 CIE(L^*,a^*,b^*) **uniform color space** [CIE79][9]. It is called a uniform color space because constant Euclidean distances between color coordinates

$$\Delta E^*_{ab} = \sqrt{(\Delta L^*)^2 + (\Delta a^*)^2 + (\Delta b^*)^2} \quad (3.35)$$

correspond to approximately equal subjective color differences. As a rule of thumb, color differences with $\Delta E^*_{ab} < 3$ are hardly perceptible, while color differences with $\Delta E^*_{ab} > 6$ are easily noticeable [Har99].

The chroma coordinates of the Munsell colors of value $V = 4$ are plotted in Figure 3.18. Although the distribution of the color points is much

[9]The 1976 CIE recommendation also defines an alternative uniform color space, called CIE(L^*,u^*,v^*), the definition of which is not discussed here.

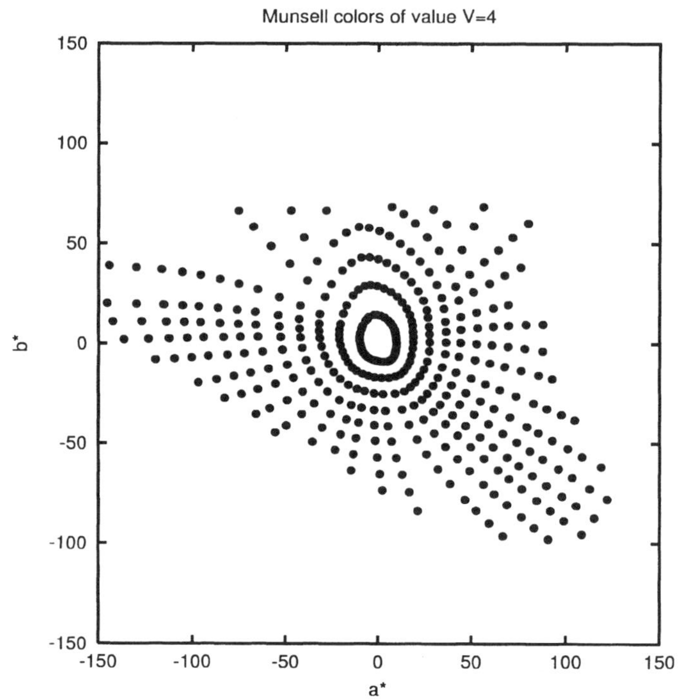

Figure 3.18. Munsell colors of value $V = 4$ are shown in the CIELAB chroma plane (a^*, b^*). The points of identical hue H lie on lines that radiate from the origin, while the points of identical chroma C lie on elliptic contours.

more uniform in Figure 3.18 than in Figure 3.16, some systematic deviations still remain. Given the low complexity of the CIELAB mappings, this is hardly surprising.

In the CIELAB recommendation, a number of color terms are introduced. The **psychometric hue** is defined as an angle

$$h_{ab} = \arctan\left(\frac{b^*}{a^*}\right) \tag{3.36}$$

in the (a^*, b^*)-plane, expressing the fact that colors of constant perceived hue have color coordinates (a^*, b^*) that aggregate along straight radial lines in the (a^*, b^*) chroma diagram. The curves in Figure 3.14 that radiate from the origin connect coordinates in (a, b)-space of constant psychometric hue. Note that, since these lines are only slightly curved, the angle in (a, b)-space also correlates well with perceived hue, especially for unsaturated colors. Indeed, for unsaturated colors, the approximation $h_{ab} \approx \arctan(0.4\, b/a)$ holds.

The quantity

$$C^*_{ab} = \sqrt{(a^*)^2 + (b^*)^2} \qquad (3.37)$$

is called the **psychometric chroma**, since it correlates well with the perceived chroma. The closed contours in Figure 3.14 connect coordinates in (a, b)-space with constant psychometric chroma. For unsaturated colors, curves of constant chroma are close to ellipses, since the approximation $C^*_{ab} \approx 500\, f'_{lab}(\ell) \cdot \sqrt{a^2 + (0.4\, b)^2}$ holds.

5.5.2 CHROMATIC ADAPTATION

The CIE(L^*,a^*,b^*) color difference formula will, because of the experimental data on which it is based, only give accurate predictions for relatively large, non-adjacent, color patches with an identical white to mid-grey surround. The observer is also assumed to be photopically adapted to an illuminant with a chromaticity that is not too different from that of average daylight. This illuminant is used as the reference color in CIELAB. One of the three following alternative procedures is usually adopted for comparing colors with different surrounds, or illuminated by a light source with a chromaticity that differs from daylight [Hun91].

The first possibility is to use the CIELAB metric as it stands, and to rely on the color adaptation that is implemented through the normalization with the reference white (X_n, Y_n, Z_n). This procedure is at best a reasonable first-order approximation.

The second procedure is usually referred to as a **von Kries transformation**, after the name of its originator. The basic idea of this approach is not to perform the normalization on the standard tristimulus values (X, Y, Z), but on the cone responses (L, M, S). The linear transformation required for mapping from tristimulus values to cone responses is for instance specified in chapter 1 and exercise 2. The assumption underlying the von Kries transformation is that the color perception of a stimulus with cone responses (L, M, S) under an illumination with reference white characterized by cone responses (L_n, M_n, S_n), is determined by the adapted cone signals $(L/L_n, M/M_n, S/S_n)$. Suppose that the actual cone responses for the stimulus and reference white are (L', M', S') and (L'_n, M'_n, S'_n), respectively. This stimulus is expected to match the color stimulus with cone responses

$$L = \frac{L_n}{L'_n}\, L', \qquad M = \frac{M_n}{M'_n}\, M', \qquad S = \frac{S_n}{S'_n}\, S' \qquad (3.38)$$

under the standard viewing circumstances where the reference white has cone responses (L_n, M_n, S_n). These adapted cone responses can subsequently be converted to standard tristimulus values (X, Y, Z) and used in the CIELAB metric.

The third alternative is to use a more complex model for describing the cone adaptation [Mal99], and to replace the simple gain-control mechanism of equation (3.38) by a more complex algorithm. Such models are referred to as color-appearance models, and one such model that is supposed to incorporate the major strong points of several existing models, has recently been standardized by the CIE [LH98b, LH98a, LLH00, Hun98].

5.5.3 1994 CIELAB

The need for a more accurate description of small color differences has led the CIE, in 1994, to introduce the following modified CIE(L^*,a^*,b^*) color-difference formula [CIE95]

$$\Delta E^*_{94} = \sqrt{\left(\frac{\Delta L^*}{k_L\, S_L}\right)^2 + \left(\frac{\Delta C^*_{ab}}{k_C\, S_C}\right)^2 + \left(\frac{\Delta H^*_{ab}}{k_H\, S_H}\right)^2}, \qquad (3.39)$$

where

$$\Delta H^*_{ab} = \sqrt{(\Delta E^*_{ab})^2 - (\Delta L^*)^2 - (\Delta C^*_{ab})^2} \qquad (3.40)$$

defines the **hue difference**, and the weighting functions

$$S_L = 1, \qquad S_C = 1 + 0.045 \cdot C^*_{ab}, \qquad S_H = 1 + 0.015 \cdot C^*_{ab} \qquad (3.41)$$

vary with the chroma C^*_{ab} of the reference color. These weighting functions incorporate the experimentally observed effect of a decreased sensitivity to chroma and hue differences with increasing chroma. If neither of the two samples can be considered as the reference, then the geometric mean $C^*_{ab} = \sqrt{C^*_1\, C^*_2}$ of the chroma values of both samples is used in the above expressions.

It is important for applications to realize the limitations that were adopted in the development of the 1996 CIELAB color-difference formula. First and foremost, the 1994 CIELAB color-difference formula is mainly intended for evaluating small to medium color differences, i.e., with $\Delta E^*_{ab} < 5$. An accurate description of such small color differences is of obvious economical interest, for instance in quality control. Large color differences, of the size observed in color ordering systems such as the Munsell system, were not considered. Second, the experimental data sets that were used for comparison against the new color-difference formula were collected in viewing conditions that approximated the following reference condition:

1 both samples are homogeneous in color and extend spatially over at least 4 degrees of visual angle;

2 they are placed in direct contact;

3 they are viewed against a uniform grey background with psychometric lightness $L^* = 50$;

4 they are illuminated with a D_{65} illuminant at 1000 lux.

The parameters factors k_L, k_C and k_H, which are all equal to one under the reference viewing conditions, may be adjusted for different viewing conditions. No guidelines are however provided for how these parameters can be assigned proper values, so that the earlier remarks on chromatic adaptation will have to serve as such.

It has been shown that the 1994 color-difference formula can be closely approximated by a Euclidean metric

$$\Delta E^{*\prime}_{ab} = \sqrt{(\Delta L^{*\prime})^2 + (\Delta a^{*\prime})^2 + (\Delta b^{*\prime})^2} \tag{3.42}$$

on the modified color coordinates $(L^{*\prime}, a^{*\prime}, b^{*\prime})$, where

$$\begin{aligned} L^{*\prime} &= \frac{L^*}{k_L} \\ a^{*\prime} &= \frac{a^* \cdot h(C^*)}{\sqrt{k_C\, k_H}} \\ b^{*\prime} &= \frac{b^* \cdot h(C^*)}{\sqrt{k_C\, k_H}}, \end{aligned} \tag{3.43}$$

and

$$h(C^*) = \frac{1}{P\,C^*} \log(1 + Q \cdot C^*), \tag{3.44}$$

with $P = 0.0500951$ and $Q = 0.0531301$ [Tho00]. Such a Euclidean metric is obviously easier to interpret and to use than the 1994 CIELAB color-difference formula itself. The modified chroma coordinates (with $k_L = k_H = k_C = 1$) of the Munsell colors of value $V = 4$ are plotted in Figure 3.19. The fact that the resulting space does not look more uniform than the one depicted in Figure 3.18 need not be surprising, since the 1994 CIELAB is not intended for accurately predicting the large color differences between the Munsell samples.

The quest for more accurate color-difference models is obviously continuing. The major experimental finding that is incorporated in more

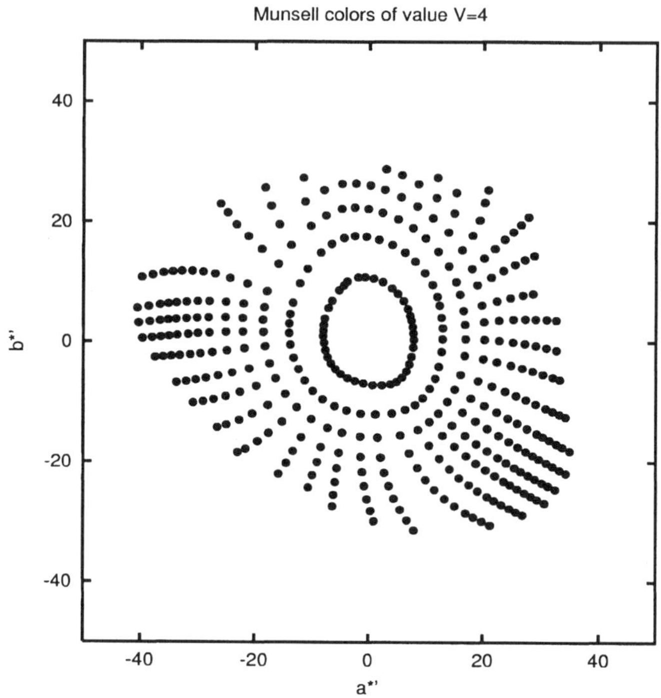

Figure 3.19. Munsell colors of value $V = 4$ are shown in the modified chroma plane $(a^{*\prime}, b^{*\prime})$. The points of identical hue H lie on lines that radiate from the origin, while the points of identical chroma C lie on elliptic contours.

recent models is the fact that the principal axes of color-difference ellipses[10] do not always point towards the neutral point, as is predicted by the CIELAB models. This effect is most pronounced in the blue region. One of the most competitive candidates for a future CIE recommendation, i.e., the CIEDE2000 formula [LCR01], for instance incorporates this effect. It remains however to be seen under which conditions the improved accuracy of such a model warrants the increased complexity.

[10]A color-difference ellips connects the chroma coordinates of colors at a constant perceived distance from a reference color.

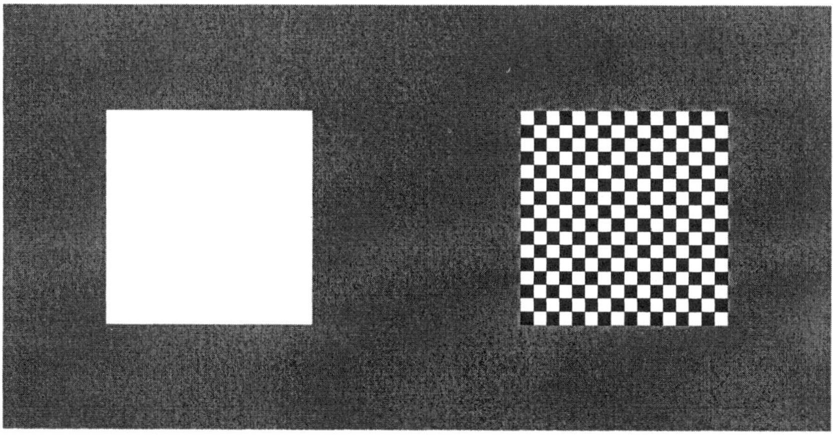

Figure 3.20. Useful patterns for display calibration in the absence of a luminance meter (see exercise 1).

EXERCISES

1 When the stimulus in figure 3.20 is viewed from a sufficiently large distance, then the flat square on the left and the checker-board square on the right will both look uniform. If the luminance of the white parts is the same in both cases, then the left square will look brighter. The grey value V_1, and hence also the luminance $L(V_1)$, of the flat square can however be decreased until its brightness (when observed from a sufficient distance for both stimuli to look uniform) matches the brightness of the checker-board square. The small black squares in the checker-board pattern are assumed to have zero luminance, while the grey value of the small white squares is denoted by V_2, with corresponding luminance $L(V_2)$. Derive how the above experimental paradigm can be used to measure some points on the grey-value-to-luminance characteristic $L(V)$ of the display. This method can hence be used to calibrate a display characteristic in the absence of a luminance meter.

2 The three cone receptors in the human eye must have spectral sensitivity characteristics

$$\begin{aligned}
\phi_L(\lambda) &= l_x \cdot \overline{x}(\lambda) + l_y \cdot \overline{y}(\lambda) + l_z \cdot \overline{z}(\lambda) \\
\phi_M(\lambda) &= m_x \cdot \overline{x}(\lambda) + m_y \cdot \overline{y}(\lambda) + m_z \cdot \overline{z}(\lambda) \\
\phi_S(\lambda) &= s_x \cdot \overline{x}(\lambda) + s_y \cdot \overline{y}(\lambda) + s_z \cdot \overline{z}(\lambda)
\end{aligned}$$

that are linear combinations of the standardized color matching functions (CMFs) $\overline{x}(\lambda)$, $\overline{y}(\lambda)$ and $\overline{z}(\lambda)$ (why?)[11]. A **dichromat** is a person that is assumed to be missing one kind of receptor (either $\phi_L(\lambda)$ missing: **protanopic**, $\phi_M(\lambda)$ missing: **deuteranopic** or $\phi_S(\lambda)$ missing: **tritanopic**).

Solve the following problems of dichromatic vision:

- Suppose that irradiance functions $L_e(\lambda)$ are represented by their tristimulus values (X, Y, Z). Prove that a dichromat will confuse two irradiances $L_{e,1}(\lambda)$ and $L_{e,2}(\lambda)$ if the respective tristimulus values (X_1, Y_1, Z_1) and (X_2, Y_2, Z_2) lie on a line with a direction (x_0, y_0, z_0) that is characteristic for the dichromat (the direction vector can be normalized such that $x_0+y_0+z_0 = 1$). Derive equations (with coefficients that are expressed in terms of $l_x, \ldots, s_z$) for these characteristic directions for protanopic $(=(x_P, y_P, z_P))$, deuteranopic $(=(x_D, y_D, z_D))$ and tritanopic $(=(x_T, y_T, z_T))$ subjects. Do not explicitly solve these equations.
- Irradiance functions $L_e(\lambda)$ can alternatively be represented by their light efficiency Y and color coordinates (x, y). Prove that a dichromat will confuse two irradiance functions $L_{e,1}(\lambda)$ and $L_{e,2}(\lambda)$ if the respective color coordinates (x_1, y_1) and (x_2, y_2) lie on a line that passes through a specific point with color coordinates (x_0, y_0), called the **color confusion center** that is characteristic for the dichromat. This is equivalent to proving that parallel lines with direction vector (x_0, y_0, z_0) (where $x_0+y_0+z_0 = 1$) in (X, Y, Z) space cross at the point (x_0, y_0) in the color diagram (x, y).
- Plot the color confusion centra

$$\begin{aligned}(x_P, y_P) &= (0.7465, 0.2535)\\ (x_D, y_D) &= (1.4, -0.4)\\ (x_T, y_T) &= (0.1748, 0.0000)\end{aligned}$$

  in the color diagram of Figure 3.5. What will a dichromat see if an irradiance function is presented with color coordinates equal to his/her color confusion center? Which of these irradiance functions can be physically realized (explain)?

[11] As was discussed in chapter 1, this statement is more accurate for the Judd-modified or Vos-modified CMFs than for the 1931 CIE standard observer curves. The values mentioned in this exercise for the color confusion centers actually assume the Judd-modified CMFs.

- Derive the receptor sensitivity curves $\phi_L(\lambda)$, $\phi_M(\lambda)$ and $\phi_S(\lambda)$, based on the specified color confusion centra and the assumption[12] that, for a normal subject, the light efficiency

$$Y = C_L + C_M + C_S$$

is the sum of the receptor outputs (where $C_L = \langle\, \phi_L(\lambda), L_e(\lambda)\, \rangle$, etc). Show that

$$\begin{bmatrix} l_x & l_y & l_z \\ m_x & m_y & m_z \\ s_x & s_y & s_z \end{bmatrix} \cdot \begin{bmatrix} x_P & x_D & x_T \\ y_P & y_D & y_T \\ z_P & z_D & z_T \end{bmatrix} = \begin{bmatrix} y_P & 0 & 0 \\ 0 & y_D & 0 \\ 0 & 0 & y_T \end{bmatrix},$$

which results in the solution[13]

$$\begin{bmatrix} l_x & l_y & l_z \\ m_x & m_y & m_z \\ s_x & s_y & s_z \end{bmatrix} = \begin{bmatrix} 0.15514 & 0.54312 & -0.03286 \\ -0.15514 & 0.45684 & 0.03286 \\ 0.000 & 0.000 & 0.01608 \end{bmatrix}.$$

For display purposes, the receptor sensitivity curves are often scaled to obtain that $C_L = C_M = C_S$ for an equal energy source. The resulting scaled solution

$$\begin{bmatrix} l_x & l_y & l_z \\ m_x & m_y & m_z \\ s_x & s_y & s_z \end{bmatrix} = \begin{bmatrix} 0.2331 & 0.8159 & -0.0494 \\ -0.4635 & 1.3650 & 0.0982 \\ 0.0000 & 0.0000 & 1.0000 \end{bmatrix}$$

has scale factors 1.5, 3.0 and 62.2 for the L, M and S cones, respectively[14]. Draw the scaled cone receptor curves (consult Appendix 1.B for the standardized CMFs for 10 nm intervals). Compare your result to Figure 1.14.

- Derive how to construct the fundamental color plane of a tritanopic observer that is shown in Figure 3.21. Find a right-handed orthogonal coordinate system $\{\varphi_1, \varphi_2\}$ such that both

[12] This assumption is due to Vos and Walraven [VW71].

[13] The numeric solution is the one provided by Smith and Pokorny [SP75], who used the Judd-Modified CMFs $\bar{x}_{51}(\lambda)$, $\bar{y}_{51}(\lambda)$ and $\bar{z}_{51}(\lambda)$ rather than the CIE standardized CMFs $\bar{x}(\lambda)$, $\bar{y}(\lambda)$ and $\bar{z}(\lambda)$, and who assumed that the light efficiency can be expressed as $Y = C_L + C_M$ rather than as $Y = C_L + C_M + C_S$. For the sake of this exercise, the numerical values provided are considered to be sufficiently accurate.

[14] Anatomical and electrophysiological knowledge about the cone receptor distribution in the visual system allows us to interpret the ratios of 1.5:3.0:62.2 (or equivalently, 1:2:41.5) between these scale factors. The L, M and S cones have a relative frequency of occurrence in the fovea of approximately 32:16:1. The L and M cones have about equal sensitivity, but since there are twice as many L cones as M cones (see chapter 1), the contribution of the L cones to the luminance is twice as large as the contribution of the M cones. The S cones do not only occur less frequently, but are moreover less sensitive (by about 30 percent).

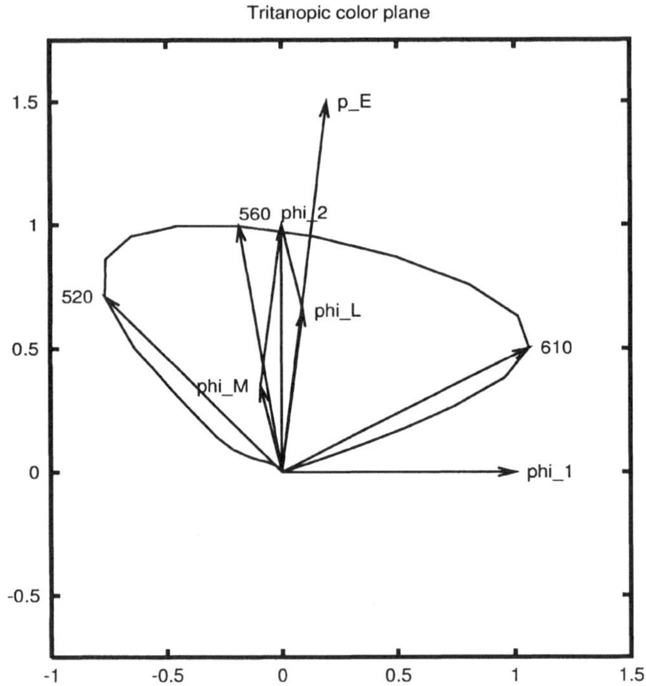

Figure 3.21. Fundamental color plane of a tritanopic observer (see exercise 2). The orthogonal coordinate system $\{\phi_1, \phi_2\}$ is such that the second coordinate corresponds to luminance. The measurement vectors for the L and M cones are represented by ϕ_L and ϕ_M, respectively. The curve corresponds to the fundamental metameres of the monochromatic spectra; those for 520 nm, 560 nm and 610 nm being indicated explicitly. The direction of the fundamental metamere of an equal energy source is indicated by p_E.

measurement vectors $\{\phi_1, \phi_2\}$ have equal length and the second measurement vector is $\phi_2(\lambda) = \phi_L(\lambda) + \phi_M(\lambda)$, i.e., the second coordinate is proportional to the spectral sensitivity (i.e., the luminance). The Gramian matrix of the CMFs of the L and M cones is

$$G_{\phi\phi} = \begin{bmatrix} \langle \phi_L, \phi_L \rangle & \langle \phi_L, \phi_M \rangle \\ \langle \phi_M, \phi_L \rangle & \langle \phi_M, \phi_M \rangle \end{bmatrix} = \frac{1}{40} \begin{bmatrix} 3.3623 & 1.6829 \\ 1.6829 & 0.9951 \end{bmatrix},$$

with inverse

$$G_{\varphi\varphi} = \begin{bmatrix} \langle \varphi_L, \varphi_L \rangle & \langle \varphi_L, \varphi_M \rangle \\ \langle \varphi_M, \varphi_L \rangle & \langle \varphi_M, \varphi_M \rangle \end{bmatrix} = 40 \cdot \begin{bmatrix} 1.9372 & -3.2761 \\ -3.2761 & 6.5453 \end{bmatrix}.$$

Draw the measurement vectors $\phi_L(\lambda)$ and $\phi_M(\lambda)$ that correspond to the L and M cones in the orthogonal coordinate system. Also depict the fundamental metameres of the equal energy source and the monochromatic spectra from 450 to 650 nm (in steps of 10 nm). Plot the measurement vectors $\phi_1(\lambda)$ and $\phi_2(\lambda)$ as a function of wavelength λ in order to observe that they indeed correspond to a double-peaked red-green opponent-color channel and a single-peaked luminance channel, respectively.

- Just for the sake of reference, the fundamental color planes for deuteranopic and protanopic observers are plotted in Figure 3.22. They can be constructed in a similar way as in case of the tritanopic observer by selecting the spectral sensitivity equal to $\phi_2(\lambda) = \phi_L(\lambda) + \phi_S(\lambda)$ and $\phi_2(\lambda) = \phi_M(\lambda) + \phi_S(\lambda)$, respectively.

3 The color coordinates (ℓ, a, b) are obtained through linear transformation of the standard color coordinates (X, Y, Z). Prove that this implies that stimuli with constant a or b coordinate are mapped into straight lines in the color diagram of Figure 3.5. Make a drawing of the color diagram in which these lines of constant a and b color coordinate are shown.

4 Suppose that a color measuring device has three sensors with respective spectral sensitivity curves $\phi_1(\lambda)$, $\phi_2(\lambda)$ and $\phi_3(\lambda)$. Prove that these sensitivity curves must be independent linear combinations of the standard CIE color matching curves $\overline{x}(\lambda)$, $\overline{y}(\lambda)$ and $\overline{z}(\lambda)$ if this sensor is to allow metameric color reproduction of an arbitrary spectrum $L_e(\lambda)$. Hence, general color reproduction puts heavy restrictions on the sensitivity curves of a color sensor. However, in many practical situations, we are only interested in reproducing reproductions, that is, the input spectra are restricted to being produced by a color device, such as an additive color source with spectra $L_e(\lambda)$ that are linear combinations of three primary spectra $p_1(\lambda)$, $p_2(\lambda)$ and $p_3(\lambda)$. Does metameric reproduction of these colors also impose strict conditions on the sensitivity curves of the sensor? Motivate the answer.

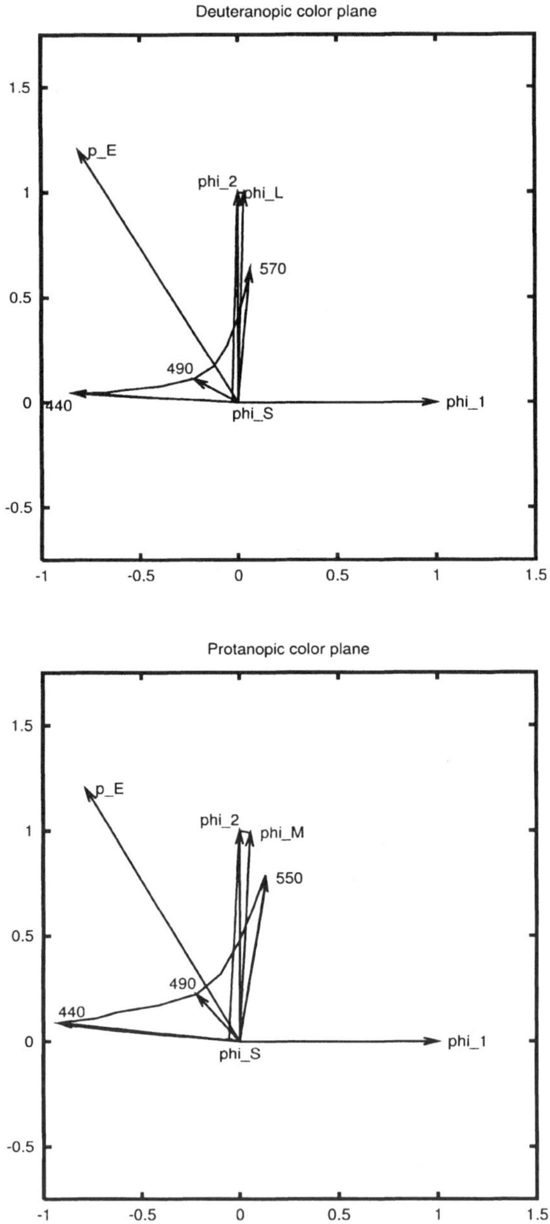

Figure 3.22. Fundamental color planes of a deuteranopic (top) and protanopic (bottom) observer.

Chapter 4

COLOR MANAGEMENT

1. INTRODUCTION

Although image communication systems such as television have been in widespread use for over half a century, the interest in the accurate reproduction of color images is much more recent. There are a number of reasons for this. First and foremost, image content is usually much more important than image quality and the available image quality is most often adequate for the purpose. For instance, the frequently resurfacing debates about the need for an improved television transmission system seem to be mostly technically driven, rather than customer-driven. Second, the fact that the reproduced image can differ substantially from the original is not necessarily objectionable, especially if this original image is not available for comparison. People are often primarily interested in a pleasing picture, even if this picture is perceived to be (slightly) unnatural. Third, until recently, it was not technically feasible to improve this situation substantially at a reasonable cost. Low-cost images are mostly photographic prints and slides, and hence produced by chemical means that are hard to control accurately. The sensors and reproducing dyes are integrated in the photographic medium, so that there are few possibilities for correcting the color reproduction by means of intermediate processing.

As in many other technological fields (such as audio), the demands on quality tend to increase with time as new (and better) systems become more commonplace. Professional photographers and graphical designers were among the first to be interested in WYSIWYG (What You See Is What You Get) systems. For example, in graphical design, where images are often computer-generated, it is customary that designers move between different workstations and from softcopy (i.e., monitor screen) to hardcopy (i.e., film or print) reproductions. It is mandatory that the colors in an image are properly managed, i.e., are not (greatly) affected by moving between these different media. With the advent of high-performance and low-cost electronic imaging systems, such as high-resolution scanners, printers and monitors, accurate color reproduction and **color management** [GM98] have now become feasible at a reasonable cost. These devices allow to combine the quality of photographic reproductions, especially the high spatial resolution, with the flexibility of computer processing. As a result of a dramatic decrease in the price of key components, professional tools such as Color Management Systems (CMS) are now also available in consumer products, such as multimedia PC's.

The goal of CMS is to enable (more) faithful color reproductions. In order to realize this, CMS use standardized color coordinate systems, such as CIELAB, to communicate color images between different devices. Such standardized color coordinate systems are called **profile connection spaces** in CMS. In addition, CMS require that the colorimetric properties of all imaging devices in a production chain are well-specified. This means that there exists a **device profile** that contains all necessary information for mapping between *device-independent* color coordinates and *device-dependent* color values [Har99] for all devices involved. Such device profiles may be determined for a specific device, or default (and probably less accurate) profiles supplied by the manufacturer, may be used. As a result, a color image delivered by an image scanner in a CMS will typically consist of the image data, most often specified in device-dependent color values, together with the profile of the scanner. In recent years, the International Color Consortium (ICC) (see http://www.color.org) has played an important role in standardizing and promoting CMS in industry.

In this chapter, we discuss image reproduction systems in which three conceptual components can be distinguished: the image sensor, the CMS for processing color images and the image display. Two common factors that most systems for the reproduction of colored images share are that the image sensor has three types of image detectors with linearly independent spectral response curves, while the image display has three types of primaries. This is of course motivated by the trichromacy of human color vision. We will however see that, especially on the display side, it may sometimes be advantageous to have more than three primaries.

The many different methods that are available for reproducing colored images can be categorized in a number of ways [Hor84]. One can distinguish between display systems that use controlled intensities of superimposed light sources with different spectral distributions (for instance, color television) and those that depend on controlled amounts of absorption by pigments or dyes (for instance, film). Superimposed light sources mix additively, while the transmission values of superimposed dye or pigment layers must be multiplied. Multiplicative color mixing methods (sometimes referred to as subtractive methods) can be subdivided into a group which requires light to be transmitted through the absorbing layers (transparencies) and a group which depends on light to be reflected from a substrate (prints). An alternative is to separate the multiplicative color reproduction techniques according to whether control is achieved by means of changes in the concentration or amount of dye (photographic transparencies and prints) or whether different colors are obtained instead by varying the fraction of the total area covered by a dye of fixed composition (lithography and ink-jet printers).

In section 2, image displays that consist of superimposed controlled light sources are discussed. The colorimetric calibration of additive color displays, which is essential to CMS, is treated in some detail. In section 3, different multiplicative color reproduction systems, based on dye concentrations and dye areas respectively, are discussed shortly. These systems are nonlinear, but an idealized linear model approximation is derived.

Some standardized color coordinate systems that are particularly important for the communication between imaging devices, such as sensors and displays, and which were originally developed for television (broadcast) applications, are described in section 4. It is shown that confusion may arise in the European PAL television system because the luminance signal is not one-to-one related to photometric luminance. Another complicating factor is that the color transformations, which are derived based on colorimetric arguments, are not performed on color coordinates (as should be the case) but on color values. The specifications by color coordinates and color values are related by nonlinear power-like transformations, and the so-called **gamma characteristics** that relate both specifications are discussed in some detail.

Image sensors can be divided into chemical (photographic) sensors and electronic (camera) sensors. A chemical sensor uses several layers that are sensitive to different spectral wavelengths. Silver halide grains are converted under the influence of light. The concentrations of converted grains are the sensor outputs. In an electronic sensor, the incoming light is split and sent through different filters before reacting with the (possibly identical) light detectors. The outputs of the detectors are continuous electrical signals. The discussion in this chapter is restricted to the case where the image sensor responses are explicitly available. In section 5, a model for such image sensors is presented and the condition under which an image sensor allows for a perfect color reproduction is discussed. In the case of photographic media, densitometers (for instance implemented in commonly available color scanners) can be used to measure the absorptions of the different photographic layers, so that the distinction between photographic and electronic sensors is becoming less important nowadays. We also discuss how electronic sensors can be characterized colorimetrically.

In section 6, attention is devoted to the important special case of reproducing reproductions. Although in many applications it is extremely difficult to guarantee that all possible input images can be exactly reproduced, it is already considered a great improvement if all reproductions are identical. This is for instance very important in applications where many duplication steps, often involving different industrial partners, intervene between the original image and the image that is eventually published (consider for instance the typical publication process of a magazine). In the case of computer generated imagery, where an 'original' image does not exist, equating the reproductions is all that is required.

Finally, in section 7, we return to the issue of image quality, as perceived by the observer, and show that a slightly unnatural reproduction of color is usually preferred over a photometrically accurate reproduction. This does not fundamentally alter the role of color management, since the problem of guaranteeing the right amount of nonlinear reproduction is obviously of similar complexity to that of guaranteeing perfect color reproduction.

2. ADDITIVE COLOR DISPLAYS

2.1 MODEL

In an additive color display, three (or more) independently controlled light sources are superimposed in order to create a **gamut of colors**. The best known examples are the color CRT (Cathode Ray Tube) monitor and the LCD (Liquid Crystal Display). The light sources in this case are the phosphor dots on the CRT screen or the LCD elements. Although the individual display elements do not actually coincide, they are so closely spaced that they cannot be resolved at a normal viewing distance. The light coming from neighboring display elements is hence effectively integrated by the optics of the eye (in an additive way).

The idealized model of figure 4.1 is assumed to describe an additive color display. More specifically, the total radiance of the light coming

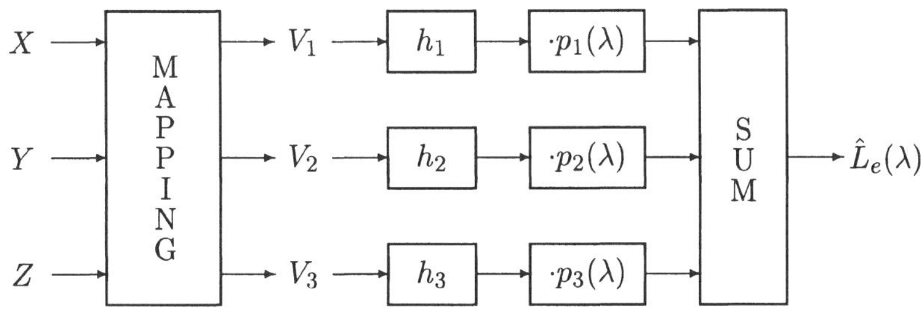

Figure 4.1. Model for an additive color display.

from a small region in the displayed image is modelled by

$$\hat{L}_e(\lambda) = h_1(V_1) \cdot p_1(\lambda) + h_2(V_2) \cdot p_2(\lambda) + h_3(V_3) \cdot p_3(\lambda). \tag{4.1}$$

The spectral distribution of the light emitted by the i-th **primary** is denoted by $p_i(\lambda)$, for $i = 1, 2, 3$. The **color values** V_i, for $i = 1, 2, 3$, specify the displayed color, while the nonlinear (electronic) processing that is performed on the color value V_i is modelled by the monotonic function h_i, for $i = 1, 2, 3$. In physical realizations, these latter functions are by necessity positive-valued and limited in amplitude. One obvious problem in display design is to determine which color values (V_1, V_2, V_3) need to be applied to the display in order to obtain an a priori desired color with tristimulus values (X, Y, Z).

In order to verify if an available display can be described by the above additive color model, the following underlying assumptions need to be tested:

1 Linearity: Fixing the signal input to two of the primaries and varying the signal to the third primary should result in colors with coordinates on a straight line in fundamental color space, for all possible color values to the fixed primaries.

2 Independence: The light output of a specific primary at a given location should be independent of the input to the other primaries, both at the same and at neighboring locations.

These assumptions have been verified for several professional color monitors, and were shown to hold with very good accuracy [Bra89]. This study also revealed that the functions h_i are typically spatially variant, but that this spatial variation could be well described by multiplying all functions h_i, for $i = 1, 2, 3$, by the same, slowly varying, spatial function. A similar correction is probably sufficient to describe the temporal

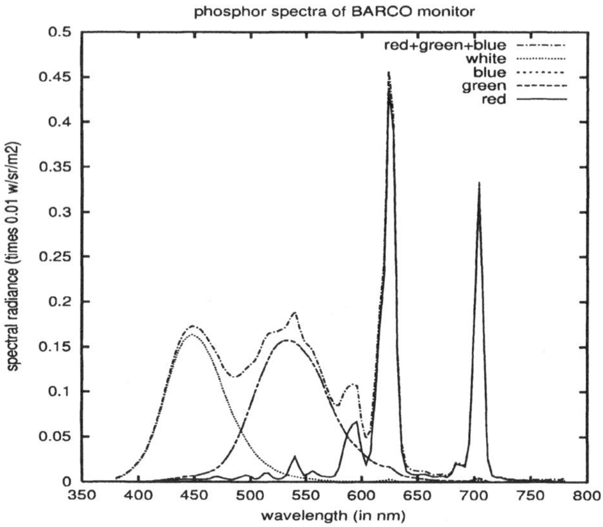

Figure 4.2. Spectral radiance as a function of wavelength for the three phosphors (red, green, blue) in a BARCO CCID 7351B color monitor. As expected for an additive color display, the spectral radiance measured for white is identical to the sum of the spectral radiances for the individual phosphors.

dependence of a typical monitor. These corrections are not overly important, since human vision is known to disregard these slow variations in space and time.

Some measured spectral radiance distributions for the red, green and blue primaries in a professional color monitor are plotted in Figure 4.2. If the additive color model is correct, then it is expected that the spectral distribution of 'white', i.e., measured with all primaries set to their maximum luminance, should be (almost) identical to the numerical sum of the spectral distributions for the individual phosphors, i.e., measured with one primary set to maximum luminance, and the two remaining primaries set to minimum luminance. This is indeed the case in Figure 4.2.

Only the fundamental metameres of the primaries are relevant for human color perception, since the tristimulus values

$$\begin{bmatrix} X \\ Y \\ Z \end{bmatrix} = D \cdot \begin{bmatrix} h_1(V_1) \\ h_2(V_2) \\ h_3(V_3) \end{bmatrix}, \tag{4.2}$$

with

$$D = \begin{bmatrix} < \overline{x}(\lambda), p_1(\lambda) > & < \overline{x}(\lambda), p_2(\lambda) > & < \overline{x}(\lambda), p_3(\lambda) > \\ < \overline{y}(\lambda), p_1(\lambda) > & < \overline{y}(\lambda), p_2(\lambda) > & < \overline{y}(\lambda), p_3(\lambda) > \\ < \overline{z}(\lambda), p_1(\lambda) > & < \overline{z}(\lambda), p_2(\lambda) > & < \overline{z}(\lambda), p_3(\lambda) > \end{bmatrix}, \tag{4.3}$$

determine the color impression of the displayed color. The relationship between the tristimulus values and the color coordinates (x_i, y_i) of the primaries

$$(X_i, Y_i, Z_i) = \left(Y_i \frac{x_i}{y_i}, Y_i, Y_i \frac{z_i}{y_i} \right), \tag{4.4}$$

with $z_i = 1 - x_i - y_i$, can be used to derive the following alternative characterization of the **display matrix**

$$D = \begin{bmatrix} x_1 \cdot Y_1/y_1 & x_2 \cdot Y_2/y_2 & x_3 \cdot Y_3/y_3 \\ Y_1 & Y_2 & Y_3 \\ z_1 \cdot Y_1/y_1 & z_2 \cdot Y_2/y_2 & z_3 \cdot Y_3/y_3 \end{bmatrix} \tag{4.5}$$

that relates the device coordinates $[h_1(V_1), h_2(V_2), h_3(V_3)]$ to the tristimulus values (X, Y, Z). The light efficiencies Y_i of the individual primaries can for instance be chosen such that the corresponding luminances $L_i = K_m \cdot Y_i$ are equal to the maximum luminances of the primaries. With this choice of normalization, the maximum value of the nonlinear functions h_i is equal to one.

The finite range of available luminances for the primaries is controlled by color values that also belong to a finite interval, i.e., $V_i \in [V_{min}, V_{max}]$. Nowadays, the color values V_i are very often stored digitally in byte format, i.e., with $V_{min} = 0$ and $V_{max} = 255$. Note that such a digital storage implies that the display as discussed here actually consists of a combination of a video board, with a digital-to-analog convertor, and an analog color CRT or LCD.

One important complicating factor of the finite luminance range of primaries is that certain colors cannot be realized, since they correspond to color values V_1, V_2 or V_3 outside the available range $[V_{min}, V_{max}]$ (see also exercise 1). The most common solution to this problem is to simply replace these non-available color values by the closest extreme value V_{min} or V_{max}. If this is not acceptable, then there are a number of possible strategies that can be adopted for increasing the gamut of colors that can be realized on a display device. The first is to take saturated primaries, such as monochromatic light sources. The main disadvantage of such primaries is that they are relatively inefficient for stimulating the visual system, so that high energies are required to obtain sufficient luminance. The primaries that are used in commercial displays are therefore a compromise between color saturation and efficiency (i.e., power consumption). A second approach is to translate, rotate and scale the entire gamut of desired colors in fundamental color space until it fits into the parallelepiped of obtainable colors [Wan87]. Since the visual system mostly judges related colors, small transformations of the entire color gamut are usually not objectionable. Larger transformations will however result in unnatural colors [Rid96b, FRB97, FBR97, FRYB97, RBF97, Rid97]. A third approach is to use more than three primaries, which of course also complicates the display. The problem of determining a mapping from standard color coordinates to color values for a display with more than three primaries can be simplified by using only three primaries at a time. The added flexibility is then only used to select the most appropriate primaries. More complex mappings may optimize additional restrictions, such as the overall power consumption.

2.2 DISPLAY CALIBRATION

A calibrated display can be defined as one for which the mapping between the applied color values (V_1, V_2, V_3) and the tristimulus values (X, Y, Z) of the displayed color is known. An important advantage of calibration is that colors can be communicated between different components in a system, such as an image sensor and an image display, by means of standard color coordinates, such as the CIE standard observer (X, Y, Z) tristimulus values. This obviously greatly simplifies the color management throughout the system, since the operating system need not be aware of the characteristics of individual devices. Another advantage is that the required calibration for these components can be developed (or even performed) by specialists that have advanced knowledge of the devices involved.

In order to accomplish calibration for a given display device, an appropriate mapping that transforms standard color coordinates into suitable color values for the image display needs to be developed. We concentrate on mappings as shown in Figure 4.3 that consist of a linear transformation D_m on the CIE tristimulus values, followed by a separate nonlinear function $\tilde{h}_i$, for $i = 1, 2, 3$, in each signal path[1].

The simplest case arises when the additive color model of figure 4.1 applies and all characteristics of the image display are known. An overall linear display [Hor84] is obtained if the nonlinearities $\tilde{h}_i$ invert the nonlinear monotonic display functions h_i. The matrix required to obtain

[1] If one such mapping is not sufficiently accurate, then we can partition the color space, and devise a separate mapping for each partition [PC89].

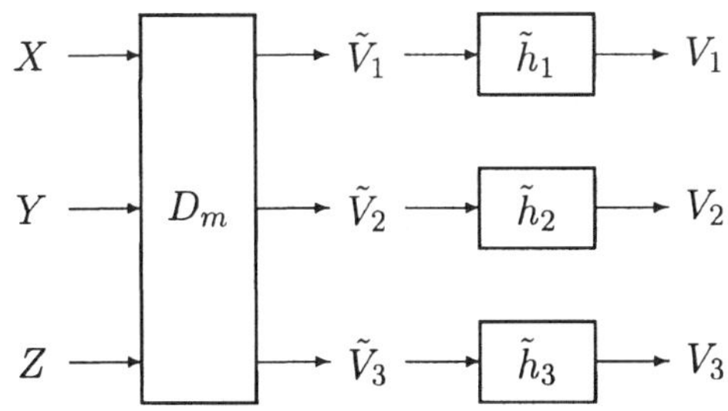

Figure 4.3. Mapping from standard color coordinates to color values for an additive color display.

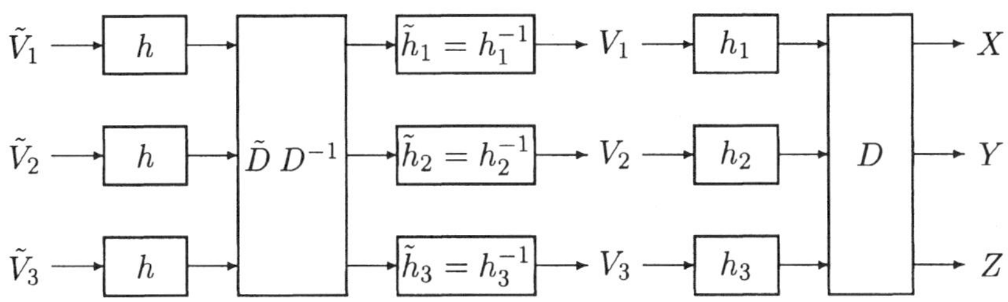

Figure 4.4. Calibration towards a non-linear display characteristic with a shared non-linearity h for all color values and a display matrix $\tilde{D}$, starting from a display with a known display matrix D and non-linearities h_i, for $i = 1, 2, 3$.

a display with color values equal to the tristimulus values is $D_m = D^{-1}$, the inverse of the display matrix D. As shown in Figure 4.4, this linearized display can easily be turned into a nonlinear display that is characterized by an arbitrary display matrix $\tilde{D}$ and a common nonlinearity h on all color values $\tilde{V}_i$, for $i = 1, 2, 3$.

If the characteristics of the display device are not known, or may vary over time, then the mapping from color coordinates to color values has to be determined experimentally. The following procedure can be used to calibrate an additive color display if a calibrated **colorimeter** is available for measuring the color coordinates of the displayed colors. A colorimeter is a linearized color sensor for which the sensor sensitivity curves are known linear combinations of the standard observer curves $\overline{x}(\lambda)$, $\overline{y}(\lambda)$ and $\overline{z}(\lambda)$, and which gives readings in (X, Y, Z)-coordinates.

It is obviously advantageous if the uncalibrated display has a characteristic that already approximates the targeted calibrated display characteristic. A setup procedure therefore exists for many display devices such as color monitors in order to approximate some standardized display characteristic (such as the standardized PAL display, described in the next section). This procedure typically involves applying the maximum value to all inputs (i.e., $V_1 = V_2 = V_3 = V_{max}$) and adjusting the video amplifier for the blue primary in order to obtain maximum luminance. Next, the video amplifiers for the red and green primaries are adjusted until the displayed color has color coordinates that are approximately equal to the color coordinates of an a priori selected white reference, such as the CIE standard illuminant D_{65}. Equations (4.2) and (4.5), with unknown display matrix D and non-linearities h_i, are assumed to model the display after this setup procedure.

The tristimulus values of the displayed color that is obtained by applying the maximum input V_{max} to the i-th primary and the minimum input V_{min} to the two remaining primaries are denoted by (X_{Mi}, Y_{Mi}, Z_{Mi}). Combining the measurements for all three primaries results in

$$\begin{bmatrix} X_{M1} & X_{M2} & X_{M3} \\ Y_{M1} & Y_{M2} & Y_{M3} \\ Z_{M1} & Z_{M2} & Z_{M3} \end{bmatrix} = D \cdot \begin{bmatrix} 1 & h_1(V_{min}) & h_1(V_{min}) \\ h_2(V_{min}) & 1 & h_2(V_{min}) \\ h_3(V_{min}) & h_3(V_{min}) & 1 \end{bmatrix}, \tag{4.6}$$

since the nonlinearities can always be chosen to satisfy $h_i(V_{max}) = 1$. The measured matrix in the left part of the above equation is hence a good (first) approximation to the display matrix D, since usually $h_i(V_{min}) \ll 1$. A more accurate estimate for the matrix D can be obtained by performing one additional measurement. One possibility is to additionally measure the tristimulus values (X_m, Y_m, Z_m) of the displayed color when all inputs are set to their minimum value. The simultaneous solution to the above matrix equation and the following set of linear equations

$$\begin{bmatrix} X_m \\ Y_m \\ Z_m \end{bmatrix} = D \cdot \begin{bmatrix} h_1(V_{min}) \\ h_2(V_{min}) \\ h_3(V_{min}) \end{bmatrix}, \tag{4.7}$$

can provide an updated estimate for D and $h_i(V_{min})$, for $i = 1, 2, 3$.

According to the additive color model, the light efficiency of any displayed color can be expressed as follows

$$Y(V_1, V_2, V_3) = Y_1 \cdot h_1(V_1) + Y_2 \cdot h_2(V_2) + Y_3 \cdot h_3(V_3) \tag{4.8}$$

in terms of the color values (V_1, V_2, V_3). It will be demonstrated below how to construct nonlinear mappings $V_i = \tilde{h}_i(\tilde{V}_i)$ such that the combined characteristics

$$h_i\left(\tilde{h}_i(\tilde{V}_i)\right) = \frac{\tilde{Y}_i}{Y_i}\, h(\tilde{V}_i), \tag{4.9}$$

are proportional to an a priori specified function h within the range $\tilde{V}_i \in [V_{min}, V_{max}]$, for arbitrary scaling factors $\tilde{Y}_i$ $(i = 1, 2, 3)$. Substitution leads to the following expression

$$\begin{aligned}
&\tilde{Y}\left(\tilde{V}_1, \tilde{V}_2, \tilde{V}_3\right) \\
&= Y_1 \cdot h_1\left(\tilde{h}_1(\tilde{V}_1)\right) + Y_2 \cdot h_2\left(\tilde{h}_2(\tilde{V}_2)\right) + Y_3 \cdot h_3\left(\tilde{h}_3(\tilde{V}_3)\right) \\
&= \tilde{Y}_1 \cdot h(\tilde{V}_1) + \tilde{Y}_2 \cdot h(\tilde{V}_2) + \tilde{Y}_3 \cdot h(\tilde{V}_3)
\end{aligned} \tag{4.10}$$

for the light efficiency of the targeted calibrated display.

If the nonlinearities $\tilde{h}_i$ are included as **look-up tables** (LUTs) $\tilde{V}_i \rightarrow V_i = \tilde{h}_i(\tilde{V}_i)$ before applying the color values V_i to the inputs of the uncalibrated display, then the overall display will be specified by the following relationship

$$\begin{bmatrix} X \\ Y \\ Z \end{bmatrix} = \tilde{D} \cdot \begin{bmatrix} h(\tilde{V}_1) \\ h(\tilde{V}_2) \\ h(\tilde{V}_3) \end{bmatrix} \tag{4.11}$$

between calibrated color values and standard tristimulus values. The display matrix

$$\tilde{D} = \begin{bmatrix} x_1 \cdot \tilde{Y}_1/y_1 & x_2 \cdot \tilde{Y}_2/y_2 & x_3 \cdot \tilde{Y}_3/y_3 \\ \tilde{Y}_1 & \tilde{Y}_2 & \tilde{Y}_3 \\ z_1 \cdot \tilde{Y}_1/y_1 & z_2 \cdot \tilde{Y}_2/y_2 & z_3 \cdot \tilde{Y}_3/y_3 \end{bmatrix} \tag{4.12}$$

can differ from the display matrix D of the uncalibrated display since the factors $\tilde{Y}_i$, for $i = 1, 2, 3$, can be freely specified as part of the calibration procedure. The color coordinates (x_i, y_i) of the primaries are obviously not altered by the calibration procedure.

As mentioned earlier, displays are usually calibrated such that colors with an a priori specified color coordinate (x_w, y_w) are obtained if the same color value is applied to all inputs, i.e., $\tilde{V}_1 = \tilde{V}_2 = \tilde{V}_3$. Obviously, if the factors $\tilde{Y}_i$ satisfy

$$\begin{bmatrix} x_1/y_1 & x_2/y_2 & x_3/y_3 \\ 1 & 1 & 1 \\ z_1/y_1 & z_2/y_2 & z_3/y_3 \end{bmatrix} \cdot \begin{bmatrix} \tilde{Y}_1 \\ \tilde{Y}_2 \\ \tilde{Y}_3 \end{bmatrix} = \tilde{Y}_w \cdot \begin{bmatrix} x_w/y_w \\ 1 \\ z_w/y_w \end{bmatrix}, \tag{4.13}$$

then the neutral color of the calibrated display will agree with the specified neutral reference. The maximum light efficiency of the white reference is equal to $\tilde{Y}_w\, h(V_{max})$, and is limited by the available luminance range of the uncalibrated display in a way that is discussed below. For example, the color coordinates of the red, green and blue phosphors of the color monitor whose luminance characteristics are shown in Figure 4.5 have been measured with a colorimeter while applying the maximum color value to the primary of interest, and the minimum color value to the two other primaries. The measurements resulted in $(x_1, y_1) = (0.614, 0.351)$, $(x_2, y_2) = (0.301, 0.593)$ and $(x_3, y_3) = (0.149, 0.070)$. These color coordinates can be used to derive that a D_{65} white reference, with $(x_w, y_w) = (0.31273, 0.32902)$, is obtained for the calibrated monitor if $\tilde{Y}_1 = 0.2460 \cdot \tilde{Y}_w$, $\tilde{Y}_2 = 0.6694 \cdot \tilde{Y}_w$ and $\tilde{Y}_3 = 0.0846 \cdot \tilde{Y}_w$.

We now return to the required procedure for determining the nonlinear transformations $\tilde{h}_i$. The starting point for the nonlinear transformations is fixed to $\tilde{h}_i(V_{min}) = V_{min}$, for $i = 1, 2, 3$. Based on this assumption and equations (4.8) and (4.10), one obtains that

$$\tilde{Y}_w \cdot h(V_{min}) = Y\left(V_{min}, V_{min}, V_{min}\right) \tag{4.14}$$

and

$$\tilde{Y}_i \cdot \left[h(\tilde{V}) - h(V_{min})\right] = Y_i \cdot \left[h_i\left(\tilde{h}_i(\tilde{V})\right) - h_i(V_{min})\right] = \overline{Y}_i\left(\tilde{h}_i(\tilde{V})\right), \tag{4.15}$$

for $i = 1, 2, 3$. The first required characteristic

$$\begin{aligned} \overline{Y}_1(V) &= Y_1 \cdot [h_1(V) - h_1(V_{min})] \\ &= Y(V, V_{min}, V_{min}) - Y(V_{min}, V_{min}, V_{min}) \end{aligned} \tag{4.16}$$

is obtained by measuring the light efficiency $Y(V, V_{min}, V_{min})$, or luminance $L = K_m \cdot Y$, for a varying color value V to the first primary, while the color values to the second and third primary are fixed to their minimum value V_{min}. The other two characteristics $\overline{Y}_2(V)$ and $\overline{Y}_3(V)$ can be obtained in a similar way by varying the color value for the second and third primary, respectively. The luminance characteristics $L_r(V) = K_m \cdot Y(V, V_{min}, V_{min})$ (for red), $L_g(V) = K_m \cdot Y(V_{min}, V, V_{min})$ (for green) and $L_b(V) = K_m \cdot Y(V_{min}, V_{min}, V)$ (for blue) are usually highly nonlinear (as in the case of a monochrome display[2]) and different for the three color channels, as is illustrated for a typical color monitor in Figure 4.5.

[2]A monochrome display was earlier characterized by how it maps input grey values V into luminances $L(V)$ on the display screen (see for instance Figure 3.7).

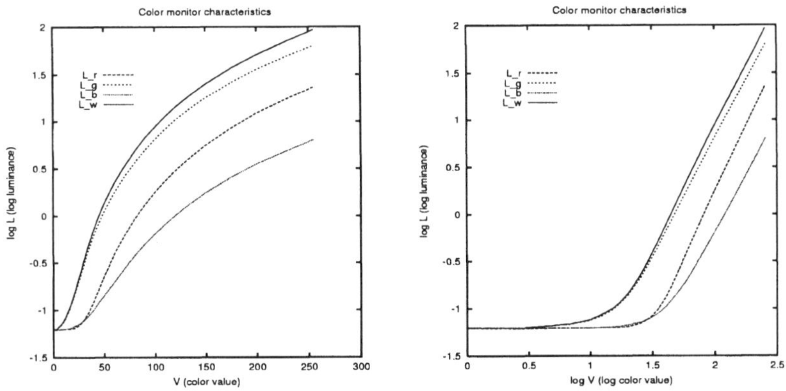

Figure 4.5. Relationships between the logarithm of the output luminance L and the input color value V (left) or $\log V$ (right) for a BARCO CCID 7351B color monitor. The curves denoted by L_r, L_g and L_b are the characteristics if only the red, green and blue color values are nonzero, while the curve denoted by L_w corresponds to an equal value applied to all inputs. For $V > 30$ ($\log V > 1.5$), the luminance characteristics can be approximated by a power law $L(V) \propto V^\gamma$, with $\gamma \approx 2.5$.

A graphical procedure for deriving the nonlinear mapping

$$\tilde{h}_i(\tilde{V}) = \overline{Y}_i^{-1}\left(\tilde{Y}_i \cdot \left[h(\tilde{V}) - h(V_{min})\right]\right) \tag{4.17}$$

from a specified function $h(\tilde{V})$ and a measured characteristic $\overline{Y}_i(V)$ is presented in Figure 4.6.

The maximum value $h(V_{max})$ that can be obtained for the nonlinearity of the calibrated display is obviously determined by the maximum luminances that can be obtained for each of the three primaries of the uncalibrated display, i.e.,

$$h(V_{max}) - h(V_{min}) = \min_i \left[\frac{\overline{Y}_i(V_{max})}{\tilde{Y}_i}\right] \tag{4.18}$$

(see also Figure 4.6). Since the minimum and maximum values $h(V_{min})$ and $h(V_{max})$ are the only boundary conditions that are imposed by the luminance range of the available display, $h(\cdot)$ can be any monotonically increasing function from $[V_{min}, V_{max}]$ to $[h(V_{min}), h(V_{max})]$. A frequently adopted characteristic in display calibration is the **gamma characteristic** with equation

$$h(V) = g(V; \gamma, s, V_o, c) = c \cdot \left[(V^s + V_o^s)^{1/s}\right]^\gamma, \tag{4.19}$$

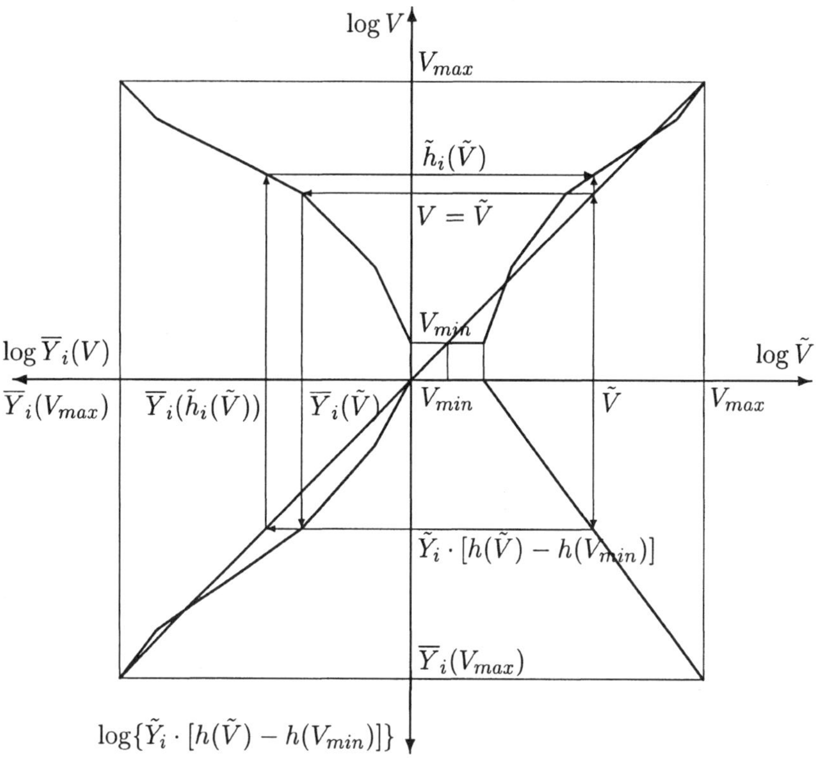

Figure 4.6. Determining the nonlinear mapping between the calibrated color value $\tilde{V}$ and the uncalibrated color value $V = \tilde{h}_i(\tilde{V})$ can be accomplished by using the difference between the measured light efficiency $\overline{Y}_i(\tilde{V})$ and the expected value $\tilde{Y}_i \cdot [h(\tilde{V}) - h(V_{min})]$. First quadrant: original ($V = \tilde{V}$) and updated ($V = \tilde{h}_i(\tilde{V})$) relation between $\tilde{V}$ and V; second quadrant: experimentally determined, and subsequently interpolated, relation between V and $\overline{Y}_i(V)$; third quadrant: targeted (straight line) and actual relation between measured and expected light efficiency; fourth quadrant: relation between $\tilde{V}$ and desired characteristic $\tilde{Y}_i \cdot [h(\tilde{V}) - h(V_{min})]$. Since most of the relationships are (approximate) power functions, it is usually advantageous to use logarithmic axes throughout.

for $V \in [V_{min}, V_{max}]$, where the constants c and V_o are derived from the known extreme values of the function h. The free parameter s controls the steepness of the transition from a constant output $c \cdot V_o^\gamma$ (for $V \ll V_o$) to the power relationship $c \cdot V^\gamma$ (for $V \gg V_o$). The special case $s = \infty$, i.e.,

$$\begin{aligned} h(V) = g(V;\gamma,\infty,V_o,c) &= c \cdot V_o^\gamma \quad \text{for} \quad V_{min} \leq V \leq V_o \\ &= c \cdot V^\gamma \quad \text{for} \quad V_o \leq V \leq V_{max} \end{aligned} \tag{4.20}$$

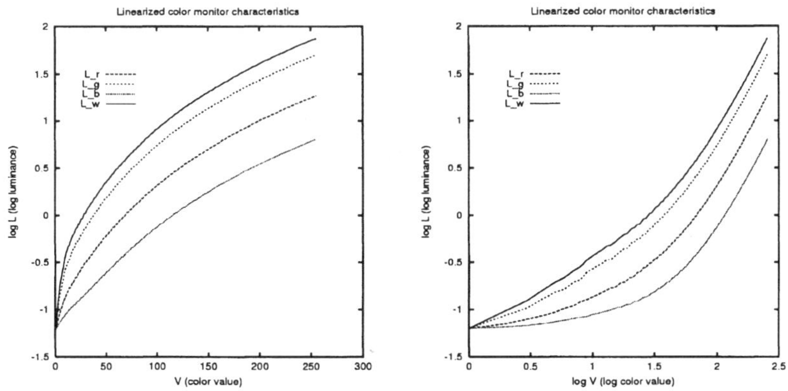

Figure 4.7. Relationships between the logarithm of the output luminance L and input color value V (left) or $\log V$ (right) for a display with a linear relationship between grey value and psychometric lightness. The curves denoted by L_r, L_g and L_b are the characteristics if only the red, green and blue color values differ from zero, while the curve denoted by L_w corresponds to an equal color value applied to all inputs.

corresponds to a discontinuous change in the slope of the gamma characteristic. The most important free parameter γ is called the **gamma of the display** [Poy93]. For instance, we will see in the next section that the standardized display in television broadcasting has a gamma characteristic with $\gamma = 2.2$.

If the display calibration is successful, then the single-channel characteristics of the calibrated display, which can be measured by putting two out of three input color values equal to V_{min}, should be equal to

$$\begin{aligned} \tilde{Y}_1(\tilde{V}) = \tilde{Y}(\tilde{V}, V_{min}, V_{min}) &= \tilde{Y}_1 \cdot [h(\tilde{V}) - h(V_{min})] + \tilde{Y}_w \cdot h(V_{min}) \\ \tilde{Y}_1(\tilde{V}) = \tilde{Y}(V_{min}, \tilde{V}, V_{min}) &= \tilde{Y}_2 \cdot [h(\tilde{V}) - h(V_{min})] + \tilde{Y}_w \cdot h(V_{min}) \\ \tilde{Y}_3(\tilde{V}) = \tilde{Y}(V_{min}, V_{min}, \tilde{V}) &= \tilde{Y}_3 \cdot [h(\tilde{V}) - h(V_{min})] + \tilde{Y}_w \cdot h(V_{min}) \end{aligned} \tag{4.21}$$

while the calibrated (white) characteristic with equal color values applied to all inputs should correspond to the light efficiency $\tilde{Y}_w \cdot h(\tilde{V})$.

It was demonstrated in chapter 3 how the psychometric lightness characteristic can be used to create a perceptually linearized monochrome display. Let $L(V)$ denote the linearized grey-value-to-luminance characteristic of equation (3.27) with minimum, maximum and reference luminance equal to L_{min}, L_{max} and L_o, respectively. A calibrated color display that behaves as this linearized monochrome display $L(\tilde{V}) =$

$K_m \cdot \tilde{Y}_w \cdot h(\tilde{V})$ in case $\tilde{V} = \tilde{V}_1 = \tilde{V}_2 = \tilde{V}_3$ can easily be created using the above calibration procedure. The resulting luminance characteristics for this calibrated display are illustrated in Figure 4.7 for $L_{min} = K_m \cdot \tilde{Y}_w \, h(V_{min}) = 0.063$ cd/m^2 and $L_{max} = L_o = K_m \cdot \tilde{Y}_w \, h(V_{max}) = 74.7$ cd/m^2 determined by the available luminance range of the uncalibrated display in figure 4.5. The corresponding range of psychometric lightness is $L^\star \in [0.76, 100]$. The white point of the calibrated display is D_{65}.

3. MULTIPLICATIVE COLOR DISPLAYS

In this section, two representative multiplicative color reproduction techniques, i.e., continuous-tone and halftone printing, are described. Idealized linear models [Hor84] that are useful for understanding the relationship between additive and multiplicative (or subtractive) color reproduction methods are also derived. These linear models only provide a crude first-order approximation. In cases where a more accurate color characterization is required, a truly nonlinear model will have to be devised. The non-linear color characterization method that will be discussed in section 5 for image sensors can for instance be adapted for this purpose [Har99].

3.1 CONTINUOUS-TONE PRINTING

Continuous-tone printing is best known from photographic reproduction. Colors are (re)produced by controlling, on a point-by-point basis, the *concentrations* of (three) different absorbing dyes in superimposed layers. For example, in case of a photographic transparency, the transmission at each point in the image can be approximated by

$$T(\lambda) = T_0(\lambda) \cdot \prod_{i=1}^{3} [p_i(\lambda)]^{C_i} , \tag{4.22}$$

where $p_i(\lambda)$ is the transmission of one unit of concentration of the i-th dye, and C_i is the spatially variant concentration of the i-th dye. The transmission of the transparency without dye is denoted by $T_0(\lambda)$. A similar expression holds for the reflectance of a photographic print.

In a photographic film, each dye layer is directly coupled to a sensitive layer. The wavelength characteristic of a sensitive layer determines how the incoming light during film exposure influences the dye concentration after film development in the corresponding (re)production layer. It is possible to accurately control the concentrations of the dyes in a photographic film, provided that the sensitive layers can be made separately sensitive to three control wavelengths. If this condition is satisfied, then one dye concentration can be varied without influencing the other two, by exposing the film through a narrow-band filter at one of the control wavelengths. This so-called lambda technique is mostly used when accurate color reproductions on film are required (see also section 6).

In a thermal printer, the dye is transferred from a ribbon cartridge containing dye to the print medium (paper or transparency material). When printing is initiated, the dye carrier and the print medium are brought into contact with a thermal print head. Spatially varying temperature profiles that reflect the image signal can be generated under computer control by means of this thermal print head. The temperature then determines the amount of dye that diffuses from the dye carrier to the print medium.

If the input irradiance of the transparency is $E_{e,s}(\lambda)$, then the output radiance becomes

$$L_e(\lambda) = T(\lambda) \cdot E_{e,s}(\lambda) = E_{e,s}(\lambda)\, T_0(\lambda) \cdot \prod_{i=1}^{3} [p_i(\lambda)]^{C_i} . \tag{4.23}$$

Since this relationship between the dye concentrations C_i and the output radiance is clearly nonlinear, accurate color calibration will require a detailed knowledge of the spectrum of the transmitted light and the absorption spectra $p_i(\lambda)$ of the dyes. We derive an additive color model approximation to the above multiplicative color model that can provide a first-order approximation to the mapping from dye concentrations to color coordinates.

Ideal dyes that satisfy a number of conditions are assumed in the simplified additive model[Hor84]. First, in order to avoid multiplicative interactions between the three layers, at most one dye should absorb at each wavelength. Second, in order to be able to produce low transmissions at all wavelengths (a necessary condition in order to produce low luminances), at least one dye should absorb at each wavelength. Hence, from both preceding conditions, one can conclude that exactly one dye should absorb at each wavelength in the visible region of the spectrum. Consequently, the range of visible wavelengths can be subdivided in three regions Λ_1, Λ_2 and Λ_3, such that $p_i(\lambda) < 1$ and $p_j(\lambda) = 1, j \neq i$, for $\lambda \in \Lambda_i$, $i = 1, 2, 3$. Hence, the transmitted radiance without any dye present can be decomposed in

$$L_0(\lambda) = T_0(\lambda) \cdot E_{e,s}(\lambda) = L_1(\lambda) + L_2(\lambda) + L_3(\lambda), \tag{4.24}$$

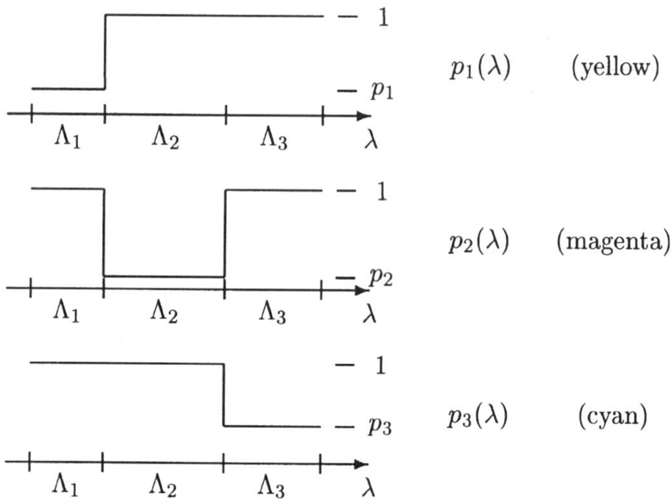

Figure 4.8. Highly idealized dye transmission curves needed to map multiplicative color models into additive color models.

where $L_i(\lambda) = T_0(\lambda) \cdot E_{e,s}(\lambda)$ for $\lambda \in \Lambda_i$, and $L_i(\lambda) = 0$ for λ outside Λ_i. Consequently, the transmitted radiance can be described by

$$L_e(\lambda) = \sum_{i=1}^{3} [p_i(\lambda)]^{C_i} \cdot L_i(\lambda). \tag{4.25}$$

This is similar to the expression for an additive light source, provided that the transmittance curves $p_i(\lambda)$ are piecewise constant, as shown in figure 4.8. The link between subtractive and additive color mixing because obvious upon realizing that the cyan, magenta and yellow (CMY) dyes are respectively red-absorbing, green-absorbing and blue-absorbing, so that the sub-spectra $L_i(\lambda)$, for $i = 3, 2, 1$, correspond to red, green and blue light sources.

It is worthwhile to note at this time that the traditional names (cyan, magenta and blue) given to the dyes in the photographic and printing industry have often been a source of confusion since they do not properly reflect the perceived colors: the cyan dye is actually perceived to be blue, rather than blue-green, while the magenta dye is perceived to be red.

3.2 HALF-TONE PRINTING

A second important multiplicative color reproduction technique is halftone printing. It is for instance used in lithography and ink-jet printers. In this case, dyes or inks of fixed concentration are used, while the *fraction of the surface* that is covered with each ink is varied. This may be attained by varying the size of dots spaced in a regular pattern (see for instance newspaper photographs). First consider a single ink. Assume that the (two-way) transmission of the ink is $p_i(\lambda)$ and that a fraction A_i of the area is covered with ink. If this dot pattern is applied to a substrate of reflectance $R_0(\lambda)$, then the average reflectance will be

$$R(\lambda) = R_0(\lambda)\left[(1 - A_i) + A_i \cdot p_i(\lambda)\right]. \tag{4.26}$$

If the input irradiance of the print is $E_{e,s}(\lambda)$, then the average output radiance becomes

$$L_e(\lambda) = R(\lambda) \cdot E_{e,s}(\lambda) = E_{e,s}(\lambda)\, R_0(\lambda) \cdot \left[(1 - A_i) + A_i \cdot p_i(\lambda)\right]. \tag{4.27}$$

The spacing between the dots is usually below the limits of visual acuity at normal viewing distances, so that this average radiance is adequate for deriving the retinal irradiance.

In the case of color prints, the dots that correspond to different inks lie on rasters that are rotated relative to one another to avoid the appearance of repeated patterns. The result is that dots overlap differently in various regions of the print. Consequently, the fraction of the area covered by no ink, one ink, two inks, or three inks, are well approximated by their respective statistical mean values such as $(1-A_1)(1-A_2)(1-A_3)$, $A_1(1-A_2)(1-A_3)$, $A_1A_2(1-A_3)$ and $A_1A_2A_3$. Consequently, the average reflectance of the completed print is obtained by multiplying the substrate reflectance with the average transmissions of the different ink layers, i.e.,

$$R(\lambda) = R_0(\lambda) \cdot \prod_{i=1}^{3} \left[(1 - A_i) + A_i \cdot p_i(\lambda)\right]. \tag{4.28}$$

The resulting relationships between the color coordinates of the output radiance and the area parameters A_i are again highly nonlinear.

The idealized dyes of figure 4.8 can again be used to map the above multiplicative model into an additive model. By splitting the maximally reflected radiance into the sum of three components, i.e.,

$$L_0(\lambda) = R_0(\lambda) \cdot E_{e,s}(\lambda) = L_1(\lambda) + L_2(\lambda) + L_3(\lambda), \tag{4.29}$$

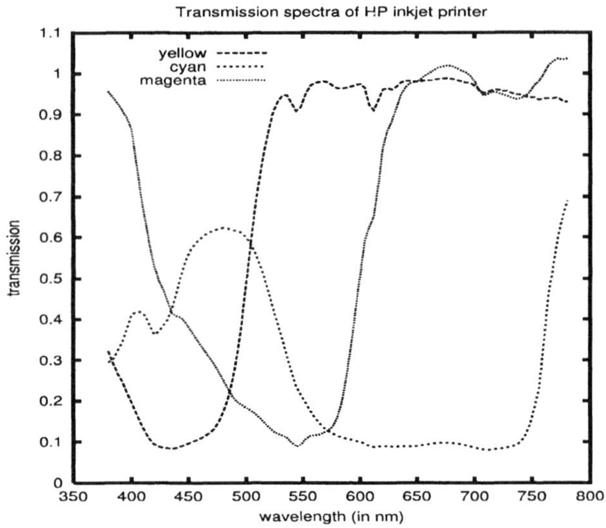

Figure 4.9. Measured transmission characteristics as a function of wavelength for the three dyes (yellow, cyan, magenta) in an HP color ink-jet printer. The characteristics are only approximate since they were obtained by dividing the radiance at a printed part of the paper by the radiance at an unprinted (white) part of the paper. The transmission curves therefore contain some residual effects of the light source used (which was non-standardized daylight).

one obtains that

$$L_e(\lambda) = \sum_{i=1}^{3} [1 - (1 - p_i) \cdot A_i] \cdot L_i(\lambda) \tag{4.30}$$

which conforms to the additive color model of the previous section.

In order to get some feeling for how crude the idealized-dye approximation is, some measured characteristics of existing dyes, as used in an HP color ink-jet printer, are plotted in Figure 4.9.

Many color printers use four (CMYK) instead of three (CMY) dyes. The black ink (K) will typically absorb light at all visible wavelengths. It is introduced to accomplish two things: to increase the contrast by allowing more absorption in dark parts of the image, and to replace some percentage of the three color primaries. This latter aspect can drastically reduce the operating cost, since large concentrations of colored inks are no longer required to produce dark areas [SCB88]. As in the case of additive color mixing with more than three primaries, there are several possible strategies for determining the amount of black ink. The most

Table 4.1. Color coordinates of the red, green and blue primaries and the white reference for standardized displays.

	NTSC	PAL	BT.709
x_r	0.67	0.64	0.64
y_r	0.33	0.33	0.33
x_g	0.21	0.29	0.30
y_g	0.71	0.60	0.60
x_b	0.14	0.15	0.15
y_b	0.08	0.06	0.06
x_w	0.310	0.313	0.313
y_w	0.316	0.329	0.329

straightforward method is to implement the neutral or grey component of a color by means of the black ink [Hun88, Kan97].

4. STANDARDIZED COLOR COORDINATES

4.1 STANDARDIZED DISPLAYS

Television (TV) broadcasting is an extreme case of color management where the same color coordinates are transmitted to many receiving systems. This only makes sense because all these receiving TV systems are assumed to approximate the same standard additive color mixing display. Amongst others, this implies that the color coordinates for the red, green and blue primaries and the white reference of this standard monitor have been agreed upon. Probably the three most widely used standardized displays are described in:

1 the NTSC (National Television Standard Committee) standard, with the Federal Communication Commission of America (FCC) primaries,

2 the PAL (Phase Alternation Lines) standard with the European Broadcasting Union (EBU) primaries, and

3 the recommendation BT.709 for High Definition Television (HDTV) from the International Telecommunications Union (ITU-R).

The color coordinates of the red, green and blue primaries and the white reference for these standardized displays are listed in Table 4.1 [NH88, PV00].

The relative light efficiencies Y_r, Y_g and Y_b of the red, green and blue primaries of the standardized display must be such that a white reference of unit light efficiency and color coordinates (x_w, y_w) is obtained if a weight of one is assigned to all three primaries, i.e.,

$$\left(\frac{x_w}{y_w}, 1, \frac{z_w}{y_w}\right) = \left(Y_r\frac{x_r}{y_r}, Y_r, Y_r\frac{z_r}{y_r}\right) + \left(Y_g\frac{x_g}{y_g}, Y_g, Y_g\frac{z_g}{y_g}\right) + \left(Y_b\frac{x_b}{y_b}, Y_b, Y_b\frac{z_b}{y_b}\right). \quad (4.31)$$

This implies that the relative light efficiencies of the primaries can be found by solving the following set of equations

$$\begin{bmatrix} x_r/y_r & x_g/y_g & x_b/y_b \\ 1 & 1 & 1 \\ z_r/y_r & z_g/y_g & z_b/y_b \end{bmatrix} \cdot \begin{bmatrix} Y_r \\ Y_g \\ Y_b \end{bmatrix} = \begin{bmatrix} x_w/y_w \\ 1 \\ z_w/y_w \end{bmatrix}. \quad (4.32)$$

For instance, for the D_{65} illuminant with $(x_w, y_w) = (0.313, 0.329)$ that is adopted as white reference in PAL and BT.709, we obtain that $Y_r = 0.222$, $Y_g = 0.707$ and $Y_b = 0.071$ for PAL and $Y_r = 0.2125$, $Y_g = 0.7154$ and $Y_b = 0.0721$ for BT.709. For NTSC systems, where the C illuminant with $(x_w, y_w) = (0.310, 0.316)$ is used as white reference, the solution is $Y_r = 0.299$, $Y_g = 0.587$ and $Y_b = 0.114$.

In section 2, it was derived that an additive mixture of primaries with known color coordinates and weights E_r, E_g and E_b has CIE standard observer tristimulus values equal to

$$\begin{bmatrix} X \\ Y \\ Z \end{bmatrix} = \begin{bmatrix} x_r \cdot Y_r/y_r & x_g \cdot Y_g/y_g & x_b \cdot Y_b/y_b \\ Y_r & Y_g & Y_b \\ z_r \cdot Y_r/y_r & z_g \cdot Y_g/y_g & z_b \cdot Y_b/y_b \end{bmatrix} \cdot \begin{bmatrix} E_r \\ E_g \\ E_b \end{bmatrix}. \quad (4.33)$$

In case of PAL, the following explicit expression is obtained

$$\begin{bmatrix} X \\ Y \\ Z \end{bmatrix} = \begin{bmatrix} 0.432 & 0.341 & 0.178 \\ 0.222 & 0.707 & 0.071 \\ 0.020 & 0.129 & 0.938 \end{bmatrix} \cdot \begin{bmatrix} E_r \\ E_g \\ E_b \end{bmatrix}, \quad (4.34)$$

or equivalently,

$$\begin{bmatrix} E_r \\ E_g \\ E_b \end{bmatrix} = \begin{bmatrix} 3.054 & -1.389 & -0.474 \\ -0.970 & 1.878 & 0.042 \\ 0.068 & -0.229 & 1.070 \end{bmatrix} \cdot \begin{bmatrix} X \\ Y \\ Z \end{bmatrix}. \quad (4.35)$$

The corresponding transformations for BT.709 are very similar, i.e.,

$$\begin{bmatrix} X \\ Y \\ Z \end{bmatrix} = \begin{bmatrix} 0.4125 & 0.3576 & 0.1804 \\ 0.2127 & 0.7152 & 0.0721 \\ 0.0193 & 0.1192 & 0.9502 \end{bmatrix} \cdot \begin{bmatrix} E_r \\ E_g \\ E_b \end{bmatrix}, \quad (4.36)$$

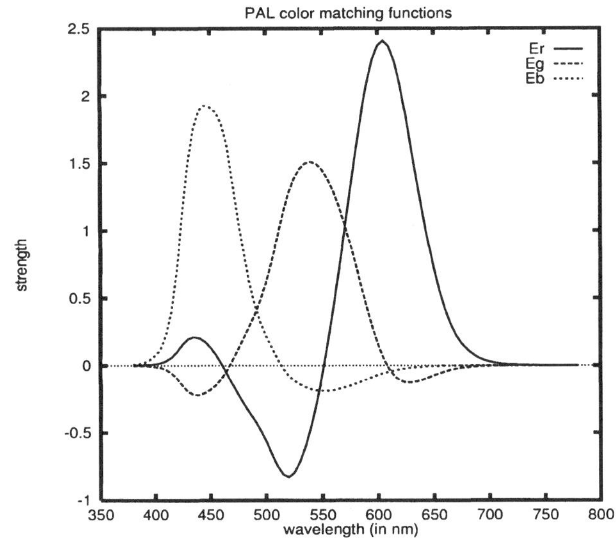

Figure 4.10. PAL color-matching functions $\overline{E}_r(\lambda)$, $\overline{E}_g(\lambda)$ and $\overline{E}_b(\lambda)$.

and

$$\begin{bmatrix} E_r \\ E_g \\ E_b \end{bmatrix} = \begin{bmatrix} 3.2405 & -1.5372 & -0.4985 \\ -0.9693 & 1.8760 & 0.0416 \\ 0.0556 & -0.2040 & 1.0573 \end{bmatrix} \cdot \begin{bmatrix} X \\ Y \\ Z \end{bmatrix}, \tag{4.37}$$

while the transformations for NTSC are somewhat different, i.e.,

$$\begin{bmatrix} X \\ Y \\ Z \end{bmatrix} = \begin{bmatrix} 0.607 & 0.174 & 0.200 \\ 0.299 & 0.587 & 0.114 \\ 0.000 & 0.066 & 1.111 \end{bmatrix} \cdot \begin{bmatrix} E_r \\ E_g \\ E_b \end{bmatrix} \tag{4.38}$$

and

$$\begin{bmatrix} E_r \\ E_g \\ E_b \end{bmatrix} = \begin{bmatrix} 1.909 & -0.532 & -0.288 \\ -0.985 & 1.997 & -0.028 \\ 0.058 & -0.119 & 0.902 \end{bmatrix} \cdot \begin{bmatrix} X \\ Y \\ Z \end{bmatrix}. \tag{4.39}$$

If the transformation matrix from (E_r, E_g, E_b) to (X, Y, Z) is denoted by D_S, then the inverse display matrix D_S^{-1} relates the standard color-matching functions $\{\overline{x}(\lambda), \overline{y}(\lambda), \overline{z}(\lambda)\}$ to the color matching functions $\{\overline{E}_r(\lambda), \overline{E}_g(\lambda), \overline{E}_b(\lambda)\}$ for the video primaries. A video camera with these color-matching functions delivers outputs E_r, E_g and E_b that can be applied directly to the primaries of the standard display. The color matching functions for the PAL primaries are plotted in Figure 4.10.

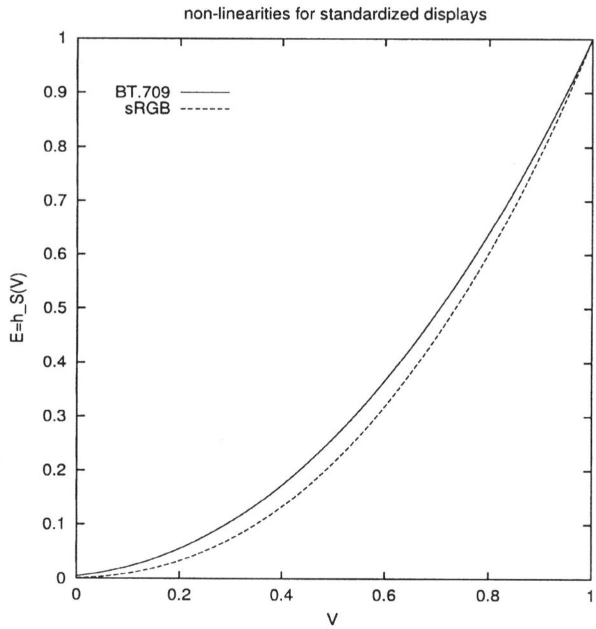

Figure 4.11. Nonlinearities for two standardized displays (BT.709 and sRGB) that correspond to typical TV and desktop viewing conditions, respectively.

The nonlinearity h_S that relates the color values (V_r, V_g, V_b) to the energies $(E_r = h_S(V_r), E_g = h_S(V_g), E_b = h_S(V_b))$ of the primaries in a standardized display is an approximate gamma characteristic with $\gamma \approx 2.2$. The following modification for small color values

$$E = h_S(V) = \begin{cases} \frac{V}{f}, & V \leq V_o \\ \left(\frac{V+V_t}{1+V_t}\right)^{\gamma_S}, & V_o \leq V \leq 1 \end{cases} \tag{4.40}$$

with $\gamma_S = 1/0.45 = 2.22$, $V_t = 0.099$, $f = 4.5$, $V_o = 0.081$ and $V \in [0,1]$, has been adopted in BT.709 to minimize the effect of noise [Poy93, PV00]. The resulting characteristic is plotted in Figure 4.11. The best approximation of the form $c \cdot (V + V_c)^\gamma$ to this standardized characteristic has a gamma $\gamma = 1.956$ that is smaller than γ_S.

It has been argued that the relatively dim viewing conditions that are assumed for broadcasting are not representative for most desktop applications. Based on this argumentation, an alternative characteristic satisfying equation (4.40), but with different parameter values $\gamma_S = 2.4$,

$V_t = 0.055$, $f = 12.92$ and $V_o = 0.03928$, has been proposed [PV00]. It is also plotted in Figure 4.11. The best approximation of the form $c\cdot(V+V_c)^\gamma$ to this alternative characteristic also has the targeted gamma $\gamma = 2.2$. This alternative characteristic underlies the recent proposal for a default color spaces (called sRGB) for internet applications [SACM97].

The mapping from primary weights E to color values V is the reverse of equation (4.40) and can be expressed as

$$V = h_S^{-1}(E) = \begin{cases} f \cdot E, & E \leq E_o \\ (1+V_t)\cdot E^{1/\gamma_S} - V_t, & E_o \leq E \leq 1 \end{cases} \tag{4.41}$$

with $E_o = V_o/f$ (for instance, equal to $E_o = 0.018$ in case of BT.709) and $V \in [0,1]$ [Poy93].

4.2 COLOR ENCODING

The (standard) weights E_r, E_g and E_b are not transmitted as such in TV broadcast systems, but are linearly transformed at the transmitter and the receiver side. The primary reason is that, by selecting the **luminance signal**

$$E_Y = 0.299 \cdot E_r + 0.587 \cdot E_g + 0.114 \cdot E_b \tag{4.42}$$

as one of these combinations, a compatibility with monochrome television could be achieved. Note that this signal is only proportional to luminance in the case of NTSC. For historical reasons (i.e., the PAL system originally used the same primaries as NTSC), this signal is also used in PAL systems. In the BT.709 recommendation, the proper luminance signal

$$E_Y = 0.2127 \cdot E_r + 0.7152 \cdot E_g + 0.0721 \cdot E_b \tag{4.43}$$

is used in the color transformations discussed below.

In the case of PAL, the color information is coded in the **color signals**

$$\begin{aligned} E_U &= 0.493\,(E_b - E_Y) \\ E_V &= 0.877\,(E_r - E_Y), \end{aligned} \tag{4.44}$$

resulting in the following overall transformation

$$\begin{bmatrix} E_Y \\ E_U \\ E_V \end{bmatrix} = \begin{bmatrix} 0.351 & 0.661 & 0.005 \\ -0.140 & -0.439 & 0.525 \\ 2.370 & -1.798 & -0.420 \end{bmatrix} \cdot \begin{bmatrix} X \\ Y \\ Z \end{bmatrix}, \tag{4.45}$$

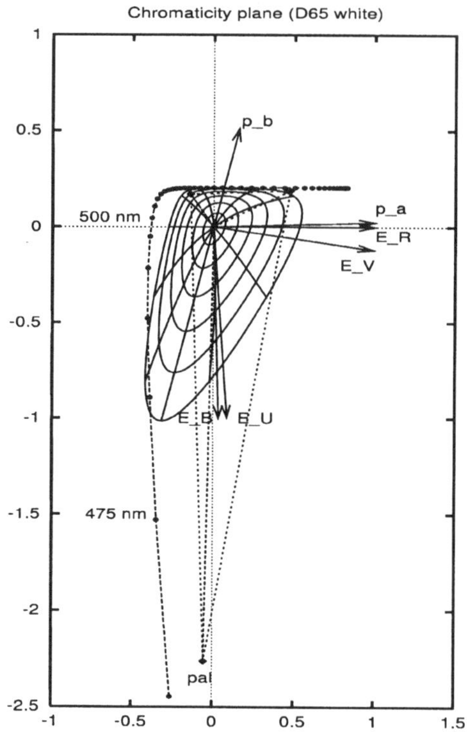

Figure 4.12. The open curve corresponds to the projections onto the chromaticity plane, parallel to the D_{65} white reference (with $X_n/Y_n = 0.951$, $Z_n/Y_n = 1.088$), of the monochromatic spectra with luminance ratio $\ell = Y/Y_n = 0.4$. The curves that radiate from the origin connect coordinates of constant psychometric hue (hue angle steps of 30°). The closed contours connect coordinates with constant psychometric chroma. Psychometric hue and chroma are according to the CIE(L^*,a^*,b^*) metric for $\ell = 0.4$. The (projected) coordinate axes (E_U, E_V) and (E_B, E_R) for PAL and BT.709, respectively, as well as the color triangle spanned by the PAL primaries, are also indicated.

or equivalently

$$\begin{bmatrix} X \\ Y \\ Z \end{bmatrix} = \begin{bmatrix} 0.951 & 0.227 & 0.294 \\ 1.000 & -0.134 & -0.156 \\ 1.088 & 1.853 & -0.052 \end{bmatrix} \cdot \begin{bmatrix} E_Y \\ E_U \\ E_V \end{bmatrix}. \tag{4.46}$$

Note that the color signals satisfy $E_U = E_V = 0$ for the white reference $E_r = E_g = E_b$.

The chromaticity plane is shown in Figure 4.12, together with the projections, parallel to the D_{65} neutral reference, of the coordinate axes

E_U and E_V. The coordinate plane (E_U, E_V) is different from the chromaticity plane, since the E_Y signal is not proportional to the luminance. More specifically, the coordinate plane (E_U, E_V) makes an angle of 13.7 degrees with the chromaticity plane. An advantage of the PAL system is that the E_V coordinate correlates well with the red-green percept. To a somewhat lesser extent, the E_U coordinate correlates with the yellow-blue percept.

In the ITU-R BT.601-4 international standard for studio quality component digital video[SE90], which is for instance used in JPEG (Joint Picture Expert Group) and MPEG (Motion Picture Expert Group) coding, the color information is coded in the color-difference signals

$$\begin{aligned} E_B &= 0.564\,(E_b - E_Y) \\ E_R &= 0.713\,(E_r - E_Y). \end{aligned} \tag{4.47}$$

This color coding is also used in recommendation BT.709, in which case it results in the following overall transformation

$$\begin{bmatrix} E_Y \\ E_B \\ E_R \end{bmatrix} = \begin{bmatrix} 0.000 & 1.000 & 0.000 \\ 0.031 & -0.679 & 0.596 \\ 2.311 & -1.809 & -0.355 \end{bmatrix} \cdot \begin{bmatrix} X \\ Y \\ Z \end{bmatrix}, \tag{4.48}$$

or equivalently

$$\begin{bmatrix} X \\ Y \\ Z \end{bmatrix} = \begin{bmatrix} 0.950 & 0.256 & 0.429 \\ 1.000 & 0.000 & 0.000 \\ 1.089 & 1.664 & -0.023 \end{bmatrix} \cdot \begin{bmatrix} E_Y \\ E_B \\ E_R \end{bmatrix}. \tag{4.49}$$

In Figure 4.12, the coordinate axes (E_B, E_R) for BT.709 are plotted in the chromaticity plane with a and b axes for the D_{65} neutral reference. Note that the correspondence of E_B and E_R with the blue-yellow and red-green axis, respectively, is even better than in the case of PAL. Since the E_Y coordinate of BT.709 is proportional to the luminance, the E_B and E_R axes lie in the chromaticity plane.

In the case of NTSC the color coordinates (E_U, E_V) from PAL are rotated over 33° to obtain the new coordinates

$$\begin{aligned} E_I &= E_V \cdot \cos 33^\circ - E_U \cdot \sin 33^\circ \\ E_Q &= E_V \cdot \sin 33^\circ + E_U \cdot \cos 33^\circ, \end{aligned} \tag{4.50}$$

which results in the following overall transformation

$$\begin{bmatrix} E_Y \\ E_I \\ E_Q \end{bmatrix} = \begin{bmatrix} 0.000 & 1.000 & 0.000 \\ 1.389 & -0.827 & -0.452 \\ 0.937 & -1.195 & 0.234 \end{bmatrix} \cdot \begin{bmatrix} X \\ Y \\ Z \end{bmatrix}, \tag{4.51}$$

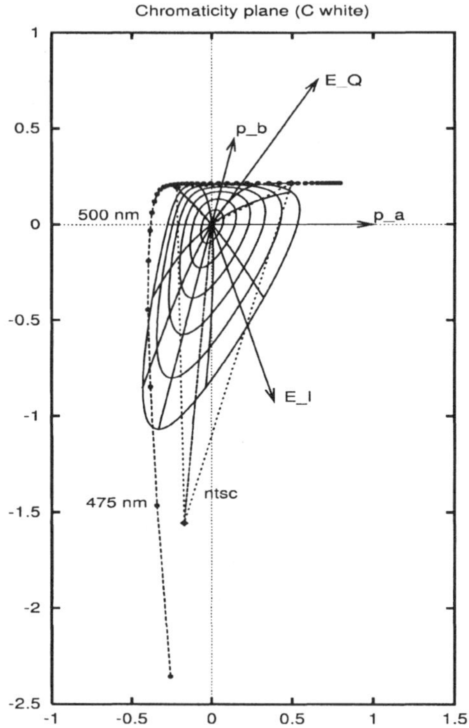

Figure 4.13. The drawn curve corresponds to the projections onto the chromaticity plane, parallel to the C white reference (with $X_n/Y_n = 0.981$, $Z_n/Y_n = 1.184$), of the monochromatic spectra with luminance ratio $\ell = Y/Y_n = 0.4$. The curves that radiate from the origin connect coordinates of constant psychometric hue (hue angle steps of 30°). The closed contours connect coordinates with constant psychometric chroma. Psychometric hue and chroma are according to the CIE(L^*,a^*,b^*) metric for $\ell = 0.4$. The coordinate axes (E_I, E_Q), as well as the color triangle spanned by the NTSC primaries, are also indicated.

or equivalently

$$\begin{bmatrix} X \\ Y \\ Z \end{bmatrix} = \begin{bmatrix} 0.981 & 0.312 & 0.606 \\ 1.000 & 0.000 & 0.000 \\ 1.184 & -1.253 & 1.858 \end{bmatrix} \cdot \begin{bmatrix} E_Y \\ E_I \\ E_Q \end{bmatrix}. \tag{4.52}$$

In Figure 4.13, the coordinate axes (E_I, E_Q) are plotted in the chromaticity plane with a and b axes for the C neutral reference. Since the E_Y coordinate of NTSC is proportional to the luminance, we are guaranteed that the E_I and E_Q axes are in the chromaticity plane. The

E_I and E_Q axes do not correlate well with the opponent color percepts, since they were optimized with another argument in mind, i.e., to reduce the transmission bandwidth required for the E_Q component.

4.3 GAMMA-CORRECTED COLOR VALUES

In actual broadcast practice, the color values V_r, V_g and V_b rather than the weights $E_r = h_S(V_r)$, $E_g = h_S(V_g)$ and $E_b = h_S(V_b)$ of the primaries are used in the above encoding. For instance, in case of BT.709 this leads to the following equations

$$\begin{aligned} V_Y &= 0.2127 \cdot V_r + 0.7152 \cdot V_g + 0.0721 \cdot V_b \\ C_B &= 0.564\,(V_b - V_Y) \\ C_R &= 0.713\,(V_r - V_Y) \end{aligned} \tag{4.53}$$

for the **luma component** V_Y and the **color components** C_B and C_R. In digital video, the original color values V_r, V_g, V_b and the luma component V_Y are quantized with 220 levels in such a way that the black level corresponds to 16 and the peak white level to 235. Similarly, the color components C_B and C_R are quantized with 225 quantization levels that are symmetrically disposed about the level 128 (i.e., from 16 to 240). Level 128 corresponds to a zero signal. The weights in the above equation guarantee that all the signals stay within their designated boundaries. The reason for encoding color values rather than primary weights is that the visible effect of such quantization to 256 levels is much reduced in the former case. We will return to this issue in the chapter on quantization.

The effect of using color values instead of primary weights on the interpretation of the luma and color components V_Y, C_B and C_R (in comparison to E_Y, E_B and E_R) is limited [Hun88]. The situation is similar to the comparison between $(a,b)-$ and (a^*,b^*)-space. The polar angle and polar distance in both spaces correlate well with the hue and chroma of the color, at least for unsaturated colors.

Note that the transformations to luma and color components are only used for encoding and transmission, so that any information of interest (such as E_r, E_g and E_b) can be recomputed whenever required.

5. IMAGE SENSORS

5.1 MODEL

In chapter 1, it was derived that the irradiance of the image target in a typical sensor (with a lens) is proportional to the object radiance $L_e(\lambda)$.

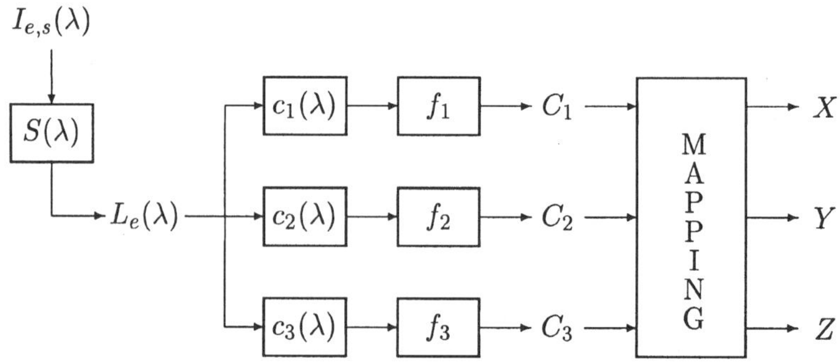

Figure 4.14. Model for color image sensing.

The wavelength dependence of this radiance can often be expressed as

$$L_e(\lambda) = I_{e,s}(\lambda) \cdot S(\lambda), \tag{4.54}$$

where $S(\lambda)$ is the spectral reflectance of the irradiated object, while $I_{e,s}(\lambda)$ describes the spectral content of the radiation source(s). The usual interpretation of this equation is that the object radiance is the product of two factors, one that varies slowly over space, that is usually attributed to the light source, and one that contains the fast spatial variations, that is usually attributed to changes in the object reflectance or orientation. Even if this physical interpretation is not valid, the human visual system seems to prefer it. For instance, an image display is a light source that varies over space and time. Nevertheless, most light variations are interpreted as being caused by a variation in surface colors, which leads to the interpretation of objects or figures being illuminated by one or more slowly varying light sources.

In a color image sensor, the irradiance is linearly integrated over the wavelength by (usually) three detectors with linearly independent spectral sensitivities. The integrated signals are available at the outputs of the sensor after, possibly nonlinear, processing. The block diagram of figure 4.14 is the resulting idealized description of an image sensor. The outputs of the sensor are given by

$$C_i = f_i(< c_i(\lambda), L_e(\lambda) >), \tag{4.55}$$

for $i = 1, 2, 3$, where $c_i(\lambda)$ denotes the spectral sensitivity of the i-th sensor, and f_i represents the nonlinear processing of the sensor signal.

In order to determine how the color values (C_1, C_2, C_3) can be related to standard color coordinates such as (X, Y, Z), it is assumed for the

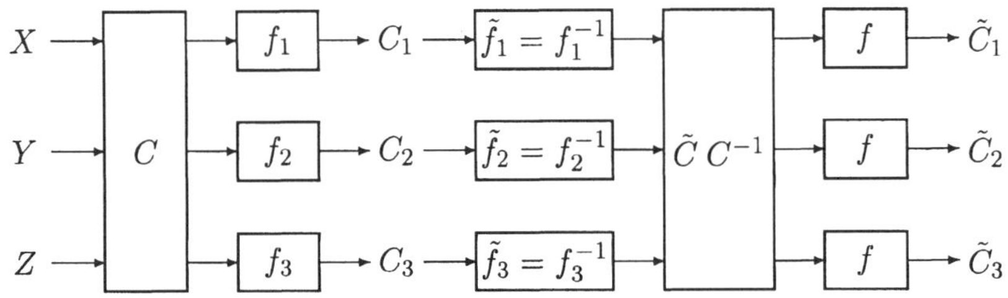

Figure 4.15. Calibration towards a non-linear sensor characteristic with a shared non-linearity f for all color values and sensor matrix $\tilde{C}$, starting from a sensor with a known sensor matrix C and non-linearities f_i, for $i = 1, 2, 3$.

time being that the characteristics of the image sensor are known, so that the nonlinearities f_i can be inverted to obtain linear sensor outputs [Hor84]. In a similar way as in chapter 3, it can be derived that the sensitivity curves $c_i(\lambda)$ span a linear subspace of the space of all radiometric functions. The image sensor maps an arbitrarily radiometric function into this subspace. If no additional information is available about the radiometric functions, then an exact color reproduction requires that fundamental color space is contained in the subspace spanned by the image sensors. This implies that the CIE standard observer sensitivity curves $\overline{x}(\lambda)$, $\overline{y}(\lambda)$ and $\overline{z}(\lambda)$ must be linear combinations of the sensor sensitivity curves $c_i(\lambda)$. In case of an image sensor with three detectors, the reverse is also true, i.e., the sensitivity curves $c_i(\lambda)$ of the sensor must be linear combinations of the standard observer curves. This is called the **Luther-Ives condition** [Ive15, Har99]. Amongst others, it implies that the spectral sensitivity curves $c_i(\lambda)$ must be zero outside the region of visible wavelengths. This condition is not always satisfied by camera systems, but may be corrected by using appropriate filters, such as infrared and/or ultraviolet blocking filters.

If the spectral sensitivities are indeed linear combinations of the CIE standard observer sensitivity curves, i.e.,

$$c_i(\lambda) = c_{ix} \cdot \overline{x}(\lambda) + c_{iy} \cdot \overline{y}(\lambda) + c_{iz} \cdot \overline{z}(\lambda), \tag{4.56}$$

then the sensor outputs can be written as

$$C_i = f_i\left(c_{ix} \cdot X + c_{iy} \cdot Y + c_{iz} \cdot Z\right), \tag{4.57}$$

for $i = 1, 2, 3$, where (X, Y, Z) are the tristimulus values of the input radiance $L_e(\lambda)$. Such a sensor can be characterized by the **sensor matrix**

$$C = \begin{bmatrix} c_{1x} & c_{1y} & c_{1z} \\ c_{2x} & c_{2y} & c_{2z} \\ c_{3x} & c_{3y} & c_{3z} \end{bmatrix}, \tag{4.58}$$

together with the nonlinearities f_i, for $i = 1, 2, 3$, and can be calibrated towards a nonlinear sensor with an arbitrary sensor matrix $\tilde{C}$ and a common nonlinearity f on all sensor outputs $\tilde{C}_i$, for $i = 1, 2, 3$, as shown in Figure 4.15. Note that this model for a calibrated sensor is identical to the model of Figure 4.4 for a calibrated additive color display, but traversed in the reverse order.

The condition that the sensor sensitivity curves must be linear combinations of the standard observer curves is no longer a necessary condition when additional information about the input radiance is available. One extreme, but important, special case will be treated in section 6, where all input radiances of interest are linear combinations of three basis functions [Wan87]. In this case, the sensitivity curves $c_i(\lambda)$ can be chosen independently of the standard observer curves.

5.2 SENSOR CALIBRATION

In most practical cases, there is very little explicit information available about the image sensor, so that it has to be calibrated experimentally. A typical approach towards such calibration consists of the following steps:

1 First, one has to decide on a structure for the calibration model. We adopt the model of Figure 4.16, where sensor output C_i is transformed non-linearly by the function $\tilde{f}_i$ to $\tilde{C}_i = \tilde{f}_i(C_i)$, before being applied to a polynomial regression model C_m. In case the sensor sensitivity curves are linear combinations of the standard observer curves, we are guaranteed that the mapping of Figure 4.16, with C_m equal to a linear transformation, can indeed accomplish the required calibration. If this condition is not satisfied, or if it cannot be verified, one can resort to higher-order regression models C_m in order to reduce the prediction errors to within specified bounds.

2 Second, standard color targets, with known colorimetric coordinates, are needed to serve as references for the calibration model. The sensor outputs (C_{1j}, C_{2j}, C_{3j}) for these color targets can then be used as the inputs into the nonlinear regression model, while the corresponding known CIELAB color coordinates (L_j^*, a_j^*, b_j^*), for $j = 1, \ldots, N$,

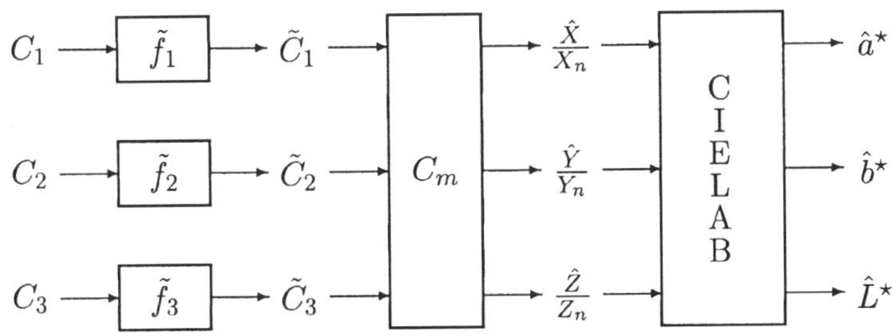

Figure 4.16. Mapping from sensor outputs (C_1, C_2, C_3) to calibrated color coordinates.

are the corresponding outputs that need to be approximated by the predicted values $(\hat{L}_j^*, \hat{a}_j^*, \hat{b}_j^*)$ from the model.

Since regression models use a minimum-mean-square-error criterium, CIELAB is the most obvious color space in which to perform this regression [Har99]. Either the 1976 or 1994 color-difference formula can be used to compare the actual to the predicted CIELAB coordinates.

There are a number of possible strategies towards obtaining color targets with known color coordinates:

1 If the spectral distribution of the light source and the spectral reflectance of a flat target are known, then the radiance can be expressed as in equation (4.54), and the corresponding color coordinates can be calculated. Experimental studies have moreover shown that the reflectance spectra of many real objects, as well as the spectra of different phases of daylight are well approximated by the sum of a limited number of 'basis' spectra [Wan87, Mal99]. If a mapping linearizes the sensor for these 'basis' spectra, then it will also be linearized for all linear combinations of them.

2 A second possibility is to measure the color coordinates of a number of test samples with a calibrated **colorimeter**. This has the additional advantage that it can be done in an illumination condition that is typical for the application. If a colorimeter is not available, but a calibrated image display (see section 2) is, then we can generate colors with controlled coordinates on the display and interactively match the test samples to displayed colors (in which case the human observer acts as colorimeter).

3 The easiest and most frequently used approach is however to present standard color charts, such as the Macbeth Color Checker [MW92, Wan95], as input to the sensor. The IT8.7/2 color transmission [ANS93a] and reflection [ANS93b] targets, which are being produced by several manufacturers, are currently the most frequently used color targets. The color coordinates of the different color patches in such a color chart are specified for a standard illuminant, such as D_{50} or D_{65}. The IT8.7 color targets, which have been standardized by ANSI and ISO, are designed to provide a fairly uniform coverage of CIELAB space (using 3 lightness values, 12 hue angles and 3 chroma values). Next to this set of 108 standard color patches, the color targets also contain characteristics of the actual dyes used in the production process of the color charts (in the form of 12-step scales for the basic colors cyan, magenta, yellow, neutral, red, green and blue and a 22-step neutral scale, all with approximately constant-lightness increments).

We have experimented with two kinds of nonlinearities in the above calibration procedure. The nonlinearities are assumed to preserve the range of sensor values, i.e., $\tilde{f}_i(0) = 0$ and $\tilde{f}_i(C_{max}) = C_{max}$, for $i = 1, 2, 3$. The first alternative is to use gamma characteristics satisfying

$$\tilde{f}_i(C) = c \cdot \left[(C^s + C_{io}^s)^{\gamma_i/s} - C_{io}^{\gamma_i} \right], \tag{4.59}$$

for $C \in [0, C_{max}]$ and s fixed (usually $s = 1$). The second alternative is to use monotonically increasing spline functions that are specified by their derivative value $\tilde{f}_i'(C_j^{(s)})$ at a number of **knot points** $C_j^{(s)}$, for $j = 0, \ldots, p$, in the interval $[0, C_{max}]$ of possible sensor outputs (see section 6.3 of chapter 5 for a more detailed description on monotonic spline functions). The monotonically increasing spline functions consist of piecewise quadratic functions and have specified continuous derivatives at the knot points. Logarithmically spaced knot points, i.e., $C_0^{(s)} = 0$ and $C_j^{(s)} = C_{max}/2^{p-j}$, for $j = 1, \ldots, p$, are used in the sensor calibration. The gamma characteristic has two optimization parameters C_{io} and γ_i, while the spline characteristic has p optimization parameters.

The polynomial regression model that is applied to the transformed color values $\tilde{C}_i = \tilde{f}_i(C_i)$ is either first order, i.e.,

$$\frac{\hat{X}}{X_n} = \sum_i^3 c_{1x}(i) \cdot \tilde{C}_i, \tag{4.60}$$

Table 4.2. Statistics of the calibration errors for an Agfa Arcus II scanner with an AGFA IT8.7/2 color chart. The average errors (ΔE_a^*, $|\Delta L^*|_a$, $|\Delta a^*|_a$ and $|\Delta b^*|_a$) and the maximum errors (ΔE_m^*, $|\Delta L^*|_m$, $|\Delta a^*|_m$ and $|\Delta b^*|_m$) are specified for both the overall color difference and the individual color coordinates. The methods C-n and L-n refer to an n-th order polynomial regression from RGB to XYZ and LAB, respectively. The original results from the Hardeberg study [Har99] are denoted by H-n.

	ΔE_a^*	ΔE_m^*	$\|\Delta L^*\|_a$	$\|\Delta L^*\|_m$	$\|\Delta a^*\|_a$	$\|\Delta a^*\|_m$	$\|\Delta b^*\|_a$	$\|\Delta b^*\|_m$
C-1	3.04	16.40	1.34	4.92	1.91	14.41	1.24	9.72
C-2	1.40	8.28	0.63	2.47	0.71	6.05	0.70	5.48
C-3	0.85	3.36	0.36	2.08	0.46	2.70	0.46	2.25
L-1	4.21	25.98	1.64	8.88	2.25	24.29	2.17	11.77
L-2	1.95	10.17	1.11	6.79	1.04	9.98	0.70	2.81
L-3	1.09	5.41	0.54	4.64	0.61	4.13	0.44	2.17
H-1	5.46	18.39	2.35	11.75	2.51	17.90	3.64	12.08
H-2	1.50	12.45	0.35	2.35	1.17	12.35	0.58	3.31
H-3	0.92	4.67	0.29	2.07	0.62	4.59	0.43	2.79

second order, i.e.,

$$\frac{\hat{X}}{X_n} = \sum_{i}^{3} c_{1x}(i) \cdot \tilde{C}_i + \sum_{i \leq j}^{3} c_{2x}(i,j) \cdot \tilde{C}_i \, \tilde{C}_j, \tag{4.61}$$

or third order

$$\frac{\hat{X}}{X_n} = \sum_{i}^{3} c_{1x}(i) \cdot \tilde{C}_i + \sum_{i \leq j}^{3} c_{2x}(i,j) \cdot \tilde{C}_i \, \tilde{C}_j + \sum_{i \leq j \leq k}^{3} c_{3x}(i,j,k) \cdot \tilde{C}_i \, \tilde{C}_j \, \tilde{C}_k, \tag{4.62}$$

with 3, 9 and 19 coefficients, respectively. Given similar equations for the predicted tristimulus values $\hat{Y}/Y_n$ and $\hat{Z}/Z_n$, we obtain that the polynomial regression model, denoted by C_m in Figure 4.16, contains 9, 27 or 57 coefficients in case of polynomial order 1, 2 or 3, respectively.

This calibration model was tested against an available data set (see [Har99], appendix C). The data were obtained by scanning an Agfa IT8.7/2 color chart containing $N = 288$ color patches with an Agfa Arcus II flatbed scanner. The nominal CIELAB values were provided by the manufacturer and the 1976 CIELAB color-difference formula was used in the non-linear optimization. The nonlinearities were restricted to gamma characteristics (with $s = 1$). The first three rows of Table 4.2 give the error statistics for polynomial regressions between transformed scanner outputs and normalized tristimulus values $(X/X_n, Y/Y_n, Z/Z_n)$, as described above. The polynomial orders are $n = 1$, $n = 2$ and $n = 3$, respectively. The corresponding error statistics for the same data set using the calibration method described by Hardeberg [Har99] are given in the last three rows of Table 4.2. This latter method differs from the presented method in three aspects. First, the nonlinearities in the Hardeberg model are not optimized over the complete data set, but only over a subset of the data (i.e., using only the 22-step neutral scale). Second, a polynomial expression is used to map the nonlinearly transformed sensor outputs directly to (L^*, a^*, b^*) rather than to the intermediate normalized tristimulus values $(X/X_n, Y/Y_n, Z/Z_n)$. Third, a constant offset is allowed in the polynomial expressions. In order to determine the effect of mapping to LAB rather than to XYZ, we also present the error statistics in case of polynomial regression between transformed sensor outputs and (L^*, a^*, b^*) coordinates in case the nonlinearities are optimized over the complete data set. These results are summarized in the middle rows of Table 4.2. The original assumption that regression towards normalized tristimulus values is to be preferred over regression directly to CIELAB is confirmed for the example data set.

Only the calibration model for polynomial order $n = 1$ is easy to interpret. For the example data set, we obtained

$$\begin{bmatrix} \hat{X}/X_n \\ \hat{Y}/Y_n \\ \hat{Z}/Z_n \end{bmatrix} = \begin{bmatrix} 0.508 & 0.293 & 0.032 \\ 0.201 & 0.662 & -0.034 \\ 0.016 & -0.071 & 0.903 \end{bmatrix} \cdot \begin{bmatrix} \tilde{f}_1(C_1) \\ \tilde{f}_2(C_2) \\ \tilde{f}_3(C_3) \end{bmatrix}. \tag{4.63}$$

The 3×3 matrix that arises after multiplying the three rows of the above matrix with the respective tristimulus values (X_n, Y_n, Z_n) of the neutral reference is an estimate of the inverse of the sensor matrix. The estimated parameters of the gamma characteristics ($\gamma_1 = 2.86$, $C_{1o} = 4.96$, $\gamma_2 = 2.69$, $C_{2o} = 1.04$ and $\gamma_3 = 2.83$, $C_{3o} = 0$) also agreed with the fact that the nominal setting of the scanner was $\gamma = 1/3$.

6. REPRODUCING REPRODUCTIONS

In this section, we discuss the important practical case that an image that is produced on one display device needs to be reproduced accurately on a second display device. It is shown that the sensor that is required to capture such a displayed image need no longer satisfy the restrictive Luther-Ives condition that had to be imposed in the previous section in order to guarantee a perfect color reproduction for arbitrary input radiances. Moreover, it is possible to calibrate such a sensor in order to directly measure the color values for a specific display device (such as the photographic film of a specific film manufacturer).

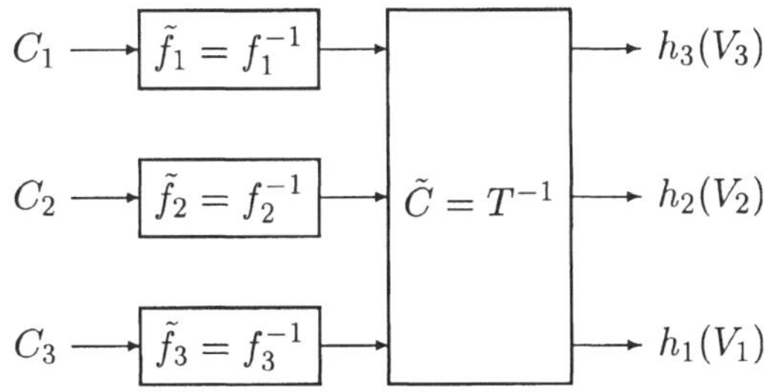

Figure 4.17. Mapping used for estimating display primary weights $(h_1(V_1), h_2(V_2), h_3(V_3))$ from sensor outputs (C_1, C_2, C_3).

Consider an image that is generated on the additive color reproduction device of Figure 4.1, i.e., the output radiance can be described by equation (4.1). Applying this radiance as an input to the image sensor of Figure 4.14 results in the sensor outputs

$$C_i = f_i \left[\sum_{j=1}^{3} h_j(V_j) < c_i(\lambda), p_j(\lambda) > \right] \tag{4.64}$$

for $i = 1, 2, 3$. The matrix T with elements $T_{ij} =< c_i(\lambda), p_j(\lambda) >$, for $i, j = 1, 2, 3$, is determined by the known spectral distributions of the sensor detectors $c_i(\lambda)$ and the display primaries $p_j(\lambda)$. If this matrix T is invertible, then the primary weights $h_j(V_j)$ can be estimated from the sensor outputs. These primary weights, together with the color coordinates of the primaries, determine the color coordinates of the displayed color. The calibration procedure of the previous section can easily be adapted to invert the sensor functions f_i, for $i = 1, 2, 3$, and the transformation matrix T. Only a limited set of displayed colors with known primary weights are needed as input for such a calibration procedure. As shown in Figure 4.17, the sensor calibration can be set up such that the weights $(h_1(V_i), h_2(V_2), h_3(V_3))$ on the display primaries are estimated directly from the sensor outputs (C_1, C_2, C_3).

In the case of multiplicative color reproduction methods, it is also possible to measure the color values of the original display. For instance, in the case of photographic transparencies, the density is equal to

$$\log T(\lambda) = \sum_{i=1}^{3} C_i \cdot \log p_i(\lambda) + \log T_o(\lambda), \tag{4.65}$$

for all wavelengths λ. The medium and dye characteristics $T_o(\lambda)$ and $p_i(\lambda)$, for $i = 1, 2, 3$, are assumed to be known. Hence, by measuring this density within (at least) three narrow spectral bands, for which $p_i(\lambda)$ can be assumed constant, the concentrations C_i can be derived by solving a set of linear equations. By proper selection of the test wavelengths λ_1, λ_2 and λ_3, the matrix of coefficients $\log p_i(\lambda_j)$, for $i, j = 1, 2, 3$, may even be diagonalized. In a similar way, the area parameters of a halftone reproduction can be derived from reflectance measurements within narrow spectral bands. Hence, in both these multiplicative color reproduction methods, an image sensor with three narrowband spectral sensitivity curves allows to derive the image production parameters. A sensor with such characteristics is called a **densitometer**.

7. COLOR REPRODUCTION AND IMAGE QUALITY

In the preceding discussion, it was implicitly assumed that a color reproduction system should aim for perfect color reproduction. However, in many cases it will be the subjective preference of the observer, rather then the physical reproduction error, that determines if a color reproduction is adequate or optimal. Clearly, in case the original image is not available as a reference, this judgement of the observer is the main determinant. However, even in cases where a perfect reproduction is available, an alternative reproduction may be preferred. Several experiments with variations in gamma [RBR91, Rou92, RKT94], chroma [Rid96b, YRFB97, FRB97, FRYB97, Rid97] and hue [FBR97, RBF97], have demonstrated that a linear reproduction is often not the preferred one. Computational theories have been devised in order to model this human preference for image reproductions that deviate (slightly) from the physically accurate reproduction. One of these theories was specifically developed for color vision [Yen98, YBR99a, YBR99b], while another one aims at a more general understanding of the phenomenon [Jan99, JB97, JB00b, JB00a]. The first theory has also been applied to optimize the color reproduction of natural scenes [YBR98c, YBR98b, YBR98a]. We will not go into these computational theories here, but will restrict ourselves to summarizing some of the experimental evidence. We refer to the above references for a more in-depth discussion.

7.1 GAMMA VARIATIONS

It was shown that many image processing chains can be approximately characterized by an exponential relationship between input- and output-luminance, i.e., $L_{out} \approx L_{in}^{\gamma}$. An obvious question to ask is hence whether there exists an optimal value for γ.

It was already demonstrated more than 30 years ago, in two independent studies [Nov69, Mar72], that the optimal relationship between input and output luminance depends on the surround light level. In the experiment by De Marsch, for example, a life scene was captured by a camera and displayed on a monitor. The task of the subjects was to adjust the "brightness" and "contrast" knobs on the monitor for optimal quality. This task was performed for three different surround light levels in the viewing room that were characterized as bright, dim and dark. The averaged optimal characteristics could be approximated by gamma characteristics with exponent $\gamma_{bright} = 1$, $\gamma_{dim} = 1.2$ and $\gamma_{dark} = 1.5$, respectively. Hence, the optimum value of γ depended on the surround light level, and exceeded one in case of a dark or dim surround.

The effect of a systematic variation of the exponent γ on the image quality has also been studied. These studies not only reveal the optimum value for γ, but also give an impression of how the quality impression varies as a function of γ. An underlying assumption is that such impressions can indeed be measured in a reliable way. It will be discussed in detail in chapter 5 how this can be accomplished by means of psychophysical measurement techniques.

We present some sample results from the "gamma" experiments described by Roufs et al. [RKT94, RKT97]. A slide scanner was used to convert eight color slides into digital images. The three color bands (V_r, V_g, V_b) contained 512 by 512 pixels and 8 bits per pixel. Look-up-tables were applied to the scanner outputs in order to calibrate all three color channels towards a common gamma-characteristic with exponent $\gamma_{scan} = 0.4$. The color values (V_r, V_g, V_b) were subsequently mapped to a luma component V_Y and two color components, according to the PAL standard. The luma component V_Y was mapped by a gamma characteristic $V_{Y,out} = g(\gamma) \cdot V_{Y,in}^{\gamma}$ with the following values for the exponent γ: 0.4, 0.7, 1.0, 1.3, 1.6, 1.9 and 2.2. The gain factor $g(\gamma)$ in the gamma characteristic was adjusted such that the average luminance of the displayed images was kept constant within each scene. The required gain factor could be determined for each image and value of γ, because the images were displayed on a calibrated color monitor with a known gamma characteristic with exponent $\gamma_{display} = 2.5$. Note that $\gamma = 1$ corresponded to a linear relationship between input and output luminances, since $\gamma_{scan} \cdot \gamma_{display} = 1$.

In a first experiment with six subjects, the color components were set equal to zero, such that the displayed images were monochrome. The experimental results, averaged across subjects, are plotted in Figure 4.18. The plot on the left in Figure 4.18 confirms that, within the dim viewing conditions used in the experiment, a value of $\gamma = 1.2$ is preferred for most

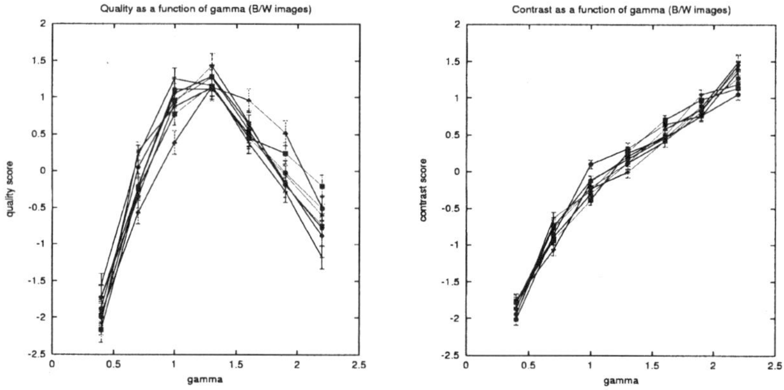

Figure 4.18. Perceived image quality (left) and contrast (right) as a function of the exponent γ of the luminance characteristic for eight monochrome images.

scenes. In order to verify that the parameter γ indeed correlates well with the impression of contrast, this contrast impression was also rated by subjects. The plot on the right in Figure 4.18 shows this perceived contrast as a function of gamma. In order to give an impression of the visual effect of the applied variation in γ, we have included some example images in Figure 4.19. Since we obviously do not have any control over the printing process used in this book, these images can only be seen as approximations to the images actually displayed in the experiment.

In a second experiment, the perceived quality was determined for colored images. The modified luma components were combined with the original color components for each scene in order to create colored images. Only four out of the previous eight scenes were used in this experiment. Four of the six subjects that took part in the first experiment also participated in this second experiment. The perceived image quality, averaged across subjects, is plotted in Figure 4.20 as a function of γ. For these colored images, a value of $\gamma = 1.2$ is preferred for all scenes.

Note that the above experiments optimize the imaging chain from color slide to displayed image. Although some efforts were undertaken to also estimate the luminance characteristic of the photographic process, i.e., the mapping from the original luminances in the scene to the optical densities in the slides, this could only be done approximately. The available measurements indicated that these photographic characteristics could be described by a gamma characteristic with an exponent

Figure 4.19. Images of the scene "Terrace" with constant average luminance and increasing values of γ, i.e., $\gamma = 0,7$ (left), $\gamma = 1.3$ (middle) and $\gamma = 2.2$ (right). The display-characteristic was assumed to have $\gamma_{display} = 2.5$ within the luminance range $[0.2, 60]$ cd/m^2.

close to one for the low densities (or high luminances), which changed into a gamma characteristic with an exponent around 1.5 for higher densities. This increased value of gamma at low luminances is not an artefact but an intended feature of the slide film, and compensates to some degree for the light adaptation that occurs in the visual system (as was discussed in section 4.3). Most film characteristics also saturated at very high densities (or low luminances). Since the gamma of the photographic film was however equal to or larger than one in the operating range of the film, the above conclusion that the optimal gamma exceeds one also remains valid for the complete imaging chain.

7.2 QUALITY: CONTRAST VERSUS NATURALNESS

Additional experiments [Jan99] have been conducted in order to better understand this preference of observers for images reproduced with a non-linear luminance characteristic. In order to obtain such insight, subjects were asked to not only rate the quality impression conveyed by the processed images, but also to rate some relevant attributes such as contrast and naturalness. The hypothesis was that a nonlinear lumi-

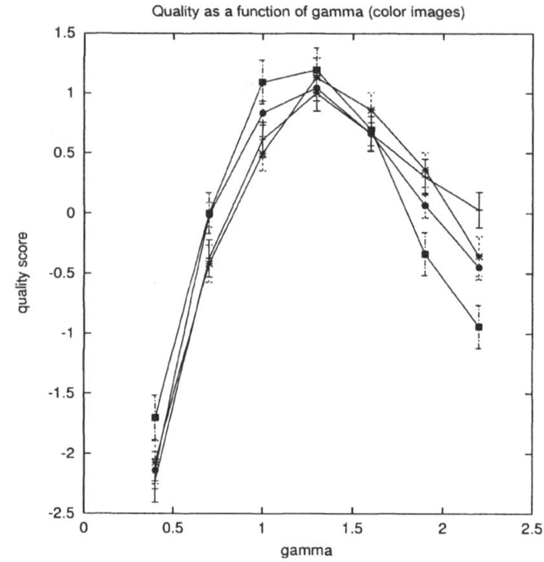

Figure 4.20. Perceived image quality as a function of the exponent γ of the luminance characteristic for four color images.

nance characteristic creates slightly unnatural colors, but at the same time increases the contrast between adjacent areas in the image, and that the overall quality impression is a compromise between these conflicting demands [JB00b, JB00a].

Two experiments were conducted with two distinct sets of processed color images. The original images of the four scenes used in the experiments were taken from a Kodak Photo CD and are shown in Figure 4.21. The spatial resolution of the images was 768 by 512 pixels, while each color was represented by an 8-bit integer value. The images were displayed on a calibrated color monitor with known color coordinates for the primaries and white point, and a gamma-characteristic with exponent $\gamma_{display} = 2.5$. By means of this calibrated monitor characteristic, the color values could be mapped into colorimetric coordinates such as the CIE-(L^*, u^*, v^*) coordinates [CIE79]. Consequently, all images were processed in the CIELUV domain. The processed color coordinates were passed through the inverse of the monitor characteristic in order to convert them back into color values that were suited for display.

The fourteen stimuli that were used in the first experiment were created in a similar way as was used in earlier experiments with colored

Figure 4.21. Original images used in the experiments addressing contrast, naturalness and quality (CNQ-16: upper left, CNQ-17: upper right, CNQ-19: lower left and CNQ-33: lower right).

images [FRB97, YRFB97]. For each scene, seven images were created by varying the color temperature of the reference white between 4650 and 10300 degrees Kelvin, 6500K corresponding to the original image. The seven color temperatures (4650K, 5150K, 5800K, 6500K, 7400K, 8650K and 10300K) were chosen such that the perceptual differences between successive cases were approximately constant. This processing was implemented using a translation of the color coordinates (u^*, v^*). Seven additional stimuli were created per scene by scaling the color coordinates (u^*, v^*) by one of the factors 0.50, 0.63, 0.79, 1.00, 1.26, 1.59 or 2.00. The psychometric lightness L^* remained constant in all these processed images.

The nine stimuli for the second experiment, on the other hand, were obtained by keeping the color coordinates (u^*, v^*) fixed, and by varying

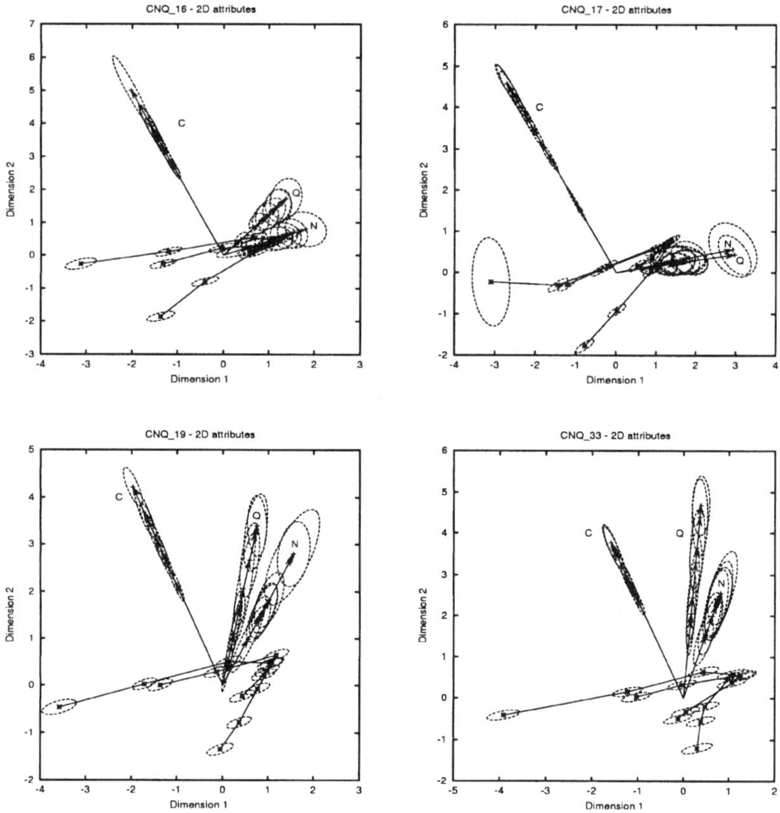

Figure 4.22. MDS models summarizing the naturalness (N), contrast (C) and quality (Q) impressions for color stimuli with varying psychometric color coordinates and fixed psychometric lightness.

the psychometric lightness by means of an S-shaped curve, i.e.,

$$L^*_{out} = \begin{cases} L^*_{min} + \left(\frac{L^*-L^*_{min}}{L^*_{ave}-L^*_{min}}\right)^{\gamma} (L^*_{ave} - L^*_{min}) & , L^*_{min} \leq L^* \leq L^*_{ave} \\ L^*_{max} - \left(\frac{L^*_{max}-L^*}{L^*_{max}-L^*_{ave}}\right)^{\gamma} (L^*_{max} - L^*_{ave}) & , L^*_{ave} \leq L^* \leq L^*_{max} \end{cases}, \tag{4.66}$$

where γ assumed one of the following values 0.25, 0.35, 0.50, 0.71, 1.00, 1.41, 2.00 or 2.82. The minimum, maximum and average value of the psychometric lightness in the original image are denoted by L^*_{min}, L^*_{max} and L^*_{ave}, respectively.

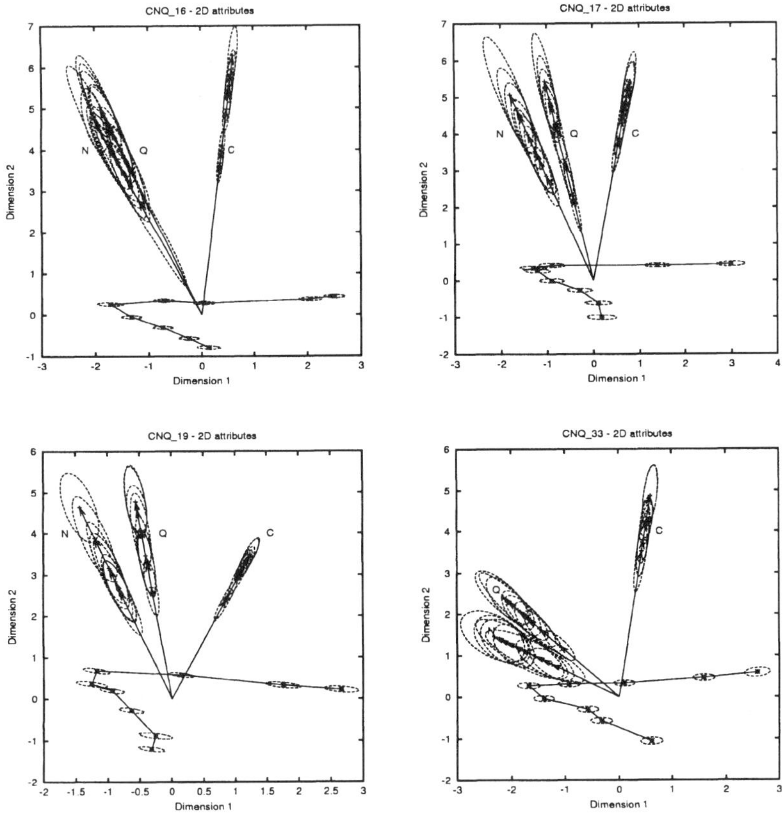

Figure 4.23. MDS models summarizing the naturalness (N), contrast (C) and quality (Q) impressions for color stimuli with varying psychometric lightness and fixed psychometric color coordinates.

In order to simultaneously visualize the different impressions or attributes (in casu, contrast, naturalness and quality) that are associated with the stimuli, we use the multidimensional scaling (MDS) approach of chapter 7. The results of such MDS analysis for the two separate experiments are presented in Figures 4.22 and 4.23, respectively. The data for the four different scenes were analyzed separately, so that there are four plots per experiment. Each plot contains a number of points, representing the (fourteen or nine) images used in the experiment. The orientations and lengths of the vectors are derived from the attribute judgements of the subjects. More precisely, they are derived by optimally correlating these attribute judgements with the projections of the stimulus points on the corresponding attribute axis (see Figure 5.8 in the next chapter for a more precise geometric interpretation). The "contrast" (C), "naturalness" (N) and "quality" (Q) vectors of all seven subjects were assumed to have the same orientation, since the corresponding sensations were assumed to be shared by all subjects. The lengths of the vectors varied across subjects, since not all subjects used the same range of numbers to express their judgements. The ellipses surrounding the end points of the vectors and the stimulus positions indicate the estimated 95 % confidence intervals for the corresponding model parameters.

The main conclusions from the experiments are as follows. Since the "quality" vectors lie in between the vectors for "naturalness" and "contrast", quality is indeed a compromise between optimal contrast and optimal naturalness. The quality vector is also more closely related to "naturalness" than to "contrast". With the exception of scene CNQ-17 in the first experiment and scene CNQ-16 in the second experiment, the subjects make a statistically significant distinction between naturalness and quality. They attribute the highest quality to images with a contrast that is higher than that of the original image, while still being aware of the fact that this is a (slightly) unnatural reproduction.

EXERCISES

1 A realistic monitor model should take into account that the luminances L_r, L_g and L_b of the three primaries are limited to finite ranges $[L_{r,min}, L_{r,max}]$, $[L_{g,min}, L_{g,max}]$ and $[L_{b,min}, L_{b,max}]$. Illustrate on the color diagram of Figure 3.5 how this practical condition limits the range of colors with a given luminance L that can be produced on a monitor. Assume a PAL monitor with $L_{r,min} = 0.25$, $L_{r,max} = 12.7$, $L_{g,min} = 0.8$, $L_{g,max} = 40.3$, $L_{b,min} = 0.08$ and $L_{b,max} = 4$ cd/m^2. Draw the gamut of obtainable colors with luminance $L = 11.5$ cd/m^2, as well as the color triangle spanned by the three primaries.

2 Derive that the angle between the chromaticity plane and the coordinate plane (E_U, E_V), spanned by the color signals of PAL, is indeed equal to 13.7 degrees.

3 Derive how the sensor calibration procedure of section 5 could be modified in order to be used for calibrating a multiplicative color display.

Chapter 5

PSYCHOPHYSICAL MEASUREMENT AND MODELLING OF IMAGE QUALITY

1. INTRODUCTION

Because of the large amount of imaging equipment being produced nowadays, there is a substantial economical interest in being able to measure and predict the effect of variations in the technical parameters of such systems on the resulting quality. Especially in the case of alternative systems with similar functionality, perceived (image) quality is one of the major discriminating factors between products from the point of view of the user. In this chapter, we will concentrate on how such quality impressions can be measured with the help of human subjects. In later chapters, more specifically, chapters 11 and 12, we will focus on approaches to predict image quality. Readers that are more interested in the bottom-up approach to image technology design may decide to skip this chapter.

People are very used to making judgements about many aspects of the things they encounter, be it objects, situations or other people. Moreover, judgements made by different people often agree remarkably well when being compared explicitly. When judging the quality of images, for instance, people especially agree on the more perceptual aspects such as the brightness, contrast and colorfulness of the images. This agreement on perceptual aspects does however not prohibit that people can have conflicting opinions about more cognitively related aspects of the same images. One aspect on which people frequently disagree is the aesthetical quality of images. There may be many different, often personal, reasons why some images (photographs or motion pictures) are considered to better convey the essence of the subject being depicted and appeal more to some people than to others. In this chapter[1], **multidimensional models** are introduced and it is demonstrated how they can reflect both such common aspects and differences between individual judgements. It should be noted that, although multidimensional models are introduced here within the context of image quality, they can also be applied to many other fields where understanding and visualizing human judgements are important.

Image technology adopts a restricted perspective on image quality. The input images are selected a priori, and the interest is in evaluating the effect of alternative image coding, processing and/or display systems on the perceived quality of these images. Even within this limited scope, one can still distinguish many different aspects, or **attributes**, to the **perceived image quality** [BHN+96]. Image attributes such as the perceived sharpness, contrast, colorfulness, naturalness of colors, etc., may all be influenced in different ways by technical systems. Alternative systems may also introduce different distortions, i.e., patterns such as noise, sampling and quantization artifacts that do not belong to the scene but are created by the imaging system. People often agree very closely in their judgements about the strengths of these image quality attributes. The importance that they assign to these individual attributes in reaching their overall judgement on the image quality may however differ.

The idea that the attribute judgements by different subjects are often very related is referred to as the **principle of homogeneity of perception** [GCJS89] and constitutes an important motivation for the **multidimensional scaling** (MDS) approach [GCJS89, YH87, CC94, Mar02] towards describing image quality that is presented in this chapter. This principle has led to the assumption that a single multidimensional configuration representing the alternative images underlies the attribute and quality judgements by all subjects. Differences in attribute judgements of subjects can be accommodated in MDS by the fact that the mapping from the joint multi-dimensional stimulus configuration to the one-dimensional attribute judgements may vary per attribute and per subject. Constructing the underlying stimulus configuration from experimental judgements by subjects and/or predicting it on the basis of instrumental (i.e., objective) measurements on the images hence becomes a major step in image-quality model building [MM01].

An **instrumental quality measure** is defined here as an algorithm for predicting the stimulus configuration that underlies human judge-

[1]This chapter is based on two earlier publications [MB01, Mar02].

ments, as well as the mapping from this stimulus configuration to quality (and attribute) judgements. The usefulness of such a measure obviously depends on how well the resulting quality predictions correlate with human judgements. Instrumental models can not only make the design of new systems more efficient, but can also be used to monitor the performance of existing systems. An advantage of multi-dimensional models [MK96] over one-dimensional quality models [EB98] in such a monitoring application is that multi-dimensional models can also provide insight into which attributes of image quality are failing when the overall quality is insufficient.

Multidimensional image-quality models are first illustrated by means of an example in section 2. Next, in section 3, some of the concepts and terminology that are needed when discussing the psychophysical measurement of image quality are summarized. Amongst others, the important distinctions between **metric** and **non-metric** scales and between **continuous** and **discrete** scales is introduced. The most frequently used experimental paradigms for measuring image quality, i.e., **single-stimulus attribute scaling**, **double-stimulus difference** (or **preference) scaling**, and **double-stimulus dissimilarity scaling**, are discussed in section 4. Next, a brief historical overview of the application of MDS in image quality measurement and modelling is presented in section 5. In section 6, we show how multidimensional model parameters can be estimated from experimental measurements for single images and/or image pairs, provided these measurements are continuous (i.e., not discrete). The program XGms is introduced as an important tool for interactively exploring alternative metric and non-metric models for continuous data obtained from one or more subjects, using any combination of the three above-mentioned experimental paradigms. In section 7, some example data sets are analyzed with XGms. Next, in section 8, the modelling of discrete data is addressed. Finally, discrete data analysis using XGms is illustrated by means of an example in section 9.

2. EXAMPLE OF MULTIDIMENSIONAL MODELLING

In order to make the discussion on multidimensional modelling more concrete, we first present an example. More specifically, we describe the multidimensional modelling of experimental data concerning the quality of images degraded by noise and blur [KM96c]. Experiments were performed for the three different scenes shown in Figure 5.1, but only the results for image Wanda will be discussed here. A detailed analysis of the complete data set is postponed until section 7.

Figure 5.1. Scenes used in experiments with blur and noise: Wanda (left), Terrace (middle) and Mondrian (right).

All combinations of four levels of blur with a binomial kernel and four levels of Gaussian additive white noise were applied to create 16 different images per scene. The blur was characterized by the standard deviation σ_b of the filter kernel (σ_b values of 0, $\sqrt{2}$, 2 and $2\sqrt{2}$ times the sampling distance), so that the perceived blur increased with σ_b. The noise was characterized by the standard deviation σ_n of the Gaussian probability density function (σ_n values of 0, 7, 10 and 14 for grey values in the range [0,255]). The noise was added after the images were blurred, and the processed images were quantized to 8 bits/pixel. All pictures contained 512×512 pixels, but only a subregion of 240×470 pixels was viewed in the experiments. This restricted region was needed in order to allow simultaneous display of two images. Some details of processed images for the Wanda scene are shown in Figure 5.2

The images were converted to 50-Hz non-interlace video and displayed on a CCID-7351B high-resolution monitor. The grey-value-to-luminance characteristic of the monitor was measured and the calibration procedure of chapter 4 was used to determine a look-up table for the grey values. After calibration, the relationship between the grey value g and the luminance L for the combined chain (lookup table, digital-to-analog

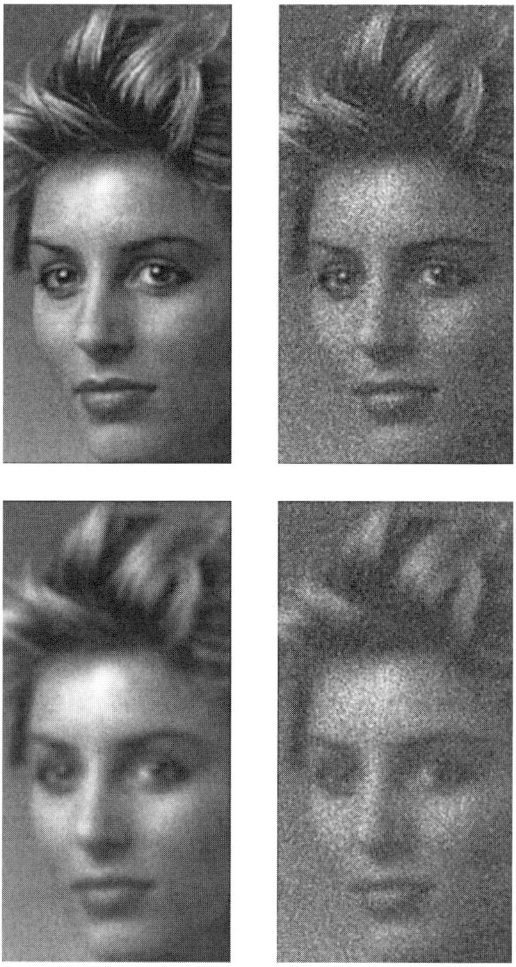

Figure 5.2. Details of processed images of scene Wanda used in experiments with blur and noise: original (upper left), only noise $(\sigma_n, \sigma_b) = (14, 0)$ (upper right), only blur $(\sigma_n, \sigma_b) = (0, 2\sqrt{2})$ (lower left) and both noise and blur $(\sigma_n, \sigma_b) = (14, 2\sqrt{2})$ (lower right).

video convertor, and monitor) became

$$L = \max\left[L_{min}, L_{max}(g/g_{max})^{\gamma}\right], \tag{5.1}$$

with $g_{max} = 255$, $L_{max} = 60$ cd/m^2, $L_{min} = 0.2$ cd/m^2, and $\gamma = 2.5$. This calibrated characteristic was verified by a second luminance measurement.

Most viewing conditions satisfied BT.500 recommendation [IR98] from the ITU-R (International Telecommunication Union - Radiocommunication). The viewing distance was 1.5 meters, which was equivalent to six times the height of the monitor. Between two successive stimuli, a uniform adaptation field was displayed during the time it took subjects to enter their response by means of a keyboard. The minimum duration of this adaptation field was 2 seconds, while its luminance of 9 cd/m^2 was approximately equal to the average luminance of the images.

Five subjects, aged between 25 and 39 years and with normal or corrected-to-normal visual acuity between 1.5 and 2, measured on a Landolt chart[2] at a distance of 5 m, participated in a double-stimulus dissimilarity experiment. All $15 \times 16/2 = 120$ different image pairs were scored once by each subject, using numbers from 0 to 10.

The same five subjects, plus two additional ones, also took part in three single-stimulus scaling experiments with the same images. In three separate experiments, subjects rated perceived noisiness, blur and quality of individual images. All images were presented 4 times, in random order, to each of the subjects (resulting in $16 \times 4 \times 3 = 192$ attribute scores per subject). Integer scores between 0 and 10 were used by the subjects to express their judgements.

The XGms program that we will introduce in detail in later sections was used to derive multidimensional models from these data. The most concise picture representing the model output for one scene (i.e., Wanda) is shown in Figure 5.3. The 16 processed images of this scene are represented by the crosses. The image in the upper left is the original, the image in the lower left contains no noise (only blur), the image in the upper right contains no blur (only noise), while the image in the lower right contains the maximum amount of blur and noise. The horizontal and vertical axes are not labelled since they don't have any physical meaning. This is a consequence of the fact that the figure can be arbitrarily translated, rotated and scaled without influencing its interpretation.

The fact that the 16 images are not arranged in a rectangle, as might be expected a priori from the independent processing of blur and noise, indicates that there are some interactions between perceived noise and perceived blur. The directions that correspond to perceived blur and perceived noise are indicated by two of the arrows. The interpretation

[2]Visual acuities of 1 or 2 on a Landolt chart imply that a subject can detect the opening in a C-ring when this opening is 1 arcmin or 0.5 arcmin of visual angle wide, respectively.

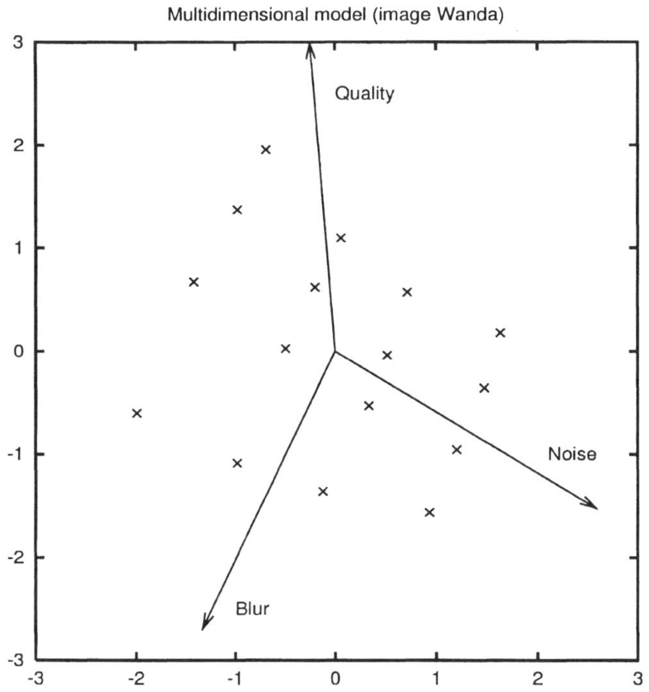

Figure 5.3. Example of a multidimensional model for a scene degraded by noise and blur. The crosses indicate the 16 processed images, while the arrows indicate the directions in which perceived blur, noise and quality increase.

of the noise axis is for instance that the orthogonal projections of two image points onto this axis indicate (on average) which of the two images is perceived to be more noisy. As expected, images with the same noise standard deviation are perceived to be (approximately) equally noisy. There is no such one-to-one relationship between the standard deviation of the filter kernel and perceived blur. More specifically, it can be derived from the model in Figure 5.3 that perceived blur increases with noise for an unblurred image, while it decreases with noise for the most heavily blurred image. The direction corresponding to overall quality is indicated by the third axis. Obviously, since both perceived noise and perceived blur degrade the overall quality, this direction of the quality vector is in between (and opposite to) the directions of the attribute vectors for noise and blur. Blur is relatively more important than noise, since the angle between the blur and the impairment direction (which is

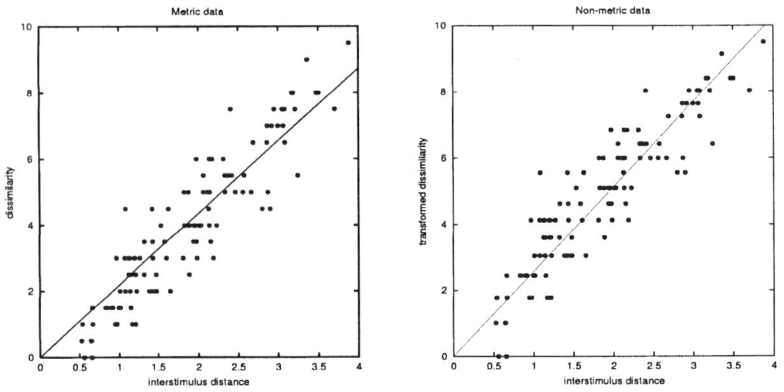

Figure 5.4. Original dissimilarity scores (left) and transformed dissimilarity scores (right) as a function of interstimulus distances.

opposite to the quality direction) is smaller than the angle between the noise and the impairment direction. The model depicted in Figure 5.3 assumes that the attribute directions are the same for all subjects. The output of an alternative model which allows for a different attribute vector per subject would look similar to Figure 5.3, but would contain one vector per attribute and per subject.

A scatter-diagram showing the relationship between the distances d_{ij} between stimulus points in Figure 5.3 and scored dissimilarities D_{ij} for one subject is shown in Figure 5.4. In the left panel, the dissimilarity data is assumed to be metric, i.e., d_{ij} is compared against D_{ij}. In the right panel, the dissimilarity data is non-metric, i.e., d_{ij} is compared against transformed dissimilarities $T(D_{ij})$. In the example, the transformation is assumed to be of the form $T(D) \propto D^q$, and the power q is determined as part of the model optimization (in the specific example $q = 0.68$). The linear regression coefficients between experimental dissimilarity data and interstimulus distances are $R = 0.982$ and $R = 0.987$ for the metric and non-metric case, respectively. These regression coefficients indicate the percentage of variation in the experimental dissimilarity data that is explained by the interstimulus distances. Obviously, the interstimulus distances correlate highly with the dissimilarity judgements.

In the previous discussion we used terms like double-stimulus dissimilarity scaling and single-stimulus attribute scaling, metric and non-metric data, etc, without providing definitions for them. In the next sections, we therefore include a short introduction to the field of psychophysics from which all these terms were borrowed.

3. PSYCHOPHYSICAL MEASUREMENT

3.1 PSYCHOPHYSICS

What is implied by the statement that we have measured subjective image quality or one of its underlying attributes? Especially in an engineering community, where (objective) measurement and (subjective) judgement are typically kept very separate, the above question may need some clarification. Therefore, let us examine in closer detail the key terms, in order to understand more precisely what is meant by *measurement* and *subjective* in the current context.

Measuring implicitly assumes that the sensation under consideration (i.e., quality) can be quantified in a consistent and meaningful way. Moreover, the process that is performed to obtain this quantification must be well-specified, and in accordance with well-accepted principles and practices. This means that the procedures that are used can be described in sufficient detail for someone else to repeat them, and that the results are not expected to change significantly in such a repeated experiment. Judging what is a significant change in the experimental results requires that the precision of the outcomes, i.e., the confidence regions on the measured sensation strengths, are also specified. Especially this latter aspect is often neglected, and we will therefore pay explicit attention to it later on.

The inclusion of the word subjective refers to the fact that subjects play an essential role in generating the responses to the presented stimuli. We have to be able to argue why subjects are sensitive to the stimulus attribute that we want to measure, and must determine how the responses that they give relate to the strengths of their attribute sensations.

Arguing why subjects are able to discriminate or rate stimuli on a given attribute, such as quality, is often very difficult, or even impossible, so that we usually postulate the existence of such a relevant mechanism. Thurstone, one of the most influential early scientists in the field of **psychophysics**, called such a mechanism a **discriminating process** [Thu27, Tor58]. It is assumed to be an inherent part of the perceptual and cognitive abilities of the subject, and may or may not be identical for all subjects. Repeated inter-subject and/or across-subject experiments with the same stimuli should at least give consistent results if this assumption is valid. For instance, for perceptual attributes such as brightness or loudness, we often assume the existence of a fairly peripheral mechanism that does not vary greatly across subjects. For such peripheral processes, attempts have been made to explicitly model the discriminating process based on known neural response mechanisms [EB98]. Knowledge about the physiology of the relevant senses, i.e, vision and hearing, often guides the construction of these models. For more cognitively influenced attributes, such as quality, we often do not have an explicit model for how the (psychological) discriminating process is derived from the presented (physical) stimuli. Obviously, much of the recent effort towards modelling image quality [EB98] is exactly aimed at generating a discriminating process that agrees with quality judgements by subjects. Especially for cognitive attributes, the relevant discriminating process may vary across subjects. Assumptions about the nature of the discriminating process have consequences for the measurement procedure, for instance, for deciding whether stimulus conditions have to be repeated within and/or across subjects.

How (external) responses given by subjects relate to (internal) sensations is more open to systematic analysis and one of the key topics of research in psychophysics. In the next subsection, we will discuss in more detail some of the most frequently used response models. Obviously, the adopted response model has a direct impact on how the subject responses need to be processed to arrive at estimates of the subjective sensations that are of interest.

The measurements described above will be referred to as **psychophysical measurements**, rather than subjective measurements, in the remainder. Psychophysics is concerned with adapting existing measurement procedures for physical entities (such as length, weight, etc) to measurement procedures for psychological sensations. It is not the intention of this chapter to offer an in-depth introduction to the field of psychophysics, but rather to provide some insight into the most important concepts and underlying models. This should help to understand why specific experimental procedures and their corresponding data analyses have been proposed for evaluating perceptual image quality and its attributes, as well as to better appreciate the limitations of such psychophysical measurements.

More detailed information on psychophysical models and procedures can for instance be found in the excellent tutorial chapter 'Measurement, Scaling, and Psychophysics' [LK88] in Stevens' Handbook of Experimental Psychology [AHLL88], as well as in the large number of references contained therein. A more theoretical treatment of how measurements can be constructed from basic axioms, and how these basic axioms give rise to different, but related, measurement procedures in physics and psychophysics, is contained in the chapter 'Psychophysical measurement and theory' [Fal86] in the Handbook of Perception and Human Performance [BKT86]. Both handbooks also contain introductory chapters on many other aspects of perception and cognition.

3.2 RESPONSE VARIABLES AND SCALES

In any psychophysical experiment, observers are required to map their sensations into responses. Usually, the valid responses are restricted by the experimenter. Obviously, imposing such response rules makes it easier to interpret results afterwards, as well as to compare results across different subjects and experiments.

A major distinction is usually made between continuous and discrete responses. In case of a **continuous response**, there is typically an upper and lower limit to the valid response values. One possible example is where subjects have to rate quality numerically between 0 and 100, but are allowed to use any numerical value in this range. Making a mark on a graphical scale [IR98] or positioning a physical slider [Rid96a] are alternative ways of generating such continuous responses. In all cases, the responses can be mapped to a finite interval of the real axis. The resulting positions represent the underlying sensations on a so-called **non-metric scale**, since equal distances on this scale do not necessarily correspond to equal differences in sensations. In general, there is a unknown nonlinear response function, which can be assumed to be monotonically increasing or decreasing, that relates the internal sensations to the continuous output scores.

Often, continuous responses are converted into discrete responses because of limitations in the accuracy of the measurement equipment. For instance, the position of a slider may determine an output voltage which is subsequently converted by an analog-to-digital convertor with a limited precision into an integer score. However, as long as the precision of the equipment is much higher than the precision with which subjects rate identical stimuli on repeated trials (i.e., the observer noise), these technical limitations are obviously inconsequential. Therefore, in the above case where subjects have to rate by means of a numerical value in the range [0,100], requesting integer responses is not expected to change the results in any substantial way.

The term **discrete response** is mostly reserved for cases in which observers have to choose from a fairly small number of discrete alternatives, called **categories**. Discrete responses for which the categories do not have a natural order are called **nominal**. Examples of nominal variables are gender (male, female), mode of transportation (bus, car, train, bicycle), etc. Although nominal response variables are often used in the social and biomedical sciences, we will not discuss them any further here since they are seldom used in the field of psychophysics.

Many categorical variables have ordered levels. Such variables are called **ordinal**. For example, the ITU-R proposes the use of the qualifications bad (1), poor (2), fair (3), good (4) and excellent (5) to designate increasing levels of quality [IR98, Teu96]. They also propose an alternative set of categories to denote decreasing amounts of image impairment, i.e, very annoying (1), annoying (2), slightly annoying (3), perceptible, but not annoying (4), and imperceptible (5). In a **category scaling** experiment on image quality, subjects are for instance required to score images in such categories.

Ordinal categories, even if they are labelled by adjectives, can be mapped to integer numbers. It is sufficient that the ordering of the categories according to the relevant sensation agrees with the smaller than (or larger than) relationship on the integer labels of the categories. Although the order of the categories is preserved by such a mapping, distances are not. The numerical distance between categories 1 and 2, and 4 and 5, for instance, is identical, while the differences in sensations are very likely unequal. For instance, for the case of the ITU-R quality categories, it has been argued that they do not divide the quality scale into intervals of equal sizes [Teu96]. We often regard ordinal variables as coarsely (and non-uniformly) quantized measures of a continuous sensation, such as quality, that is not itself observable. It is often reasonable to assume that the ordered categories correspond to non-overlapping and exhaustive intervals on a real axis that describes the amplitude of the non-observable discriminating process [AP81]. This idea will be made explicit in section 8.

In conclusion, for psychophysical experiments that use either a continuous response or a response by means of discrete ordinal categories, one can establish a mapping from responses to real or integer numbers, respectively. These numbers belong to a non-metric scale and correspond to sensations, such as image quality, that are not directly observable. The monotonic response function that relates internal sensations to external responses is usually unknown. This process of assigning numbers to sensations is usually referred to as **scaling** [You84]. **Psychometrics** is the subfield within psychophysics that is especially concerned with modelling scaling data.

An **interval** variable is a special case of either a non-metric continuous or a discrete ordinal variable for which the distances between the continuous scores or the category number, respectively, do have absolute meaning. Many of the physical measurement scales (for length, temperature, income, age) have this interval property, i.e., are **metric**. An interval scale that has a well-defined origin, such as in the case of length and the Kelvin temperature scale, is referred to as a **ratio** scale.

Most statistics, such as mean and variance, have been developed to capture important aspects of ratio (or interval) data. They should therefore not be applied to the majority of (non-metric) scaling data. One exception is the median, which is only based on the ordering of the categories or continuous responses. Since scientific inferences are most often based on statistics, it is important to realize that many of the most frequently used data analysis techniques (such as regression, analysis of variance, etc) can not be applied directly to non-metric scaling data. Although this may seem obvious, current practice shows differently. For example, the often used **mean opinion score** (MOS) arises by attributing the scores 1 to 5 to the categories bad, poor, fair, good, excellent, and subsequent averaging of these scores across repetitions and/or subjects. Fortunately, many non-metric data analyses techniques, implemented in freely available software programs [You84], have been developed to handle continuous non-metric scaling data. Also in the field of categorical data analysis [Agr90], much progress has been made to generalize many of the concepts from continuous data analysis. In the specific case of ordinal category response data, extensive theory, supported by procedures in popular statistical packages, is available [McC80, AP81, Agr89, Wic98]. We will also present a new interactive program XGms that can be used to analyze most response data of interest within a common framework.

3.3 NON-METRIC SCALES AND RESPONSE FUNCTIONS

As was discussed in the previous section, it is usually assumed in psychophysics that there is an unknown, monotonically increasing or decreasing, nonlinear function that relates the internal sensation strengths to the (non-metric) subject responses [Fal86]. For example, let us denote the quality of the images i used in an experiment by x_i. Fechner assumed that the probability that image i is preferred over image j in a two-alternative-forced-choice experiment can be expressed as follows

$$P(i > j) = R[x_i - x_j], \tag{5.2}$$

for some strictly increasing (and continuous) **response function** R [Fal86]. This model has a geometric interpretation in one dimension. The images i are represented by their coordinates x_i in one dimension and the differences between stimulus coordinates are monotonically related to the observed responses, in casu, the probabilities $P(i > j)$. The responses only belong to a metric scale if the function R can be assumed to be linear.

The observed subject responses can be used to estimate the response function R and the sensation strengths x_i. Usually, the robustness of these estimates is increased by putting some constraints on the function R that relates (differences in) sensation strengths to subject responses. It may for instance be chosen to belong to a parameterized family of monotonically increasing functions, such as the class of power functions.

Substituting $x_i^\star = \alpha \cdot x_i + \beta$ and $R^\star(s) = R(s/\alpha)$ in the above expression confirms that the scale x_i has the interval property, i.e., it is determined up to an arbitrary linear transformation, since

$$P(i > j) = R[x_i - x_j] = R^\star[x_i^\star - x_j^\star] \tag{5.3}$$

are two equivalent descriptions. Amongst others, this implies that psychophysical measurements cannot be used to obtain absolute quality judgements, but only relative judgements (i.e., one can only conclude that the quality difference between images i_1 and j_1 is greater or smaller than the quality difference between images i_2 and j_2).

4. PSYCHOPHYSICAL PROCEDURES

Within the limits of this chapter it is not possible to discuss in detail all available psychophysical procedures, let alone illustrate them by means of examples. Therefore, after giving a general overview, we will concentrate on some specific procedures that we have found most useful for measuring image quality and its attributes.

Luce and Krumhansl [LK88] make a distinction between **local** and **global** psychophysical procedures. In local psychophysical methods, the focus is on stimulus differences that are small enough to cause confusions. The goal of the experiment is to measure probabilities of detection or discrimination. Quality differences between stimuli are typically expressed in terms of number of **just noticeable differences (JNDs)**. The developments in this area have been very much influenced by the advent of detection theory [Tre68].

Global psychophysical measurements deal with a substantial part of a sensory dimension all at once. The differences between the stimuli in a global psychophysical experiment are sufficiently large, so that there is little or no confusion between stimuli, especially not for the most extreme stimuli in the range. The probability of detection or discrimination carries little information in this case, so that alternative techniques such as category scaling and magnitude estimation are needed to obtain substantial information.

Ideally, both approaches should give similar results for identical stimuli. More specifically, sensation strengths derived from global approaches (scaling) should be linearly related to JNDs from a local approach. Unfortunately, this is not necessarily the case. Although the differences will be limited in cases where the range of sensation strengths is not too large, it should be realized that results of psychophysical experiments can and will be influenced by the experimental procedure [GRK+97]. Therefore, in the end, we will have to choose which procedure we consider to capture at best the sensations that are of interest.

4.1 LOCAL PSYCHOPHYSICS

Local psychophysical methods aim at determining detection and discrimination thresholds. It is found empirically that the relation between a physical magnitude (or magnitude difference) and the probability of detection (or discrimination) is often described by a **psychometric function**, i.e., this probability increases continuously from 0 (or 0.5) to 1 as the amplitude (difference) increases. This behavior is usually accounted for by detection models [Tre68], which assume that the relevant discrimination process is distorted by internal noise.

Detection (and discrimination) experiments are often performed using a **forced choice** paradigm. In such a case, the subject is presented with two stimuli, either simultaneously or sequentially, and has to select the stimulus which gives rise to the largest sensation strength. In this case, the psychometric curve varies from 0.5 to 1, since there is a 50 % chance of guessing correctly. The stimuli presented to the observer can be selected a priori, such as in the **method of constant stimuli**, or adaptively, such as in the **staircase (up-down) method**.

The method of constant stimuli can for instance be used to determine the threshold in bit rate for which a coded image is distinguishable from the original. A number of coded images of decreasing bit rate are generated (for instance, by the same coding algorithm with an increasing quantization stepsize) and combined with the original to form stimulus pairs. The order of the original and coded image in a pair is varied randomly. The subject's task is to indicate the image with the lowest quality on each trial. The stimulus pairs are randomized and repeated a sufficient number of times to allow for an estimate of the detection probability at each level of bit rate. These probabilities are subsequently plotted as a function of the bit rate, and, somewhat arbitrarily, the bit rate for which the coded image is selected correctly in 75 % of the presentations is chosen as the **threshold** and corresponds to 1 JND. Typically, a cumulative probability function with a limited number of parameters, such as an error function with unknown mean and standard deviation, is fitted through the data points in order to obtain a more stable estimate of the threshold. The constant stimulus procedure has the obvious disadvantage that many measurements are required to derive a single number (i.e., the bitrate at threshold). Moreover, a fairly accurate a priori knowledge of the threshold is required in order to avoid using images in the experiment that are either not or trivially distinguishable from the original. For these reasons, thresholds nowadays are mostly measured using adaptive methods.

In an adaptive method, we assume that the stimuli are ordered according to increasing sensation strength. For instance, we may assume that the probability of detecting a difference between the original and a coded image increases with decreasing bit rate. Instead of estimating the complete psychometric curve, an adaptive method aims at finding the stimulus level (in our case, the bit rate) for which the detection probability is equal to an a priori specified value. The only mechanism used in the experiment is to either increase or decrease the stimulus strength under influence of the subject responses, hence the name staircase or up-down procedure. For instance, a typical so-called "up-down" rule is to increase the stimulus strength after each wrong answer, but to only decrease stimulus strength after n correct answers [Lev71]. This up-down procedure is continued until the presented stimuli oscillate around a constant stimulus strength. Typically, the procedure is stopped when any one of the stimuli has been used for a prescribed number of times. If the probability for correct detection of stimuli at this (limit) strength is p, then the probability for n successive correct detections is p^n, at least if all observations are performed independently. In all other cases, with a total probability of $1-p^n$, the stimulus strength is increased. If the stimulus strength oscillates around a constant level, then the probabilities for increase and decrease must be equal, i.e, $1-p^n = p^n$ or $p = (0.5)^{1/n}$. For instance, $p = 0.707, 0.794, 0.841$ for $n = 2, 3, 4$, respectively. The main assumption in this adaptive procedure is that successive judgements are performed independently. The typical way to guarantee that this is approximately true in an actual experiment is to time-multiplex several up-down procedures (for instance for different scenes), such that successive judgements for the same staircase are well separated in time. More advanced adaptive methods, in which the step size is also controlled, have been proposed [Lev71].

4.2 GLOBAL PSYCHOPHYSICS

In global psychophysics, the probability of discriminating stimuli is close to 1 for many of the stimulus pairs. Measuring such probabilities close to 1, even with a large number of observations, is very troublesome and error-prone. Therefore, supra-threshold measurement techniques are required to obtain information about the (differences in) stimulus sensations. Next to this technical argument, there are additional reasons for considering global measurements. Being able to perceive a complete range of stimulus strengths is more representative for typical (natural) circumstances in which we have an idea about the complete range of sensation strengths. Learning effects, i.e., systematic changes in responses of subjects to identical stimuli in the course of an experiment, therefore tend to level off more quickly in global experiments. Also, some local methods, such as the staircase method, require that the stimuli can be ordered according to increasing strength. Especially in cases where the stimulus variations are complex (for example, images coded by different image coding algorithms at variable bit-rates), it may be very difficult, or even impossible, to find the required a priori ordering. Last but not least, the same number of trials typically provide much more information when using a global measurement procedure rather than a local measurement procedure. For instance, in an unpublished study related to the scaling experiments described in section 7 [RM90], it was shown that comparable (and more stable) results for the detection threshold of coding artifacts are obtained using categorical scaling than using two-alternative-forced-choice detection with a constant stimulus method. The bit rate up to which the coded images received approximately the same quality score as the original image agreed well with the detection threshold, at least within the error of the measurement for the detection experiment. Of course, in addition, the scaling experiment also provided quality scores over a complete range of bit rates.

One global method that has been used to assess image quality is the **tradeoff method**. For example, in [MNP77], the task of the subject was to equate coded images to images with additive Gaussian noise (in terms of overall quality). The output of a single measurement for a coded image was the equivalent noise standard deviation. This method is very closely related to the more general method of **magnitude estimation**. This method of magnitude estimation has been advocated by S.S. Stevens [Ste57]. In this method, subjects are asked to report, usually in the form of a number, the ratio of the sensation strengths conveyed by two stimuli, called the test stimulus and the reference stimulus. The reference stimulus may even be implicit, in the sense that subjects are allowed to imagine a reference stimulus that corresponds to the unit of their numerical scale. Instead of a numerical scale, a graphical scale or a slider may also be used to report the responses. Stevens claimed that the responses thus obtained by the subjects were in one-to-one relationship with the internal sensations, i.e., had ratio scale properties. He applied this magnitude estimation method to estimating the sensation strengths for many different perceptual modalities (loudness, brightness, coarseness, etc) and found that in all cases a power-law relationship between the reported sensation and the physical parameter describing the stimuli was adequate. The power did however vary across modalities. For instance, in the case of the relation between stimulus luminance L (in cd/m^2) and reported brightness B, he found

$$B \propto (L - L_o)^\beta \tag{5.4}$$

for $\beta \approx 0.3$ and $L > L_o$, where L_o is the luminance required for detection. A different value for the exponent (i.e., $\beta = 0.56$) was found in case the stimuli were presented on a background (with luminance around 20 cd/m^2) [RBR91]. The latter study also applied a **paired comparison** method to the stimuli in order to verify Stevens' assumption of a ratio scale.

In paired comparison, the use of numbers is avoided by comparing two pairs of stimuli. The only task of the subjects is to report that pair for which the difference in sensation is the smallest. Analysis of the rank order of these pairs can be used to obtain a scale for the stimulus sensation strengths [She62, Kru64a]. Although this method is generally accepted as the most stable method to obtain a linear sensation scale, it is very time-consuming, and therefore seldom used. In the above application to the luminance-brightness relation, it revealed that this relation is better expressed by a power-law with exponent $\beta = 0.13$. This exponent is considerably smaller than the one found in case of magnitude estimation, which implies that the relationship between brightness sensation and response is most likely not linear, as assumed by Stevens, but better approximated by a power law with exponent $4.3 = 0.56/0.13$. It has been argued [Wag75] that this may be due to the limited ability of subjects to handle large number ranges in a linear way.

The use of large numbers is avoided in **category scaling**. This is one of the reasons why we have almost exclusively used category scaling for our experiments on image quality. Since giving names to categories may also introduce a bias in the use of these categories, we most often use **numerical category scaling**, in which successive categories are identified by increasing integer numbers. For the above example of the luminance-brightness relationship, it was verified that a numerical category scaling experiment using 10 categories did indeed result in a power law with the same exponent $\beta = 0.13$ as in the case of non-metric scaling [RBR91]. Hence, in this case of numerical category scaling, the response was linearly related to the sensation strength.

Scaling can be applied in both single and double stimulus experiments. In **single-stimulus scaling**, the images are presented one-by-one to the observer. A (fixed) reference image may potentially be shown together with the test image. The task of the subject is to rate the quality of the test image. In the ITU-R BT.500 recommendation for subjective assessment [IR98], a graphical scale, potentially implemented using a physical slider, is proposed as a continuous rating scale. This scale is divided into five equal intervals that are denoted by the quality or impairment categories described earlier. This method hence combines aspects of both continuous and category scaling. In the case of numerical category scaling, on the other hand, one can for instance use integer numbers from 1 to 10 to rate the test stimuli.

In a double-stimulus experiment, pairs of images are presented to the observer. The task of the subject in to either scale both stimuli, in **double-stimulus scaling**, to scale the dissimilarity between both stimuli, in **dissimilarity scaling**, or to scale the attribute (e.g., quality) difference, in **difference scaling**.

In double-stimulus scaling, both presented stimuli are scaled individually and the numerical difference between the responses for the individual stimuli is retained as the final response. Any response mechanism discussed above for single-stimulus scaling can be used to express the responses for the individual stimuli. In the ITU-R BT.500 recommendation, it is proposed to perform double-stimulus scaling by putting marks for both images on a graphical scale, and retaining the distance between both markings as the response. Again, quality or impairment categories are indicated alongside these graphical scales. Obviously, the ratings for both images have to be on a metric scale, in order for the distance between the marks to be monotonically related to the difference in quality[3]. There is no obvious reason, in view of the foregoing discussion, why this assumption should be valid. We will therefore not promote this method any further here.

In dissimilarity scaling, the subjects' task is to indicate how dissimilar or different they perceive two images to be [MM98]. Any aspect that contributes to the dissimilarity can be taken into account. The order of presentation of the stimuli in the pair should be irrelevant. Again, continuous scaling or numerical category scaling is most often used to express the response to a stimulus pair. Dissimilarity scores are by definition positive or zero.

In difference scaling, subjects can for instance respond to a stimulus combination by means of a numerical category from -5 to +5, or using a mark on a continuous interval that is symmetric around the origin. If

[3]In the ITU-R BT.500 recommendation, one of the two stimuli to be scaled is always the original image. This implies that the difference between both scale values can also be interpreted as single-stimulus scaling, where zero corresponds to the quality of the original image. The numerical difference in the scale values can be assumed to increase monotonically with the quality difference between the original and the test image.

the first (or leftmost) stimulus has the largest attribute strength, then a negative score is given, while a positive score corresponds to the second (or rightmost) stimulus having the largest attribute strength. Hence, the order of the stimuli does matter in this case. The absolute value of the response is monotonically related to the strength of the attribute difference. The zero category (or origin) can be used in case both stimuli are judged to have equal attribute strength.

One important advantage of difference scaling over single and double stimulus scaling is that the latter methods are more easily influenced by contextual effects [PW86, DMW95, Rid96a, DM96, DM97b]. For instance, in an experiment described by de Ridder [Rid96a], image sets with skewed distributions (either positively or negatively skewed, i.e., containing a relatively large proportion of high and low quality images, respectively) were rated using all three scaling techniques. Some images were however identical in both experiments. Only in the case of difference scaling was there no significant effect of the stimulus distribution on the quality scores.

A **trial session** is usually advisable for scaling experiments. In such a session, the subject is shown a representative set of the stimuli, so that he/she can get acquainted with the range over which quality varies. This is done in order to help subjects decide on how to use the available categories, or graphical scale, to cover the entire range of sensations. If such a trial session is not used, then it is more likely that subjects will change the use of their categories, or scale, in the course of the experiment. An additional measure that is often taken to reduce the effect of learning on the analyzed data is **counter balancing**. This means that the stimuli, or stimulus pairs, are shown both in forward and reverse order in repeated experiments with the same subject or with different subjects. We assume that in this way the effect of possible learning effects on the estimated sensation strengths is further reduced by the averaging within the data processing.

5. MULTIDIMENSIONAL MODELLING OF IMAGE QUALITY

If subjects are requested to evaluate image quality in a psychophysical experiment, they are able to analyze and justify their judgements, especially if they are somewhat experienced in making such judgements. They can for instance report different kinds of distortions in coded images [YW98], and are aware of the fact that their overall quality judgement is determined by the relative weight that they attribute to these individual impairments. They are also able to report the sensation strengths for individual impairments in a similar way as they can report their overall quality sensations.

In order to simultaneously model the results from different experiments with the same stimuli, multi-dimensional geometrical models can be adopted. In such a model, images are represented by points in a multidimensional space. All observations, obtained using one or several of the above-mentioned experimental paradigms, are related to geometrical properties of these points, such as distances between points and coordinates of point projections onto selected axes. This multi-dimensional model can be viewed as an extension to the one-dimensional geometrical model adopted in section 3.3 for modelling judgements on a single attribute.

Based on the above considerations, we subdivide the task of image quality modelling into:

1 establishing the stimulus configuration (i.e., the discriminating process) that underlies all attribute judgements by different subjects, and

2 determining how this stimulus configuration relates to the judgements for different attributes and/or individuals.

In case of an instrumental measure for image quality, the stimulus configuration is supplied by the measure and the remaining task (2) is to determine if this configuration can indeed be used to predict the subject responses. The XGms program that we will introduce below supports both options, i.e., derivation of a stimulus configuration from experimental data, which is called **multidimensional scaling (MDS)** [KW78, YH87, CC94], and comparison of a given stimulus configuration against experimental data, which is called **regression**.

A number of algorithms have been developed within the field of mathematical psychology (i.e., psychometrics) to position stimuli based on experimental data. Most of these algorithms model dissimilarity data. The stimuli are positioned in space according to metric or non-metric models. In non-metric models, the distances between stimulus positions are only monotonically related to the judged dissimilarities, i.e., only the rank order of the experimental data is important [She62, Kru64a, Kru64b]. Metric models, in contrast, maintain a linear relationship between the experimental dissimilarities and the distances between stimulus positions. The Euclidean distance is the metric most frequently used, although more general Minkowski metrics and weighted distances have also been used. In the latter case, the multidimensional psychological space in which the stimuli are positioned is not identical for all subjects (i.e., does not conform strictly to the principle of homogeneity of perception); the spaces for individual subjects are linked by linear (affine) transformations [CC94]. An example of a freely available MDS program for modelling dissimilarity data that implements most of the above metric and non-metric options is ALSCAL[4].

Marmolin & Nyberg [MN75] and Goodman & Pearson [GP79] have used MDS to study image quality. Dissimilarity judgements for pairs of impaired images were used in both studies. The dimensions of the multidimensional spaces thus established were labelled based on an examination of the positions of the impaired images. This labeling was however not verified by separate experiments. Escalante-Ramírez et al. [ER92, ERRM94, ERMR95] studied the perceptual quality of noise-reduced computer tomography images using MDS techniques. In addition to the space obtained using dissimilarity data, these authors also obtained a second space through a principal component analysis of the scaling data for the main attributes: noise, blur, and visibility of structures. They hence assumed the scaling data to be metric in this case. Both stimulus configurations found in this study were very similar and could be related by a linear transformation. The attribute data could also be used to identify the attribute directions in the multi-dimensional space that was obtained from dissimilarity data. A similar study was undertaken by Kayargadde and Martens [KM96c] to model images degraded by noise and blur (see also sections 2 and 7), while Teunissen and Westerink used MDS to study the quality differences between television sets [TW96]. A problem encountered in all the above studies is that no program is available for finding a single stimulus configuration, based on all available experimental data. Although a program for the joint analysis of single-stimulus ratings, pairwise preferences and dissimilarities has been proposed [Ram80], the available implementation[5] does not seem to function properly for the most general case.

Existing MDS programs are mostly offline programs. This implies that the user must enter his/her parameter choices, such as the dimension of the space or the selection between metric or non-metric analysis, and that the results are returned in a file, to be examined after the program has finished. This makes it very difficult and cumbersome to appreciate the impact of alternative model choices, and hence to choose the model that most adequately describes the data. Also, since MDS programs are nonlinear optimization programs, the reported solution may correspond to a local optimum instead of a global one. The inter-

[4]http://forrest.psych.unc.edu/research/ALSCAL.html
[5]http://www.psych.mcgill.ca/faculty/ramsay/ramsay.html

active MDS program XGvis [BSL+02] has been developed to help remedy such problems. The program XGvis allows to interactively control the main parameters in the MDS model, and exchanges the calculated stimulus configuration with a second program, called XGobi [SCB98]. This latter program is an interactive dynamic data visualization tool for the X Window environment. By combining the functionality of both programs, the user is not only able to dynamically alter the parameters of the MDS model within XGvis, but to also view and manipulate the stimulus configuration in XGobi. This interactivity for instance allows the user to assist the optimization algorithm in avoiding sub-optimum solutions that correspond to local minima. The XGvis program however only implements a metric model for the dissimilarity data of a single subject, and hence has too limited functionality to be very useful for modelling image quality. In order to overcome these limitations, XGvis has been extended to include the joint analysis of data from single-stimulus scaling, double-stimulus difference scaling and dissimilarity scaling for multiple subjects. The resulting program XGms supports both metric and non-metric modelling of (parts of) the data.

6. MDS MODELLING OF CONTINUOUS DATA

In this section, we describe the class of multidimensional models that is implemented in the program XGms for the case that the response data is continuous. The case of discrete response data will be treated in section 8. We first describe the maximum-likelihood optimization criterium that is used in the program to estimate the model parameters from the experimental data. Next, we show how optimized monotonic transformations can be used to replace the response data, which may be non-metric, by transformed data that is approximately metric. Many parameters of the multidimensional model (such as the dimension n of the space) can be controlled by the user at run-time, so that their effect on the stimulus configuration and on the relationship between experimental data and model predictions can be explored. In order to give a better impression of how a user can influence the construction of multidimensional models, the user interface to the program XGms is also described shortly.

6.1 MAXIMUM-LIKELIHOOD (ML) OPTIMIZATION

6.1.1 DISSIMILARITY DATA

The experimental dissimilarity data are denoted by $D_{k,i,j}$, for subject $k = 1, \ldots, K_d$ and stimulus pair $(i,j) \in I_d(k)$. The goal of the XGms program is to construct a stimulus configuration $\mathbf{x}_1, \ldots, \mathbf{x}_N$ in an n-dimensional vector space, such that the experimentally observed dissimilarity scores $D_{k,i,j}$ are monotonically related to the interstimulus distances

$$d_{ij} = \|\mathbf{x}_i - \mathbf{x}_j\|_l = \left[\sum_{m=1}^{n} |x_{im} - x_{jm}|^l \right]^{1/l}, \tag{5.5}$$

where the distance is computed according to a **Minkowski norm** with power l. The default value $l = 2$ corresponds to a Euclidean (or length) norm[6].

More precisely, a linear relationship is pursued between transformed dissimilarity scores $TD_{k,i,j} = T_{dk}(D_{k,i,j})$ and interstimulus distances d_{ij}, i.e., $TD_{k,i,j} \approx d_k \cdot d_{ij}$, where the regression factor d_k and the transformation T_{dk} need to be derived from the data. An important implication of this pursued relationship is that the transformed dissimilarities $TD_{k,i,j}$ are assumed to be metric (i.e., have ratio properties), since they are compared to the metric distances d_{ij}. We hence implicitly assume that there exists a monotonic transformation T_{dk} that maps the non-metric observed dissimilarities $D_{k,i,j}$ into metric transformed dissimilarities $TD_{k,i,j}$. Note that this assumed transformation can be regarded as the inverse of a response function R_{dk} that relates internal sensation strengths on a metric scale to external responses (as in section 3.3).

According to the principle of homogeneity of perception, the stimulus configuration $\mathbf{x}_1, \ldots, \mathbf{x}_N$ should be shared by all subjects, while the transformations T_{dk} and regression factors d_k can be subject-dependent. In order to preserve the physical meaning of dissimilarity zero, we require that the applied transformations satisfy $T_{dk}(0) = 0$. We also impose that the maximum dissimilarity value $M_d(k)$ for subject k is mapped onto itself, i.e., $T_{dk}[M_d(k)] = M_d(k)$, for $k = 1, \ldots, K_d$. For the time being, we assume that the transformations T_{dk} on the experimental data

[6]The use of a Minkowski norm with $l \neq 2$ implies that coordinate differences in horizontal and vertical directions are treated differently from coordinate differences in oblique directions. Such Minkowski norms should therefore be handled with care. Amongst others, the optimum stimulus configuration will often depend critically on the initial configuration, and convergence problems may occur, especially for $l < 2$ [She74, Ram77a].

$D_{k,i,j}$ are known. If the input data can be assumed to be metric, no transformations need to be applied, i.e. $TD_{k,i,j} = D_{k,i,j}$, for all subjects $k = 1, \ldots, K_d$ and all stimulus combinations (i, j). This is the default choice of the program XGms at initialization. If the input data is non-metric, then transformations T_{dk} that map the input dissimilarities $D_{k,i,j}$ to transformed dissimilarities $TD_{k,i,j}$ that have ratio properties can be derived from the data in a way to be discussed later.

Note that the index set

$$I_d(k) = \{\ (i,j) \mid i \neq j,\ D_{k,i,j} \neq NA,\ TD_{k,i,j} < U_d(k),\ \ldots\} \tag{5.6}$$

for subject k can be a subset of all possible stimulus combinations, so that missing data can be handled (a missing dissimilarity is indicated by NA). We can also exclude large dissimilarities (for example, because they are judged to be inaccurate) by setting $U_d(k)$ to a value smaller than the maximum dissimilarity value $M_d(k)$ for subject k, where $k = 1, \ldots, K_d$. The number of observations $N_d(k)$ for subject k is therefore usually substantially smaller than the N^2 possible stimulus combinations.

The assumption that the stimulus configuration models the experimental data can be mathematically reformulated by demanding that the probability $p(TD_{k,i,j})$ that subject k assigns transformed response $TD_{k,i,j}$ to the dissimilarity for stimulus pair (i, j) can be expressed in terms of the model parameters, i.e.,

$$p(TD_{k,i,j}) = \phi\left[\ TD_{k,i,j} - \widehat{TD}_{k,i,j}; \sigma_d(k)\ \right], \tag{5.7}$$

where the average transformed response to stimulus pair (i, j) is

$$\widehat{TD}_{k,i,j} = d_k \cdot \|\mathbf{x}_i - \mathbf{x}_j\|_l. \tag{5.8}$$

The standard deviation $\sigma_d(k)$ is a measure for the variability in the transformed responses $TD_{k,i,j}$ of subject k on repeated presentations of the same stimulus combination (i, j). This standard deviation is assumed to only depend on the subject k, and not on the stimulus pair[7]. The **probability density function (PDF)** $\phi[x; \sigma_d(k)]$ is also called the **link function** for subject k.

[7]XGms also allows for a noise standard deviation of the form $\sigma_d(k) \cdot \exp[\tau_d(k) \cdot \widehat{TD}_{k,i,j}]$, where the parameter $\tau_d(k)$ determines the stimulus dependence of the noise [McC80]. This model option is not discussed any further here.

The available link functions in the XGms program are the **logistic PDF**

$$\phi(x;\sigma) = \frac{1}{\rho}\left[\exp\left(\frac{x}{2\rho}\right) + \exp\left(-\frac{x}{2\rho}\right)\right]^{-2}, \tag{5.9}$$

with $\rho = \sigma\,\frac{\sqrt{3}}{\pi}$, and the **generalized Gaussian PDF**

$$\phi(x;\sigma) = \frac{r}{2\rho\,\Gamma(1/r)}\,\exp\left[-\left|\frac{x}{\rho}\right|^{r}\right], \tag{5.10}$$

with $\rho = \sigma\,\sqrt{\frac{\Gamma(1/r)}{\Gamma(3/r)}}$, where

$$\Gamma(a) = \int_0^\infty z^{a-1}\,e^{-z}\,dz \tag{5.11}$$

denotes the gamma function. The **Gaussian** (or **normal**) **PDF**

$$\phi(x;\sigma) = \frac{1}{\rho\,\sqrt{\pi}}\,\exp\left[-\left|\frac{x}{\rho}\right|^{2}\right], \tag{5.12}$$

with $\rho = \sigma\sqrt{2}$, and the **exponential PDF**

$$\phi(x;\sigma) = \frac{1}{2\rho}\,\exp\left[-\left|\frac{x}{\rho}\right|\right], \tag{5.13}$$

with $\rho = \sigma/\sqrt{2}$, are special cases of the generalized Gaussian PDF that correspond to $r = 2$ and $r = 1$, respectively. Some example PDFs with unit standard deviation are plotted in figure 5.5.

The overall model for dissimilarity data is summarized in Figure 5.6. The stimulus configuration is assumed to be shared by all subjects, while the response mechanism is subject-dependent through the proportionality factor d_k, the noise standard deviation $\sigma_d(k)$ and the non-linear response mapping $R_{dk} = T_{dk}^{-1}$, for $k = 1, \ldots, K_d$.

The transformation T_{dk} is monotonic, so that the probability $p(D_{k,i,j})$ of obtaining the original dissimilarity $D_{k,i,j}$ can be derived from the probability $p(TD_{k,i,j})$ of obtaining the transformed dissimilarity $TD_{k,i,j}$, i.e.,

$$\begin{aligned} p(D_{k,i,j}) &= p(TD_{k,i,j}) \cdot \left|T'_{dk}(D_{k,i,j})\right| \\ &= \phi\left[\, TD_{k,i,j} - d_k \cdot \|\mathbf{x}_i - \mathbf{x}_j\|_l ; \sigma_d(k)\,\right] \cdot \left|T'_{dk}(D_{k,i,j})\right|, \end{aligned} \tag{5.14}$$

where T'_{dk} denotes the derivative function of T_{dk}. If all subject responses are independent, then

$$P_d = \prod_{k=1}^{K_d} \prod_{(i,j)\in I_d(k)} p(D_{k,i,j}) \tag{5.15}$$

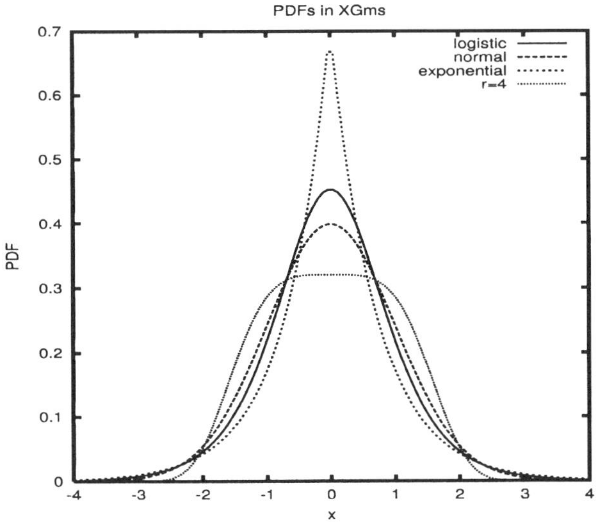

Figure 5.5. Plots of PDFs with standard deviation $\sigma = 1$. The different curves correspond to the logistic PDF and the generalized Gaussian PDFs with exponent $r = 2$ (normal), $r = 1$ (exponential) and $r = 4$.

is the overall probability, according to the model, of finding the specific dissimilarity responses $D_{k,i,j}$ that are observed in the experiment. Finding the model parameters that maximize this probability P_d, or equivalently, that minimize the inverse of the **log-likelihood function**, i.e.,

$$L_d = -\log P_d = -\sum_{k=1}^{K_d} \sum_{(i,j)\in I_d(k)} \log p(D_{k,i,j}), \tag{5.16}$$

corresponds to adjusting the model parameters such that the experimentally observed responses are the most likely outcomes [Tre68, SO91]. The regression parameters d_k and standard deviations $\sigma_d(k)$, for $k = 1, \ldots, K_d$, and the stimulus positions $\mathbf{x}_i$, for $i = 1, \ldots, N$, that minimize L_d are called the **maximum-likelihood (ML) estimates** for the model. In appendix 5.A, we show that the stress optimization that is used for parameter estimation in many existing MDS programs [BSL$^+$02, Kru64a, CC94] is a special case of the more general ML optimization that is proposed here.

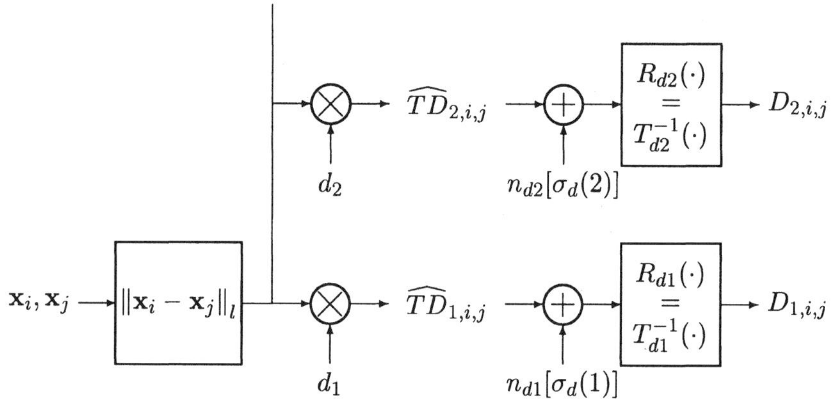

Figure 5.6. MDS model relating interstimulus distances $\|\mathbf{x}_i - \mathbf{x}_j\|_l$ to subject dissimilarity ratings $D_{k,i,j}$ for stimulus pair (i,j) and subject k. The stimulus configuration is shared by all subjects, while the response mechanism $R_{dk} = T_{dk}^{-1}$ is subject-dependent. The response variability for subject k is modelled by the noise source n_{dk} with standard deviation $\sigma_d(k)$.

Once the stimulus positions $\mathbf{x}_i$, for $i = 1, \ldots, N$, are known, minimization of the individual terms

$$L_d(k) = - \sum_{(i,j) \in I_d(k)} \log p(D_{k,i,j}) \tag{5.17}$$

in the ML criterium can be used to determine the noise standard deviations $\sigma_d(k)$ and the regression factors d_k, for $k = 1, \ldots, K_d$. These regression factors can be used to define the *normalized transformed dissimilarities*

$$TD^{\star}_{k,i,j} = \frac{1}{d_k} TD_{k,i,j}. \tag{5.18}$$

If all subjects behave similarly, then the normalized transformed dissimilarities for an identical stimulus pair (i,j) should have similar values across subjects, i.e., be approximately equal to the interstimulus distance $d_{ij} = \|\mathbf{x}_i - \mathbf{x}_j\|_l$. We will see in the next section how these normalized scores can be used to test the hypothesis that a model can describe the average subject behavior when judging dissimilarity.

Note that the stimulus positions $\mathbf{x}_i$ only occur in the average transformed dissimilarities $d_k \cdot \|\mathbf{x}_i - \mathbf{x}_j\|_l$. Consequently, a uniform dilation of the stimulus configuration, i.e., $\mathbf{x}_i \to \delta_k \cdot \mathbf{x}_i$, will have no influence on the ML criterium, provided that the regression parameters are adjusted accordingly, i.e., $d_k \to d_k/\delta_k$. In order to avoid such undetermined model behavior, a condition is imposed on the stimulus configuration in order to uniquely determine its scale factor. Similarly, a translation of the stimulus configuration has no influence on the interstimulus distances either, so that an additional condition is imposed to uniquely determine this translation. In summary, the ML criterium is invariant under linear translations and uniform dilation of the stimulus coordinates $\mathbf{x}_i = [x_{im}; m = 1, \ldots, n]$, i.e., mappings of the form

$$x_{im} \to d \cdot x_{im} + e_m, \tag{5.19}$$

for $i = 1, \ldots, N$ and $m = 1, \ldots, n$. A priori conditions on the stimulus configuration are therefore required in order to guarantee a unique stimulus configuration. The translation vector $\mathbf{e} = [e_m; m = 1, \ldots, n]$ is determined by requiring that the configuration is centered at the origin, i.e., $\sum_{i=1}^{N} x_{im} = 0$, for $m = 1, \ldots, n$. The dilation factor d is derived from the condition that $\sum_{i=1}^{N} \sum_{m=1}^{n} x_{im}^2$ is constant (in our case, equal to $N \cdot n$). Stimulus configurations that satisfy these conditions are called normalized. Because of this normalization of the stimulus configuration, the number of **degrees of freedom (DOF)** in this configuration is equal to

$$F_{\mathbf{x}} = (N-1) \cdot n - 1. \tag{5.20}$$

If the distance metric is Euclidean, i.e., if $l = 2$, then an arbitrary orthogonal transformation (rotation or mirroring around the origin) of the stimulus configuration has no effect on the interstimulus distances either. This further reduces the DOF in the stimulus configuration to

$$F_{\mathbf{x}|l=2} = (N-1) \cdot n - 1 - \frac{n(n-1)}{2} = \left(N - \frac{n+1}{2}\right) \cdot n - 1. \tag{5.21}$$

In the default setting of the XGms program, no measures are taken to uniquely select the orientation of the stimulus configuration. This implies that the orientation of the output stimulus configuration will depend on the initial stimulus configuration. The default settings of the XGms program can however be modified such that an orthogonal transformation is selected that aligns the stimulus configuration along its principal axes [Ram77a].

6.1.2 PREFERENCE DATA

The ML criterium for double-stimulus preference data[8] can be derived in a similar way as in the case of dissimilarity data, and is equal to

$$L_p = \sum_{k=1}^{K_p} L_p(k) = -\sum_{k=1}^{K_p} \sum_{(i,j)\in I_p(k)} \log p(P_{k,i,j}), \quad (5.22)$$

where

$$p(P_{k,i,j}) = \phi \left[TP_{k,i,j} - \widehat{TP}_{k,i,j}; \sigma_p(k) \right] \cdot \left| T'_{pk}(P_{k,i,j}) \right| \quad (5.23)$$

is the probability, according to the MDS model[9], that the preference rating $P_{k,i,j}$ is given by subject k, with $k = 1, \ldots, K_p$, for stimulus pair (i, j), with $i, j = 1, \ldots, N$. The transformed preference scores $TP_{k,i,j} = T_{pk}(P_{k,i,j})$ are compared against their expected values

$$\widehat{TP}_{k,i,j} = m_k \cdot ([\mathbf{p}_k, \mathbf{x}_i] - [\mathbf{p}_k, \mathbf{x}_j]) \,. \quad (5.24)$$

These predictions for subject k are derived from the stimulus positions $\mathbf{x}_i$ and $\mathbf{x}_j$ and the *preference vector* $\mathbf{p}_k$ in ways to be described below. The regression parameter m_k expresses the linear relationship between the transformed scores and the predictions. Again, a subset

$$I_p(k) = \{ \, (i,j) \mid i \neq j, \; P_{k,i,j} \neq NA, \; |TP_{k,i,j}| < U_p(k), \; ...\} \quad (5.25)$$

of all possible stimulus combinations can be selected, so that the number of observations $N_p(k)$ for subject k is usually substantially smaller than the N^2 possible stimulus combinations. While dissimilarities are always positive numbers, preferences can be both negative or positive, depending on whether the i-th stimulus is preferred over the j-th stimulus, or vice versa. Therefore, the upper limit $U_p(k)$ works on the absolute value of the transformed preferences. It is smaller than the maximum amplitude $M_p(k)$ of the transformed preferences of subject k, where $k = 1, \ldots, K_p$.

The overall model for preference data is summarized in Figure 5.7. The stimulus configuration is assumed to be shared by all subjects, while the response mechanism is subject-dependent through the prediction vector $\mathbf{p}_k$, the proportionality factor m_k, the noise standard

---

[8]Preference refers to a difference in quality. If another attribute is used, then attribute differences are scaled. We use the term preference in a generic sense here to also refer to such attribute difference judgements.

[9]XGms also allows for a noise standard deviation of the form $\sigma_p(k) \cdot \exp[\tau_p(k) \cdot |\widehat{TP}_{k,i,j}|]$, where the parameter $\tau_p(k)$ controls the stimulus dependence of the noise.

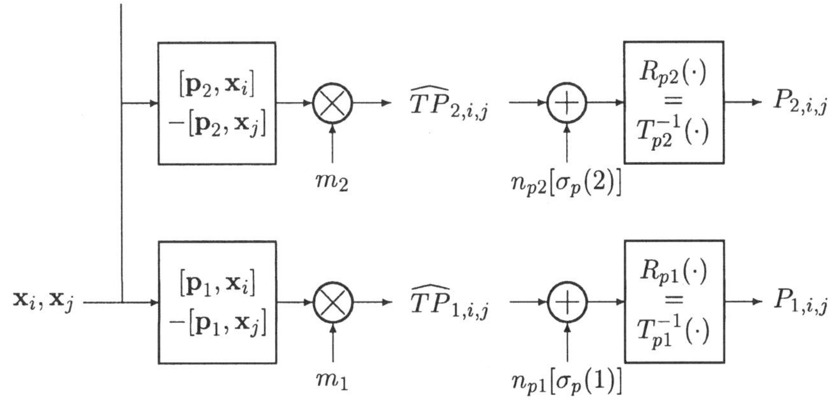

Figure 5.7. MDS model relating stimulus positions $(\mathbf{x}_i, \mathbf{x}_j)$ to subject preference ratings $P_{k,i,j}$ for stimulus pair (i, j) and subject k. The stimulus configuration is shared by all subjects, while the preference vector $\mathbf{p}_k$ and response mechanism $R_{pk} = T_{pk}^{-1}$ are subject-dependent. The response variability for subject k is modelled by the noise source n_{pk} with standard deviation $\sigma_p(k)$.

deviation $\sigma_p(k)$ and the non-linear response mapping $R_{pk} = T_{pk}^{-1}$, for $k = 1, \ldots, K_p$.

XGms allows to choose between two prediction models. The user can (interactively) switch between both models in order to decide which model best describes the data.

According to the **vector-product** or **inner-product model**, the prediction is equal to the vector product

$$\begin{aligned} [\mathbf{p}_k, \mathbf{x}_i] - [\mathbf{p}_k, \mathbf{x}_j] &= \langle \mathbf{p}_k, \mathbf{x}_i \rangle - \langle \mathbf{p}_k, \mathbf{x}_j \rangle = \langle \mathbf{p}_k, \mathbf{x}_i - \mathbf{x}_j \rangle \\ &= \sum_{m=1}^{n} p_{km} \cdot (x_{im} - x_{jm}) \end{aligned} \tag{5.26}$$

between the difference vector $\mathbf{x}_i - \mathbf{x}_j$, pointing from stimulus j to stimulus i, and the preference vector $\mathbf{p}_k$. The corresponding geometrical interpretation is illustrated in Figure 5.8. The difference between the orthogonal projections (or coordinates) of the stimulus positions $\mathbf{x}_i$ and $\mathbf{x}_j$ onto a one-dimensional axis, with direction specified by the vector $\mathbf{p}_k$, determines the average preference between both stimuli for subject k. If the same stimulus combinations are scored very differently by different subjects, then this should be reflected in preference vectors $\mathbf{p}_k$ with distinct orientations. Such different orientations are obviously only possible in case the dimension n of the space exceeds one.

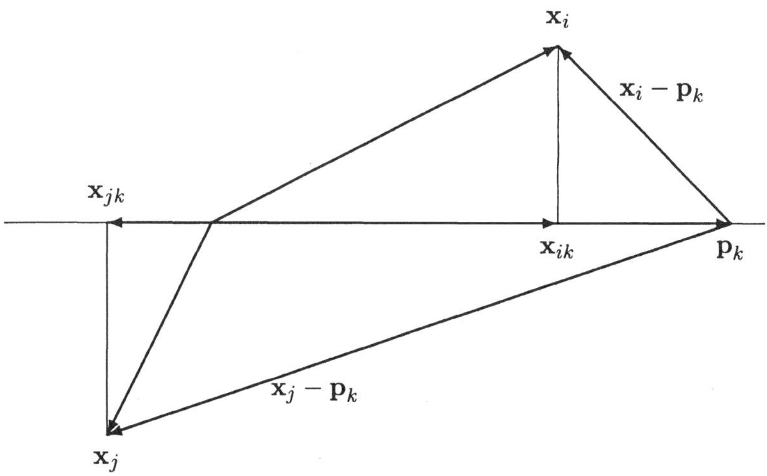

Figure 5.8. Geometric interpretation of the preference prediction models. If $\mathbf{x}_{ik}$ is the orthogonal projection of the vector $\mathbf{x}_i$ onto the axis with direction $\mathbf{p}_k$, then the inproduct definition implies that $\langle \mathbf{p}_k, \mathbf{x}_i \rangle = \langle \mathbf{p}_k, \mathbf{x}_{ik} \rangle$. The projected vectors can be expressed as $\mathbf{x}_{ik} = x_{ik} \cdot \mathbf{p}_k / \|\mathbf{p}_k\|_2$ and $\mathbf{x}_{jk} = x_{jk} \cdot \mathbf{p}_k / \|\mathbf{p}_k\|_2$ since they are aligned with the preference vector $\mathbf{p}_k$. The numbers x_{ik} and x_{jk} can be interpreted as coordinates on the preference axis, while the inner-product model prediction $\langle \mathbf{p}_k, \mathbf{x}_i - \mathbf{x}_j \rangle = \langle \mathbf{p}_k, \mathbf{x}_{ik} - \mathbf{x}_{jk} \rangle = \|\mathbf{p}_k\|_2 \cdot (x_{ik} - x_{jk})$ is proportional to the difference in the coordinates of the projected vectors. The ratio of the lengths of the vectors $\mathbf{x}_i - \mathbf{p}_k$ and $\mathbf{x}_j - \mathbf{p}_k$ determines the preference prediction in case of the ideal-point model.

A consequence of the ratio property of the transformed preferences is that only the directions of the preference vectors $\mathbf{p}_k$ are uniquely determined; their amplitudes may be scaled arbitrarily. We can hence adopt $\|\mathbf{p}_k\|_2 = 1$ and conclude that the vector-product model contains n DOF for each preference k, i.e., $n-1$ DOF for the direction of the vector $\mathbf{p}_k$ and 1 DOF for the regression factor m_k.

The alternative **ideal-point model** or **unfolding model**

$$[\mathbf{p}_k, \mathbf{x}_i] - [\mathbf{p}_k, \mathbf{x}_j] = -\log(\|\mathbf{x}_i - \mathbf{p}_k\|_l) + \log(\|\mathbf{x}_j - \mathbf{p}_k\|_l) \qquad (5.27)$$

is based on the ratio of the distances of the stimulus positions $\mathbf{x}_i$ and $\mathbf{x}_j$ from the "ideal point" $\mathbf{p}_k$ [Ram80, CC94]. This ideal point cannot coincide with a stimulus point since $[\mathbf{p}_k, \mathbf{x}_i]$ must remain finite[10]. If the same stimulus pairs are scored very differently by different subjects,

[10] This is avoided in XGms by using a slightly modified prediction formula, i.e., $[\mathbf{p}_k, \mathbf{x}_i] = -\log(r_o + \|\mathbf{x}_i - \mathbf{p}_k\|_l)$, where r_o is a small offset, such as $r_o = 10^{-3}$.

then this should be reflected by ideal points $\mathbf{p}_k$ at some distance apart. Unlike the vector-product model, the ideal-point model can also model different subject behaviors in case the space is one-dimensional. The model contains $n+1$ model parameters for each preference k, i.e., n DOF for the position of the ideal point $\mathbf{p}_k$ and 1 DOF for the regression factor m_k.

The predictions from both models can be related in case of a Euclidean metric, with $l = 2$. More specifically, if $\|\mathbf{p}_k\|_2 \gg \|\mathbf{x}_i\|_2, \|\mathbf{x}_j\|_2$, then the first-order Taylor series expansion

$$-\log(\|\mathbf{x}_i - \mathbf{p}_k\|_2) \approx -\log(\|\mathbf{p}_k\|_2) + \frac{1}{\|\mathbf{p}_k\|_2^2} \langle \mathbf{p}_k, \mathbf{x}_i \rangle, \tag{5.28}$$

can be used to derive the following approximation

$$-\log\left(\frac{\|\mathbf{x}_i - \mathbf{p}_k\|_2}{\|\mathbf{x}_j - \mathbf{p}_k\|_2}\right) \approx \frac{1}{\|\mathbf{p}_k\|_2^2} \langle \mathbf{p}_k, \mathbf{x}_i - \mathbf{x}_j \rangle. \tag{5.29}$$

Hence, if the ideal-point $\mathbf{p}_k$ is far removed from the stimulus positions, then only the direction of the vector $\mathbf{p}_k$ is relevant for the predictions, and the ideal-point model becomes equivalent to a vector-product model. In practice, this will imply that estimation of the amplitude $\|\mathbf{p}_k\|_2$ becomes ill-determined in case of the ideal-point model, and that a vector-product model, with one less DOF, should be preferred.

Very often, the available preference data can be divided into groups. For example, the indices $k = 1, \ldots, K_p/2$ may refer to $K_p/2$ different subjects rating preference (differences in image quality), while indices $k = K_p/2+1, \ldots, K_p$ may refer to the same subjects rating another attribute difference (such as the difference in amount of perceived blur). In such a case, it often makes sense to look for a common prediction model for all subjects within a group. This corresponds to estimating a single prediction vector $\mathbf{p}_k = \mathbf{p}$ for all indices k in a group. In case of a single group with K_p subjects, this reduces the number of parameters from $K_p n$ to K_p+n-1 for the vector-product model, and from $K_p(n+1)$ to K_p+n for the ideal-point model. XGms allows to define such groups and can also export the *normalized transformed preferences*

$$TP^{\star}_{k,i,j} = \frac{1}{m_k} TP_{k,i,j} \tag{5.30}$$

that allow for an easy comparison of experimental data across subjects within a group. Indeed, for stimulus pair (i, j), the repeated judgements $TP^{\star}_{k,i,j}$ across subjects k in a group should aggregate around the 'group' prediction $[\mathbf{p}, \mathbf{x}_i] - [\mathbf{p}, \mathbf{x}_j]$.

6.1.3 ATTRIBUTE DATA

The ML criterium for single-stimulus attribute data is equal to

$$L_a = \sum_{k=1}^{K_a} L_a(k) = -\sum_{k=1}^{K_a} \sum_{(i,j)\in I_a(k)} \log p(A_{k,i,j}), \tag{5.31}$$

where

$$p(A_{k,i,j}) = \phi\left[\, TA_{k,i,j} - \widehat{TA}_{k,i}; \sigma_a(k) \,\right] \cdot \left|T'_{ak}(A_{k,i,j})\right| \tag{5.32}$$

is the probability, according to the MDS model[11], that the attribute rating $A_{k,i,j}$ is given by subject k for stimulus i on the j-th repetition ($k = 1, \ldots, K_a$; $i = 1, \ldots, N$ and $j = 1, \ldots, R_a$). The transformed attribute scores $TA_{k,i,j} = T_{ak}(A_{k,i,j})$ are compared against their expected values

$$\widehat{TA}_{k,i} = c_k + f_k \cdot [\mathbf{a}_k, \mathbf{x}_i]. \tag{5.33}$$

These predictions for subject k are derived from the stimulus positions $\mathbf{x}_i$ and the *attribute vector* $\mathbf{a}_k$. The regression parameters c_k and f_k determine the linear relationship between transformed attribute scores and predictions for subject k. Again, a subset

$$I_a(k) = \{\ (i,j) \mid A_{k,i,j} \neq NA,\ \ |TA_{k,i,j} - B_a(k)| < U_a(k),\ \ldots\} \tag{5.34}$$

off all possible attribute scores can be used in the ML criterium, so that the number of observations $N_a(k)$ for subject k may be smaller than $N \cdot R_a$. The attribute values for subject k lie in the interval $[m_a(k), M_a(k)]$, and we impose that the transformations T_{ak} preserve the extreme values, i.e., $T_{ak}[m_a(k)] = m_a(k)$ and $T_{ak}[M_a(k)] = M_a(k)$, for $k = 1, \ldots, K_a$. The upper limit $U_a(k)$ and the base value $B_a(k)$ can be used to indicate that only the transformed attribute scores in the range $[B_a(k) - U_a(k), B_a(k) + U_a(k)]$ contribute to the ML criterium.

The overall model for attribute data is summarized in Figure 5.9. The stimulus configuration is assumed to be shared by all subjects, while the response mechanism is subject-dependent through the prediction vector $\mathbf{a}_k$, the proportionality factor f_k, the offset factor c_k, the noise standard deviation $\sigma_a(k)$ and the non-linear response mapping $R_{ak} = T_{ak}^{-1}$, for $k = 1, \ldots, K_a$.

---

[11] XGms also allows for a noise standard deviation of the form $\sigma_a(k) \cdot \exp[\tau_a(k) \cdot |\widehat{TA}_{k,i} - B_a(k)|]$, where $B_a(k)$ is a bias value. The free parameter $\tau_a(k)$ controls the stimulus dependence of the noise.

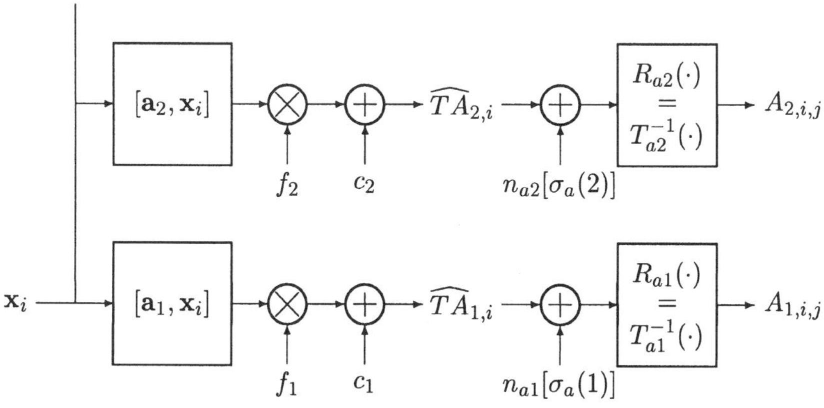

Figure 5.9. MDS model relating stimulus positions $\mathbf{x}_i$ to subject attribute ratings $A_{k,i,j}$ for stimulus i, repetition j and subject k. The stimulus configuration is shared by all subjects, while the attribute vector $\mathbf{a}_k$ and response mechanism $R_{ak} = T_{ak}^{-1}$ are subject-dependent. The response variability for subject k is modelled by the noise source n_{ak} with standard deviation $\sigma_a(k)$.

A requirement on the prediction rule $[\mathbf{a}_k, \mathbf{x}_i]$ is that it should allow to compensate for a dilation and/or translation $\mathbf{x}_i \to d \cdot \mathbf{x}_i + \mathbf{e}$ of the stimulus configuration by a change in the regression parameters. This implies that the prediction rule must satisfy

$$[\mathbf{a}_k, d \cdot \mathbf{x}_i + \mathbf{e}] = h_0(d, \mathbf{e}) + h_1(d, \mathbf{e}) \cdot [\mathbf{a}_k(d, \mathbf{e}), \mathbf{x}_i] \tag{5.35}$$

for all stimulus positions $\mathbf{x}_i$. The required mapping of the regression parameters is then given by $\mathbf{a}_k \to \mathbf{a}_k(d, \mathbf{e})$, $c_k \to c_k(d, \mathbf{e}) = c_k + f_k \cdot h_0(d, \mathbf{e})$ and $f_k \to f_k(d, \mathbf{e}) = f_k \cdot h_1(d, \mathbf{e})$. The same two prediction models as in the case of preference data are available for modelling attribute data in XGms. Both models can be shown to satisfy the above condition. In case of a Euclidean metric, i.e., $l = 2$, both prediction models are also invariant under a rotation of the stimulus configuration, provided of course that the attribute vectors $\mathbf{a}_k$ and the stimulus points $\mathbf{x}_i$ are rotated in the same way.

The vector-product model compares the transformed attribute scores $TA_{k,i,j}$ with the linear prediction

$$c_k + f_k \cdot [\mathbf{a}_k, \mathbf{x}_i] = c_k + f_k \cdot \langle\, \mathbf{a}_k, \mathbf{x}_i \,\rangle = c_k + f_k \cdot \sum_{m=1}^{n} a_{km} \cdot x_{im}. \tag{5.36}$$

The average strength of attribute k for stimulus i hence increases with the coordinate of the stimulus projection on a one-dimensional axis with direction indicated by the attribute vector $\mathbf{a}_k$. The offset value c_k, the scale factor f_k and the direction of the attribute vector $\mathbf{a}_k$ comprise a total of $n+1$ parameters. We again adopt the convention that $\|\mathbf{a}_k\|_2 = 1$ in this case.

The alternative to this vector-product model is the ideal-point (or unfolding) model, in which the transformed attribute scores for stimulus i are related to the distance $\|\mathbf{x}_i - \mathbf{a}_k\|_l$ of the stimulus at position $\mathbf{x}_i$ from an 'ideal' image at position $\mathbf{a}_k$ [Ram80], i.e.,

$$c_k + f_k \cdot [\mathbf{a}_k, \mathbf{x}_i] = c_k + f_k \cdot [\alpha_k - \log(\|\mathbf{x}_i - \mathbf{a}_k\|_l)], \tag{5.37}$$

where the offset

$$\alpha_k = \frac{1}{N} \sum_{i=1}^{N} \log(\|\mathbf{x}_i - \mathbf{a}_k\|_l) \tag{5.38}$$

is included to obtain that the attribute predictions are centered around the origin. The offset value c_k, the scale factor f_k and the ideal position $\mathbf{a}_k$ comprise a total of $n+2$ parameters.

As in the case of preference data, the attribute data may be subdivided into groups. This corresponds to estimating a single prediction vector $\mathbf{a}_k = \mathbf{a}$ for all indices k in a group. In case of a single group with K_a subjects, this reduces the number of parameters from $K_a\,(n+1)$ to $2K_a+n-1$ for the vector-product model, and from $K_a\,(n+2)$ to $2K_a+n$ for the ideal-point model. The *normalized transformed attributes*

$$TA^{\star}_{k,i,j} = \frac{1}{f_k}\,(TA_{k,i,j} - c_k), \tag{5.39}$$

across subjects k and repetitions j in a group are expected to aggregate around the 'group' prediction $[\mathbf{a}, \mathbf{x}_i]$ for stimulus i.

6.2 POWER-LIKE DATA TRANSFORMATIONS

The monotonically increasing transformations from $D_{k,i,j}$, $P_{k,i,j}$ and $A_{k,i,j}$ to $TD_{k,i,j}$, $TP_{k,i,j}$ and $TA_{k,i,j}$ are fixed in the above discussion of the ML criterium. These transformations can however be modified interactively by the user. In order to limit the possible variations, the available nonlinear transformations are based on the asymmetric power-like function[12]

$$P(x; q, x_t) = \mathrm{sgn}(x) \cdot \frac{(|x| + x_t)^q - x_t^q}{q}, \tag{5.40}$$

[12]The sign function is denoted by $\mathrm{sgn}(x)$, i.e., $\mathrm{sgn}(x)$=1 if $x > 0$, $\mathrm{sgn}(x)$=-1 if $x < 0$, and $\mathrm{sgn}(x)$=0 if $x = 0$.

that can be characterized by two parameters q and x_t. This function varies from 0, for values $|x| \ll x_t$ well below the threshold, to a power function with exponent q for values $|x| \gg x_t$ well above the threshold. Note that the often-used power-law relationship [Ram77a]

$$P(x; q, 0) = \text{sgn}(x) \cdot \frac{|x|^q}{q} \tag{5.41}$$

is included as a special case. The above definition needs to be modified to a logarithmic relationship

$$P(x; 0, x_t) = \text{sgn}(x) \cdot [\log(|x| + x_t) - \log x_t] \tag{5.42}$$

in case $q = 0$. The continuity in this transition from a power-law to a logarithmic relationship is most easily observed by verifying that the derivative to the function is

$$P'(x; q, x_t) = (|x| + x_t)^{q-1} \tag{5.43}$$

in all cases. The fact that this derivative is strictly positive implies that $P(x; q, x_t)$ is indeed monotonically increasing.

The nonlinear transformations that are used in the XGms program are designed such that they preserve the range of the input data. It is easily verified that, if the input data x is within the range $[m, M]$, that

$$T(x; q, x_t, b) = \frac{(M - m) \cdot P(x - b) - M \cdot P(m - b) + m \cdot P(M - b)}{P(M - b) - P(m - b)} \tag{5.44}$$

where $b \in [m, M]$ is the origin of the power function and $P(x) = P(x; q, x_t)$, is also limited to this same range. The linear relationship

$$T(x; 1, x_t, b) = x \tag{5.45}$$

is used as the default transformation by the program XGms. It corresponds to assuming that the input data is metric.

In case of dissimilarity and preference data, we choose $m = -M$, where M is the maximum (absolute) score, and $b = 0$, since the zero value is the natural origin for the transformation. The resulting simplified transformation

$$T(x; q, x_t) = \frac{M \cdot P(x; q, x_t)}{P(M; q, x_t)} \tag{5.46}$$

satisfies $T(x; q, x_t) = 0$ and $T(M; q, x_t) = M$, and contains two parameters, i.e., q and x_t. In XGms, the threshold value $x_t = t \cdot M$ is specified by t as a fraction of M.

In the case of attribute data, there is no natural origin for the transformation, so that the bias b is also a parameter. The transformed bias value is denoted by $B = T(b; q, x_t, b)$. In this case, the threshold $x_t = t \cdot (M - m)$ is specified by t as a fraction of the overall range.

6.3 NON-METRIC DATA

The model described above is essentially metric, since the nonlinear transformations on the experimental data are assumed to be known a priori (i.e., specified by an external action of the user). Rather than fitting the geometric model to the experimentally obtained dissimilarities $D_{k,i,j}$, preferences $P_{k,i,j}$ and attribute scores $A_{k,i,j}$, the fit is made to the transformed values $TD_{k,i,j}$, $TP_{k,i,j}$ and $TA_{k,i,j}$ that are assumed to have ratio or interval properties. In non-metric scaling, the monotonic transformations on the data are not specified a priori, but are determined as part of the optimization process. The XGms program allows for such non-metric optimizations under user control. The user can select to replace some, or all, of the experimental data by optimally-transformed data in the above minimization of the ML criterium.

The ML criterium contains expressions of the form

$$L = -\sum_{i=1}^{I} \left\{ \log \phi \left[T(o_i) - \hat{o}_i \right] + \log \left| T'(o_i) \right| \right\}, \tag{5.47}$$

where $T(o_i)$ denotes the transformed observation o_i for case i and $\hat{o}_i$ is the corresponding prediction according to the stimulus configuration and regression parameters in the current model. These predictions $\hat{o}_i$ are for instance linearly related to the distances between the (known) stimulus positions in case of dissimilarity and to the coordinates or coordinate differences of the stimulus positions along known attribute or preference directions. The noise standard deviation that characterizes the link function ϕ is also assumed to be known.

In non-metric MDS [Kru64a, CC94], the monotonic transformation T that minimizes L for known predictions $\hat{o}_i$ and link function ϕ is selected as the transformation for the experimental data.

6.3.1 OPTIMUM POWER-LIKE TRANSFORMATION

The first possibility is to assume that the optimum transformation can be closely approximated by expression (5.44). A non-linear optimization over the parameters q, t (and b in case of attribute scores) can then be performed in XGms. The parameters are restricted to the range $q \in [-6, 6]$ and $t \in [0.02, 4]$ in order to avoid possible run-away arguments in the optimization[13]. In case of attribute scores, the origin b for the transformation is limited to the range $[m, M]$ of scores that occur. The number of DOF in an optimized power-like transformation is hence $F_t = 2$ or $F_t = 3$, depending on whether the transformed data is ratio (dissimilarity and preference data) or interval (attribute data), respectively.

Although this optimization over a restricted set of power-like transformations often gives reasonable results, it may be interesting to compare the obtained optimum power-like transformation with the optimum transformation out of all possible monotonic transformations. For example, in case of attributes, knowledge of the optimum monotonic transformation often helps to select an appropriate bias parameter b for the power-like transformations (it is usually advantageous to put this bias b at the attribute score for which this optimum transformation is most asymmetric). Therefore, XGms also allows for an optimization across all possible monotonic transformations.

6.3.2 OPTIMUM MONOTONIC (KRUSKAL) TRANSFORMATION

An algorithm for determining the optimum monotonic transformation in non-metric MDS was originally developed by Kruskal [Kru64a, CC94]. We use an alternative and more flexible method based on spline interpolation introduced by Ramsay [Ram77b].

Suppose that the observations $\{o_i; i = 1, \ldots, I\}$ contain $J \leq I$ distinct values. Without loss of generality, we can assume that the data are sorted such that the first J observations $\{o_j; j = 1, \ldots, J\}$ are all distinct and in increasing order. Such sorting has no effect on the value of the ML criterium. A monotonically increasing transformation function

$$T(o) = T(o_1) + \int_{o_1}^{o} T'(x)\, dx \tag{5.48}$$

can be obtained by integrating a positive-valued derivative function $T'(o)$. We take a derivative function $T'(o)$ that linearly interpolates between J positive values $T'(o_j)$, for $j = 1, \ldots, J$. The corresponding integrated function will be piecewise quadratic with a continuous derivative, as shown in the example of Figure 5.10. A non-linear optimization of the ML criterium of equation (5.47) as a function of $T'(o_j)$, for $j = 1, \ldots, J$, can hence be used to find the optimum monotonically

[13] Non-linear optimizations for bounded parameters are performed using the DMNFB routine from the Netlib library, available through http://www.netlib.org.

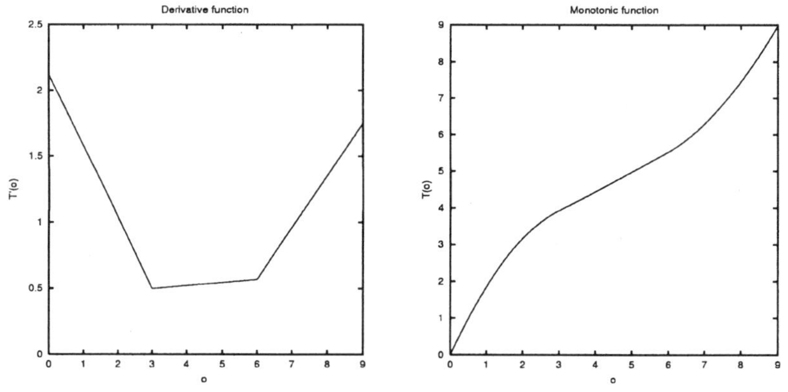

Figure 5.10. The figure on the left shows a piecewise linear derivative function $T'(o)$ that is strictly positive. The corresponding monotonically increasing function $T(o)$ on the right is piecewise quadratic with a continuous derivative.

increasing transformation on the data. The numbers $T'(o_j)$ have to be rescaled at each iteration step to guarantee that the transformation T preserves the data range $[o_1, o_J]$, i.e., the following normalization conditions

$$T(o_J) - T(o_1) = \int_{o_1}^{o_J} T'(x)\,dx = o_J - o_1 \tag{5.49}$$

and $T(o_1) = o_1$ are imposed. Since only the $J-2$ intermediate values are modified from o_j to $T(o_j)$, for $j = 2, \ldots, J-1$, we obtain that the number of DOF in the above *Kruskal optimization* is $F_t = J - 2$.

6.3.3 OPTIMUM SPLINE TRANSFORMATION

The Kruskal approach can be easily generalized to **spline transformations**, in which case the derivatives $T'(\tilde{o}_j)$ are only specified for a limited number of **knot points** $\tilde{o}_j$, for $j = 1, \ldots, s+2$ (with $s+2 < J$). The first and last knot point are equal to the minimum and maximum value, i.e., $\tilde{o}_1 = o_1$ and $\tilde{o}_{s+2} = o_J$, respectively, and the number of internal knot points is s. All required derivatives $T'(o)$ are obtained by linear interpolation of the values at the knot points, while the monotonic transformation $T(o)$ is again obtained by integrating the normalized derivative function. XGms allows for *spline optimization* with $0 \leq s < J-2$ equally-spaced internal knot points. Especially when J is large, such a spline optimization of a reduced order is a practical alternative to the general Kruskal optimization. The number of DOF in the optimized monotonic transformation is $F_t = s + 1$ in case of a spline optimization with s internal knot points. One DOF is added by the integration but two DOFs are consumed by the normalization conditions that guarantee that the range of the data is preserved.

6.4 MAXIMUM-LIKELIHOOD PARAMETER ESTIMATION

6.4.1 ITERATIVE OPTIMIZATION

The overall ML criterium

$$L = L_d + L_p + L_a = \sum_{k=1}^{K_d} L_d(k) + \sum_{k=1}^{K_p} L_p(k) + \sum_{k=1}^{K_a} L_a(k), \qquad (5.50)$$

with

$$F_O = F_{dO} + F_{pO} + F_{aO} = \sum_{k=1}^{K_d} N_d(k) + \sum_{k=1}^{K_p} N_p(k) + \sum_{k=1}^{K_a} N_a(k) \qquad (5.51)$$

degrees of freedom (DOF) in the data, needs to be minimized as a function of the stimulus positions $\mathbf{x}_1, ..., \mathbf{x}_N$, the regression parameters d_k (for dissimilarity), m_k and $\mathbf{p}_k$ (for preference), c_k, f_k and $\mathbf{a}_k$ (for attribute scaling), and the PDF noise standard deviations $\sigma_d(k)$, $\sigma_p(k)$ and $\sigma_a(k)$. In case of non-metric MDS, the monotonic transformations T_{dk}, T_{pk} and T_{ak} must also be optimized per subject. All required optimizations are performed iteratively[14]. If the stimulus configuration needs to be optimized (as in MDS), then one iteration step involves three stages. If the stimulus configuration is fixed (as in regression analysis), then one iteration step involves only two stages.

In the first stage, the stimulus positions are optimized, assuming fixed values for the regression parameters, the standard deviations and the monotonic transformations. The stimulus configuration is normalized after each optimization step, i.e., its translation and dilation factor are determined by a priori conditions (see above). In case of regression analysis, the stimulus configuration is fixed (after normalization), and the first stage in the iteration is skipped.

In the second stage, the regression parameters and standard deviations are optimized for a fixed stimulus configuration and known monotonic transformations on the data. As discussed above, this latter optimization involves only individual terms in the ML criterium (i.e., $L_d(k)$, $L_p(k)$

[14]Non-linear optimizations in XGms are performed using the iterative routine DMNF from the Netlib library.

and $L_a(k)$) and is therefore fairly simple and efficient, especially in case of a vector-product model and a Gaussian PDF, with $r = 2$, where it reduces to solving a set of simultaneous linear equations. In all other cases, a (slower) non-linear optimization is required to find the regression parameters and standard deviations.

Non-metric MDS or regression adds a third stage to each iteration step in the minimization of the ML criterium. In this third stage, the monotonic transformations that minimize the ML criterium for the current stimulus configuration, regression parameters and noise standard deviations are updated. The optimizations of the monotonic transformations can again be performed separately for each individual term in the ML criterium (i.e., $L_d(k)$, $L_p(k)$ and $L_a(k)$).

6.4.2 DEGREES OF FREEDOM

One linear regression parameter d_k is required to specify the dissimilarity predictions for subject k, with $k = 1, \ldots, K_d$. Hence, the total number of independent dissimilarity parameters is

$$F_d = K_d + J_d + \sum_{k=1}^{K_d} F_{dt}(k), \tag{5.52}$$

where $J_d = K_d$ in case the standard deviations of all subjects are estimated separately, and $J_d = 1$ in case of stress optimization (see Appendix 5.A). We assume that there are $F_{dt}(k)$ DOF in the transformation T_{dk} of the non-metric dissimilarities $D_{k,i,j}$ into the metric dissimilarities $TD_{k,i,j}$ for subject k.

One linear regression parameter m_k is required per preference. If the K_p preferences are split into K_{pv} groups using a vector-product model, and K_{pu} groups using an unfolding model (with $K_{pv} + K_{pu} \leq K_p$), then the total number of independent preference parameters is

$$F_p = K_p + (n-1) \cdot K_{pv} + n \cdot K_{pu} + J_p + \sum_{k=1}^{K_p} F_{pt}(k), \tag{5.53}$$

where $J_p = K_p$ in case the standard deviations of all subjects are estimated separately, and $J_p = 1$ in case of stress optimization (see Appendix 5.A). We assume that there are $F_{pt}(k)$ DOF in the transformation T_{pk} of the non-metric preferences $P_{k,i,j}$ into the metric preferences $TP_{k,i,j}$ for subject k.

Similarly, two linear regression parameters c_k and f_k are required per attribute to specify the predictions. If the K_a attributes are split into K_{av} groups using a vector-product model, and K_{au} groups using an unfolding model (with $K_{av} + K_{au} \leq K_a$), then the total number of independent attribute parameters is

$$F_a = 2 \cdot K_a + (n-1) \cdot K_{av} + n \cdot K_{au} + J_a + \sum_{k=1}^{K_a} F_{at}(k), \qquad (5.54)$$

where $J_a = K_a$ in case the standard deviations of all subjects are estimated separately, and $J_a = 1$ in case of stress optimization (see Appendix 5.A). We assume that there are $F_{at}(k)$ DOF in the transformation T_{ak} of the non-metric attributes $A_{k,i,j}$ into the metric attributes $TA_{k,i,j}$ for subject k.

6.4.3 HYPOTHESIS TESTING

The total number of parameters[15] in the MDS model is

$$F_M = F_{\mathbf{x}} + F_d + F_p + F_a, \qquad (5.55)$$

so that there are $F_O - F_M$ remaining DOFs for judging the residual errors between the actual data and the model predictions.

The following general result on ML estimation [Ram77a, SO91] can be used to compare the goodness-of-fit of alternative multidimensional models. Suppose that L_1 is the optimized ML criterium value for a model with F_1 DOFs, and that L_2 ($> L_1$) is the optimized ML criterium value for a reduced model with F_2 ($< F_1$) DOFs[16]. It can be shown that the statistic

$$G_{12}^2 = 2 \cdot (L_2 - L_1) \qquad (5.56)$$

satisfies a χ^2-distribution (*chi-squared* distribution) with $F_{12} = F_1 - F_2$ DOFs, i.e., the probability that G_{12}^2 exceeds the value χ^2 is given by

$$P(G_{12}^2 > \chi^2; F_{12}) = \frac{\Gamma\left(\frac{F_{12}}{2}, \frac{\chi^2}{2}\right)}{\Gamma\left(\frac{F_{12}}{2}\right)}, \qquad (5.57)$$

in case both models apply and the number of observations F_O is large (in theory, F_O goes to infinity). We have introduced the notation

$$\Gamma(a, x) = \int_x^{\infty} z^{a-1}\, e^{-z}\, dz \qquad (5.58)$$

---

[15]Note that the number of DOF in the stimulus configuration is $F_{\mathbf{x}} = 0$ in case of regression analysis and given by equation (5.20) (if $l \neq 2$) or (5.21) (if $l = 2$) in case of MDS.

[16]For example, the first model may assume non-metric data, while the second model assumes metric data. Or alternatively, the first model may include one preference and/or attribute vector per subject, while the second model may divide the preferences and/or attributes in groups that share a common prediction vector. The more general model may also be the original data with ML criterium $L = 0$ and $F_M = F_O$ DOFs.

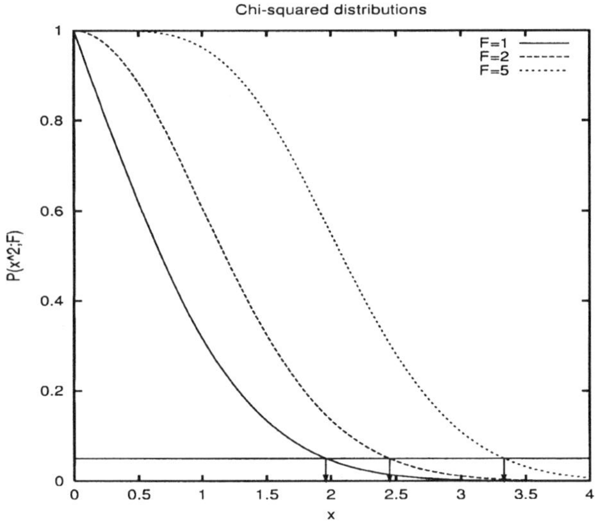

Figure 5.11. Plots of the χ^2-distribution for a number of degrees of freedom F equal to 1, 2, and 5. The horizontal line corresponds to a probability of $\alpha = 0.05$.

for the **incomplete gamma function** and $\Gamma(a) = \Gamma(a, 0)$ for the complete gamma function. Suppose that $\chi^2_\alpha(F_{12})$ is the value for which $P(G^2_{12} > \chi^2_\alpha(F_{12}); F_{12}) = \alpha$, then an observed value $G^2_{12} > \chi^2_\alpha(F_{12})$ is an indication that both models are not equivalent. Indeed, the probability that such a value occurs in case both models are equivalent is less than α. Most often, a probability $\alpha = 0.05$ is used in such a χ^2-test. The chi-squared distributions for DOF equal to 1, 2 and 5 are plotted in Figure 5.11. From this figure, we can for instance derive that $\chi_{0.05}(5) = 3.33$.

6.4.4 CONFIDENCE REGIONS

An important advantage of ML estimation is that it cannot only be used to obtain an estimate for the optimum model parameters themselves, but also for the accuracy with which these parameters are determined by the data. We now discuss in some more detail how some general results from ML estimation theory can be applied to derive such so-called confidence regions for the model parameter estimates. The reader that is not interested in the mathematical details can safely skip the rest of this section.

We can assemble the P_M parameters of a multidimensional model in a (column) vector $\boldsymbol{\theta}$. Note that, because of the normalization conditions applied to the stimulus configuration and the monotonic transformations, P_M will be larger than the number of DOF F_M mentioned in equation (5.55), since F_M is equal to the number of *independent* parameters in the MDS model. The ML estimate $\hat{\boldsymbol{\theta}}$ is the parameter vector that minimizes the ML criterium $L(\boldsymbol{\theta})$ and that satisfies the normalization conditions. The dependence of the ML criterium $L(\boldsymbol{\theta})$ on the available experimental data and the model parameters was discussed in the previous section.

Suppose that the model with parameter vector $\overline{\boldsymbol{\theta}}$ is the true description of the data. Since the ML estimate is based on a finite number of observations (i.e., F_O), the ML estimate $\hat{\boldsymbol{\theta}}$ need not be identical to $\overline{\boldsymbol{\theta}}$. There exists however a general theorem [Tre68] that states that the ML estimates obtained on repeated trials[17] will observe a normal distribution if the number of observations F_O is large. The average vector of this normal distribution is the true parameter vector $\overline{\boldsymbol{\theta}}$, while the variance-covariance matrix is equal to the **Fisher matrix** or **information matrix**

$$H = E\left[(\hat{\boldsymbol{\theta}} - \overline{\boldsymbol{\theta}}) \cdot (\hat{\boldsymbol{\theta}} - \overline{\boldsymbol{\theta}})^t\right] = E\left[\nabla^2 L(\hat{\boldsymbol{\theta}})\right], \tag{5.59}$$

where

$$\nabla^2 L(\boldsymbol{\theta}) = \left[\frac{\partial^2}{\partial\theta_i\,\partial\theta_j} L(\boldsymbol{\theta});\ i,j = 1,\ldots,P_M\right] \tag{5.60}$$

is the symmetric matrix of size $P_M \times P_M$ containing all possible second derivatives of the ML criterium $L(\boldsymbol{\theta})$ [Tre68, Ram78, SO91]. The expectation symbol E indicates that H is the average value obtained for $\nabla^2 L(\hat{\boldsymbol{\theta}})$ in case ML estimates are performed for (infinitely) many trials. Since only one data set is available in an actual experiment, our best estimate for the information matrix H is $\hat{H} = \nabla^2 L(\hat{\boldsymbol{\theta}})$, where $\hat{\boldsymbol{\theta}}$ is equal to the ML estimate for our actual data set. If the number of observations F_O is large, we expect that $H \approx \hat{H}$.

Because the ML estimates are normally distributed, the probability of finding estimates $\hat{\boldsymbol{\theta}}$ within the ellipsoid with equation

$$(\hat{\boldsymbol{\theta}} - \overline{\boldsymbol{\theta}})^t \cdot H^{\#} \cdot (\hat{\boldsymbol{\theta}} - \overline{\boldsymbol{\theta}}) = \chi^2_\alpha(F_M) \tag{5.61}$$

[17] Although the same MDS model with parameter vector $\overline{\boldsymbol{\theta}}$ describes these repeated trials, the F_O data values in each trial are different, because of the statistical nature of the model. As a result, the derived ML estimate $\hat{\boldsymbol{\theta}}$ will also vary on each trial.

is equal to $1-\alpha$, provided that $\chi^2_\alpha(F_M)$ is the value of the chi-squared distribution with F_M DOF for which $P(\chi^2 > \chi^2_\alpha(F_M); F_M) = \alpha$ [SO91]. Such an ellipsoid is called an **asymptotic confidence region**. The matrix H is not invertible since its size P_M is larger than its rank F_M, so that the Moore-Penrose pseudo-inverse $H^\#$ of the matrix H is used in the above expression [Ram78]. In practice, where only one experimental data set is available, we construct confidence regions by applying equation (5.61) with H replaced by its estimate $\hat{H}$.

In MDS, we are mostly interested in **conditional confidence regions**. They are obtained by splitting the total parameter vector into two parts, i.e., $\boldsymbol{\theta} = [\boldsymbol{\theta}_r\ \boldsymbol{\theta}_R]$, where the first component $\boldsymbol{\theta}_r$ contains only a small number r of parameters, and the second component $\boldsymbol{\theta}_R$ contains the remaining parameters. For example, the first component may contain the n coordinates of one stimulus point $\mathbf{x}_i$ or just a single parameter, such as the noise standard deviation of the link function for a specific observer (e.g., $\sigma_a(k)$). The corresponding subdivision of the information matrix is

$$H = \begin{bmatrix} H_{rr} & H_{rR} \\ H_{Rr} & H_{RR} \end{bmatrix}, \tag{5.62}$$

where H_{rr} is usually an invertible matrix of size $r \times r$. Suppose that we now restrict our search to MDS models for which most parameters are fixed to their ML estimate, i.e., $\overline{\boldsymbol{\theta}}_R = \hat{\boldsymbol{\theta}}_R$. These restricted models are characterized by the parameter vector $\boldsymbol{\theta}_r$. We can repeat the above reasoning to derive that the asymptotic confidence region for this parameter vector satisfies

$$(\hat{\boldsymbol{\theta}}_r - \overline{\boldsymbol{\theta}}_r)^t \cdot H_{rr}^{-1} \cdot (\hat{\boldsymbol{\theta}}_r - \overline{\boldsymbol{\theta}}_r) = \chi^2_\alpha(r), \tag{5.63}$$

where $\chi^2_\alpha(r)$ is the value of the chi-squared distribution with r DOF for which the probability $P(\chi^2 > \chi^2_\alpha(r); r) = \alpha$. Such a confidence region is called a conditional confidence region, since it expresses the discrepancy between the ML parameter estimate $\hat{\boldsymbol{\theta}}_r$ and the true parameter vector $\overline{\boldsymbol{\theta}}_r$ under the condition that all remaining parameters $\overline{\boldsymbol{\theta}}_R$ are assumed equal to their ML estimate $\hat{\boldsymbol{\theta}}_R$ [Ram78].

XGms can estimate conditional confidence regions for individual regression parameters (d_k, m_k, c_k and f_k), noise standard deviations ($\sigma_d(k)$, $\sigma_p(k)$ and $\sigma_a(k)$), stimulus positions $\mathbf{x}_i$, attribute and preference vectors ($m_k \cdot \mathbf{p}_k$ and $f_k \cdot \mathbf{a}_k$ or $\mathbf{p}_k$ and $\mathbf{a}_k$, depending on whether an vector-product model or an ideal-point model is used, respectively).

Conditional confidence regions can be graphically represented in case $r = 1$ or $r = 2$. If $r = 1$, then $\hat{H}_{rr} = \hat{\sigma}_1^2$ and equation (5.63) is equivalent to stating that there is an estimated $100 \cdot (1-\alpha)$ percentage chance that the true parameter value $\overline{\theta}_1$ falls within the interval

$$[\hat{\theta}_{1-}, \hat{\theta}_{1+}] = [\hat{\theta}_1 - \hat{\sigma}_1 \cdot \chi_\alpha(1), \hat{\theta}_1 + \hat{\sigma}_1 \cdot \chi_\alpha(1)],$$

(for example, $\chi_{0.05}(1) = 1.96$). In case $r = 2$, the estimated variance-covariance matrix can be written in the form

$$\hat{H}_{rr} = \begin{bmatrix} \hat{\sigma}_1^2 & \hat{\sigma}_1\,\hat{\sigma}_2\,\hat{\rho} \\ \hat{\sigma}_1\,\hat{\sigma}_2\,\hat{\rho} & \hat{\sigma}_2^2 \end{bmatrix}, \tag{5.64}$$

with $\hat{\rho} \in [-1, 1]$, so that the boundary of the $(1-\alpha)$ probability confidence ellips is specified by the following equations

$$\begin{aligned} \overline{\theta}_1(t) &= \hat{\theta}_1(t) + \hat{\sigma}_1 \cdot \chi_\alpha(2)\, \frac{\sqrt{1+\hat{\rho}} \cdot \cos t + \sqrt{1-\hat{\rho}} \cdot \sin t}{\sqrt{2}} \\ \overline{\theta}_2(t) &= \hat{\theta}_2(t) + \hat{\sigma}_2 \cdot \chi_\alpha(2)\, \frac{\sqrt{1+\hat{\rho}} \cdot \cos t - \sqrt{1-\hat{\rho}} \cdot \sin t}{\sqrt{2}}, \end{aligned} \tag{5.65}$$

for $t \in [-\pi, +\pi]$ (for example, $\chi_{0.05}(2) = 2.45$).

6.5 GRAPHICAL USER INTERFACE TO XGMS

The graphical user interface to XGms is depicted in Figure 5.12 for an experimental data set containing both dissimilarity data and attribute data. This interface was derived from the graphical user interface of the XGvis program [BSL+02].

The panel at the top contains the major action buttons. The user can either specify an initial stimulus configuration at startup or can load a new initial stimulus configuration (using "Load CONF") at any time during the XGms session. He/she can return at any time to the last-specified initial configuration using "Init CONF", or can add a random component to the current stimulus configuration using the "Scramble CONF" button. The current stimulus configuration can be kept fixed, and used for regression analysis against the experimental data. A single iteration step in the regression can be triggered with the "Step REG" button, while multiple iterations can be started and stopped with the "Run REG" button. Alternatively, the current stimulus configuration can be used as the initial configuration in an MDS analysis. Starting and stopping of the iterations in such an MDS analysis is controlled by the "Step MDS" and "Run MDS" buttons.

Either the original data (Di and Ai), power-transformed data (Dt and At) or spline-transformed data (Ds and As) can be used in the

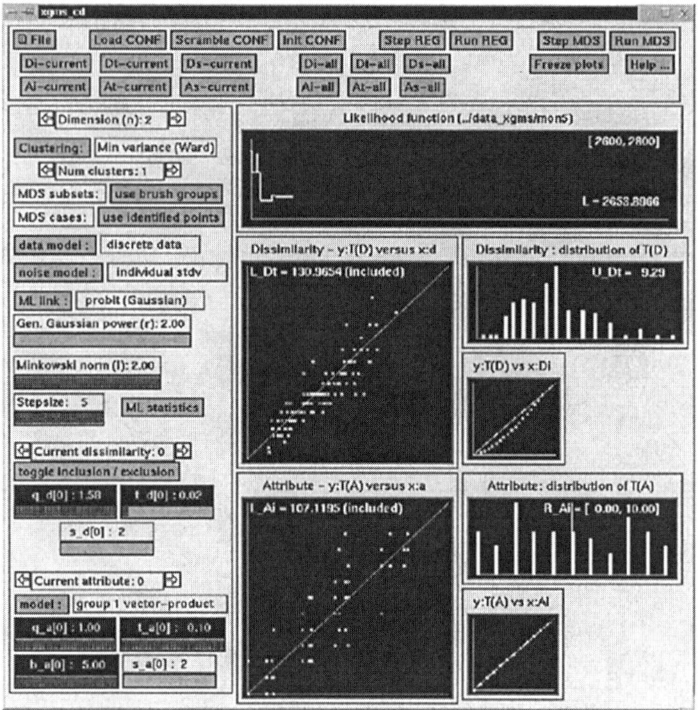

Figure 5.12. XGms graphical user interface.

regression or MDS analysis. Dissimilarities and/or attribute scores from individual subjects can be selected for optimum transformation, or the scores from all subjects can be transformed simultaneously (Di-current, Dt-current or Ds-current versus Di-all, Dt-all or Ds-all). In the example, the original data are used for the attributes, while the dissimilarity data are transformed using a power-like function.

In order to allow processing by other data analysis or visualization programs, most intermediate data in the XGms program can also be output to ASCII files using the options in the "file" menu. Amongst others, the normalized transformed scores $TD^{\star}_{k,i,j}$, $TP^{\star}_{k,i,j}$ and/or $TA^{\star}_{k,i,j}$ and their model predictions $\|\mathbf{x}_i - \mathbf{x}_j\|_l$, $[\mathbf{p}_k, \mathbf{x}_i] - [\mathbf{p}_k, \mathbf{x}_j]$ and $[\mathbf{a}_k, \mathbf{x}_i]$ can all be exported in this way.

The bottom left panel[18] allows to control the parameters of the algorithm, such as

1 the dimension n of the stimulus space,

2 the selection between continuous and discrete input data,

3 the noise model used (individual noise standard deviations or derived from the variations in the data through a common factor as in stress optimization),

4 the link function (logistic or generalized Gaussian),

5 the power r of the generalized Gaussian link function,

6 the Minkowski power l of the distance metric,

7 the number of iteration steps in one call to the optimization routines DMNF and DMNFB (controls speed of convergence),

8 the prediction model used (either vector product or ideal point),

9 the parameters q, t (and b in case of attributes) that control the nonlinear power-like transformations on the individual subsets of the data, and

10 the number s of internal knot points in a spline transformation[19].

The user can obtain ML statistics, including the current number of DOF in the model and the value L of the ML criterium, at any time in the XGms session by hitting the "ML statistics" button. XGms also compares the outcomes on two successive calls to "ML statistics". In this way, χ^2-tests between alternative models can be made.

If data for several subjects and/or attributes are available, then the 'current' selection for the dissimilarity or attribute allows to view the settings and model fits for the current subset of the data. The user can for instance select which parameters of the power-like transformation are fixed to a user-specified value, and which can be optimized by XGms. In the case of preference and attribute data, a choice can be made between a vector-product model and an unfolding model. Attributes and preferences can also be grouped so that they share a common prediction vector. It is moreover possible to (temporarily) exclude the subset of the data currently being viewed from the ML criterium.

[18]Xgms also allows to perform clustering on the stimulus configuration, an option that is not discussed any further here. The left panel therefore also contains two action buttons that allow to interactively control this clustering.

[19]Kruskal transformation is considered as a special case of spline transformation, indicated by s equal to max.

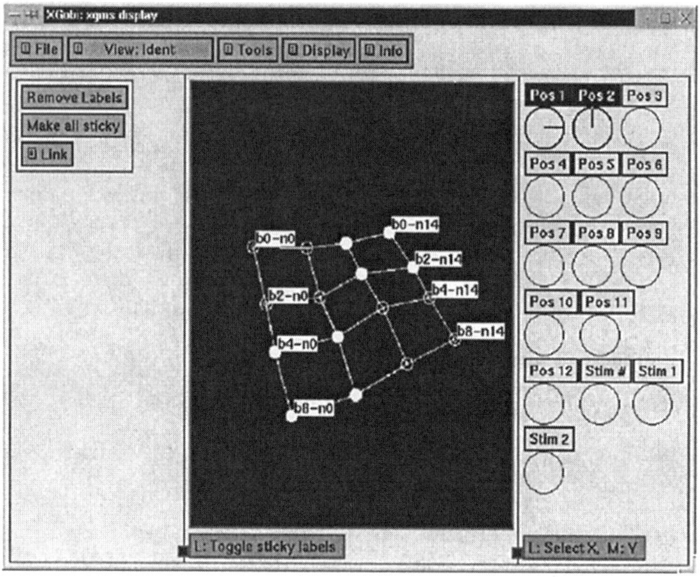

Figure 5.13. XGobi graphical user interface.

Another way of selecting a subset of the available data is through the "use brush groups" and "use identified points" options [BSL+02]. The XGobi program that is used to visualize the stimulus configuration allows to partition the stimuli into non-overlapping subsets by means of brushing[20]. In case "use brush groups" is selected, then only those dissimilarities and preferences for which both stimuli belong to the same subgroup are used in the regression and/or MDS analysis. Similarly, XGobi also allows to identify a subset of the stimulus points. In case "use identified stimuli" is selected, then only those data that involve the identified stimuli are used in the analysis. Both options can be active at the same time, in which case only the dissimilarities (and preferences) between identified stimuli within the same brush group are used in the ML optimization.

The graph at the top in the bottom right panel of the XGms window shows how the ML criterium has varied as a function of time during the XGms session. In addition, scatterplots of transformed experimental

[20]Brushing means that the attributes (such as color, shape, ...) of the stimulus points can be changed. Stimulus points with the same attributes are assumed to belong to the same brush group.

data versus predicted model data, and barplots representing the histograms of the transformed experimental data are shown for the 'current' dissimilarity and attribute. The data in the scatterplots can also be output to the visualization program XGobi in order to examine them more closely. The small panels depict the overall transformations from the input data to the transformed data used in the ML minimization. In order to speed up program execution, these scatterplots can be temporarily frozen by the "freeze plots" action button. The limits ($U_d(k)$, $U_p(k)$ or $[B_a(k) - U_a(k), B_a(k) + U_a(k))$ that can be used to define yet another subset of the 'current' data can be changed by clicking at the desired position in the histogram of the transformed data.

As mentioned before, the stimulus configuration is exchanged between the MDS program XGms and the visualization program XGobi. The graphical user interface to the XGobi program is shown in Figure 5.13. The (two-dimensional) stimulus configuration in the example is displayed as a point plot, and the user can view and/or alter this configuration. In Figure 5.13, two brush groups with 8 stimuli each have been created using open and filled circles, respectively, while 8 of the 16 available stimuli have been identified. The preference and attribute vectors (or ideal points) can also be visualized in XGobi. XGobi allows the visualization of higher-dimensional stimulus configurations by dynamically moving a 2-D projection plane through the stimulus space [SCB98].

7. EXAMPLES OF CONTINUOUS DATA ANALYSIS

In this section, multidimensional model descriptions are illustrated for two separate experimental data sets. The two data sets will require one- and two-dimensional spaces, respectively. In both experiments, numerical category scaling was used to collect subject responses, so that, strictly speaking, the subject responses are not continuous but discrete. We will nevertheless treat the data as continuous in this section. An alternative way of analyzing discrete data will be treated in section 8, and will be applied in section 9 to re-analyze some of the data in this section in order to compare the outcomes of continuous and discrete data analyses.

7.1 CODED IMAGES

7.1.1 EXPERIMENTAL SET-UP

The first data set, collected for images coded at varying bit-rate [RM90], is used to illustrate one-dimensional MDS. Two natural images, a head-and-shoulder scene (Wanda) and an outdoor scene (Terrace),

Figure 5.14. Scenes used in experiment with coded images: Wanda (left) and Terrace (right).

were used in the experiment. The original digital pictures contained 512 × 512 pixels, and 8 bits/pixel. Additional stimuli were created by scale-space coding [MM89] of the images. The bit-rate of the coded images was controlled by the quantization stepsize Q of a uniform quantizer. The quantization stepsize Q varied between 1 and $Q_m = 127$. The case $Q = 1$ resulted in an output image that was identical to the input image, while the choice $Q = 127$ resulted in an image with noticeable blur. Only a subregion of the coded images, containing 240 × 470 pixels, was viewed in the experiments. This restricted region was needed in order to allow simultaneous display of two images. The original images (after cropping) are shown in Figure 5.14.

The images were displayed on a CONRAC 2400 high-resolution monochrome monitor. The grey-value-to-luminance characteristic of the monitor was measured and a lookup table was determined such that the relationship between the grey value g and the luminance L for the combined chain (lookup table, digital-to-analog convertor, and monitor) satisfied the gamma characteristic of equation (5.1) with $g_{max} = 255$, $L_{max} = 110$ cd/m^2, $L_{min} = 1$ cd/m^2, and $\gamma = 2.5$. This calibrated characteristic was verified by a second luminance measurement.

Most viewing conditions satisfied the ITU-R BT.500 recommendation [IR98]. The viewing distance was 1.7 meters, which was equivalent to six times the height of the monitor. At this viewing distance, each pixel on the monitor extended 1 min arc $\times$ 1 min arc of retinal angle. Stimuli were displayed for a fixed duration of 5 seconds. Between two successive stimuli, a uniform adaptation field was displayed during the time it took subjects to enter their response by means of a keyboard. The minimum duration of this adaptation field was 2 seconds, while its luminance of 40 cd/m^2 was approximately equal to the average luminance of the images.

The two subjects, aged 22 and 35 years, had normal or corrected-to-normal visual acuity of 2, measured on a Landolt chart at a distance of 5 m. The subjects performed two different experiments, one using single-stimulus numerical category scaling and one using double-stimulus preference scaling.

In the single-stimulus numerical category scaling experiment, subjects assessed the impairment of the images. There were 15 stimuli per scene. Each stimulus was presented 8 times, in random order, to each of the subjects. Stimuli for the two scenes were alternated. Each subject hence rated $2 \times 15 \times 8 = 240$ stimuli in total. An 11-point numerical category scale, with categories labelled 0 to 10, was used by the subjects to express their judgements.

The double-stimulus preference scaling was performed in a separate experiment. In order to limit the duration of this second experiment, only a subset of 11 out of the 15 stimuli used in the first experiment was included. All 110 combinations of the stimuli were rated twice by the subjects (in two separate sessions). Each subject hence rated $2 \times 110 \times 2 = 440$ stimulus pairs in total. The subjects were instructed to rate differences in the perceived strength of impairment on a 21-point numerical category scale, ranging from -10 (the rightmost image is much more impaired than the leftmost image) to +10 (the leftmost image is much more impaired than the rightmost image).

7.1.2 STIMULUS CONFIGURATIONS

The stimuli were analyzed separately per scene and per subject, resulting in four XGms sessions. Per session, a single one-dimensional stimulus configuration was derived from the attribute data (with 8 repetitions) and the preference data (with 2 repetitions). Gaussian noise distributions were assumed to describe the variations in the experimental

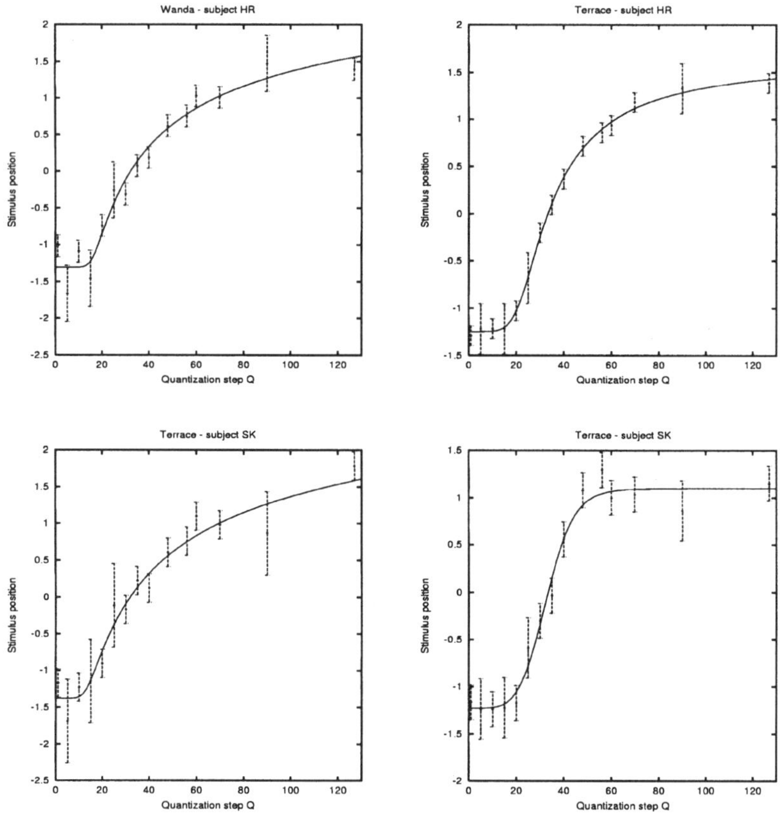

Figure 5.15. Stimulus positions obtained with XGms from experimental data with scale-space coded images. The top and bottom rows correspond to subjects HR and SK, respectively, while the left and right columns correspond to the scenes 'Wanda' and 'Terrace', respectively. Each graph contains the stimulus configuration obtained using MDS, and an approximation of this MDS configuration by a non-linear function $F(Q)$ of the quantization stepsize Q.

data, while inner-product models were used for predicting the average attribute and preference scores. The one-dimensional stimulus configurations were determined for different transformations on the experimental data, and the resulting model fits were compared using a χ^2-test (with probability $\alpha = 0.05$) on the log likelihood. The results of the statistical tests are presented in Tables 5.1 and 5.2, for subjects HR and SK, respectively. In all cases, there was statistical evidence that the general (Kruskal) transformation (denoted by Ps,As) was superior to both

Table 5.1. Likelihood ratio statistics used to compare different alternative model descriptions for the continuous impairment scores of coded images for subject HR. Accepted model simplifications are indicated by (*).

Case	Wanda - HR	Terrace - HR
Pi,Ai|Ps,As	$G^2(23) = 138.3 > 35.2$	$G^2(25) = 130.2 > 37.7$
Pi,Ai|Pt,At	$G^2(7) = 101.5 > 14.1$	$G^2(7) = 68.1 > 14.1$
Pt,At|Ps,As	$G^2(16) = 36.8 > 26.3$	$G^2(18) = 62.1 > 28.9$
Ps,Ai|Ps,As	$G^2(9) = 60.0 > 16.9$	$G^2(9) = 71.3 > 16.9$
Ps,Ai|Ps,At	$G^2(3) = 44.5 > 7.8$	$G^2(3) = 59.1 > 7.8$
Ps,At|Ps,As	$G^2(6) = 15.6 > 12.6$	$G^2(6) = 12.2 < 12.6(*)$

Table 5.2. Likelihood ratio statistics used to compare different alternative model descriptions for the continuous impairment scores of coded images for subject SK. Accepted model simplifications are indicated by (*).

Case	Wanda - SK	Terrace - SK
Pi,Ai|Ps,As	$G^2(26) = 108.6 > 38.9$	$G^2(25) = 108.6 > 40.1$
Pi,Ai|Pt,At	$G^2(7) = 49.8 > 14.1$	$G^2(7) = 55.5 > 14.1$
Pt,At|Ps,As	$G^2(19) = 58.8 > 30.2$	$G^2(20) = 53.1 > 31.4$
Ps,Ai|Ps,As	$G^2(8) = 17.2 > 15.5$	$G^2(9) = 8.8 < 16.9(*)$
Ps,Ai|Ps,At	$G^2(3) = 17.7 > 7.8$	$G^2(3) = 4.0 < 7.8(*)$
Ps,At|Ps,As	$G^2(5) = -0.5 < 11.1(*)$	$G^2(6) = 4.8 < 12.6(*)$

the power-like transformation (denoted by Pt,At) and no transformation (denoted by Pi,Ai). For the attribute data, the general (Kruskal) transformation (Ps,As) and the power-like transformation (Ps,At) were mostly equivalent, except in the case Wanda-HR. In the case Terrace-SK, the attribute data could even be considered to be metric (i.e., a monotonic transformation on the attribute data did not significantly improve the model fit).

Although, in the majority of the cases, transformations on the experimental data had a significant effect on the fit between (transformed) experimental data and model predictions, they only had a very limited influence on the stimulus configuration. We therefore only report the derived stimulus configurations for the case of Kruskal transformations on all data (Ps,As in the above notation). Figure 5.15 not only shows the obtained stimulus configurations, but also the estimated 95% conditional confidence regions on the individual stimulus coordinates. Note that the confidence regions are significantly larger for the four stimuli (corresponding to $Q = 5$, 15, 25 and 90) that were only presented in the single-stimulus scaling experiment.

We expected a priori that the relationship between the image impairment (i.e., the stimulus configuration) and the quantization stepsize Q could be approximated by an expression of the form

$$x_i \approx \hat{x}_i = F(Q_i) = f \cdot P(Q_i; q, Q_t, s) + c = f \cdot \frac{(Q_i^s + Q_t^s)^{q/s} - Q_t^q}{q} + c, \tag{5.66}$$

where Q_i and x_i denote the known quantization stepsize and estimated MDS coordinate for stimulus i, respectively. The three parameters q, Q_t and s determine a power-like transformation, while f and c are two linear regression parameters. Note that the power-like transformation $P(Q_i; q, Q_t, s)$ is very similar to the power-like transformation used in the XGms program; the only difference being the inclusion of an additional parameter s. This parameter controls the steepness of the transition from below to above threshold. As can be observed from figure 5.15, the above non-linear relationship provides a good description of the experimentally-determined MDS configurations. The optimal parameter values for the transformations, which have been obtained by minimization of a quadratic error criterium, are specified in Table 5.3, together with the linear correlation coefficients between the experimental and predicted stimulus configurations. The relatively large values for s indicate a steep transition from below to above threshold in all cases, while the negative values for q imply a saturation in perceived impairment for large values of the quantization stepsize Q. Since the parameters of the power-law functions are difficult to interpret, we have also included an alternative specification by means of three characteristic quantization step sizes Q_v, for $v = 0.1$, $v = 0.5$ and $v = 0.9$, where Q_v is defined by

$$F(Q_v) = (1 - v) \cdot F(0) + v \cdot F(Q_m), \tag{5.67}$$

with $Q_m = 127$ equal to the maximum value of the quantization stepsize. Note from Table 5.3 that the values for $Q_{0.1}$ and $Q_{0.5}$ are very close for the four cases, indicating a similar behavior near threshold, but that there are some distinct differences in the saturation behavior (which are also reflected in the exponent q).

7.1.3 ANALYSIS OF VARIANCE

One of the applications of XGms is to generate normalized transformed data ($TA^\star_{k,i,j}$, $TP^\star_{k,i,j}$, $TD^\star_{k,i,j}$ in the earlier notation) that are approximately metric and that can hence be used as inputs into more traditional statistical analyses such as analysis of variance (ANOVA)

Table 5.3. Parameters of the power-law transformations that are used to derive the predicted stimulus positions from the known quantization step size Q. The rightmost column contains the correlations R between the predicted stimulus configurations and the MDS stimulus configurations derived from the experimental data.

Scene/subject	Q_t	s	q	$Q_{0.1}$	$Q_{0.5}$	$Q_{0.9}$	R
Wanda - HR	16.5	10.	-0.56	18.2	34.8	90.0	0.9827
Wanda - SK	14.4	6.9	-0.39	15.7	34.3	92.2	0.9806
Terrace - HR	23.8	6.0	-1.61	20.8	34.7	74.3	0.9994
Terrace - SK	39.9	4.8	-10.0	21.5	32.8	46.1	0.9918

and linear regression. These statistical methods assume that the measured data are continuous and metric, and that variations in the data can be described by Gaussian noise with a fixed standard deviation[21]. Since analysis of variance and linear regression are used so frequently, we describe them shortly in Appendix 5.B and illustrate them here on our example data. We refer to the excellent book by Draper and Smith [DS98] for a more in-depth discussion.

We specifically discuss how ANOVA can be used to test if predictions following equation (5.66) can indeed describe the experimental data. More specifically, we use predictions $\hat{x}_i = F(Q_i)$, where Q_i is the known quantization stepsize for stimulus i and F is a power-like transformation satisfying equation (5.66). The regression factors c and f in this equation are chosen such that the predictions form a normalized configuration, i.e., $\sum_i \hat{x}_i = 0$ and $\sum_i \hat{x}_i^2 = N$, where N is the number of stimuli ($N = 15$ in our case). This is the same normalization that was also applied to the MDS stimulus configurations. Since the data for different scenes and subjects are analyzed separately, the transformation F is allowed to vary across scene/subject. We use the transformation parameters of Table 5.3 to characterize these power-like transformations.

The results of the ANOVA for the attribute data are summarized in Table 5.4. Application of the general results in appendix 5.B to the specific case at hand leads to the conclusion that the total number of DOF in the normalized transformed attribute scores for one subject and one scene is DOF(total)=120-6-2=112. The number of DOF in the original attribute data is 120, since the 15 stimuli were scored 8 times. A spline transformation with 5 internal knot points, and 6 DOF, is used to transform the non-metric attribute scores to a met-

[21] Although this is very often omitted, it is up to these statistical methods to test the validity of the Gaussian noise assumption for a specific data set [DS98].

Table 5.4. The third column contains the correlation coefficients R between the predicted stimulus positions and the experimentally observed impairment scores for the coded images. The lack of fit statistic in the last column indicates whether or not this correlation R is significantly smaller than the pure-error correlation R_p. Accepted models are indicated by (*).

Scene/Subject	R_p	R	*Lack of fit statistic*
Wanda - HR	0.906	0.891	$F(13, 96) = 1.103 < 1.824$ (*)
Wanda - SK	0.893	0.840	$F(13, 96) = 3.292 > 1.824$
Terrace - HR	0.968	0.953	$F(13, 96) = 3.444 > 1.824$
Terrace - SK	0.950	0.940	$F(13, 96) = 1.363 < 1.824$ (*)

ric scale, while two linear-regression parameters (denoted by c_k and f_k in the previous section) are needed in the definition of the normalized scores. The predictions $\hat{x}_i = F(Q_i)$ are derived from the known quantization stepsizes Q_i by means of a power-like transformation with three parameters, so that DOF(residue)=DOF(total)-3=109 is the number of DOF in the residuals (or prediction errors). Optimal prediction of the $N = 15$ attribute scores involves DOF(lack of fit)=N-2=13 parameters (2 DOF are deduced because of the normalization), so that DOF(pure error)=109-13=96 are left for testing the goodness of fit, based on the distribution $F(13, 96)$. More specifically, an observed ratio $F = \text{MS(lack of fit)}/\text{MS(pure error)} > F_\alpha = 1.824$ is interpreted as a lack of fit (with a probability exceeding 95 percent if $\alpha = 0.05$).

From Table 5.4, we can deduce that the regression models describe most of the variance in the experimental data, since $R > 0.84$ in all cases. The linear correlation coefficients R are larger for the Terrace scene than for the Wanda scene. Since this trend is also observed in the pure-error correlations R_p, this only reflects the fact that the latter scene was harder to judge consistently. An improved model can potentially be pursued in the cases Wanda-SK and Terrace-HR where the probability that the model fits the data to within the experimental error is very small (since the observed value F is larger than $F_{0.05} = 1.824$). In Figure 5.16, we have plotted the (transformed) experimental data as a function of the regression variable $\hat{x}_i = F(Q_i)$. The response data for a given stimulus i are represented by the mean with error bars. The length of the error bars is equal to two times the estimated standard deviation of the scores. If linear regression is an adequate model, then the regression line should pass sufficiently close to all data points. We observe that the regression line does not intersect all error bars in Figure 5.16 for those cases where a lack of fit is reported in Table 5.4.

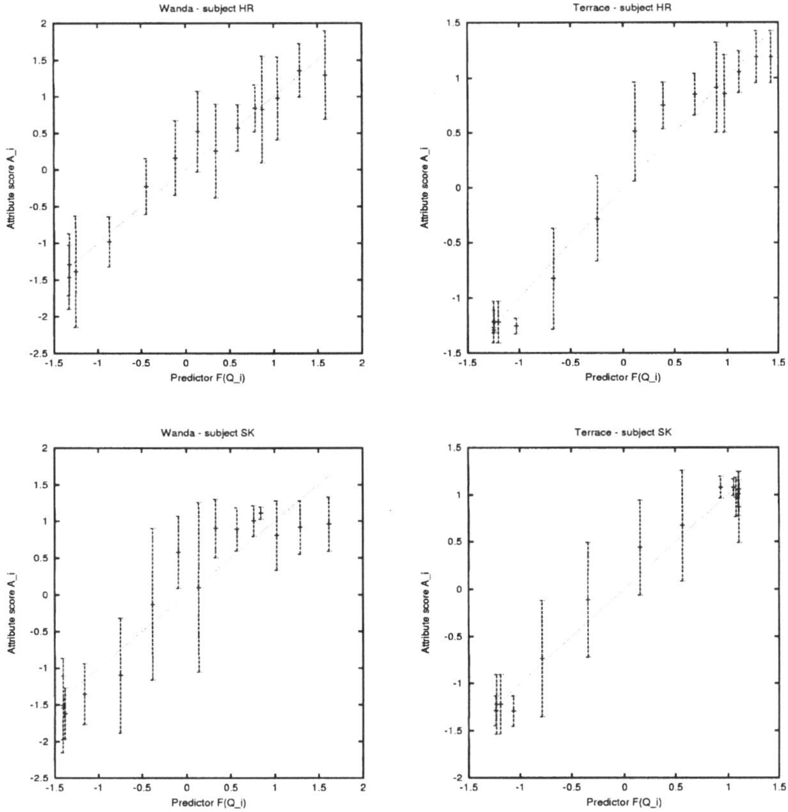

Figure 5.16. Linear predictions for single-stimulus impairment scores given to scale-space coded images. The top and bottom rows correspond to subjects HR and SK, respectively, while the left and right columns correspond to scenes 'Wanda' and 'Terrace', respectively. Each graph contains the experimental data, represented by the mean scores with error bars, as well as the linear regression line. The predictor for image i is a power-like function $\hat{x}_i = F(Q_i)$ of the quantization stepsize Q_i used in the coding of that image, as specified in equation (5.66). The function F varies across scene and subject.

The results of the ANOVA for the preference data are summarized in Table 5.5. There are 110 stimulus pairs that have each been scored twice. Since the preference scores were collected in separate sessions, a separate transformation and normalization was performed on the two sets of preference scores. Each spline transformation involved 5 internal knot points with 6 DOF, while each normalization required one regression

Table 5.5. The third column contains the correlation coefficients R between differences in the predicted stimulus positions and the experimentally observed preference scores for the coded images. The lack of fit statistic in the last column indicates whether or not this correlation R is significantly smaller than the pure-error correlation R_p. Accepted models are indicated by (*).

Scene/Subject	R_p	R	*Lack of fit statistic*
Wanda - HR	0.942	0.911	$F(54, 149) = 1.456 > 1.424$
Wanda - SK	0.925	0.890	$F(54, 149) = 1.193 < 1.424$ (*)
Terrace - HR	0.974	0.965	$F(54, 149) = 0.871 < 1424$ (*)
Terrace - SK	0.922	0.893	$F(54, 149) = 0.937 < 1.424$ (*)

parameter (m_k in the notation of the previous section). Consequently, DOF(total)=220-2(6+1)=206 is the number of DOF in the normalized transformed preference scores. The three non-linear parameters in the power-like function $\hat{x}_i = F(Q_i)$ are the same as in the case of the attribute data. The preference prediction for stimulus pair (i, j) is equal to $\hat{\gamma}_{ij} = \hat{x}_i - \hat{x}_j$. This leaves DOF(residue)=206-3=203 for judging the residual errors. The optimal predictor for the subset of $N = 11$ stimuli that were used in the preference experiments requires DOF(lack of fit)=$N \cdot (N-1)/2-1$=54 parameters, which leaves DOF(lack of fit)=203-54=149 for testing the goodness of fit. More specifically, a $F(54, 149)$ statistical test is applied to the ratio of the mean squares between 'lack of fit' and 'pure error' to determine if the regression model is an adequate description of the data. In Figure 5.17, we have plotted the preference data as a function of the predictions $\hat{\gamma}_{ij}$ for all four combinations of scene and subject. The linear regression model describes the experimental data in three out of the four cases.

7.2 IMAGES WITH BLUR AND NOISE

The second data set, which was already considered shortly in section 2, concerns images degraded by noise and blur [KM96c]. The measurements performed with the 16 different images for each of the three scenes were as follows: double-stimulus dissimilarity scaling of all $15 \times 16/2 = 120$ different image pairs by 5 subjects, and single-stimulus attribute scaling with 4 repetitions on three attributes (perceived noisiness, blur and quality) by 7 subjects (resulting in $16 \times 4 \times 3 = 192$ attribute scores per subject).

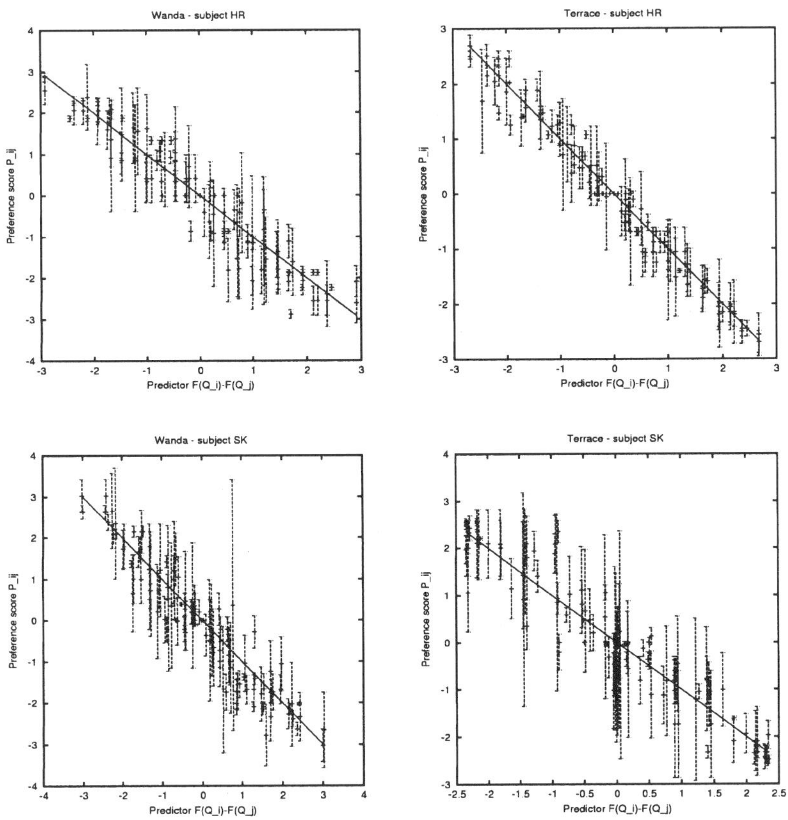

Figure 5.17. Linear predictions for double-stimulus preference scores given to pairs of scale-space coded images. The top and bottom rows correspond to subjects HR and SK, respectively, while the left and right columns correspond to scenes 'Wanda' and 'Terrace', respectively. Each graph contains the experimental data, represented by the mean score with error bars, as well as the linear regression line. The predictor for image pair (i, j) is $\hat{\gamma}_{ij} = \hat{x}_i - \hat{x}_j$, where $\hat{x}_i = F(Q_i)$ is a power-like function of the quantization stepsize Q_i used in the coding of image i. The function F varies across scene and subject.

7.2.1 STIMULUS CONFIGURATIONS

We used XGms with a vector-product prediction model for the attribute scores to find two-dimensional stimulus configurations for each of the scenes. We assumed that the subjects constituted a homogeneous group, so that the attribute vectors that modelled noise, blur and quality were shared by all 7 subjects. A Gaussian noise distribution with subject- and attribute-dependent standard deviation was assumed for all data. The stimulus configurations were determined for different transformations on the experimental data. There was no statistical evidence that a general (Kruskal) transformation on the experimental data performed better than a spline interpolation with two internal knots, so that we will use the latter spline transformations in the subsequent data analysis. The resulting stimulus configurations, together with their conditional confidence regions, are shown in Figure 5.18. Similarly as in the previous example, the stimulus configurations changed only slightly when the transformations on the experimental dissimilarity and attribute data were varied (from metric to non-metric, for instance). It should be noted that comparing multi-dimensional stimulus configurations is more complicated than comparing one-dimensional stimulus configurations, and Appendix 5.C therefore describes in more detail how such comparisons can be performed.

The stimuli in our experiment can be objectively characterized by the coordinate pair (σ_b, σ_n). For each scene, we use the following prediction model

$$\begin{bmatrix} \hat{x} & \hat{y} \end{bmatrix} = \begin{bmatrix} P_b(\sigma_b) & P_n(\sigma_n) \end{bmatrix} \cdot \begin{bmatrix} f_{xx} & f_{yx} \\ f_{xy} & f_{yy} \end{bmatrix} + \begin{bmatrix} c_x & c_y \end{bmatrix}, \quad (5.68)$$

to approximate the stimulus coordinates in the MDS configuration, where P_b and P_n are optimized power-law transformations on the image blur (σ_b) and noise (σ_n) standard deviations, respectively. The transformations P_b and P_n are optimized per scene. Since these transformations preserve the end points, and since there are only four levels of noise and blur, they can be characterized by how they transform the two intermediate values for the blur and noise standard deviations, as shown in Table 5.6. The comparisons between the predicted stimulus configurations and the MDS stimulus configurations are shown in Figure 5.19, while some derived statistics, such as the inner-product correlations, are mentioned in Table 5.6. The correlations between both configurations are very high in all cases.

In order to assist in the interpretation, the transformation matrix of dimension 2 can be decomposed as follows

$$\begin{aligned} F &= \begin{bmatrix} 1 & 0 \\ 0 & \lambda \end{bmatrix} \begin{bmatrix} 1 & 0 \\ \cos\phi & \sin\phi \end{bmatrix} \begin{bmatrix} f & 0 \\ 0 & f \end{bmatrix} \begin{bmatrix} \cos\omega & \sin\omega \\ -\sin\omega & \cos\omega \end{bmatrix} \\ &= \begin{bmatrix} f\cdot\cos\omega & f\cdot\sin\omega \\ f\lambda\cdot\cos(\omega+\phi) & f\lambda\cdot\sin(\omega+\phi) \end{bmatrix} = \begin{bmatrix} f_{xx} & f_{yx} \\ f_{xy} & f_{yy} \end{bmatrix}. \quad (5.69) \end{aligned}$$

The important parameters are λ, the ratio of the weights attributed to the coordinates $P_b(\sigma_b)$ and $P_n(\sigma_n)$, respectively, and ϕ, the angle

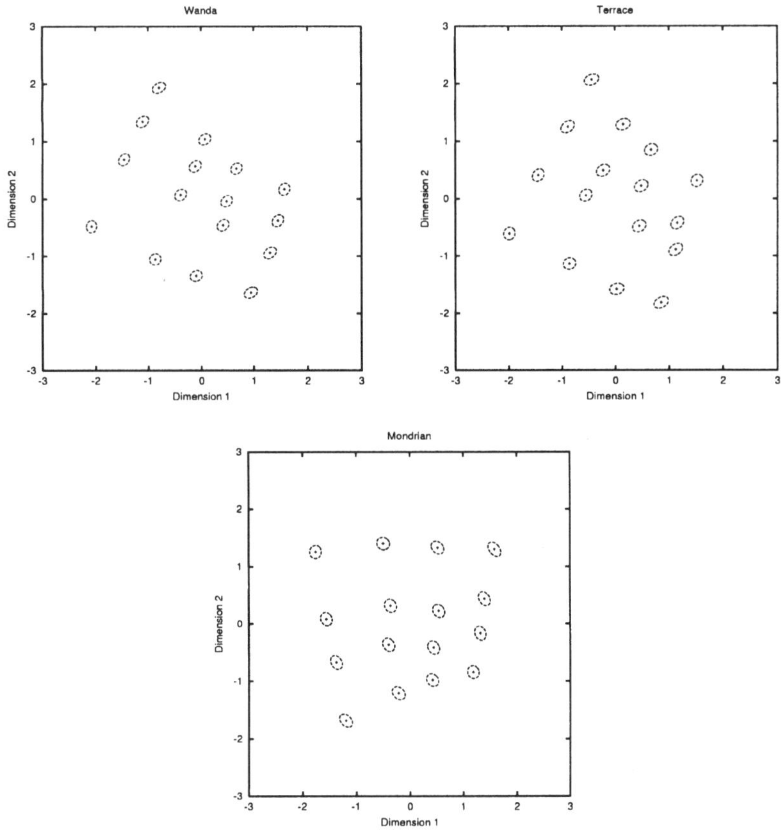

Figure 5.18. Two-dimensional stimulus configurations from the blur/noise experiment for scenes 'Wanda' (upper left), 'Terrace' (upper right) and 'Mondrian' (lower middle). The ellipses indicate the conditional confidence regions for the individual stimulus points.

between the transformed coordinate axes. The dilation factor f and the rotation ω are again irrelevant because of the indeterminacy in the MDS stimulus configurations.

7.2.2 ANALYSIS OF VARIANCE

The XGms program can export transformed normalized scores that are approximately metric (i.e., $TD^{\star}_{k,i,j}$ and $TA^{\star}_{k,i,j}$ in the earlier notation), and that can hence be used for performing ANOVA. In this section, we investigate if the model of equation (5.68), using the transformed parameter values of Table 5.6, can describe the experimental

Table 5.6. The columns on the left specify the transformed parameter values used for predicting the stimulus configuration in the experiments with blur and noise. The three parameters on the right result from the comparison between the predicted stimulus configurations and the MDS configurations: they are the angles ϕ (in degrees) and relative weights λ on the transformed coordinate axes $P_b(\sigma_b)$ and $P_n(\sigma_n)$, and the inner-product correlations ρ between both configurations.

	$P_b(\sigma_b)$				$P_n(\sigma_n)$						
Scene	0	$\sqrt{2}$	2	$2\sqrt{2}$	0	7	10	14	ϕ	λ	ρ
Wanda	0	0.73	1.46	$2\sqrt{2}$	0	5.65	9.46	14	83.1	0.280	0.9937
Terrace	0	0.89	1.60	$2\sqrt{2}$	0	5.10	9.52	14	79.8	0.216	0.9915
Mondrian	0	0.87	1.64	$2\sqrt{2}$	0	5.50	9.58	14	83.6	0.232	0.9913

data. More specifically, we want to know if the distances between the model coordinates are proportional to the transformed dissimilarities, and if specific linear combinations of these coordinates are proportional to the different transformed attribute data (for blur, noise and quality). Since the coordinates of equation (5.68) are themselves linear combinations of $P_b(\sigma_b)$ and $P_n(\sigma_n)$, we can alternatively also question if the transformed attribute scores can be described by linear combinations of $P_b(\sigma_b)$ and $P_n(\sigma_n)$.

The normalized transformed dissimilarity scores have DOF(total)= $5 \times (120 - 3 - 1) = 580$, which can be motivated as follows: 120 stimulus pairs were judged by 5 subjects, an optimized spline function with 2 internal knot points (and DOF=3) was applied to the dissimilarity scores of each individual subject, while one linear regression parameter d_k that defines the ratio between distances and transformed dissimilarities is used in the normalization. There are DOF(regression)=6 *relevant* parameters[22] in the model of equation (5.68), so that DOF(residue)=574. The optical predictor for the dissimilarities between $N = 16$ stimuli contains DOF(lack of fit)= $N \cdot (N-1)/2 - 1 = 119$, which leaves DOF(pure error)=455. An $F(119, 455)$ statistical test on the ratio of the mean squares between 'lack of fit' and 'pure error' hence determines the goodness of fit for the dissimilarity scores.

The data for a single attribute (blur, noise or quality) has DOF(total)= $7 \times (16 \cdot 4 - 3 - 2) = 413$, which can be motivated as follows: 16 stimuli were judged 4 times by 7 subjects, an optimized spline function with 2 internal knot points (and DOF=3) was applied to the attribute scores of each individual subject, while two linear regression parameters c_k

[22]The relevant parameters in the linear part of equation (5.68) are ϕ and λ, while each of the non-linear transformations $P_b(\sigma_b)$ and $P_n(\sigma_n)$ is characterized by two parameters.

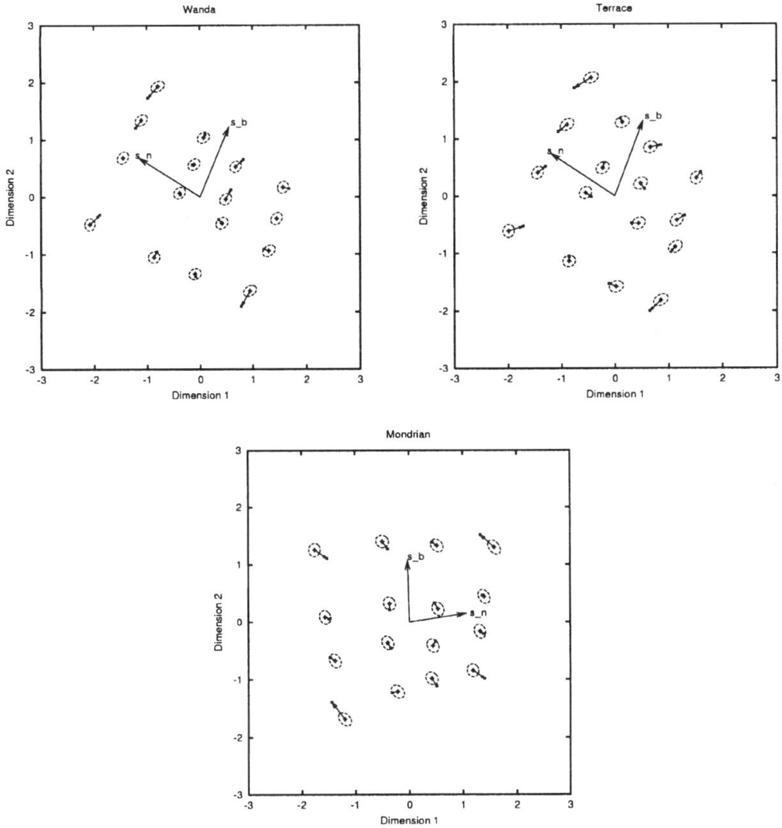

Figure 5.19. Two-dimensional stimulus configurations from the blur/noise experiment for scenes 'Wanda' (upper left), 'Terrace' (upper right) and 'Mondrian' (lower middle). The start and end points of the arrows indicate the stimulus positions according to the MDS model and the predictive model of equation (5.68), respectively. The directions corresponding to the (transformed) blur and noise coordinates $P_b(\sigma_b)$ and $P_n(\sigma_n)$ are indicated by s_b and s_n, respectively.

and f_k are used in the normalization. The number of regression parameters DOF(regression)=7, and consequently DOF(residue)=406. In addition to the six parameters that define the predicted 2-D stimulus configuration, one extra parameter is needed to specify the orientation of the attribute vector. The optimal attribute predictor requires DOF(lack of fit)=14, so that DOF(pure error)=392. An $F(14, 392)$ statistical test on the ratio of the mean squares between 'lack of fit' and 'pure error' hence determines the goodness of fit for the attribute scores.

Table 5.7. The fourth column contains the correlation coefficients R between the predicted stimulus positions and the experimentally observed scores for the images with noise and blur. The lack of fit statistic in the last column indicates whether or not this correlation R is significantly smaller than the pure-error correlation R_p. Accepted models are indicated by (*).

Attribute	*Scene*	R_p	R	*Lack of fit measure*
Dissimilarity	Wanda	0.988	0.981	$F(119, 455) = 2.147 > 1.259$
	Terrace	0.987	0.979	$F(119, 455) = 2.390 > 1.259$
	Mondrian	0.988	0.975	$F(119, 455) = 3.901 > 1.259$
Blur	Wanda	0.878	0.863	$F(14, 392) = 3.150 > 1.717$
	Terrace	0.895	0.884	$F(14, 392) = 2.791 > 1.717$
	Mondrian	0.889	0.882	$F(14, 392) = 1.583 < 1.717$ (*)
Noise	Wanda	0.955	0.953	$F(14, 392) = 1.314 < 1.717$ (*)
	Terrace	0.964	0.954	$F(14, 392) = 7.849 > 1.717$
	Mondrian	0.962	0.957	$F(14, 392) = 3.532 > 1.717$
Quality	Wanda	0.919	0.894	$F(14, 392) = 8.215 > 1.717$
	Terrace	0.939	0.927	$F(14, 392) = 5.329 > 1.717$
	Mondrian	0.911	0.897	$F(14, 392) = 4.003 > 1.717$

The results of the ANOVA using repetitions across subjects are reported in Table 5.7. They indicate that the predictive models describe most of the variance in the dissimilarity data and the attribute data ($R > 0.85$ in all cases). For the attribute data, the linear correlation coefficient R is largest for perceived noise, and smallest for perceived blur. Since this trend is also observed in the pure-error correlations R_p, this only reflects the fact that blur is harder to judge consistently than quality, while noise is most easy to judge. In almost all cases, there is statistical evidence that the linear prediction model can be improved, i.e., there are some effects (mostly the interactions between noise and blur) that are not fully accounted for by the model. These deviations between the linear predictions and the experimental data are illustrated for the case of quality in Figure 5.20. Especially the quality score for the original image deviates substantially from the regression line.

8. MDS MODELLING OF DISCRETE DATA

In the MDS modelling of section 6 it was assumed that the experimental data are continuous. Although discrete data, such as obtained through numerical category scaling, can be entered as continuous data in XGms (as was done in the examples of the previous section), this constitutes an obvious approximation that does not make optimal use of the discrete nature of the data. If the quantization noise that is introduced by the categorical scaling can be assumed to be substantially

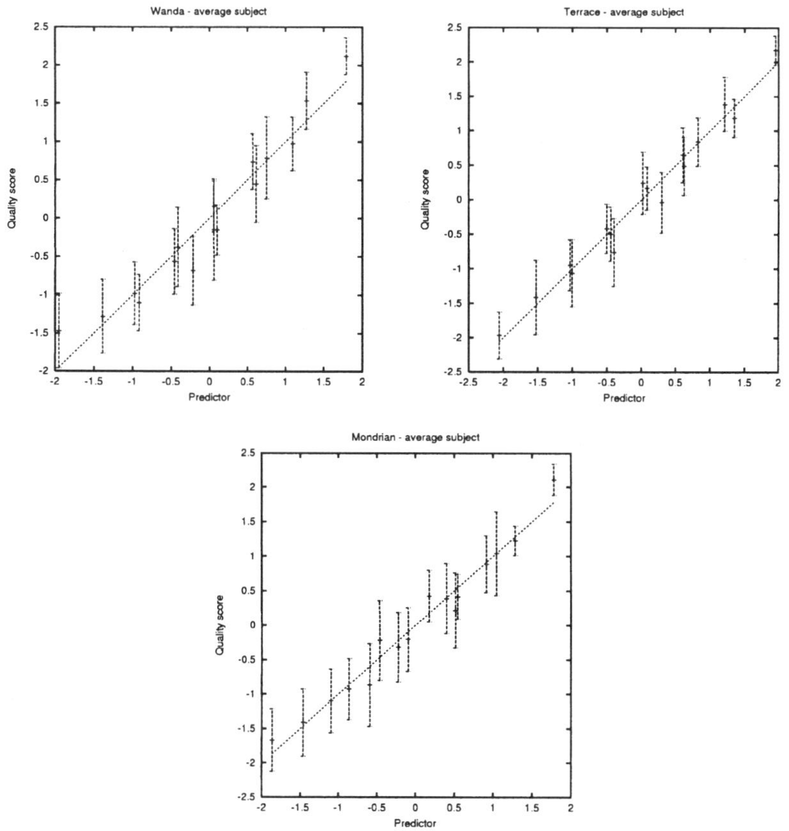

Figure 5.20. Quality predictions versus experimental data in the blur/noise experiment for scenes 'Wanda' (upper left), 'Terrace' (upper right) and 'Mondrian' (lower middle). Each graph contains the experimental data, represented by the mean score with error bars (plus/minus one standard deviation), as well as the linear regression line.

smaller than the internal noise underlying the judgements (i.e., if many categories are used), this approximation may however be sufficiently accurate.

An alternative approach towards analyzing discrete data is discussed in this section. The main implication of using discrete data is that the maximum-likelihood (ML) criterium used within XGms must be modified. The required modification for the case of attribute data will be discussed in detail. The necessary modifications for dissimilarity data and preference data [Bos01, MB01] are very similar and will not be discussed explicitly.

8.1 THURSTONE MODELLING OF ATTRIBUTE DATA

In the following discussion, it is assumed that subjects can judge stimuli on a specified attribute in a consistent way. Based on the arguments discussed in section 3, Thurstone [Thu27] proposed to model this by a two-step process: the perception of a stimulus is modelled by a mapping of this stimulus to a sensation strength on an (attribute-specific) one-dimensional axis, and the response of the subject on a numerical category scale is assumed to be monotonically related to this sensation strength.

Applied to the case at hand, it is assumed that stimulus i triggers a sensation strength $u_i(k)$ for the k-th attribute (or k-th subject). The assumption that a stimulus need not have the same sensation strength on successive trials is modelled by assuming that $u_i(k)$ is a stochastic variable [Thu27, Tor58]. More specifically, the cumulative probability $P[u_i(k) < u]$ that the sensation strength $u_i(k)$ is smaller than u is a monotonically increasing function of u with values between 0 and 1, i.e.,

$$P[u_i(k) < u] = \Phi\left[u - \beta_i(k); \sigma_a(k)\right], \qquad (5.70)$$

where $\sigma_a(k)$ denotes the standard deviation in sensation strengths for attribute k. This standard deviation is assumed to be stimulus-independent. The average sensation strength for stimulus i and attribute k is denoted by $\beta_i(k)$, for $i = 1, \ldots, N$ and $k = 1, \ldots, K_a$. The **cumulative distribution** Φ is also referred to as the link function and most often chosen to be a logistic distribution [Agr89]

$$\Phi(u;\sigma) = \frac{1}{1+\exp\left(-\frac{u}{\rho}\right)}, \qquad (5.71)$$

with $\rho = \sigma\,\sqrt{3}/\pi$, or a generalized Gaussian distribution

$$\Phi(u;\sigma) = \frac{1}{2}\left[1 + \operatorname{sgn}(u)\,\left(1 - \frac{\Gamma\left(\frac{1}{r}, \left|\frac{u}{\rho}\right|^r\right)}{\Gamma\left(\frac{1}{r}\right)}\right)\right], \qquad (5.72)$$

with $\rho = \sigma\sqrt{\Gamma(1/r)/\Gamma(3/r)}$, where $\Gamma(a,x)$ denotes the incomplete gamma function. The generalized Gaussian distribution for $r = 2$ is the cumulative normal distribution

$$\Phi(u;\sigma) = \frac{1}{2}\left[1 + \operatorname{erf}\left(\frac{u}{\rho}\right)\right], \qquad (5.73)$$

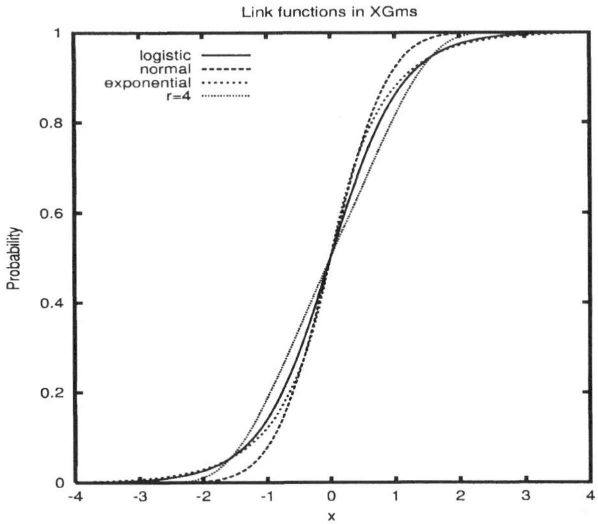

Figure 5.21. Plots of link functions with unit standard deviation. The different curves correspond to the logistic link and the generalized Gaussian links with exponent $r = 2$ (normal), $r = 1$ (exponential) and $r = 4$.

with $\rho = \sigma\sqrt{2}$ and

$$\mathrm{erf}(x) = \frac{2}{\sqrt{\pi}} \int_0^x \exp(-t^2)\, dt, \tag{5.74}$$

while the generalized Gaussian distribution for $r = 1$ is the cumulative exponential distribution

$$\Phi(u; \sigma) = \frac{1}{2}\left[1 + \mathrm{sgn}(u) \cdot \left(1 - \exp\left(-\left|\frac{u}{\rho}\right|\right)\right)\right], \tag{5.75}$$

with $\rho = \sigma/\sqrt{2}$. Some example link functions with unit standard deviation are drawn in figure 5.21.

The sensation strengths $u_i(k)$ cannot be observed directly. In the case of continuous responses, the mapping from internal sensation strengths to observed responses was modelled by a non-linear but monotonic transformation. In the current case of discrete responses, it is assumed that the mapping from the internal sensation strength $u_i(k)$ to the discrete response category $\tilde{u}_i(k)$ reported by the subject can be described by a (non-uniform) quantization process[23] with decision boundaries $\alpha_q(k)$, for $q = 2, \ldots, Q_k$ [Tor58, McC80]. Mathematically, this is expressed by the fact that the probability $P[\tilde{u}_i(k) < q]$ that a response category $\tilde{u}_i(k)$ smaller than q is selected for stimulus i is identical to the probability that the internal sensation strength $u_i(k)$ is smaller than the category boundary $\alpha_q(k)$, i.e.,

$$P[\tilde{u}_i(k) < q] = P\left[u_i(k) < \alpha_q(k)\right] = \Phi\left[\, \alpha_q(k) - \beta_i(k); \sigma_a(k) \,\right], \quad (5.76)$$

or equivalently

$$L_{q|ki} = \Phi^{-1}\left(P[\tilde{u}_i(k) < q]\right) = \frac{\alpha_q(k) - \beta_i(k)}{\sigma_a(k)}, \quad (5.77)$$

for $i = 1, \ldots, N$ and $q = 2, \ldots, Q_k$. Note that the Q_k available categories are denoted by integer numbers $q = 1, 2, \ldots, Q_k$. Since $\alpha_1(k) = -\infty$ and $\alpha_{Q_k+1}(k) = +\infty$, only $Q_k - 1$ category boundaries need to be specified.

The experimental data for attribute k can be summarized in a so-called **two-way contingency table** [Agr89, Agr90], in which cell (i, q) contains the number of times n_{kiq} that stimulus i is assigned to category q when questioned about attribute k. If n_{ki} is the total number of times that stimulus i is questioned about attribute k, then $f_{kiq} = n_{kiq}/n_{ki}$ is the frequency with which category q is selected for stimulus i on attribute k.

The Thurstone model assumes that the $N \cdot (Q_k - 1)$ observed numbers

$$\hat{L}_{q|ki} = \Phi^{-1}\left(\sum_{p=1}^{q-1} f_{kip}\right) \quad (5.78)$$

for $i = 1, \ldots, N$ and $q = 2, \ldots, Q_k$, approximate the model quantities $L_{q|ki}$ of equation (5.77), and can hence be decomposed into 2 independent factors with N parameters $\beta_i(k)/\sigma_k$, for $i = 1, \ldots, N$, representing the (normalized) stimulus strengths and $Q_k - 1$ parameters $\alpha_q(k)/\sigma_k$, for $q = 2, \ldots, Q_k$, representing the (normalized) category boundaries for attribute k. The observed numbers $\hat{L}_{q|ki}$ are called **logits** or **probits** if a logistic or a cumulative normal distribution Φ is assumed for the internal noise, respectively [McC80, Agr89].

Equation (5.77), with $L_{q|ki}$ replaced by the corresponding observations $\hat{L}_{q|ki}$, looks like a classical analysis-of-variance (ANOVA) decomposition

[23] A more extensive discussion on quantization will be presented in chapter 10.

[DS98], with the exception that the numbers $\hat{L}_{q|ki}$ cannot be assumed to be normally distributed with a constant variance. That is the reason why an iterative maximum likelihood estimation procedure will be required to estimate the model parameters $\alpha_q(k)/\sigma_k$, and $\beta_q(k)/\sigma_k$ instead of a minimum-mean-square error procedure. This ML estimator, which is discussed in more detail in the next section, will approximate the experimental frequency f_{kiq} of scoring stimulus i in category q by the model probability

$$p_{kiq} = \Phi\left[\, \alpha_{q+1}(k) - \beta_i(k); \sigma_a(k) \,\right] - \Phi\left[\, \alpha_q(k) - \beta_i(k); \sigma_a(k) \,\right], \quad (5.79)$$

for $i = 1, \ldots, N$ and $q = 1, \ldots, Q_k$. The number of expected observations in category q for stimulus i and attribute k is then equal to $\hat{n}_{kiq} = n_{ki} \cdot p_{kiq}$ and should be approximately equal to n_{kiq}, which is the number of observations in the actual experiment.

8.2 ML OPTIMIZATION FOR A SINGLE ATTRIBUTE

The **likelihood function** describes the probability of finding the specific frequency distribution f_{kiq} that is obtained in an actual experiment. This likelihood function depends on the model probabilities p_{kiq}. More specifically, the probability of finding the frequency distribution $f_{kiq} = n_{kiq}/n_{ki}$ for stimulus i is described by a **multinomial distribution** [Agr89]

$$P_{ki} = n_{ki}! \prod_{q=1}^{Q_k} \frac{p_{kiq}^{\,n_{kiq}}}{n_{kiq}!}. \quad (5.80)$$

Assuming that the observations for the different stimuli are performed independently, we obtain that the probability of the entire experimental data set for attribute k is equal to the likelihood function

$$P_a(k) = \prod_{i=1}^{N} P_{ki} = \prod_{i=1}^{N} n_{ki}! \prod_{q=1}^{Q_k} \frac{p_{kiq}^{\,n_{kiq}}}{n_{kiq}!} \quad (5.81)$$

which needs to be maximized with respect to the $N + Q_k$ model parameters that determine the probabilities p_{kiq} in a way specified by equation (5.79). These parameters are: the N stimulus positions $\beta_i(k)$, for $i = 1, \ldots, N$, the $Q_k - 1$ category boundaries $\alpha_q(k)$, for $q = 2, \ldots, Q_k$, and the noise standard deviation $\sigma_a(k)$. This maximization of $P_a(k)$ is equivalent to minimizing the inverse of the log-likelihood function

[Agr90], i.e.,

$$L_a(k) = -\log P_a(k) = -\sum_{i=1}^{N}\sum_{q=1}^{Q_k} n_{kiq}\log(p_{kiq}) - \sum_{i=1}^{N}\sum_{q=1}^{Q_k}\log\left(\frac{n_{ki}!}{n_{kiq}!}\right), \tag{5.82}$$

where the second term can be omitted in the optimization since it does not depend on the model parameters[24].

Note that any translation and scaling of the stimulus strengths $\beta_i(k)$ can be compensated by an identical translation of the category boundaries $\alpha_q(k)$ and a change in the noise standard deviation $\sigma_a(k)$. Stimulus strengths hence belong to an interval scale and are only determined up to a linear transformation. In order to remove this indeterminacy, we can put normalization conditions on the stimulus strengths, i.e., $\sum_{i=1}^{N}\beta_i(k) = 0$ and $\sum_{i=1}^{N}\beta_i^2(k) = N$. This normalization reduces the number of free parameters in the Thurstone model to $N + Q_k - 2$.

8.3 ML OPTIMIZATION IN MDS

Based on the above results, we argue that the MDS program XGms can be adapted to numerical category scaling data by replacing the minimization of the log-likelihood function L_a of equation (5.31) by a minimization of an alternative log-likelihood function

$$L_a = \sum_{k=1}^{K_a} L_a(k), \tag{5.83}$$

where $L_a(k)$ is now given by equation (5.82). In the expression for the probabilities p_{kiq}, the stimulus strengths $\beta_i(k)$ for the different attributes k have to be replaced by their predictions according to the MDS model, i.e.,

$$\beta_i(k) = c_k + f_k \cdot [\mathbf{a}_k, \mathbf{x}_i], \tag{5.84}$$

for $i = 1, \ldots, N$, where $[\cdot,\cdot]$ denotes either a vector-product or an unfolding model. This modification to the model of subsection 8.2 is caused by the fact that the stimulus strengths are not estimated separately

[24]In order to avoid numerical problems with zero (or very small) probabilities, the above theoretical probabilities p_{kiq} can be replaced by stabilized probabilities $p^*_{kiq} = (p_{kiq}+p_o)/(1+p_o \cdot Q_k)$, where p_o is a small constant (such as $p_o = 10^{-4}$). The physical interpretation is that we assume a very small, but nonzero, minimum probability $p_o/(1+p_o \cdot Q_k) \approx p_o$ for every category q to be used when scoring stimulus i on attribute k.

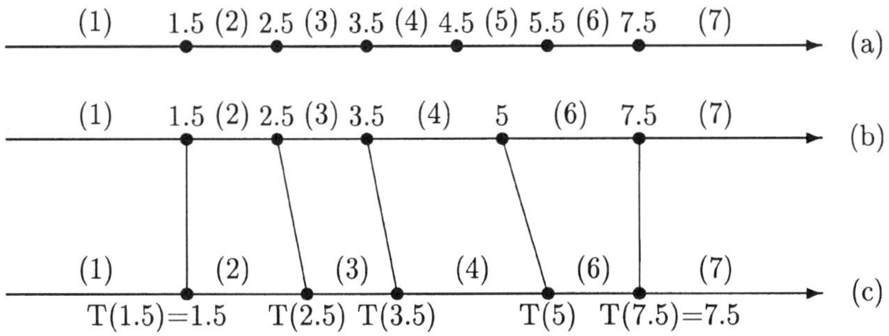

Figure 5.22. Quantizer decision boundaries for metric and non-metric MDS. In (a), the categories range from (1) to (7) and all categories are used, in which case the category boundaries for a uniform quantizer are located at half-integer values. In (b), it is assumed that category (5) is not used by the observer, and should hence be removed: the category boundaries 4.5 and 5.5 merge into a new category boundary equal to 5. In (c), the internal category boundaries are mapped by a nonlinear transformation T that preserves the outer category boundaries; the result is a non-linear quantizer.

for each attribute k, but are derived from a common stimulus configuration $\mathbf{x}_i$, for $i = 1, \ldots, N$, in an n-dimensional space (according to the homogeneity of perception principle). The number of DOF in this stimulus configuration is denoted by $F_{\mathbf{X}}$ and is identical to what was previously obtained for continuous data (see equations (5.20) and (5.21) for an MDS space with a general Minkowski metric and with a Euclidean metric, respectively).

For each of the K_a attributes, two linear regression parameters c_k and f_k and one attribute vector $\mathbf{a}_k$ are used to specify the predictions. The attribute vector is normalized in case of the vector-product model. In addition, there is one noise standard deviation $\sigma_a(k)$ per attribute. Similarly as in the case of continuous data, we may choose to estimate a separate noise standard deviation $\sigma_a(k)$ for each attribute k, or derive the noise standard deviations from the variations in the data using a common factor (as in appendix 5.A). The number of DOF in the attribute data F_a is hence again expressed by equation (5.54).

Metric and non-metric MDS in the case of discrete data corresponds to assuming a uniform or a non-uniform quantizer between the internal sensations strengths $u_i(k)$ and the categorical subject responses $\tilde{u}_i(k)$. In XGms, the categorical responses $\tilde{u}_i(k)$ are represented by integer numbers. Since there are only a limited number of categories, we can assume that these integers are restricted to the range $[m, M]$. In case of metric MDS, the decision boundaries for attribute k are uniformly distributed in the interval $[m, M]$, and will by default be positioned at half-integer values. This is illustrated in part (a) of Figure 5.22 for the case $m = 1$ and $M = 7$. Not all categories in the admissible range $[m, M]$ need however be used by the subject, i.e., the number of used categories Q_k may be smaller than $M - m + 1$. A category needs to be removed if it is not used, and the category boundaries $\alpha_q(k)$, $q = 2, \ldots, Q_k$, adjusted accordingly. This is illustrated in part (b) of Figure 5.22 for the case that one category (5) is not used, and $Q_k = 6$ is smaller than $M - m + 1 = 7$. In case of non-metric MDS, the $Q_k - 3$ internal category boundaries may be mapped by a monotonic transformation that preserves the outer category boundaries, hence resulting in a non-uniform quantizer, as illustrated in part (c) of Figure 5.22. XGms allows for the same monotonic transformations that were also used in the case of continuous data, i.e., the power-like transformation described by equation (5.44), which is controlled by three parameters q, x_t and b, or a monotonically increasing spline function. The most general monotonic transformation arises by assigning all category boundaries of the uniform quantizer to be knot points of the spline function. The category boundaries $\alpha_q(k)$, $q = 2, \ldots, Q_k$ that minimize the log-likelihood $L_a(k)$ are hence determined indirectly by optimizing the transformation T_{ak} from uniform to non-uniform category boundaries for each subject k. The number of DOF in the category boundaries is either $F_{at}(k) = 0$, in case of a uniform quantizer, $F_{at}(k) = 3$, in case of a power-like transformation[25], $F_{at}(k) = Q_k - 3$ in case of a Kruskal transformation, or $F_{at}(k) = s_a(k) + 1$ in case of a spline interpolation with $s_a(k)$ internal knot points (with $0 \leq s_a(k) < Q_k - 3$).

Note that the MDS model of this subsection and the single-attribute model of subsection 8.2 are identical in case there is only one attribute which is modelled by a vector-product model applied to a stimulus configuration with dimension $n = 1$ and quantized by a general quantizer with $Q_k - 3$ internal category bounds (in which case $F_{\mathbf{X}} = N - 2$, $F_a = 3 + (Q_k - 3) = Q_k$ and $F_M = N + Q_k - 2$ is the total number of DOF in the model).

The only modification that must be made to the model for continuous attribute data, presented in Figure 5.9, to adapt it for discrete data is that a uniform quantizer must be added after the nonlinear response mapping $R_{ak}(\cdot) = T_{ak}^{-1}(\cdot)$. The response mapping $R_{ak}(\cdot)$ and this uni-

[25]Note that for dissimilarities and preferences, $F_{dt}(k) = F_{pt}(k) = 2$ in case of an optimized power-like transformation.

form quantizer together constitute the non-uniform quantizer that was discussed above (and illustrated in Figure 5.22). A similar modification can be made to the models of Figure 5.6 and Figure 5.7 to adapt them for discrete dissimilarity and preference data, respectively.

8.4 ORDINAL REGRESSION AND INSTRUMENTAL MODELS

The previous discussion should have made obvious that (linear) regression, as discussed in appendix 5.B, is also not the most appropriate way of testing if a given stimulus configuration $\mathbf{x}_i$, for $i = 1, \ldots, N$, can describe numerical category scaling data. Rather, using a multi-dimensional model with a fixed stimulus configuration is more in line with the categorical nature of the experimental data. An important difference with the regression method for continuous data is that we also allow for optimization of the category boundaries $\alpha_q(k)$ that determine the quantization process between the model predictions and the predicted response categories. The resulting procedure is called **ordinal regression** [McC80, Agr89] and can be performed using a similar ML optimization procedure as used in XGms. The only difference between MDS and ordinal regression is that MDS optimizes the stimulus configuration $\mathbf{x}_i$, for $i = 1, \ldots, N$, while this stimulus configuration is specified a priori (and fixed) in case of ordinal regression. We called such a model that is based on a fixed stimulus configuration (i.e., with $F_{\mathbf{X}} = 0$) an **instrumental model**. Since XGms can be initialized with an externally supplied stimulus configuration, it can also be used to perform ordinal regression and hence to investigate if such an instrumental model provides an adequate description of the available experimental data.

8.5 CHI-SQUARED STATISTICS

The model predictions $\hat{n}_{kiq}$ according to any of the aforementioned models (i.e., optimization per attribute as in section 8.2, simultaneous MDS optimization for all attributes as in section 8.3, or ordinal regression based on a given stimulus configuration as in section 8.4) can be compared with the observed numbers n_{kiq} in the following way. Either the **Pearson chi-squared statistic**

$$X^2 = \sum_{k=1}^{K_a} \sum_{i=1}^{N} \sum_{q=1}^{Q_k} \frac{(n_{kiq} - \hat{n}_{kiq})^2}{\hat{n}_{qik}} \tag{5.85}$$

or the **Wilks likelihood ratio statistic**

$$G^2 = 2 \sum_{k=1}^{K_a} \sum_{i=1}^{N} \sum_{q=1}^{Q_k} n_{kiq} \cdot \log \left(\frac{n_{kiq}}{\hat{n}_{kiq}} \right) \tag{5.86}$$

can be used to infer if the model predictions provide an adequate fit to the data[Agr90]. If the number of (independent) observations is sufficiently large, and the predictor model applies, then both statistics can be shown to have a χ^2-distribution with $F_O - F_M$ degrees of freedom, where F_O and F_M denote the number of DOF in the experimental data and the model, respectively. If $X^2 < \chi^2_\alpha(F_O - F_M)$ and/or $G^2 < \chi^2_\alpha(F_O - F_M)$, where most often $\alpha = 0.05$, then there is no compelling evidence for rejecting the model as an adequate description of the data.

Per stimulus i and attribute k, there are $Q_k - 1$ degrees of freedom in the experimental data. This is one less than the number of categories Q_k since the sum of the observations n_{kiq} over all categories is known a priori since it is equal to

$$n_{ki} = \sum_{q=1}^{Q_k} n_{kiq}, \tag{5.87}$$

which is the number of times that stimulus i has been questioned on attribute k. Hence, the total number of DOF in the experimental attribute data[26] is equal to

$$F_{aO} = N \cdot \sum_{k=1}^{K_a} (Q_k - 1). \tag{5.88}$$

Note that this is the only expression for the DOF that is different from the case of continuous data (see equation (5.51)).

Similarly as in the case of continuous data, the likelihood ratio statistic G^2 can be used to compare results from different models. Suppose that G_1^2 is the statistic for a general model with F_1 DOF and ML criterium L_1, and that G_2^2 is the statistic for a reduced model with F_2 DOF and ML

[26]This number of DOF is often unrealistically large. This is a consequence of the fact that there is often a strong coupling between the stimuli and the categories that are used. For example, a bad image will never be scored in the category "excellent". This problem can be addressed by turning some combinations of stimuli and categories into "structural zeros" [Agr90], and to deduce the number of structural zeros from the total number of DOF. Since we will be mainly interested in comparing alternative models with a number of DOF that is much smaller than the total number of DOF, this theoretical issue usually causes very few practical problems.

criterium L_2. Obviously, the reduced model uses fewer parameters such that $F_1 > F_2$. Applying the hypothesis-testing theorem of section 6.4 to the likelihood function for discrete data reveals that $G_{12}^2 = G_2^2 - G_1^2 = 2 \cdot (L_2 - L_1)$ should observe a χ^2-distribution with $F_1 - F_2$ DOFs if the reduced model is a specific case of the general model and both models are equivalent descriptions of the data. This difference statistic can hence be used to infer if the reduced and general model are both equally suited models for describing the experimental data.

The log-likelihood function can also be used to derive confidence regions for the parameters in the MDS model. The relevant theory in section 6.4 can be applied without modifications.

We refer to the tutorial papers [McC80, Agr89] or chapters 8 and 9 of the book by Agresti [Agr90] for a more in-depth coverage on modelling of ordered categorical response data.

9. EXAMPLES OF DISCRETE DATA ANALYSIS

We now re-analyze some of the example data of section 7.1. The main difference is that we respect the discrete nature of the data in this new analysis.

9.1 STIMULUS CONFIGURATIONS

A separate XGms analysis was performed on the experimental data for the coded images per scene and per subject. These data consisted of eight impairment ratings for all stimuli and two impairment-difference ratings for a subset of the stimulus pairs. A Gaussian link function was assumed, and nonlinear quantizers based on a general (Kruskal) transformation were used to relate stimulus positions to impairment ratings and differences in positions to impairment-difference ratings. The resulting stimulus configurations are shown in Figure 5.23, together with the stimulus configurations of Figure 5.15 that were obtained using a continuous data analysis. Only in case of the Wanda scene and subject SK are the differences between the stimulus configurations large enough to be noticeable. The linear correlation coefficients between the stimulus configuration for continuous and discrete data are 0.99985, 0.99988, 0.99953 and 0.99996 for the four cases Wanda-HR, Terrace-HR, Wanda-SK and Terrace-SK, respectively. This result confirms our expectation that the differences between a continuous and a discrete data analysis remain small if the number of categories used is sufficiently large.

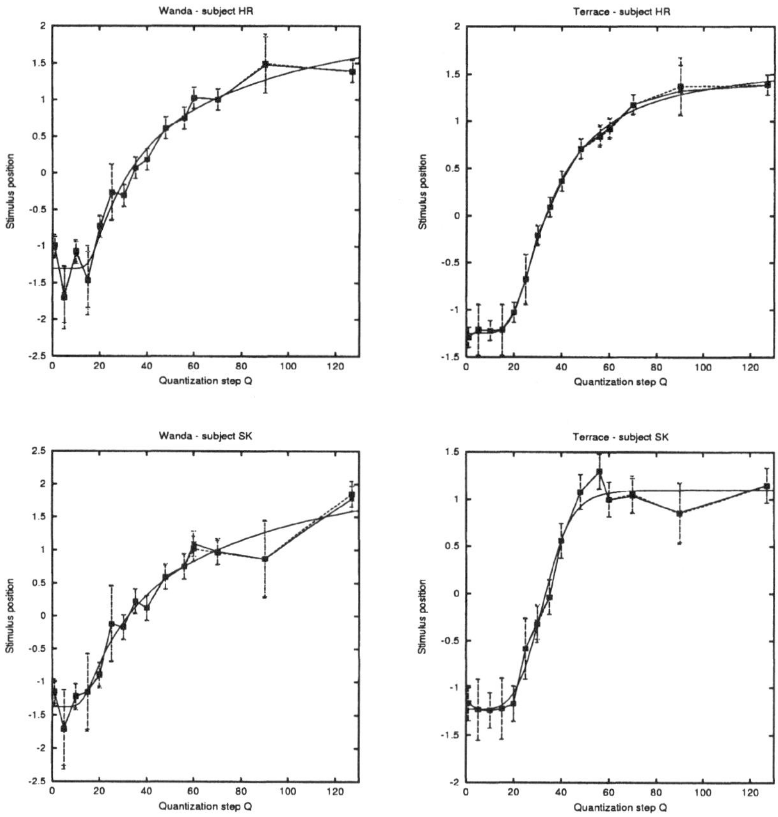

Figure 5.23. Stimulus positions from experimental data with scale-space coded images. The top and bottom rows correspond to subjects HR and SK, respectively, while the left and right columns correspond to the scenes 'Wanda' and 'Terrace', respectively. Each graph contains three curves: the stimulus configuration obtained by treating the data as continuous (see also Figure 5.15), an approximation of this configuration by a power-like function $F(Q)$ of the quantization stepsize Q, and the stimulus configuration obtained by treating the data as discrete.

9.2 ORDINAL REGRESSION OF ATTRIBUTE SCORES

We consider four possible models for describing the attribute data. The first possible variation in the models is whether the stimulus positions are supplied by an external model, as in section 8.4, or whether these stimulus positions are estimated from the data, as in sections 8.2 and 8.3. These alternatives are referred to as P (predicted) and E (es-

Table 5.8. Likelihood ratio statistics used to compare different alternative model descriptions for the impairment scores of coded images (subject HR). Accepted model simplifications are indicated by (*).

Case	Wanda - HR	Terrace - HR
P-QN|E-QN	$G^2(10) = 16.2 < 18.3$ (*)	$G^2(10) = 13.9 < 18.3$ (*)
E-QL|E-QN	$G^2(8) = 48.7 > 15.5$	$G^2(8) = 34.0 > 15.5$
P-QL|E-QN	$G^2(18) = 60.1 > 28.9$	$G^2(18) = 81.6 > 28.9$
P-QL|E-QL	$G^2(10) = 11.4 < 18.3$ (*)	$G^2(10) = 47.6 > 18.3$
P-QL|P-QN	$G^2(8) = 43.9 > 15.5$	$G^2(8) = 67.7 > 15.5$

Table 5.9. Likelihood ratio statistics used to compare different alternative model descriptions for the impairment scores of coded images (subject SK). Accepted model simplifications are indicated by (*).

Case	Wanda - SK	Terrace - SK
P-QN|E-QN	$G^2(10) = 36.3 > 18.3$	$G^2(10) = 15.3 < 18.3$ (*)
E-QL|E-QN	$G^2(7) = 14.9 > 14.1$	$G^2(8) = 12.2 < 15.5$ (*)
P-QL|E-QN	$G^2(17) = 49.5 > 27.6$	$G^2(18) = 23.6 < 28.9$ (*)
P-QL|E-QL	$G^2(10) = 34.6 > 18.3$	$G^2(10) = 11.4 < 18.3$ (*)
P-QL|P-QN	$G^2(7) = 13.2 < 14.1$ (*)	$G^2(8) = 8.3 < 15.5$ (*)

timated), respectively. We use a non-linear power function $P(Q)$ of the quantization stepsize Q for predicting the positions of the coded images on a 1-D axis. The parameters for the non-linear functions are taken from Table 5.3. The second possible variation in the models is whether a non-linear or a linear quantizer is used to relate the stimulus positions to the subject ratings. These alternatives are referred to as QN and QL, respectively. The two models E-QL and P-QN are both special cases of the most general model P-QN, while the model P-QL is a special case of all the other models. This implies for instance that we can test the statistical evidence for a model simplification like P-QN|E-QN, i.e., the probability that the more specific model P-QN is valid, assuming that the more general model E-QN is valid.

The conditional tests between different models are summarized in Tables 5.8 and 5.9, for subjects HR and SK, respectively. The accepted model simplifications are indicated by $(*)$ in the tables, and the resulting relationships are represented graphically in Figure 5.24. In case of subject HR, we conclude that the predictor can model the data, since P-QN is not rejected, but that a non-linear quantizer is required to relate predicted positions to subject responses, since E-QL is rejected. In case of subject SK, a linear quantizer is probably sufficient to describe

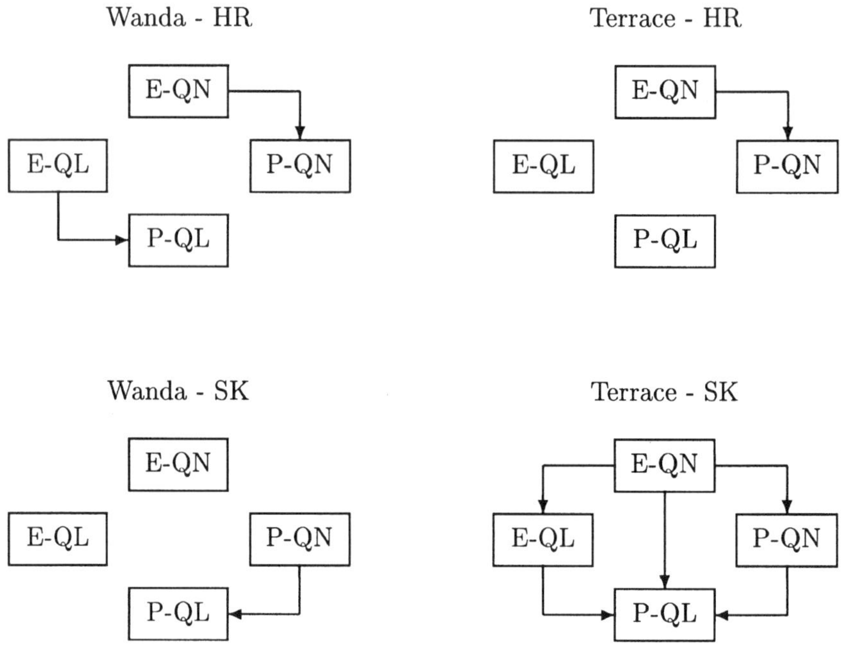

Figure 5.24. Accepted conditional tests between models for the category scaling data in the coding experiment are represented by arrows. The top and bottom rows correspond to subjects HR and SK, respectively, while the left and right columns correspond to the scenes 'Wanda' and 'Terrace', respectively.

the category scaling, since E-QL is accepted in case of the Terrace scene and only marginally rejected in case of the Wanda scene. The predictor provides only an approximate description of the data in case of the Wanda scene.

10. SUMMARY

In this chapter, we have shown how experimental data on image quality and its attributes can be integrated using MDS models and/or compared against instrumental model predictions. The interactive program XGms has been introduced as an interesting extension to available MDS and regression programs. It has been shown that the geometric models implemented in XGms can indeed provide adequate descriptions for two available experimental data sets.

DATA ANALYSIS SOFTWARE

XGms was developed by the author, starting from the existing open-source program XGvis. The analysis of dissimilarity data for a single subject, as provided by XGvis, has been extended by XGms to the joint analysis of dissimilarity, preference and scaling data for multiple subjects. The input data need no longer be continuous and metric, but may be continuous or discrete with an arbitrary combination of metric and non-metric scores. Replacing the stress optimization that is used in XGvis by a maximum-likelihood optimization has also made it possible to perform statistical tests between alternative models and to derive confidence intervals for most estimated model parameters.

Two other programs have also been developed by the author in order to assist in performing the analyses presented in this chapter:

1 Mapreg: performs non-linear regression between an independent characterization of stimuli (i.e., a known stimulus configuration in one or more dimensions) and a dependent characterization of the stimuli (for example, an MDS stimulus configuration derived from experimental data). A separate power-like transformation with three parameters q, t and s, following equation (5.66), is allowed on all dimensions in the independent characterization. Mapreg was used in section 7 to map from the quantization stepsizes Q, in case of coded images, and the blur and noise standard deviations (σ_b, σ_n), in case of blurred/noisy images, to MDS stimulus positions.

2 Matreg: determines the optimal matrix transformation required for mapping one stimulus configurations onto another, as described in appendix 5.C. This was used in section 7.2 to find the optimum linear transformation between the predicted stimulus configurations and the experimentally determined MDS configurations in case of the blurred/noisy images.

Requests for a copy of the XGms programs can be made to the author at J.B.O.S.Martens@tue.nl. The XGvis and XGobi programs are available via the home page of Deborah Swayne[27].

ACKNOWLEDGEMENT

The author wishes to thank the many people that have contributed to the public software in XGobi, XGvis and Netlib. Without this freely available software, realization of the above programs would not have been feasible.

[27] http://www.research.att.com/~dfs

APPENDIX 5.A: STRESS OPTIMIZATION

In this appendix, the relationship between the ML optimization proposed in section 6 and the stress optimization that is used in many existing MDS programs is established. This requires a more in-depth analysis of the ML optimization for the specific case of the generalized Gaussian PDF. We summarize the results for the different types of experimental data.

DISSIMILARITY DATA

In case of a generalized Gaussian PDF, the ML criterium of equation (5.16) for the dissimilarity data, which is equal to the inverse of the log-likelihood function, can be expressed as

$$\begin{aligned} L_d &= \sum_{k=1}^{K_d} \sum_{(i,j)\in I_d(k)} \left\{ \left| \frac{TD_{k,i,j} - d_k \cdot \|\mathbf{x}_i - \mathbf{x}_j\|_l}{\rho_d(k)} \right|^r - \log \left| T'_{dk}(D_{k,i,j}) \right| \right\} \\ &\quad + \sum_{k=1}^{K_d} N_d(k) \cdot \log \rho_d(k) + \sum_{k=1}^{K_d} N_d(k) \cdot \log \left[\frac{2}{r} \Gamma\left(\frac{1}{r}\right) \right], \end{aligned}$$

where $\rho_d(k) = \sigma_d(k) \sqrt{\frac{\Gamma(1/r)}{\Gamma(3/r)}}$, and $N_d(k)$ denotes the number of observations in the set $I_d(k)$, for $k = 1, \ldots, K_d$. The ML estimates for the standard deviations can be solved explicitly in this case, i.e.,

$$\rho_d(k) = \left[\frac{r}{N_d(k)} \sum_{(i,j)\in I_d(k)} \left| TD_{k,i,j} - d_k \cdot \|\mathbf{x}_i - \mathbf{x}_j\|_l \right|^r \right]^{1/r},$$

for $k = 1, \ldots, K_d$.

A relationship with stress optimization can be accomplished by assuming a model with a generalized Gaussian PDF in which the standard deviation $\sigma_d(k)$ (and derived parameter $\rho_d(k)$) for subject k is proportional to the variation in the data of this subject, i.e.,

$$\rho_d(k) = g_d \cdot \left[\frac{r}{N_d(k)} \sum_{(i,j)\in I_d(k)} \left| TD_{k,i,j} \right|^r \right]^{1/r},$$

where the proportionality factor g_d is assumed to be the same for all subjects. The ML estimate of this factor g_d is the total weighted stress

$$S_d = \left[\frac{\sum_{k=1}^{K_d} N_d(k) \cdot S_d^r(k)}{\sum_{k=1}^{K_d} N_d(k)} \right]^{1/r},$$

where

$$S_d(k) = \left[\frac{\sum_{(i,j)\in I_d(k)} \left| TD_{k,i,j} - d_k \cdot \|\mathbf{x}_i - \mathbf{x}_j\|_l \right|^r}{\sum_{(i,j)\in I_d(k)} \left| TD_{k,i,j} \right|^r} \right]^{1/r}$$

denotes the **stress** between the experimental data and the model predictions for subject k. In case the data consists solely of dissimilarity scores, then the ML optimization of the stimulus configuration for this choice of the standard deviations reduces to a minimization of $\sum_k N_d(k) \cdot \log S_d$, or equivalently, a minimization of S_d itself. This generalizes the stress function used in many existing MDS programs, such as XGvis [BSL+02], to the case of multiple subjects. Once the standard deviations $\sigma_d(k)$, $k = 1, \ldots, K_d$, and the stimulus positions $\mathbf{x}_i$, for $i = 1, \ldots, N$, are known, minimization of the individual terms $S_d(k)$, or equivalently,

$$\min_{d_k} \sum_{(i,j)\in I_d(k)} \mid TD_{k,i,j} - d_k \cdot \|\mathbf{x}_i - \mathbf{x}_j\|_l \mid^r,$$

can be used to determine the regression factors d_k, for $k = 1, \ldots, K_d$. In case $r = 2$, the above minimization reduces to a simple **linear regression**, i.e.,

$$d_{k|r=2} = \frac{\sum_{(i,j)\in I_d(k)} TD_{k,i,j} \cdot \|\mathbf{x}_i - \mathbf{x}_j\|_l}{\sum_{(i,j)\in I_d(k)} \|\mathbf{x}_i - \mathbf{x}_j\|_l^2}$$

PREFERENCE DATA

A relation to stress optimization can also be established for the preference data in case the link function is a generalized Gaussian PDF and the standard deviation $\sigma_p(k)$ (and derived parameter $\rho_p(k)$) for subject k is proportional to the variation in the data of this subject, i.e.,

$$\rho_p(k) = g_p \cdot \left[\frac{r}{N_p(k)} \sum_{(i,j)\in I_p(k)} \mid TP_{k,i,j} \mid^r \right]^{1/r},$$

where $N_p(k)$ denotes the number of preference scores for subject k. The ML estimate for this proportionality factor g_p is the total weighted stress

$$S_p = \left[\frac{\sum_{k=1}^{K_p} N_p(k) \cdot S_p^r(k)}{\sum_{k=1}^{K_p} N_p(k)} \right]^{1/r},$$

where

$$S_p(k) = \left[\frac{\sum_{(i,j)\in I_p(k)} \mid TP_{k,i,j} - m_k \cdot ([\mathbf{p}_k, \mathbf{x}_i] - [\mathbf{p}_k, \mathbf{x}_j]) \mid^r}{\sum_{(i,j)\in I_d(k)} \mid TP_{k,i,j} \mid^r} \right]^{1/r}$$

is the stress between the experimental preference data and the model predictions for subject k.

In case of the vector-product model and a Gaussian link function with $r = 2$, the scaled preference vector $\mathbf{p}_k^\star = m_k \cdot \mathbf{p}_k$ can be determined by minimization of the stress $S_p(k)$, or equivalently, through

$$\min_{\mathbf{p}_k^\star} \sum_{(i,j)\in I_p(k)} \mid TP_{k,i,j} - < \mathbf{p}_k^\star, \mathbf{x}_i - \mathbf{x}_j > \mid^2.$$

This is equivalent to finding the optimal **multiple regression** [DS98] coefficients between the transformed preferences $TP_{k,i,j}$ for subject k and the difference vectors $\mathbf{x}_i - \mathbf{x}_j$, and can be reduced to solving a set of linear equations.

ATTRIBUTE DATA

A relation to stress optimization can also be established for the attribute data in case the link function is a generalized Gaussian PDF and the standard deviation $\sigma_a(k)$ (and derived parameter $\rho_a(k)$) for subject k is proportional to the variation in the data of this subject, i.e.,

$$\rho_a(k) = g_a \cdot \left[\frac{r}{N_a(k)} \sum_{(i,j)\in I_a(k)} \mid TA_{k,i,j} - TA_{k..} \mid^r \right]^{1/r},$$

where $N_a(k)$ is the number of observations in the index set $I_a(k)$, and

$$TA_{k..} = \frac{1}{N_a(k)} \sum_{(i,j)\in I_a(k)} TA_{k,i,j},$$

denotes the average transformed attribute score for subject k. The ML estimate for the proportionality factor g_a is the weighted total stress

$$S_a = \left[\frac{\sum_{k=1}^{K_a} N_a(k) \cdot S_a^r(k)}{\sum_{k=1}^{K_a} N_a(k)} \right]^{1/r}$$

where

$$S_a(k) = \left[\frac{\sum_{(i,j)\in I_a(k)} \mid TA_{k,i,j} - (c_k + f_k \cdot [\mathbf{a}_k, \mathbf{x}_i]) \mid^r}{\sum_{(i,j)\in I_a(k)} \mid TA_{k,i,j} - TA_{k..} \mid^r} \right]^{1/r}$$

is the stress between the experimental attribute data and the model predictions for subject k.

The regression parameters can be determined by minimizing the stress $S_a(k)$, or equivalently, by

$$\min_{c_k, f_k, \mathbf{a}_k} \sum_{(i,j)\in I_a(k)} \mid TA_{k,i,j} - (c_k + f_k \cdot [\mathbf{a}_k, \mathbf{x}_i]) \mid^r .$$

In case of the vector-product model and a Gaussian link function, with $r = 2$, this optimization is equivalent to finding the optimal multiple regression coefficients between the transformed attribute scores and the known stimulus positions $\mathbf{x}_i$ [DS98]. The unknown regression parameters c_k and $\mathbf{a}_k^\star = f_k \cdot \mathbf{a}_k$ can be determined through solving a set of linear equations.

COMBINED DATA

The ML optimization of the stimulus configuration involves all the data and is equivalent to the optimization of

$$\sum_{k=1}^{K_d} N_d(k) \cdot \log S_d + \sum_{k=1}^{K_p} N_p(k) \cdot \log S_p + \sum_{k=1}^{K_a} N_a(k) \cdot \log S_a$$

under the above assumptions that lead to stress optimization (i.e., generalized Gaussian PDFs and noise standard deviations that are proportional to the variations in the data).

APPENDIX 5.B: ANALYSIS OF VARIANCE

Analysis of variance (ANOVA) is concerned with attributing the variation in observed data to several (potential) causes, and with determining the statistical evidence for these causes. The application of ANOVA requires that the data varies continuously on an absolute scale, and that the variation in the response data for identical stimuli can be described by a Gaussian noise distribution (with a constant variance).

In this appendix, we discuss how ANOVA can be used to verify if attribute, preference and/or dissimilarity data can be described by a regression model. This is equivalent to testing the assumption that the data is the sum of a stochastic (noise) component and a deterministic component that can be derived (through regression) from a given stimulus configuration $\hat{\mathbf{x}}_i$. We will use the example data set of section 7.1 to illustrate some of the concepts and definitions in this appendix.

ATTRIBUTE DATA

The observed attribute scores $A_{k,i,j}$ for different subjects k cannot be used directly in an ANOVA for at least two reasons. Firstly, the original data $A_{k,i,j}$ can usually not be assumed to be metric. Secondly, even if the data is metric it is at best interval (or ratio) data. This means that the data is only determined up to a linear transformation, which may differ across subjects. XGms can however export normalized transformed attribute scores $TA^{\star}_{k,i,j}$ that do not have these drawbacks, and which are hence better suited for comparing responses across subjects that belong to a homogeneous group (i.e., that are assumed to share a common attribute vector). Because ANOVA assumes a Gaussian noise model, it seems most logical to also use an ML criterium based on a Gaussian PDF (with exponent $r = 2$) when creating the normalized transformed attribute scores with XGms[28].

The most frequently-used statistic for testing the goodness-of-fit of a model to (repeated) measurements is the **correlation coefficient** R, which is defined[29] as

$$R^2 = 1 - \frac{\text{SS(residue)}}{\text{SS(total)}} = 1 - \frac{\sum_{k=1}^{k_a} \sum_{(i,j)\in I_a(k)} \left| TA^{\star}_{k,i,j} - \hat{\beta}_i \right|^2}{\sum_{k=1}^{k_a} \sum_{(i,j)\in I_a(k)} \left| TA^{\star}_{k,i,j} \right|^2}$$

in case of k_a subjects in a homogeneous group. Note that stimulus i is characterized by its prediction $\hat{\beta}_i$. In case of XGms, $\hat{\beta}_i = [\mathbf{a}, \hat{\mathbf{x}}_i]$, where $\hat{\mathbf{x}}_i$ is the supplied position for the i-th stimulus. The attribute vector $\mathbf{a}$ is optimized using the attribute data of all k_a subjects. In our example case of a one-dimensional stimulus configuration and an inner-product prediction model with $||\mathbf{a}|| = 1$, we simply obtain $\hat{\beta}_i = \hat{x}_i$.

The number of degrees of freedom (DOF) in the normalized transformed attribute scores is equal to

$$n_T = DOF(\text{total}) = \sum_{k=1}^{k_a} [N_a(k) - F_a(k)] - 2 \cdot k_a.$$

[28]Strictly speaking, one should also check if the noise standard deviations $\sigma_a(k)/|f_k|$ of the normalized transformed scores are (approximately) constant across the different subjects k.

[29]This definition of the correlation coefficient R relies on the fact that the normalized scores are centered on the origin, i.e., that $TA^{\star}_{k,\cdot,\cdot} = (1/N_a(k)) \sum_{(i,j)\in I_a(k)} TA^{\star}_{k,i,j} = 0$.

The total number $N_a(k)$ of DOF in the data of subject k must be reduced by the number of independent parameters $F_a(k)$ in the transformation that is applied to the original attribute scores in order to map non-metric data $A_{k,i,j}$ to metric data $TA_{k,i,j}$. In addition, two DOF are used per subject for obtaining the regression coefficients c_k and f_k that are involved in defining the normalized attribute scores $TA^{\star}_{k,i,j} = (TA_{k,i,j} - c_k)/f_k$.

If n_P parameters are required for generating the predictions $\hat{\beta}_i$, then the number of DOF in the prediction errors is $DOF(\text{residue}) = n_R = n_T - n_P$. For instance, in case of our example data set, the predictions $\hat{x}_i = F(Q_i)$ are derived from known quantization step sizes Q_i through a power-like function F with three parameters. The number of DOF in the prediction vector **a** must also be added to n_P. This number is equal to $n-1$ or n, depending on whether an inner-product or an ideal-point model is used, respectively. Only in case of a one-dimensional stimulus configuration and an inner-product prediction rule does this not require any extra parameters, since $\hat{\beta}_i = \hat{x}_i$ in this case. Hence, in our example case, $n_P = 3$.

Correlation coefficients indicate the percentage of variance in the data that can be described by a prediction model, but do not allow to judge whether such a model can potentially be improved. This requires that the variance in the prediction error is compared against the inherent variance that is in the experimental data. We expect a smaller correlation between the actual data and the model predictions in case the responses to identical stimuli on repeated trials vary more. This accuracy of the experimental data can be estimated if the stimuli have been scored more than once (either within or across subjects). In such case, the best possible predictor[30] for stimulus p is the average score

$$\beta_p = \frac{\sum_{k=1}^{k_a} \sum_{(i,j)\in I_a(k)} TA^{\star}_{k,i,j}\, \delta_{pi}}{\sum_{k=1}^{k_a} \sum_{(i,j)\in I_a(k)} \delta_{pi}} = \frac{A(p)}{N(p)},$$

where $\delta_{pi} = 1$ if $i = p$, and $\delta_{pi} = 0$ if $i \neq p$. The sum of all attribute scores for stimulus p is denoted by $A(p)$, while $N(p)$ is equal to the total number of times that stimulus p is scored.

Using these best predictors, SS(residue) can be split into

$$\text{SS(residue)} = \text{SS(pure error)} + \text{SS(lack of fit)},$$

where

$$\text{SS(pure error)} = \sum_{k=1}^{k_a} \sum_{(i,j)\in I_a(k)} \left(TA^{\star}_{k,i,j} - \beta_i \right)^2$$

accumulates the variances on successive trials around the average score β_i, and

$$\text{SS(lack of fit)} = \sum_{k=1}^{k_a} \sum_{(i,j)\in I_a(k)} \left(\hat{\beta}_i - \beta_i \right)^2$$

[30]Obviously, an even better fit to the data can be obtained by using predictors of the form $\beta_{k,i}$ that depend in a more general way on the subject k and stimulus i. Such a predictor is however not in agreement with the assumption that the subjects form a homogeneous group.

is the SS of the differences between the average scores β_i and the predictions $\hat{\beta}_i$ according to the multidimensional model. The maximum value R_p that can be obtained for the linear correlation coefficient is hence

$$R_p^2 = 1 - \frac{\text{SS(pure error)}}{\text{SS(total)}} = 1 - \frac{\sum_{k=1}^{k_a} \sum_{(i,j)\in I_a(k)} \mid TA^\star_{k,i,j} - \beta_i \mid^2}{\sum_{k=1}^{k_a} \sum_{(i,j)\in I_a(k)} \left| TA^\star_{k,i,j} \right|^2},$$

since SS(residue) cannot become smaller than SS(pure error).

If the normalized scores $TA^\star_{k,i,j}$ are derived from minimizing an ML criterium based on a Gaussian PDF, then the following relationships exist between the normalized attribute scores and the predictions:

$$\sum_{(i,j)\in I_a(k)} (TA^\star_{k,i,j} - \hat{\beta}_i) = 0 \quad \text{and} \quad \sum_{(i,j)\in I_a(k)} (TA^\star_{k,i,j} - \hat{\beta}_i) \cdot \hat{\beta}_i = 0$$

for all subjects k, so that one obtains

$$\sum_{k=1}^{k_a} \sum_{(i,j)\in I_a(k)} (TA^\star_{k,i,j} - \hat{\beta}_i) = 0 \quad \text{and} \quad \sum_{k=1}^{k_a} \sum_{(i,j)\in I_a(k)} (TA^\star_{k,i,j} - \hat{\beta}_i) \cdot \hat{\beta}_i = 0$$

by summing over all subjects. On the other hand, the optimal attribute predictors satisfy

$$\sum_{k=1}^{k_a} \sum_{(i,j)\in I_a(k)} (TA^\star_{k,i,j} - \beta_i) \cdot \delta_{pi} = 0$$

for any stimulus p, and hence also (by summing over all stimuli)

$$\sum_{k=1}^{k_a} \sum_{(i,j)\in I_a(k)} (TA^\star_{k,i,j} - \beta_i) \cdot \alpha_i = 0$$

for any choice of the factors α_i. Applying the latter expression for $\alpha_i = 1$ and $\alpha_i = \hat{\beta}_i$, and comparing with the above relationships between the normalized attribute scores and the predictions $\hat{\beta}_i$, leads to the following two boundary conditions on the optimal predictors

$$\sum_{k=1}^{k_a} \sum_{(i,j)\in I_a(k)} \beta_i = \sum_{k=1}^{k_a} \sum_{(i,j)\in I_a(k)} \hat{\beta}_i \quad \text{and} \quad \sum_{k=1}^{k_a} \sum_{(i,j)\in I_a(k)} \beta_i \cdot \hat{\beta}_i = \sum_{k=1}^{k_a} \sum_{(i,j)\in I_a(k)} \hat{\beta}_i^2,$$

so that we can conclude that there are only n_f = DOF(lack of fit) = $N - 2$ independent values β_i.

If the regression model 'fits' the data, then the standard deviation of the regression errors in case of the model predictors $\hat{\beta}_i$ should not exceed the standard deviation of the regression errors in case of the best predictors β_i. Unbiased estimates for these standard deviations are the mean squares (MS), i.e.,

$$\begin{aligned} \text{MS(lack of fit)} &= \text{SS(lack of fit)/DOF(lack of fit)} \\ \text{MS(pure error)} &= \text{SS(pure error)/DOF(pure error)}, \end{aligned}$$

where DOF(pure error) $= n_R - n_f = n_e$ is the number of DOF in SS(pure error). The ratio

$$F = \frac{\text{MS(lack of fit)}}{\text{MS(pure error)}},$$

is hence a good statistic for testing the goodness of fit of the multidimensional model. In case of equal standard deviations, we expect $F \approx 1$. A large value of F, on the other hand, indicates that the prediction error for the regression model is significantly larger than the prediction error for the best predictor, and hence that R is substantially smaller than R_p. A better model than the regression model under test can be pursued in such cases. A lack of fit does however not prohibit that a large part of the variance in the attribute data may be explained by the available model, i.e., that both R and R_p are well above 0.9, for instance, so that the model may still be a very useful (but not perfect) description of the data.

It has been shown that the ratio F satisfies an $F(n_f, n_e)$-distribution under the hypothesis that *the standard deviations of the underlying (Gaussian) distributions are equal* [DS98]. If this is the case, then the probability that a value greater than F_α occurs is equal to

$$P(F > F_\alpha | n_f, n_e) = \frac{I_x\left(\frac{n_e}{2}, \frac{n_f}{2}\right)}{B\left(\frac{n_e}{2}, \frac{n_f}{2}\right)} = \alpha,$$

with $x = n_e/(n_e + n_f \cdot F_\alpha)$, where we have introduced the notation

$$I_x(a,b) = \int_0^x z^{a-1}(1-z)^{b-1}\, dz$$

for the **incomplete beta function**, and $B(a,b) = I_1(a,b)$ for the complete beta function. This result can be used to set up a quantitative **F-test** for the goodness of fit. If the observed ratio F is larger than F_α, where most often $\alpha = 0.05$, then it is considered very unlikely that both standard deviations are equal, and hence that all the variance in the data is accounted for by the model.

PREFERENCE DATA

The ANOVA that is presented above for single-stimulus attribute data must be modified in some aspects in case of double-stimulus preference data. The correlation coefficient R for preference data is defined by

$$R^2 = 1 - \frac{\text{SS(residue)}}{\text{SS(total)}} = 1 - \frac{\sum_{k=1}^{k_p} \sum_{(i,j)\in I_p(k)} \left| TP^\star_{k,i,j} - \hat{\gamma}_{ij} \right|^2}{\sum_{k=1}^{k_p} \sum_{(i,j)\in I_p(k)} \left| TP^\star_{k,i,j} \right|^2},$$

where $\hat{\gamma}_{ij} = [\mathbf{p}, \hat{\mathbf{x}}_i] - [\mathbf{p}, \hat{\mathbf{x}}_j]$ is the preference prediction for stimulus pair (i,j). The stimulus positions $\hat{\mathbf{x}}_i$ are supplied as input into the regression, while the shared prediction vector $\mathbf{p}$ is optimized by XGms. In our example case of a one-dimensional stimulus configuration and an inner-product prediction model, the preference prediction for stimulus pair (i,j) is equal to $\hat{\gamma}_{ij} = \hat{x}_i - \hat{x}_j$.

The number of DOF in the normalized transformed preference scores is

$$n_T = \text{DOF(total)} = \sum_{k=1}^{k_p} [N_p(k) - F_p(k)] - k_p$$

in case k_p subjects are assumed to form a homogeneous group, and a transformation with $F_p(k)$ DOF is applied on the preference data of subject k in order to map the non-metric data $P_{k,i,j}$ to the metric data $TP_{k,i,j}$. If n_P parameters are involved in generating the predictions $\hat{\gamma}_{ij}$, then $n_R = \text{DOF(residue)} = n_T - n_P$ is the number of DOF in SS(residue), i.e., the sum of the squared prediction errors.

There are two models that are worth considering for comparison against the prediction model. The most general "asymmetric" model only assumes that prediction scores are asymmetric, i.e., $\gamma_{pq} = -\gamma_{qp}$ and $\gamma_{pp} = 0$. Such a model is specified by $N \cdot (N-1)/2$ values which can be estimated as follows from the data

$$\gamma_{pq} = -\gamma_{qp} = \frac{P(p,q) - P(q,p)}{N(p,q) + N(q,p)},$$

for $p, q = 1, \ldots, N$. The sum of all observed preference scores for stimulus pair (p, q) is denoted by

$$P(p,q) = \sum_{k=1}^{k_p} \sum_{(i,j)\in I_p(k)} TP^{\star}_{k,i,j} \cdot \delta_{pi}\, \delta_{qj},$$

while

$$N(p,q) = \sum_{k=1}^{k_p} \sum_{(i,j)\in I_p(k)} \delta_{pi}\, \delta_{qj}$$

is equal to the number of repetitions for this stimulus pair. In case an ML criterium based on a Gaussian PDF is used, then the optimal asymmetric predictors satisfy the condition

$$\sum_{k=1}^{k_p} \sum_{(i,j)\in I_p(k)} \hat{\gamma}_{ij} \cdot \gamma_{ij} = \sum_{k=1}^{k_p} \sum_{(i,j)\in I_p(k)} \hat{\gamma}^2_{ij},$$

so that the number of DOF in this "asymmetric" model is $n_f = N \cdot (N-1)/2 - 1$.

The alternative "difference" model has a difference structure that is similar to the prediction model, i.e., the preference score for stimulus pair (i, j) is assumed to be of the form $\beta_i - \beta_j$. This model has the obvious advantage that β_i can be interpreted as an estimate for the quality of stimulus i. Such a model is specified by $N - 1$ values[31] that can be obtained by solving the following set of linear equations

$$\beta_p \sum_{q\neq p} [N(p,q) + N(q,p)] - \sum_{q\neq p} \beta_q \cdot [N(p,q) + N(q,p)] = \sum_{q=1}^{N} [P(p,q) - P(q,p)],$$

for $p = 1, \ldots, N$. In case an ML criterium based on a Gaussian PDF is used, then the optimal predictors satisfy the condition

$$\sum_{k=1}^{k_p} \sum_{(i,j)\in I_p(k)} \hat{\gamma}_{ij} \cdot (\beta_i - \beta_j) = \sum_{k=1}^{k_p} \sum_{(i,j)\in I_p(k)} \hat{\gamma}^2_{ij},$$

[31]Since only differences between quality scores β_i are considered, the origin of these quality scores is clearly undetermined. The normalization condition $\sum_{i=1}^{N} \beta_i = 0$ can for instance be used to remove this indeterminacy.

so that the number of DOF in this "difference" model is $n_d = N - 2$.

The prediction model is a special case of the "difference" model, which is in turn a specific case of the "asymmetric" model. The total SS can hence be split in the following three components

$$\mathrm{SS(total)} = \mathrm{SS(pure\ error)} + \mathrm{SS(asym\text{-}diff)} + \mathrm{SS(lack\ of\ fit)},$$

where

$$\mathrm{SS(pure\ error)} = \sum_{k=1}^{k_p} \sum_{(i,j)\in I_p(k)} | TP^{\star}_{k,i,j} - \gamma_{ij} |^2$$

accumulates the variances on successive trials around the optimal preference scores γ_{ij} according to the "asymmetric" model,

$$\mathrm{SS(asym\text{-}diff)} = \sum_{k=1}^{k_p} \sum_{(i,j)\in I_p(k)} | \gamma_{ij} - (\beta_i - \beta_j) |^2$$

expresses the discrepancy between the "asymmetric" model and the "difference" model, while

$$\mathrm{SS(lack\ of\ fit)} = \sum_{k=1}^{k_p} \sum_{(i,j)\in I_p(k)} | (\beta_i - \beta_j) - \hat{\gamma}_{ij} |^2$$

expresses the discrepancy between the "difference" model and the prediction model.

The maximum possible correlation coefficient R_d for any model with a difference structure is equal to

$$R_d^2 = 1 - \frac{\mathrm{SS(diff\ error)}}{\mathrm{SS(total)}} = 1 - \frac{\sum_{k=1}^{k_p} \sum_{(i,j)\in I_p(k)} | TP^{\star}_{k,i,j} - (\beta_i - \beta_j) |^2}{\sum_{k=1}^{k_p} \sum_{(i,j)\in I_p(k)} \left| TP^{\star}_{k,i,j} \right|^2},$$

where SS(diff error) = SS(residue) − SS(lack of fit). Since DOF(lack of fit) $= n_d$ is the number of DOF in the "difference" model, we can derive, in a similar way as in the case of attribute data, that an $F(n_d, n_R - n_d)$-test on the ratio

$$F = \frac{\mathrm{MS(lack\ of\ fit)}}{\mathrm{MS(diff\ error)}} = \frac{\mathrm{SS(lack\ of\ fit)}/n_d}{\mathrm{SS(diff\ error)}/(n_R - n_d)}$$

can be used to test the goodness of fit of the prediction model. A value of F that is significantly larger than one indicates that the prediction model and the "difference" model are not equivalent.

The maximum possible correlation coefficient R_p for any model satisfying asymmetry is equal to

$$R_p^2 = 1 - \frac{\mathrm{SS(pure\ error)}}{\mathrm{SS(total)}} = 1 - \frac{\sum_{k=1}^{k_p} \sum_{(i,j)\in I_p(k)} | TP^{\star}_{k,i,j} - \gamma_{ij} |^2}{\sum_{k=1}^{k_p} \sum_{(i,j)\in I_p(k)} \left| TP^{\star}_{k,i,j} \right|^2}.$$

Since the "asymmetric" model has DOF(asym-diff) $= n_f - n_d$ more parameters than the "difference" model, we can derive that an $F(n_f - n_d, n_R - n_f)$-test on the ratio

$$F = \frac{\mathrm{MS(asym\text{-}diff)}}{\mathrm{MS(pure\ error)}} = \frac{\mathrm{SS(asym\text{-}diff)}/(n_f - n_d)}{\mathrm{SS(pure\ error)}/(n_R - n_f)}$$

can be used to test if the preference data can indeed be described by a "difference" model.

There are hence two ways in which the regression model can fail the data. A first possibility is that the preference data cannot be described by a "difference" model. This implies that an adequate fit to the normalized preference data cannot be found within the class of models implemented in XGms. The second possibility is that a "difference" model is sufficient to describe the data, but that such a "difference" model cannot be based on the supplied stimulus configuration.

DISSIMILARITY DATA

The ANOVA that is required for dissimilarity data resembles the one presented above for preference data. The correlation coefficient R for dissimilarity data is defined by

$$R^2 = 1 - \frac{\text{SS(residue)}}{\text{SS(total)}} = 1 - \frac{\sum_{k=1}^{K_d} \sum_{(i,j)\in I_d(k)} \left| TD^{\star}_{k,i,j} - \hat{\tau}_{ij} \right|^2}{\sum_{k=1}^{K_d} \sum_{(i,j)\in I_d(k)} \left| TD^{\star}_{k,i,j} \right|^2},$$

where the distance $\hat{\tau}_{ij} = \|\hat{\mathbf{x}}_i - \hat{\mathbf{x}}_j\|_l$ is the dissimilarity prediction for stimulus pair (i,j). In our one-dimensional example, this simply reduces to $\hat{\tau}_{ij} = |\hat{x}_i - \hat{x}_j|$. The number of DOF in the normalized transformed dissimilarity scores is

$$n_T = \text{DOF(total)} = \sum_{k=1}^{K_d} [N_d(k) - F_d(k)] - K_d$$

in case a transformation with $F_d(k)$ DOF is applied on the dissimilarity data of subject k, i.e., to map the non-metric data $D_{k,i,j}$ to the metric data $TD_{k,i,j}$. If n_P parameters are involved in generating the stimulus positions $\hat{\mathbf{x}}_i$, then $n_R = \text{DOF(residue)} = n_T - n_P$ is the number of DOF in SS(residue).

The maximum correlation coefficient R_p is

$$R_p^2 = 1 - \frac{\text{SS(pure error)}}{\text{SS(total)}} = 1 - \frac{\sum_{k=1}^{K_d} \sum_{(i,j)\in I_d(k)} \left| TD^{\star}_{k,i,j} - \tau_{ij} \right|^2}{\sum_{k=1}^{K_d} \sum_{(i,j)\in I_d(k)} \left| TD^{\star}_{k,i,j} \right|^2},$$

and is obtained for the dissimilarity predictors

$$\tau_{pq} = \tau_{qp} = \frac{D(p,q) + D(q,p)}{N(p,q) + N(q,p)},$$

for $p, q = 1, \ldots, N$, that minimize SS(pure error). Note that this "symmetric" predictor is assumed to satisfy $\tau_{pq} = \tau_{qp}$ and $\tau_{pp} = 0$, so that it is specified by $N \cdot (N-1)/2$ values. The sum of all dissimilarity scores for stimulus pair (p,q) is denoted by

$$D(p,q) = \sum_{k=1}^{K_d} \sum_{(i,j)\in I_d(k)} TD^{\star}_{k,i,j} \cdot \delta_{pi}\, \delta_{qj},$$

while

$$N(p,q) = \sum_{k=1}^{K_d} \sum_{(i,j)\in I_d(k)} \delta_{pi}\, \delta_{qj}$$

is equal to the number of times that stimulus pair (p, q) is scored.

The lack-of-fit measure

$$\text{SS(lack of fit)} = \sum_{k=1}^{K_d} \sum_{(i,j)\in I_p(k)} (\hat{\tau}_{ij} - \tau_{ij})^2$$

sums the squared differences between the "symmetric" dissimilarity predictions and the dissimilarity predictions according to the regression model. In case an ML criterium based on a Gaussian PDF is used, then the best predictors satisfy the condition

$$\sum_{k=1}^{K_d} \sum_{(i,j)\in I_d(k)} \tau_{ij} \cdot \hat{\tau}_{ij} = \sum_{k=1}^{K_d} \sum_{(i,j)\in I_d(k)} \hat{\tau}_{ij}^2,$$

so that $n_f = \text{DOF(lack of fit)} = N \cdot (N-1)/2 - 1$. An $F(n_f, n_R - n_f)$-test on the ratio

$$F = \frac{\text{MS(lack of fit)}}{\text{MS(pure error)}} = \frac{\text{SS(lack of fit)}/n_f}{\text{SS(pure error)}/(n_R - n_f)}$$

can be used to test the goodness of fit of the regression model.

APPENDIX 5.C: COMPARING STIMULUS CONFIGURATIONS

There is a complicating factor that must be considered when comparing XGms stimulus configurations with dimension $n > 1$. More specifically, the stimulus positions arising from an XGms analysis can be arbitrarily translated, rotated/mirrored (in case of a Euclidean metric) and dilated by a factor without influencing the model predictions for the stimuli. If we collect the stimulus positions into an $N \times n$ matrix X, then the above remarks can be mathematically summarized by stating that the transformed stimulus configuration

$$Y = X \cdot (s\, U) + C,$$

is an equally valid stimulus configuration to describe the experimental data. In this formula s is a central dilation factor, U is an $n \times n$ orthonormal matrix (i.e., $U^t \cdot U$ is equal to the identity matrix) and C is an $N \times n$ translation matrix (with all identical rows specifying the translation vector). The program XGms adds normalization conditions in order to guarantee a unique solution for the stimulus position matrix X.

When comparing two stimulus configurations X and Y, for instance arising from two separate model fits to the same data, then the transformation parameters for one of the configurations, say X, must be optimized before determining the distance from and/or correlation with the second configuration Y [Sch70]. The optimum translation can be easily determined by requiring that the column averages after transformation become equal. If the translation matrices C_x and C_y are needed to make the column averages equal to zero for X and Y, respectively, then the optimum overall translation matrix is $C = C_x \cdot s\, U - C_y$. In XGms, where the stimulus configurations are centered on the origin, we are guaranteed that $C_x = C_y = 0$.

The central dilation factor s and the transformation matrix U required for the optimum transformation from $\tilde{X} = X + C_x$ to $\tilde{Y} = Y + C_y$ are determined by minimizing the sum of the squared distances between the stimulus positions, i.e.,

$$d^2 = \text{trace}\left[(\tilde{X} \cdot s\, U - \tilde{Y})^t\, (\tilde{X} \cdot s\, U - \tilde{Y})\right]$$

(where the trace of a square matrix is equal to the sum of its diagonal elements). Substituting the optimum dilation factor

$$s = \frac{\text{trace}(\tilde{Y}^t\, \tilde{X}\, U)}{\text{trace}(U^t\, \tilde{X}^t\, \tilde{X}\, U)}$$

into this expression results in

$$d^2 = \text{trace}(\tilde{Y}^t\, \tilde{Y})\, (1 - \rho^2),$$

so that minimizing d^2 is equivalent to maximizing the inner-product correlation

$$\rho = \frac{\text{trace}[\tilde{Y}^t\, \tilde{X}\, U]}{\left\{\text{trace}[U^t\, \tilde{X}^t\, \tilde{X}\, U] \cdot \text{trace}[\tilde{Y}^t\, \tilde{Y}]\right\}^{1/2}}.$$

Maximizing this inner-product correlation is in turn equivalent to maximizing the inner product trace$[\tilde{Y}^t\, \tilde{X}\, U]$, since the denominator in the latter expression is constant if U is a orthonormal matrix. The solution to the inner-product maximization problem is well-known [Sch70, RtBS84, Ber77]. More precisely, the optimum **Procrustes transformation** is given by

$$U = P\,Q^t,$$

where $\tilde{X}^t\,\tilde{Y} = P\,S\,Q^t$ is the singular value decomposition of the $n \times n$ matrix $\tilde{X}^t\,\tilde{Y}$, and S is a diagonal matrix of singular values.

Next to comparing alternative XGms stimulus configurations, we may also want to compare an MDS stimulus configuration Y with an a priori supplied stimulus configuration X. In such a case, the matrix transformation between both stimulus configurations is usually not restricted, but can be an arbitrary linear transformation. The optimum linear transformation matrix F between $\tilde{X}$ and $\tilde{Y}$ in this general (unconstrained) case (i.e., $Y \approx X \cdot F + C$ or $\tilde{Y} \approx \tilde{X} \cdot F$) is

$$F = (\tilde{X}^t\,\tilde{X})^{-1} \cdot \tilde{X}^t\,\tilde{Y},$$

which is a direct consequence of the projection theorem in Euclidean space (see chapter 2). The columns of the $n \times n$ matrix F constitute vectors of multiple regression coefficients between the columns of $\tilde{X}$ and the columns of $\tilde{Y}$ [RtBS84].

Chapter 6

DISCRETE PERIODIC SIGNALS AND FOURIER TRANSFORMATIONS

1. INTRODUCTION

In the first four chapters, we have concentrated on the wavelength description of electromagnetic radiation. It was shown that three numbers, derived by proper weighting of the electromagnetic spectrum, can carry all the information needed for human color vision. These three numbers however only specify the color for one point (or uniform patch) in the visual field. Meaningful images only arise if these signals are allowed to vary in space and time. In the following chapters, we therefore concentrate on how such spatially and temporally varying visual information can be processed by technical systems in a way that is optimized for a human observer.

We first impose some reasonable restrictions on what we imply by spatially and/or temporally varying visual information [NH88]. First, we will only consider images of finite size. We hence assume that the viewer has the impression of looking at the outside world through a (rectangular) window of finite dimensions. Second, we assume that the images are either static, i.e., do not vary in time, or belong to a sequence of finite duration. Third, we assume that the images do not contain explicit depth information, and that the viewer cannot participate actively in acquiring the visual information. The viewer can hence not use binocular vision to derive the distance of objects, and has no control over the viewing position and focal depth. This does not prohibit that he infers depth by indirect means. For instance, he can derive the (relative) depth of objects from their relative motions. Often, he can also infer the observer (camera) position and focal depth with respect to the objects in view.

The above assumptions are valid for most technical systems. For instance, in broadcast applications, the images are captured by an image sensor, such as a camera target, of finite size. The images can be broken up into sequences of finite duration (between scene cuts). The camera position and focal depth are controlled in the studio, and are hence beyond control of the viewer. The images are displayed on the fixed, approximately flat, surface of a monitor or projection screen, so that no explicit depth information is available.

If the above restrictions are valid, then the visual information can be represented by three color signals that vary over a finite interval in space and time. For instance, the luminance of an image (sequence) can be represented by $L(x, y, t)$. The spatial coordinates (x, y) belong to a finite window, while the time coordinate t is either absent or belongs to a finite time interval.

The continuous variation of color signals in space, time and amplitude prohibits a representation by a finite amount of data, even if the signals are limited to finite intervals. We therefore want to approximate these continuously varying signals in space and time by discrete signals, i.e., signals that are specified by their values at discrete positions in space and time. In chapter 10 on quantization, we will also discuss how the remaining continuous variation in amplitude can be handled.

In this chapter, it is discussed how spatio-temporal signals can be processed by **linear space-invariant (and/or time-invariant) systems**. Harmonic basis signals are shown to be ideally suited for studying such systems, and it is derived how their introduction leads to the concept of **Fourier transformations**.

We first consider discrete signals and Fourier transformations for one-dimensional signals in section 2, and subsequently extend the theory to multi-dimensional signals in section 4. In order to make this possible, the concept of **periodicity**, which is fairly trivial in one dimension, has to be generalized to the concept of **lattices** in multiple dimensions, in section 3.

2. ONE-DIMENSIONAL SIGNAL TRANSFORMATIONS

In this section, we present a mathematical theory for handling one-dimensional signals that have finite size and/or are discrete. This theory is a straightforward application of the general linear system theory, discussed in chapter 2. It will be extended to multi-dimensional signals in section 4. This extension to multiple dimensions will necessitate a more complex notation, which obviously makes the theory less transparent. It is hoped that a parallel presentation of the one-dimensional case helps to better understand the general multi-dimensional case.

2.1 PERIODIC AND DISCRETE SIGNALS

A one-dimensional signal $f_p(x)$ is **periodic** with period T if it is fully specified by the values $f_p(x), x \in [0, T[$, in an interval of length T, and if $f_p(x) = f_p(x + kT)$, for $x \in [0, T[$ and $k \in \mathbb{Z}$. A non-periodic signal $f(x)$ can be converted into a periodic signal

$$f_p(x) = \sum_{k \in \mathbb{Z}} f(x + kT), \tag{6.1}$$

for $x \in \mathbb{R}$, by summing shifted versions of the signal $f(x)$. We have to impose restrictions on the input signal $f(x)$ if we want to be able to retrieve $f(x)$ from $f_p(x)$. For instance, if we can assume that $f(x)$ is zero outside an a priori known interval of length smaller than or equal to T, then $f(x)$ can be uniquely recovered from $f_p(x)$. Periodic signals will turn out to be a mathematical convenience for handling such signals of finite extent.

A one-dimensional signal $f_d(x)$ is **discrete** or **sampled** if it is fully specified by its values $f_d(i\,\Delta)$ at positions $x = i\,\Delta$ that are integer multiples of the discretization distance Δ. Since there are usually infinitely many continuous signals that pass through the same sample values, we will again have to impose a priori restrictions in order to derive the unique continuous signal of interest from the sample values. This inverse process of constructing a continuous signal from sample values is called **interpolation**.

If the period T is a multiple of the discretization distance Δ, i.e., if $T = N\,\Delta$, then we can define signals $f_{dp}(x)$ that are both **discrete and periodic**, i.e., signals that are fully specified by $[f_{dp}(i\,\Delta), i = 0, \ldots, N-1]$ and for which $f_{dp}(i\,\Delta) = f_{dp}((i + k\,N)\,\Delta)$, for $i = 0, \ldots, N-1$ and $k \in \mathbb{Z}$.

There is no unique way of constructing a continuous periodic signal that passes through the N sample values $[f_{dp}(i\Delta), i = 0, \ldots, N-1]$. The approach that we will pursue here is to select the interpolated signal from a linear vector space of order N that is spanned by an a priori selected basis

$$B_\psi = \{\psi_n(x); n = 0, \ldots, N-1\} \tag{6.2}$$

of signals that are periodic with period T. All signals of interest are then of the form

$$f_{dp}(x) = \sum_{n=0}^{N-1} \hat{f}_{dp}(n) \cdot \psi_n(x), \tag{6.3}$$

for $x \in \mathbb{R}$. By defining a vector product that only depends on the available sample values, i.e.,

$$\langle f_{dp}(x), g_{dp}(x) \rangle = \Delta \sum_{i=0}^{N-1} f_{dp}^{\star}(i\,\Delta) \cdot g_{dp}(i\,\Delta), \tag{6.4}$$

we can make this vector space into a Hilbert space, provided that we consider only discrete periodic signals $f_{dp}(x)$ that are square-summable, i.e., for which the squared norm

$$\| f_{dp}(x) \|^2 = \langle f_{dp}(x), f_{dp}(x) \rangle = \Delta \sum_{i=0}^{N-1} | f_{dp}(i\,\Delta) |^2 \tag{6.5}$$

is finite. If N is finite, this corresponds to the demand that all sample values have a finite amplitude $|f_{dp}(i\,\Delta)|$. The resulting Hilbert space is denoted by $\mathcal{L}^2(\{0, \ldots, N-1\})$.

The Gramian matrix $G_{\psi\psi}$ of the basis B_ψ will be non-singular, so that the dual basis

$$B_\chi = \{\chi_n(x); n = 0, \ldots, N-1\} \tag{6.6}$$

is guaranteed to exist, at least if N is finite. We also assume that the dual basis exists and is known in case N is infinite. The representation $[f_{dp}(x)]_\psi = [\hat{f}_{dp}(n), n = 0, \ldots, N-1]^T$ of the interpolated signal $f_{dp}(x)$ has coordinates

$$\hat{f}_{dp}(n) = \langle \chi_n(x), f_{dp}(x) \rangle = \Delta \sum_{i=0}^{N-1} \chi_n^{\star}(i\,\Delta) \cdot f_{dp}(i\,\Delta), \tag{6.7}$$

for $n = 0, \ldots, N-1$, and can hence also be derived from the sample values $[f_{dp}(i\,\Delta), i = 0, \ldots, N-1]$. The overal signal interpolation is then given by

$$\boxed{f_{dp}(x) = \sum_{n=0}^{N-1} \psi_n(x) \cdot \Delta \sum_{i=0}^{N-1} \chi_n^{\star}(i\,\Delta) \cdot f_{dp}(i\,\Delta),} \tag{6.8}$$

for $x \in \mathbb{R}$.

If interpolation followed by discretization is applied to the basis signals themselves, then we obtain that

$$\psi_m(j\,\Delta) = \sum_{i=0}^{N-1} \psi_m(i\,\Delta)\,\Delta \sum_{n=0}^{N-1} \chi_n^\star(i\,\Delta) \cdot \psi_n(j\,\Delta), \tag{6.9}$$

for $j, m = 0, \ldots, N-1$, since it is reasonable to assume that the basis signals should be mapped onto themselves. These equations imply that

$$\Delta \sum_{n=0}^{N-1} \chi_n^\star(i\,\Delta) \cdot \psi_n(j\,\Delta) = \delta_{ij}, \tag{6.10}$$

for $i, j = 0, \ldots, N-1$, where δ_{ij} denotes Kronecker's delta function. The latter condition can in turn be used to show that interpolated signals $f_{dp}(x)$, for $x \in \mathbb{R}$, indeed pass through the N sample values $[f_{dp}(i\,\Delta), i = 0, \ldots, N-1]$.

We now illustrate the above discretization and interpolation procedure for two important special cases.

2.2 CANONICAL BASIS

In the spatial, or **canonical**, approach, we consider the vector space of signals spanned by the orthogonal basis

$$B_\psi = \left\{ \psi_n(x) = \sum_{k \in \mathbb{Z}} p_d\,(x - (n + k\,N)\,\Delta)\,;\; n = 0, \ldots, N-1 \right\} \tag{6.11}$$

for $x \in \mathbb{R}$, of periodic pulse trains. In the above expression

$$p_d(x) = \begin{cases} 1 & \text{if} \quad x \in [p_o, p_o + \Delta[\\ 0 & \text{if} \quad x \notin [p_o, p_o + \Delta[\end{cases} \tag{6.12}$$

for $x \in \mathbb{R}$, denotes the unit pulse. The real parameter p_o can be chosen freely within the range $]-\Delta, 0]$, but $p_o = -\Delta/2$ will be the most frequent choice. Note that $\psi_n(i\,\Delta) = \delta_{ni}$, for $n, i = 0, \ldots, N-1$, so that $\psi_n(x)$ is equal to 1 at the sample positions $x = n\,\Delta + k\,T$, for $k \in \mathbb{Z}$, and zero at all other sample positions.

Since the canonical basis is orthogonal, the test signals are proportional to the basis signals, i.e., $\chi_n(x) = \frac{1}{\Delta}\,\psi_n(x)$. The periodic interpolated signal

$$f_{dp}(x) = \sum_{n=0}^{N-1} f_{dp}(n\,\Delta) \cdot \psi_n(x), \tag{6.13}$$

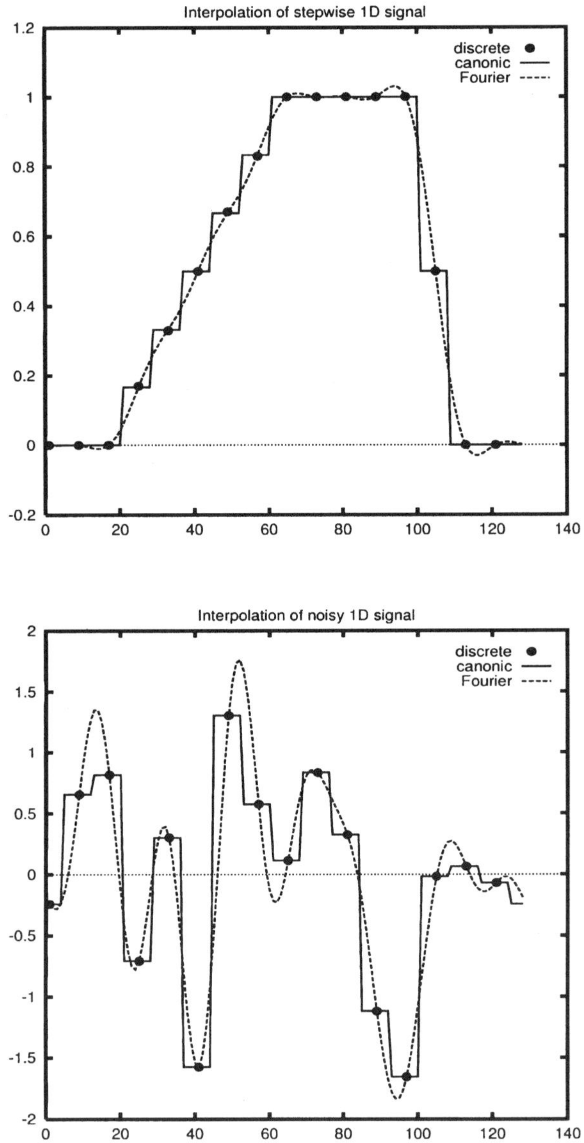

Figure 6.1. Continuous periodic signals obtained by interpolating discrete samples $f_{dp}(n\,\Delta)$, for $n = 0, \dots, N-1$ ($N = 16$, $\Delta = 8$), of a piecewise linear and a noisy signal. The stepwise curves are the canonical interpolations (with $p_o = -\Delta/2$), while the smooth curves are the Fourier interpolations (with $n_o = -\lfloor N/2 \rfloor$).

for $x \in \mathbb{R}$, is equal to the input values at the sample coordinates $x = i\,\Delta$ and piecewise constant. The canonical interpolation with $p_o = -\Delta/2$ is called the *nearest-neighbor* or *zero-order* interpolation and is illustrated for two example signals in Figure 6.1.

2.3 FOURIER BASIS

In the frequency, or **Fourier**, approach, we consider the vector space of signals that is spanned by the orthogonal basis of harmonic signals, i.e.,

$$B_\psi = \left\{ \psi_n(x) = \frac{1}{T} \cdot \exp\left(-j2\pi \frac{n\,x}{T}\right); n = n_o, \ldots, n_o + N - 1 \right\}, \tag{6.14}$$

for $x \in \mathbb{R}$. The integer parameter n_o can be chosen freely, but $n_o = -\lfloor N/2 \rfloor$ will be the most frequent choice[1].

The biorthogonality between the basis functions $\psi_n(x)$ and the test functions $\chi_n(x) = T \cdot \psi_n(x)$, for $x \in \mathbb{R}$ and $n = n_o, \ldots, n_o + N - 1$, can be verified using *Poisson's summation formula*

$$\sum_{i=0}^{N-1} \exp\left(j2\pi \frac{n\,i}{N}\right) = \begin{cases} N & \text{if} \quad n = k \cdot N,\ k \in \mathbb{Z} \\ 0 & \text{if} \quad n \neq k \cdot N \end{cases} \tag{6.15}$$

The interpolated signals are of the form

$$\boxed{f_{dp}(x) = \frac{1}{T} \sum_{n=n_o}^{n_o+N-1} \hat{f}_{dp}(n) \cdot \exp\left(-j2\pi \frac{n\,x}{T}\right),} \tag{6.16}$$

for $x \in \mathbb{R}$, and are indeed periodic with period T. The representation $[f_{dp}(x)]_\psi = [\hat{f}_{dp}(n), n = n_o, \ldots, n_o + N - 1]$ of the interpolated signal can be derived from the original sequence of numbers through the **discrete Fourier series**

$$\boxed{\hat{f}_{dp}(n) = \langle \chi_n(x), f_{dp}(x) \rangle = \Delta \sum_{i=0}^{N-1} f_{dp}(i\,\Delta) \cdot \exp\left(j2\pi \frac{n\,i}{N}\right),} \tag{6.17}$$

for $n = n_o, \ldots, n_o + N - 1$. The discrete Fourier series is a discrete and periodic signal with period N since $\hat{f}_{dp}(n + k \cdot N) = \hat{f}_{dp}(n)$, for $k \in \mathbb{Z}$.

[1] The largest integer that is smaller than or equal to the real number a is notated by $\lfloor a \rfloor$.

It is easily verified that the vector product between two signals $f_{dp}(x)$ and $g_{dp}(x)$ can also be expressed as follows

$$\langle f_{dp}(x), g_{dp}(x) \rangle = \frac{1}{T} \sum_{n=n_o}^{n_o+N-1} \hat{f}^\star_{dp}(n) \cdot \hat{g}_{dp}(n) \tag{6.18}$$

in terms of the discrete Fourier series of the signals (*Parseval's theorem*).

If the input function $f_{dp}(x)$ has only real values, then the discrete Fourier series satisfies $\hat{f}^\star_{dp}(n) = \hat{f}_{dp}(-n)$. This implies that in this case the real and imaginary parts of the discrete Fourier series form symmetric and anti-symmetric series, respectively. Since the signals that we consider are mostly real-valued, we obtain that the harmonic basis functions often occur in the following combinations

$$\begin{aligned} \frac{1}{2}\,(\psi_n(x) + \psi_{-n}(x)) &= \cos\frac{2\pi n x}{T} \\ \frac{1}{2j}\,(\psi_n(x) - \psi_{-n}(x)) &= \sin\frac{2\pi n x}{T} \end{aligned} \tag{6.19}$$

Such combinations are called **gratings** with **frequency** $|n|/T$. Hence, only frequencies that are integer multiples of $1/T$ occur in a periodic signal of length T.

Even if the input samples are real, the interpolated signal will usually be complex-valued for positions x that are in between the sampling positions. Only in specific cases, such as $n_o = -\lfloor N/2 \rfloor$ and N odd, does the interpolation procedure result in a real-valued signal. However, in all cases we can retain the real part of the interpolated signal, i.e.,

$$\Re f_{dp}(x) = \frac{1}{T} \sum_{n=n_o}^{n_o+N-1} \left(\hat{a}_{dp}(n) \cdot \cos\frac{2\pi n\, x}{T} + \hat{b}_{dp}(n) \cdot \sin\frac{2\pi n\, x}{T} \right) \tag{6.20}$$

as the signal interpolation, where $\hat{f}_{dp}(n) = \hat{a}_{dp}(n) + j \cdot \hat{b}_{dp}(n)$, for $n = n_o, \dots, n_o + N - 1$, denotes the decomposition of the complex-valued Fourier coefficients in a real and imaginary part. These real-valued Fourier interpolations (with $n_o = -\lfloor N/2 \rfloor$) of the example signals are also shown in Figure 6.1.

2.4 CONVOLUTION

Discrete Fourier series descriptions are mainly used for studying discrete linear position-invariant systems. Such systems can be described as **discrete convolution** operators that map an input signal with sample values $f_{dp}(i\,\Delta)$ into an output signal with sample values

$$\boxed{g_{dp}(i\,\Delta) = h_d(i\,\Delta) * f_{dp}(i\,\Delta) = \Delta \sum_{j\in\mathbb{Z}} h_d((i-j)\,\Delta)\cdot f_{dp}(j\,\Delta),} \tag{6.21}$$

for $i = 0, \ldots, N-1$. The sample values $h_d(j\,\Delta)$, for $j \in \mathbb{Z}$, completely characterize the system. It is called a linear position-invariant system because shifting the input values by a number of sample positions simply shifts the output values by the same number of sample positions.

A linear position-invariant system is uniquely characterized by its **impulse response**. This is the response of the system to the discrete periodic input signal with samples $f_{dp}(0) = 1$ and $f_{dp}(i\,\Delta) = 0$, for $i = 1, \ldots, N-1$. The discrete periodic impulse response signal has sample values

$$h_{dp}(i\,\Delta) = \sum_{k\in\mathbb{Z}} h_d\left((i + k\,N)\Delta\right), \tag{6.22}$$

for $i = 0, \ldots, N-1$, and is hence equal to the discrete periodic signal that arises by summing shifted versions of the non-periodic discrete signal $h_d(j\,\Delta)$, for $j \in \mathbb{Z}$. The input-output relation for a discrete convolution can be rewritten as

$$\boxed{g_{dp}(i\,\Delta) = \Delta \sum_{j=0}^{N-1} h_{dp}((i-j)\,\Delta)\cdot f_{dp}(j\,\Delta) = \langle h^{\star}_{dp}(i\,\Delta - x), f_{dp}(x)\rangle} \tag{6.23}$$

for $i = 0, \ldots, N-1$. This shows that a linear space-invariant system is a measurement system in which all test vectors are displaced versions of $h^{\star}_{dp}(-x)$, i.e., the complex conjugate of the mirrored impulse response.

The discrete harmonic signals $\psi_n(i\,\Delta)$ are *eigenvectors* of any linear position-invariant system, i.e., they are mapped proportionally into $\hat{h}_{dp}(n)\cdot\psi_n(i\,\Delta)$, for $i \in \mathbb{Z}$ and $n = 0, \ldots, N-1$. The *eigenvalues* $[\hat{h}_{dp}(n), n = 0, \ldots, N-1]^T$ are an alternative way of characterizing a linear position-invariant system and are obtained by taking the discrete Fourier series of the impulse response. The discrete Fourier series of the input and output signals are related through

$$\boxed{\hat{g}_{dp}(n) = \hat{h}_{dp}(n)\cdot\hat{f}_{dp}(n),} \tag{6.24}$$

for $n = 0, \ldots, N-1$.

An extensive discussion on linear position- (and/or time-) invariant systems and Fourier transformations is contained in most introductory textbooks on digital signal processing, so that we refer to these textbooks for more detailed information (see for instance [Ell87, NH88, Pra91, GW93]). Some important properties of discrete Fourier series are also discussed in exercise 1.

2.5 FOURIER TRANSFORMATIONS AND FOURIER SERIES

It is worthwhile to examine some limit cases of the above Hilbert space of discrete square-summable periodic signals.

2.5.1 CONTINUOUS PERIODIC SIGNALS

Discretization of a periodic signal $f_p(x)$ at a very high density, which corresponds to choosing $\Delta \to 0$, leads to the following expressions for the **continuous Fourier series**

$$\hat{f}_p(n) = \int_0^T f_p(x) \cdot \exp\left(j2\pi \frac{n\,x}{T}\right)\, dx, \tag{6.25}$$

for $n \in \mathbb{Z}$, and its inverse

$$f_p(x) = \frac{1}{T} \sum_{n \in \mathbb{Z}} \hat{f}_p(n) \cdot \exp\left(-j2\pi \frac{n\,x}{T}\right), \tag{6.26}$$

for $x \in [0, T[$. The resulting space of square-integrable continuous periodic signals, with vector product

$$\langle f_p(x), g_p(x) \rangle = \int_0^T f_p^\star(x) \cdot g_p(x)\, dx, \tag{6.27}$$

and (dual) Fourier basis

$$B_\chi = \left\{ \chi_n(x) = \exp\left(-j2\pi \frac{n\,x}{T}\right);\ n \in \mathbb{Z} \right\}, \tag{6.28}$$

is denoted by $\mathcal{L}^2([0, T[)$. Note that the Fourier basis B_χ is countable, so that $\mathcal{L}^2([0, T[)$ is a separable vector space.

The **continuous convolution** operator for such densely sampled signals can be expressed as

$$g_p(x) = \int_{-\infty}^{+\infty} h(x - y) \cdot f_p(y)\, dy \tag{6.29}$$

for $x \in \mathbb{R}$, in terms of the non-periodic impulse response $h(x)$, or as

$$g_p(x) = \int_0^T h_p(x-y) \cdot f_p(y)\, dy \tag{6.30}$$

in terms of the periodic impulse response

$$h_p(x) = \sum_{k \in \mathbb{Z}} h(x + k\,T), \tag{6.31}$$

for $x \in \mathbb{R}$.

2.5.2 DISCRETE APERIODIC SIGNALS

If we consider discrete signals $f_d(x)$, for $x = i\,\Delta$, with a very large periodicity, which corresponds to choosing $T \to \infty$, we obtain the following expressions for the **discrete Fourier transformation**

$$\hat{f}_d(u) = \Delta \sum_{i \in \mathbb{Z}} f_d(i\,\Delta) \cdot \exp\left(j2\pi u\, i\,\Delta\right), \tag{6.32}$$

for $u \in [-1/(2\,\Delta), 1/(2\,\Delta)[$, and its inverse

$$f_d(i\,\Delta) = \int_{-1/(2\,\Delta)}^{1/(2\,\Delta)} \hat{f}_d(u) \cdot \exp\left(-j2\pi u\, i\,\Delta\right)\, du, \tag{6.33}$$

for $i \in \mathbb{Z}$. The resulting space of square-summable discrete signals, with vector product

$$\langle f_d(x), g_d(x) \rangle = \Delta \sum_{i \in \mathbb{Z}} f_d^{\star}(i\,\Delta) \cdot g_d(i\,\Delta), \tag{6.34}$$

and (dual) Fourier basis

$$B_\chi = \{\chi_u(x) = \exp(-j2\pi u\, x);\; u \in [-1/(2\,\Delta), 1/(2\,\Delta)[\}, \tag{6.35}$$

is denoted by $\mathcal{L}^2(\mathbb{Z})$. Although the Fourier basis is not countable in this case, the canonical basis is, so that $\mathcal{L}^2(\mathbb{Z})$ is also a separable vector space.

2.5.3 CONTINUOUS APERIODIC SIGNALS

If we consider signals that have both a very large periodicity and a very small discretization distance, then we obtain the following expressions for the **continuous Fourier transformation**

$$\hat{f}(u) = \int_{\mathbb{R}} f(x) \cdot \exp(j2\pi u\, x)\, dx \tag{6.36}$$

for $u \in \mathbb{R}$, and its inverse

$$f(x) = \int_{\mathbb{R}} \hat{f}(u) \cdot \exp(-j2\pi u\, x)\, du, \tag{6.37}$$

for $x \in \mathbb{R}$. The resulting space of square-integrable continuous signals, with vector product

$$\langle f(x), g(x) \rangle = \int_{\mathbb{R}} f^{\star}(x) \cdot g(x)\, dx, \tag{6.38}$$

and (dual) Fourier basis

$$B_{\chi} = \{\chi_u(x) = \exp(-j2\pi u\, x); u \in \mathbb{R}\}, \tag{6.39}$$

is denoted by $\mathcal{L}^2(\mathbb{R})$. In this case, neither the canonical nor the Fourier basis is countable. This does however not imply that $\mathcal{L}^2(\mathbb{R})$ is not separable; as was shown in chapter 2, where a countable basis (of orthogonal polynomials multiplied by a window) was derived for $\mathcal{L}^2(\mathbb{R})$.

2.5.4 ALIASING

The well-known relationship between the discrete Fourier transformation $\hat{f}_d(u)$ of a sampled signal $[f_d(i\,\Delta) = f(i\,\Delta), i \in \mathbb{Z}]$ and the continuous Fourier transformation $\hat{f}(u)$ of the original continuous signal $[f(x), x \in \mathbb{R}]$, i.e.,

$$\hat{f}_d(u) = \sum_{k \in \mathbb{Z}} \hat{f}\left(u + k\,\frac{1}{\Delta}\right), \tag{6.40}$$

can easily be derived from the above expressions. This latter expression is the Fourier equivalent of the spatial relationship between the original non-periodic signal $f(x)$ and the derived periodic signal $f_p(x)$ in equation (6.1). It implies that a signal with a Fourier transformation that is zero outside a frequency interval of length at most $1/\Delta$ can be uniquely recovered from its sample values. If this condition is not satisfied, then the interpolated signal is said to be an **aliased** approximation to the original signal. Figure 6.2 illustrates the comparison between the Fourier transformations of an original (densely sampled) signal and a discrete (coarsely sampled) signal. The aliased Fourier interpolation in the upper part of Figure 6.1 arose by only retaining the $N = 16$ Fourier components that belong to the peak closest to the origin in the frequency domain.

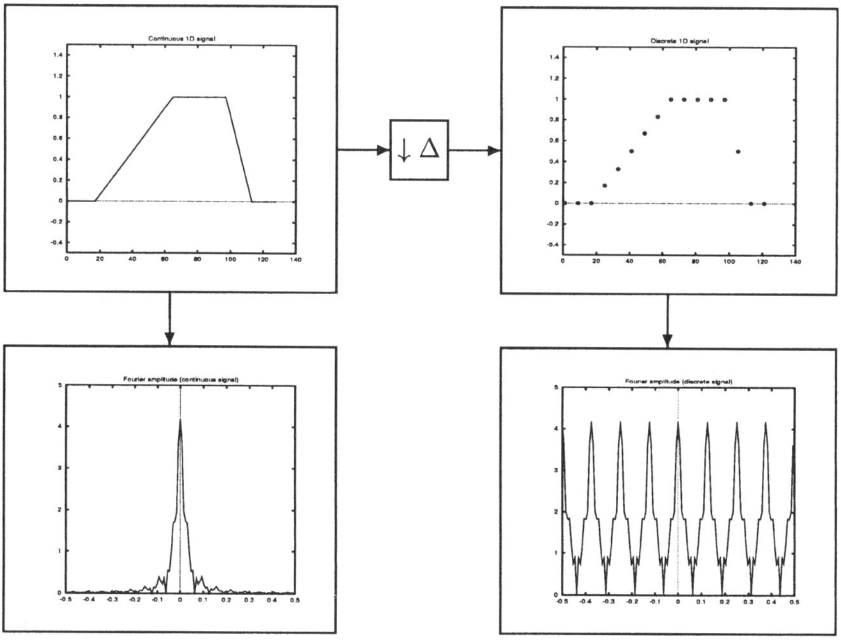

Figure 6.2. Comparison between an original (densely sampled) signal and a discrete (coarsely sampled) signal. The upper panels show the space/time variations of the original and discrete signals. The lower panels show the Fourier amplitude functions $\log(1 + |\hat{f}(u)|)$ and $\log(1 + |\hat{f}_d(u)|)$.

3. LATTICES

3.1 DEFINITIONS

In the previous section, we discussed signals that were periodic and/or discrete in one dimension. The definitions of these signals were based on actions that had to be repeated at regular intervals. Either the signal itself has to be repeated at regular intervals, in the case of periodic signals, or the signal has to be discretized at regular intervals, in the case of discrete signals. A regular interval simply implies a fixed distance in one dimension. Defining regular intervals in multi-dimensional Euclidean space is somewhat more complicated. A simplified treatment is possible because most regular intervals that have been considered up to now for applications in image technology can be described by lattice theory [Dub85]. We therefore briefly review lattice theory in this section. We introduce the theory for an arbitrary number of dimensions, although most applications will be limited to one, two or three dimensions (1D, 2D or 3D).

We assume a basis of real vectors $\{\mathbf{v}_1, \ldots, \mathbf{v}_D\}$ in the Euclidean space $\mathbb{R}^D$. A **lattice** Λ in $\mathbb{R}^D$ is defined as the set of all linear vector combinations

$$\Lambda = \{n_1 \cdot \mathbf{v}_1 + \ldots + n_D \cdot \mathbf{v}_D; n_i \in \mathbb{Z}, i = 1, \ldots, D\}, \tag{6.41}$$

with *integer* coefficients n_i. Suppose that the vector $\mathbf{v}_i$ has representation $[v_{1i} \ldots v_{Di}]^T$ in the standard orthonormal basis of $\mathbb{R}^D$, and let us denote the vector of coefficients by $\mathbf{n} = [n_1 \ldots n_D]^T (\in \mathbb{Z}^D)$. The $D \times D$ matrix $V = [v_{ij}]$ then describes the transformation between the representation in the lattice basis and the representation in the standard basis, so that the lattice Λ is equal to the set of vectors with a representation of the form $V \cdot \mathbf{n}$ in the standard basis, where $\mathbf{n} \in \mathbb{Z}^D$,.

The basis for a given lattice is not unique. If E is any matrix of integers such that $\det E = \pm 1$, then the matrix $E \cdot V$ provides another basis for Λ. However, the **determinant of the lattice** $d(\Lambda) = |\det V|$ is unique and independent of the particular choice of basis. The inverse of $d(\Lambda)$ is called the **lattice density**. If E is any matrix of integers such that $\det E \neq \pm 1$, then the matrix $E \cdot V$ provides a basis for a **sublattice** Λ_s of Λ. The determinant of this sublattice is $d(\Lambda_s) = |\det E| \cdot d(\Lambda)$.

In Figure 6.3, we illustrate two important 2D lattices, i.e., the **rectangular lattice**, corresponding to

$$V = \begin{bmatrix} x_1 & 0 \\ 0 & x_2 \end{bmatrix}, \tag{6.42}$$

and the **quincunx lattice**, corresponding to

$$V = \begin{bmatrix} x_1 & x_1/2 \\ 0 & x_2 \end{bmatrix}, \tag{6.43}$$

both having determinant $d(\Lambda) = x_1 \cdot x_2$. If $x_2 = x_1$, then the rectangular lattice is called **square**. If $x_2 = (\sqrt{3}/2) \cdot x_1$, then the quincunx lattice is called **hexagonal**. If $x_2 = (1/2) \cdot x_1$, then the quincunx lattice is called **diagonal**.

We define a **unit cell** of a lattice Λ as a set $\wp \subset \mathbb{R}^D$ such that $\mathbb{R}^D$ is the disjoint union of copies of $\wp$ centered on each lattice point, i.e., the sets $\mathbf{x} + \wp$ and $\mathbf{y} + \wp$ do not overlap for $\mathbf{x}, \mathbf{y} \in \Lambda, \mathbf{x} \neq \mathbf{y}$, and the union $\bigcup_{\mathbf{x} \in \Lambda} (\mathbf{x} + \wp) = \mathbb{R}^D$. The (hyper)volume of a unit cell of a lattice Λ is equal to $d(\Lambda)$. There are many possible choices for the unit cell of a lattice. One that is very easy to specify is the **fundamental parallelepiped**, given by

$$\wp = \{f_1 \cdot \mathbf{v}_1 + \ldots + f_D \cdot \mathbf{v}_D; 0 \leq f_i < 1, i = 1, \ldots, D\}. \tag{6.44}$$

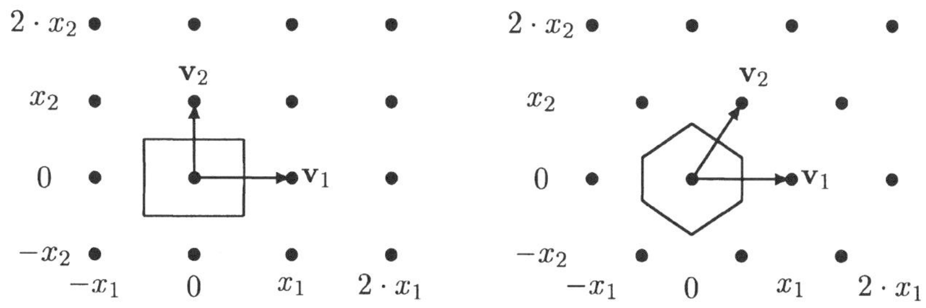

Figure 6.3. Rectangular and quincunx lattice, with corresponding Voronoi cells.

Probably the most frequently used unit cell is the **Voronoi cell**, defined as the set of all points in $\mathbb{R}^D$ that are closer to the origin (which belongs to all lattices) than to any other lattice point. The Voronoi cells for the lattices in Figure 6.3 are indicated.

3.2 RECIPROCAL LATTICE

A concept that is essential for the Fourier description of signals that are periodic and/or discrete on a lattice is that of the reciprocal lattice. Given a lattice Λ, the set of all vectors $\mathbf{u}$ such that $\mathbf{u}^T\mathbf{x}$ is an integer for all $\mathbf{x} \in \Lambda$ is called the **reciprocal lattice** $\Lambda^\star$ of Λ. Suppose that the matrix $U = [u_{ij}]$ satisfies $U^T V = I$, with I the $D \times D$ identity matrix. The set of vectors $\mathbf{u}_1, \ldots, \mathbf{u}_D$, with representations $[u_{1i} \ldots u_{Di}]^T$ in the standard basis of $\mathbb{R}^D$, is then a basis of $\Lambda^\star$. The determinant of the reciprocal lattice hence satisfies $d(\Lambda^\star) = 1/d(\Lambda)$.

The reciprocal lattices of Figure 6.3 are depicted in Figure 6.4. The reciprocal lattice of the rectangular lattice is specified by the matrix

$$U = \begin{bmatrix} 1/x_1 & 0 \\ 0 & 1/x_2 \end{bmatrix}, \tag{6.45}$$

while the reciprocal lattice of the quincunx lattice is specified by the matrix

$$U = \begin{bmatrix} 1/x_1 & 0 \\ -1/(2x_2) & 1/x_2 \end{bmatrix}, \tag{6.46}$$

while both reciprocal lattices have determinant $d(\Lambda^\star) = 1/(x_1 \cdot x_2)$.

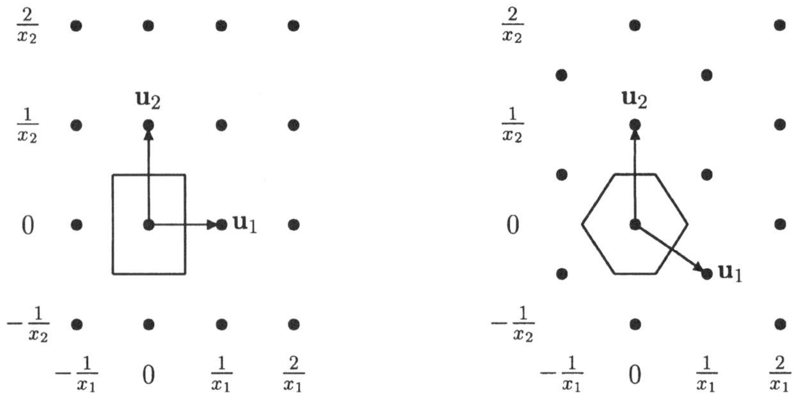

Figure 6.4. Reciprocal lattices of the rectangular and quincunx lattices of Figure 6.3, with corresponding Voronoi cells.

4. MULTI-DIMENSIONAL SIGNAL TRANSFORMATIONS

In this section, we extend some well-known results on one-dimensional signals and Fourier transformations, that were reviewed in section 2, to D-dimensional signals.

4.1 PERIODIC AND DISCRETE SIGNALS

A D-dimensional signal $f_p(\mathbf{x})$ is **periodic** on the lattice Λ_p if it is fully specified by the values $[f_p(\mathbf{x}), \forall \mathbf{x} \in \wp_p]$, where $\wp_p$ is a unit cell of the *periodicity lattice* Λ_p, and if $f_p(\mathbf{x}) = f_p(\mathbf{x}+\mathbf{y}), \forall \mathbf{x} \in \wp_p$ and $\forall \mathbf{y} \in \Lambda_p$. A non-periodic signals $f(\mathbf{x})$ can be converted into a periodic signal

$$f_p(\mathbf{x}) = \sum_{\mathbf{y} \in \Lambda_p} f(\mathbf{x}+\mathbf{y}), \tag{6.47}$$

for $\mathbf{x} \in \mathbb{R}^D$, by summing shifted versions of the signal $f(\mathbf{x})$. The original signal $f(\mathbf{x})$ cannot always be recovered from the periodic signal $f_p(\mathbf{x})$. If the signal $f(\mathbf{x})$ is however zero outside the unit cell $\wp_p$, then $f_p(\mathbf{x})$ and $f(\mathbf{x})$ are equivalent.

A D-dimensional signal $f_d(\mathbf{x})$ is **discrete** or **sampled** on the *discretization lattice* Λ_d if it is fully specified by the values $[f_d(\mathbf{x}), \forall \mathbf{x} \in \Lambda_d]$. There are infinitely many signals that pass through the same sample values $[f_d(\mathbf{x}), \mathbf{x} \in \Lambda_d]$, so that we must assume a priori knowledge in order to select the unique one that is of interest to us. This inverse process of constructing a continuous signal that passes through a given set of sample values $[f_d(\mathbf{x}), \mathbf{x} \in \Lambda_d]$ is called **interpolation**.

If the lattice Λ_p is a sublattice of the lattice Λ_d, then we can define signals $f_{dp}(\mathbf{x})$ that are both discrete and periodic, i.e., signals that are fully specified by the values $[f_{dp}(\mathbf{x}), \forall \mathbf{x} \in (\wp_p \cap \Lambda_d)]$, and for which $f_{dp}(\mathbf{x}) = f_{dp}(\mathbf{x}+\mathbf{y}), \forall \mathbf{x} \in (\wp_p \cap \Lambda_d)$ and $\forall \mathbf{y} \in \Lambda_p$. The set $\wp_p \cap \Lambda_d$ contains the $N = d(\Lambda_p)/d(\Lambda_d)$ positions on the discretization lattice Λ_d that are within the unit cell $\wp_p$ of the periodicity lattice Λ_p.

There is no unique way of constructing a periodic signal that passes through the sample values $[f_{dp}(\mathbf{x}), \forall \mathbf{x} \in (\wp_p \cap \Lambda_d)]$. The approach that we pursue here is to select the interpolated signal from a linear vector space of order N that is spanned by an a priori selected basis

$$B_\psi = \{\psi_n(\mathbf{x}); n = 0, \dots, N-1\} \tag{6.48}$$

of signals that are periodic on the lattice Λ_p. All signals of interest are then of the form

$$f_{dp}(\mathbf{x}) = \sum_{n=0}^{N-1} \hat{f}_{dp}(n) \cdot \psi_n(\mathbf{x}), \tag{6.49}$$

for $\mathbf{x} \in \mathbb{R}^D$. By defining a vector product that only depends on the sample values, i.e.,

$$\langle f_{dp}(\mathbf{x}), g_{dp}(\mathbf{x}) \rangle = d(\Lambda_d) \sum_{\mathbf{x} \in (\wp_p \cap \Lambda_d)} f_{dp}^\star(\mathbf{x}) \cdot g_{dp}(\mathbf{x}), \tag{6.50}$$

with $d(\Lambda_d)$ equal to the determinant of the lattice Λ_d, we can make this vector space into a Hilbert space, provided that we only consider periodic signals $f_{dp}(\mathbf{x})$ that are square-summable over the discrete unit cell $\wp_p \cap \Lambda_d$, i.e., for which the squared norm

$$\| f_{dp}(\mathbf{x}) \|^2 = \langle f_{dp}(\mathbf{x}), f_{dp}(\mathbf{x}) \rangle = d(\Lambda_d) \sum_{\mathbf{x} \in (\wp_p \cap \Lambda_d)} | f_{dp}(\mathbf{x}) |^2 \tag{6.51}$$

is finite. The resulting Hilbert space is denoted by $\mathcal{L}^2(\wp_p \cap \Lambda_d)$.

The basis B_ψ is assumed to have a dual basis

$$B_\chi = \{\chi_n(\mathbf{x}); n = 0, \dots, N-1\}. \tag{6.52}$$

so that the representation $[f_{dp}(\mathbf{x})]_\psi = [\hat{f}_{dp}(n), n = 0, \dots, N-1]^T$ of the interpolated signal $f_{dp}(\mathbf{x})$ consists of the coefficients

$$\hat{f}_{dp}(n) = \langle \chi_n(\mathbf{x}), f_{dp}(\mathbf{x}) \rangle = d(\Lambda_d) \sum_{\mathbf{x} \in (\wp_p \cap \Lambda_d)} \chi_n^\star(\mathbf{x}) \cdot f_{dp}(\mathbf{x}), \tag{6.53}$$

for $n = 0, \ldots, N-1$, and can hence be derived from $[f_{dp}(\mathbf{x}), \mathbf{x} \in (\wp_p \cap \Lambda_d)]$. The overal signal interpolation is then given by

$$\boxed{f_{dp}(\mathbf{x}) = \sum_{n=0}^{N-1} \psi_n(\mathbf{x}) \cdot d(\Lambda_d) \sum_{\mathbf{x} \in (\wp_p \cap \Lambda_d)} \chi_n^\star(\mathbf{x}) \cdot f_{dp}(\mathbf{x}),} \tag{6.54}$$

for $\mathbf{x} \in \mathbb{R}^D$.

If interpolation followed by discretization is applied to the basis signals themselves, then we obtain that

$$\psi_m(\mathbf{y}) = \sum_{\mathbf{x} \in (\wp_p \cap \Lambda_d)} \psi_m(\mathbf{x}) \cdot d(\Lambda_d) \sum_{n=0}^{N-1} \chi_n^\star(\mathbf{x}) \cdot \psi_n(\mathbf{y}), \tag{6.55}$$

for $\mathbf{y} \in (\wp_p \cap \Lambda_d)$ and $m = 0, \ldots, N-1$, since it is reasonable to assume that the basis signals are mapped onto themselves. These equations imply that

$$d(\Lambda_d) \sum_{n=0}^{N-1} \chi_n^\star(\mathbf{x}) \cdot \psi_n(\mathbf{y}) = \delta_{\mathbf{x}-\mathbf{y}}, \tag{6.56}$$

for $\mathbf{x}, \mathbf{y} \in (\wp_p \cap \Lambda_d)$. The latter condition in turn implies that all interpolated signals $f_{dp}(\mathbf{x})$, for $\mathbf{x} \in \wp_p$, indeed pass through the N sample values $[f_{dp}(\mathbf{y}), \mathbf{y} \in \wp_p \cap \Lambda_d)]$.

We now illustrate this interpolation procedure for two important special cases.

4.2 CANONICAL BASIS

In the spatial, or **canonical**, approach, we consider the vector space of signals spanned by the orthogonal basis

$$B_\psi = \left\{ \psi_{\mathbf{u}}(\mathbf{x}) = \sum_{\mathbf{y} \in \Lambda_p} p_d(\mathbf{x} - (\mathbf{u} + \mathbf{y})); \mathbf{u} \in (\wp_p \cap \Lambda_d) \right\}, \tag{6.57}$$

for $\mathbf{x} \in \mathbb{R}^D$, of periodic pulse trains. In the above expression

$$p_d(\mathbf{x}) = \begin{cases} 1 & \text{if} \quad \mathbf{x} \in \wp_d \\ 0 & \text{if} \quad \mathbf{x} \notin \wp_d \end{cases} \tag{6.58}$$

for $\mathbf{x} \in \mathbb{R}^D$, denotes a unit pulse. The basis B_ψ has the required number of basis vectors.

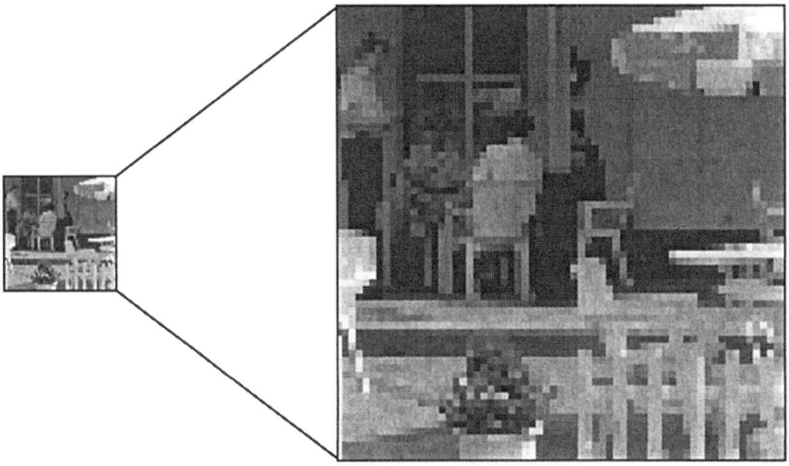

Figure 6.5. Image interpolation with a canonical basis.

The basis is orthogonal, so that the test signals $\chi_{\mathbf{u}}(\mathbf{x}) = \psi_{\mathbf{u}}(\mathbf{x})/d(\Lambda_d)$, for $\mathbf{x} \in \mathbb{R}^D$ and $\mathbf{u} \in (\wp_p \cap \Lambda_d)$, are proportional to the corresponding basis signals. The periodic piecewise-constant interpolated signal is given by

$$f_{dp}(\mathbf{x}) = \sum_{\mathbf{u} \in (\wp_p \cap \Lambda_d)} f_{dp}(\mathbf{u}) \cdot \psi_{\mathbf{u}}(\mathbf{x}), \tag{6.59}$$

for $\mathbf{x} \in \mathbb{R}^D$, and has the required values at the sample coordinates $\mathbf{x} \in (\wp_p \cap \Lambda_d)$, since

$$\psi_{\mathbf{u}}(\mathbf{x}) = \begin{cases} 1 & \text{if} \quad \mathbf{x} - \mathbf{u} \in \Lambda_p \\ 0 & \text{if} \quad \mathbf{x} - \mathbf{u} \notin \Lambda_p \end{cases} \tag{6.60}$$

for $\mathbf{x} \in \Lambda_d$ and $\mathbf{u} \in (\wp_p \cap \Lambda_d)$. Zero-order image interpolation with a canonical basis is illustrated in Figure 6.5. The most noticeable degradation introduced by this interpolation is *blocking* or *staircasing*.

4.3 FOURIER BASIS

In the frequency, or **Fourier**, approach, we consider the vector space of signals that is spanned by the orthogonal basis of harmonic signals, i.e.,

$$B_\psi = \left\{ \psi_{\mathbf{u}}(\mathbf{x}) = d\left(\Lambda_p^\star\right) \cdot \exp(-j2\pi \mathbf{u}^T \mathbf{x}); \mathbf{u} \in \left(\wp_d^\star \cap \Lambda_p^\star\right) \right\}, \tag{6.61}$$

for $\mathbf{x} \in \mathbb{R}^D$, where $\Lambda_p^\star$ is the reciprocal lattice of the periodicity lattice Λ_p, and $\wp_d^\star$ is a unit cell of the reciprocal lattice $\Lambda_d^\star$ of the discretization lattice Λ_d. This basis also contains $d(\Lambda_d^\star)/d(\Lambda_p^\star) = d(\Lambda_p)/d(\Lambda_d)$ basis vectors.

The biorthogonality between the basis functions $\psi_{\mathbf{u}}(\mathbf{x})$ and the test functions $\chi_{\mathbf{u}}(\mathbf{x}) = d(\Lambda_p) \cdot \psi_{\mathbf{u}}(\mathbf{x})$, for $\mathbf{x} \in \mathbb{R}^D$ and $\mathbf{u} \in (\wp_d^\star \cap \Lambda_p^\star)$, can be verified using *Poisson's summation formula*

$$d(\Lambda_d) \sum_{\mathbf{x} \in (\wp_p \cap \Lambda_d)} \exp(j2\pi \mathbf{u}^T \mathbf{x}) = \begin{cases} d(\Lambda_p) & \text{if} \quad \mathbf{u} \in \Lambda_d^\star \\ 0 & \text{if} \quad \mathbf{u} \notin \Lambda_d^\star \end{cases} \tag{6.62}$$

The interpolated signals are of the form

$$\boxed{f_{dp}(\mathbf{x}) = d\left(\Lambda_p^\star\right) \sum_{\mathbf{u} \in (\wp_d^\star \cap \Lambda_p^\star)} \hat{f}_{dp}(\mathbf{u}) \cdot \exp(-j2\pi \mathbf{u}^T \mathbf{x}),} \tag{6.63}$$

for $\mathbf{x} \in \mathbb{R}^D$, and are hence periodic on the lattice Λ_p. The representation $[f_{dp}(\mathbf{x})]_\psi = [\hat{f}_{dp}(\mathbf{u}), \mathbf{u} \in (\wp_d^\star \cap \Lambda_p^\star)]^T$ of the interpolated signal can be derived from the original sequence of numbers through the **discrete Fourier series**

$$\boxed{\hat{f}_{dp}(\mathbf{u}) = \langle \chi_{\mathbf{u}}(\mathbf{x}), f_{dp}(\mathbf{x}) \rangle = d(\Lambda_d) \sum_{\mathbf{x} \in (\wp_p \cap \Lambda_d)} f_{dp}(\mathbf{x}) \cdot \exp(j2\pi \mathbf{u}^T \mathbf{x}),} \tag{6.64}$$

for $\mathbf{u} \in (\wp_d^\star \cap \Lambda_p^\star)$. The discrete Fourier series is a discrete periodic signal with periodicity lattice $\Lambda_d^\star$ and discretization lattice $\Lambda_p^\star$. It is easily verified that the vector product between two signals $f_{dp}(\mathbf{x})$ and $g_{dp}(\mathbf{x})$ can be expressed as follows

$$\langle f_{dp}(\mathbf{x}), g_{dp}(\mathbf{x}) \rangle = d\left(\Lambda_p^\star\right) \sum_{\mathbf{u} \in (\wp_d^\star \cap \Lambda_p^\star)} \hat{f}_{dp}^\star(\mathbf{u}) \cdot \hat{g}_{dp}(\mathbf{u}) \tag{6.65}$$

in terms of the discrete Fourier series of the signals (*Parseval's theorem*).

The discrete Fourier series of a real-valued function $f_{dp}(\mathbf{x})$ satisfies $\hat{f}_{dp}^\star(\mathbf{u}) = \hat{f}_{dp}(-\mathbf{u})$. Since the signals that we consider are mostly real-valued, we obtain that the harmonic basis signals often occur in the following combinations

$$\begin{aligned} \frac{1}{2} \left(\psi_{\mathbf{u}}(\mathbf{x}) + \psi_{-\mathbf{u}}(\mathbf{x})\right) &= \cos(2\pi \mathbf{u}^T \mathbf{x}) \\ \frac{1}{2j} \left(\psi_{\mathbf{u}}(\mathbf{x}) - \psi_{-\mathbf{u}}(\mathbf{x})\right) &= \sin(2\pi \mathbf{u}^T \mathbf{x}) \end{aligned} \tag{6.66}$$

Such combinations are called **gratings** with **frequency vector u**. Only frequency vectors $\mathbf{u}$ that belong to the dual lattice $\Lambda_p^\star$ occur in a signal

Figure 6.6. Two-dimensional gratings with frequency vectors $(16/T, 0)$ (left) and $(16/T, 8/T)$ (right). The discretization and periodicity lattices are square with respective determinant 1 and T^2 ($T = 256$ in the example). The y-axis is pointing downwards in the figures, so that the orientation θ that the grating makes with the x-axis is measured clockwise.

with periodicity lattice Λ_p. Two examples of two-dimensional (cosine) gratings are shown in Figure 6.6. Note that a two-dimensional grating with frequency vector $\mathbf{u} = (u{\cdot}\cos\theta, u{\cdot}\sin\theta)$ is oriented along the direction θ.

Even if the input samples are real, the interpolated signal can be complex-valued for positions $\mathbf{x}$ in between the sampling positions. We can however retain the real part of the interpolated signal, i.e.,

$$\Re f_{dp}(\mathbf{x}) = d\left(\Lambda_p^\star\right) \sum_{\mathbf{u}\in(\wp_d^\star\cap\Lambda_p^\star)} \left(\hat{a}_{dp}(\mathbf{u})\cdot\cos(2\pi\mathbf{u}^T\mathbf{x}) + \hat{b}_{dp}(\mathbf{u})\cdot\sin(2\pi\mathbf{u}^T\mathbf{x})\right) \tag{6.67}$$

for $\mathbf{x} \in \mathbb{R}^D$, as the signal interpolation, where $\hat{f}_{dp}(\mathbf{u}) = \hat{a}_{dp}(\mathbf{u}) + j\cdot\hat{b}_{dp}(\mathbf{u})$ denotes the decomposition of the complex-valued Fourier coefficients in a real and an imaginary part. Image interpolation of a real-valued image with a Fourier basis is illustrated in Figure 6.7. The most noticeable degradation introduced by this interpolation is *ringing*.

4.4 CONVOLUTION

Discrete Fourier series descriptions are mainly used for studying discrete linear position-invariant systems. Such systems can be described as **discrete convolution** operators that map an input signal with sample

Figure 6.7. Image interpolation with a Fourier basis.

values $f_{dp}(\mathbf{x})$, into an output signal with sample values

$$\boxed{g_{dp}(\mathbf{x}) = h_d(\mathbf{x}) * f_{dp}(\mathbf{x}) = d(\Lambda_d) \sum_{\mathbf{y}\in\Lambda_d} h_d(\mathbf{x}-\mathbf{y}) \cdot f_{dp}(\mathbf{y}),} \tag{6.68}$$

for $\mathbf{x} \in (\wp_p \cap \Lambda_d)$. The sample values $h_d(\mathbf{x})$, for $\mathbf{x} \in \Lambda_d$, characterize the system.

A linear position-invariant system is uniquely characterized by its **impulse response**. This is the response of the system to the discrete periodic input signal with samples $f_{dp}(\mathbf{0}) = 1$ and $f_{dp}(\mathbf{x}) = 0$, for $\mathbf{x} \neq \mathbf{0} \in (\wp_p \cap \Lambda_d)$. The discrete periodic impulse response signal has sample values

$$h_{dp}(\mathbf{x}) = \sum_{\mathbf{y}\in\Lambda_p} h_d(\mathbf{x}+\mathbf{y}), \tag{6.69}$$

for $\mathbf{x} \in (\wp_p \cap \Lambda_d)$, and arises by summing shifted versions of the non-periodic signal $h_d(\mathbf{x})$, $\mathbf{x} \in \Lambda_d$. The input-output relation for a discrete convolution can hence be rewritten as

$$\boxed{g_{dp}(\mathbf{x}) = d(\Lambda_d) \sum_{\mathbf{y}\in(\wp_p\cap\Lambda_d)} h_{dp}(\mathbf{x}-\mathbf{y}) \cdot f_{dp}(\mathbf{y}) = \langle h^{\star}_{dp}(\mathbf{x}-\mathbf{y}), f_{dp}(\mathbf{y})\rangle} \tag{6.70}$$

for $\mathbf{x} \in (\wp_p \cap \Lambda_d)$. This shows that a linear space-invariant system is a measurement system in which all test vectors are displaced versions of $h^\star_{dp}(-\mathbf{x})$, i.e., the complex conjugate of the mirrored impulse response.

The discrete harmonic signals $\psi_{\mathbf{u}}(\mathbf{x})$, for $\mathbf{x} \in \Lambda_d$, are *eigenvectors* of any linear position-invariant system, i.e., they are mapped proportionally into $\hat{h}_{dp}(\mathbf{u}) \cdot \psi_{\mathbf{u}}(\mathbf{x})$, for $\mathbf{x} \in \Lambda_d$ and $\mathbf{u} \in (\wp^\star_d \cap \Lambda^\star_p)$. The set of *eigenvalues* $[h_{dp}(\mathbf{x})]_\psi = [\hat{h}_{dp}(\mathbf{u}), \mathbf{u} \in (\wp^\star_d \cap \Lambda^\star_p)]$ is an alternative way of characterizing the linear position-invariant system and is obtained by taking the discrete Fourier series of the periodic impulse response. The discrete Fourier series of the input and output signals of a discrete linear position-invariant system are related through

$$\boxed{\hat{g}_{dp}(\mathbf{u}) = \hat{h}_{dp}(\mathbf{u}) \cdot \hat{f}_{dp}(\mathbf{u}),} \tag{6.71}$$

for $\mathbf{u} \in (\wp^\star_d \cap \Lambda^\star_p)$.

4.5 FOURIER TRANSFORMATIONS AND FOURIER SERIES

It is worthwhile to examine some limit cases of the above Hilbert space of discrete periodic square-summable signals.

4.5.1 CONTINUOUS PERIODIC SIGNALS

Discretization of a periodic signal $f_p(\mathbf{x})$ at a very high density, which corresponds to choosing $d(\Lambda_d) \to 0$ or $\Lambda_d = \wp^\star_d = \mathbb{R}^D$, gives rise to the following expressions for the **continuous Fourier series**

$$\hat{f}_p(\mathbf{u}) = \int_{\wp_p} f_p(\mathbf{x}) \cdot \exp(j2\pi\mathbf{u}^T\mathbf{x})\, d\mathbf{x}, \tag{6.72}$$

for $\mathbf{u} \in \Lambda^\star_p$, and its inverse

$$f_p(\mathbf{x}) = d\left(\Lambda^\star_p\right) \sum_{\mathbf{u} \in \Lambda^\star_p} \hat{f}_p(\mathbf{u}) \cdot \exp(-j2\pi\mathbf{u}^T\mathbf{x}), \tag{6.73}$$

for $\mathbf{x} \in \wp_p$. The resulting space of square-integrable continuous periodic signals, with vector product

$$\langle f_p(\mathbf{x}), g_p(\mathbf{x})\rangle = \int_{\wp_p} f^\star_p(\mathbf{x}) \cdot g_p(\mathbf{x})\, d\mathbf{x}, \tag{6.74}$$

and (dual) Fourier basis

$$B_\chi = \left\{\chi_{\mathbf{u}}(\mathbf{x}) = \exp(-j2\pi\mathbf{u}^T\mathbf{x}); \mathbf{u} \in \Lambda^\star_p\right\}, \tag{6.75}$$

is denoted by $\mathcal{L}^2(\wp_p)$.

The **continuous convolution** operator for such densely sampled signals can be expressed as

$$g_p(\mathbf{x}) = \int_{\mathbb{R}^D} h(\mathbf{x}-\mathbf{y}) \cdot f_p(\mathbf{y})\, d\mathbf{y} \tag{6.76}$$

for $\mathbf{x} \in \mathbb{R}^D$, in terms of the non-periodic impulse response $h(\mathbf{x})$, or as

$$g_p(\mathbf{x}) = \int_{\wp_p} h_p(\mathbf{x}-\mathbf{y}) \cdot f_p(\mathbf{y})\, d\mathbf{y} \tag{6.77}$$

in terms of the periodic impulse response

$$h_p(\mathbf{x}) = \sum_{\mathbf{y}\in\Lambda_p} h(\mathbf{x}+\mathbf{y}), \tag{6.78}$$

for $\mathbf{x} \in \mathbb{R}^D$.

4.5.2 DISCRETE APERIODIC SIGNALS

If we consider discrete signals $f_d(\mathbf{x})$, for $\mathbf{x} \in \Lambda_d$, with a very large periodicity, which corresponds to choosing $d(\Lambda_p) \to \infty$ or $\Lambda_p^\star = \wp_p = \mathbb{R}^D$, we obtain the following expressions for the **discrete Fourier transformation**

$$\hat{f}_d(\mathbf{u}) = d(\Lambda_d) \sum_{\mathbf{x}\in\Lambda_d} f_d(\mathbf{x}) \cdot \exp(j2\pi\mathbf{u}^T\mathbf{x}), \tag{6.79}$$

for $\mathbf{u} \in \wp_d^\star$, and its inverse

$$f_d(\mathbf{x}) = \int_{\wp_d^\star} \hat{f}_d(\mathbf{u}) \cdot \exp(-j2\pi\mathbf{u}^T\mathbf{x})\, d\mathbf{u}, \tag{6.80}$$

for $\mathbf{x} \in \Lambda_d$. The resulting space of square-summable discrete signals, with vector product

$$\langle f_d(\mathbf{x}), g_d(\mathbf{x})\rangle = d(\Lambda_d) \sum_{\mathbf{x}\in\Lambda_d} f_d^\star(\mathbf{x}) \cdot g_d(\mathbf{x}), \tag{6.81}$$

and (dual) Fourier basis

$$B_\chi = \left\{\chi_\mathbf{u}(\mathbf{x}) = \exp(-j2\pi\mathbf{u}^T\mathbf{x}); \mathbf{u} \in \wp_d^\star\right\}, \tag{6.82}$$

is denoted by $\mathcal{L}^2(\Lambda_d)$.

4.5.3 CONTINUOUS APERIODIC SIGNALS

If we consider the case of a periodicity lattice with a very large determinant and a discretization lattice with a very small determinant, then we obtain the following expressions for the **continuous Fourier transformation**

$$\hat{f}(\mathbf{u}) = \int_{\mathbb{R}^D} f(\mathbf{x}) \cdot \exp(j2\pi\mathbf{u}^T\mathbf{x})\, d\mathbf{x} \tag{6.83}$$

for $\mathbf{u} \in \mathbb{R}^D$, and its inverse

$$f(\mathbf{x}) = \int_{\mathbb{R}^D} \hat{f}(\mathbf{u}) \cdot \exp(-j2\pi\mathbf{u}^T\mathbf{x})\, d\mathbf{u}, \tag{6.84}$$

for $\mathbf{x} \in \mathbb{R}^D$. The resulting space of square-integrable continuous signals, with vector product

$$\langle f(\mathbf{x}), g(\mathbf{x})\rangle = \int_{\mathbb{R}^D} f^\star(\mathbf{x}) \cdot g(\mathbf{x})\, d\mathbf{x}, \tag{6.85}$$

and (dual) Fourier basis

$$B_\chi = \left\{\chi_{\mathbf{u}}(\mathbf{x}) = \exp(-j2\pi\mathbf{u}^T\mathbf{x}); \mathbf{u} \in \mathbb{R}^D\right\}, \tag{6.86}$$

is denoted by $\mathcal{L}^2(\mathbb{R}^D)$.

4.5.4 ALIASING

The well-known relationship between the discrete Fourier transformation $\hat{f}_d(\mathbf{u})$ of a sampled signal $[f_d(\mathbf{x}) = f(\mathbf{x}), \mathbf{x} \in \Lambda_d]$ and the continuous Fourier transformation $\hat{f}(\mathbf{u})$ of the original continuous signal $[f(\mathbf{x}), \mathbf{x} \in \mathbb{R}^D]$, i.e.,

$$\hat{f}_d(\mathbf{u}) = \sum_{\mathbf{v}\in\Lambda_d^\star} \hat{f}(\mathbf{u}+\mathbf{v}), \tag{6.87}$$

for $\mathbf{u} \in \wp_d^\star$, can be derived from the above expressions. This latter expression is the Fourier equivalent of the spatial relationship between an original non-periodic signal $f(\mathbf{x})$ and the derived periodic signal $f_p(\mathbf{x})$ in equation (6.47). It implies that a signal with a Fourier transformation that is zero outside a frequency region that is a unit cell of the dual lattice $\Lambda_d^\star$ can be uniquely recovered from its sample values. If this condition is not satisfied, then the interpolated signal is said to be an **aliased** approximation to the original signal. Figure 6.8 illustrates the comparison between the Fourier transformations of an original (densely sampled) signal and a discrete (coarsely sampled) signal. The aliased Fourier interpolation in Figure 6.7 arose by only retaining the $N \times N$ (with $N = 64$) Fourier components that belong to the peak closest to the origin in the frequency domain.

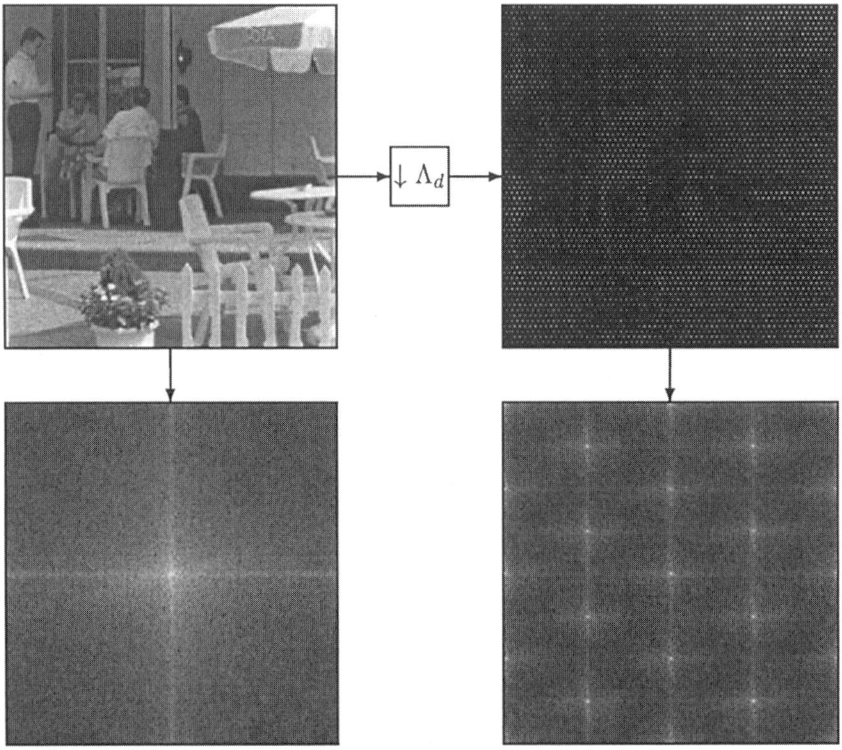

Figure 6.8. Comparison between an original (densely sampled) image and a discrete (coarsely sampled) image. The upper panels show the space variations of the original and discrete images. The lower panels show the amplitude functions $(|\hat{f}(\mathbf{u})|)^{0.1}$ and $(|\hat{f}_d(\mathbf{u})|)^{0.1}$. The discretization lattice is diagonal with determinant $d(\Lambda_d) = 16$.

EXERCISES

1 The following symmetries can be distinguished for a signal

$$\begin{array}{ll} \text{even} & f(-\mathbf{x}) = f(\mathbf{x}) \\ \text{odd} & f(-\mathbf{x}) = -f(\mathbf{x}) \\ \text{Hermitian} & f(-\mathbf{x}) = f^*(\mathbf{x}) \\ \text{anti-Hermitian} & f(-\mathbf{x}) = -f^*(\mathbf{x}) \\ \text{real} & f^*(\mathbf{x}) = f(\mathbf{x}) \\ \text{imaginary} & f^*(\mathbf{x}) = -f(\mathbf{x}) \end{array}$$

Prove the following symmetry relationships between a signal $f(\mathbf{x})$ and its Fourier series (or transformation) $\hat{f}(\mathbf{u})$, i.e.,

$f(x)$	$\hat{f}(\mathbf{u})$
even	even
odd	odd
Hermitian	real
anti-Hermitian	imaginary
real	Hermitian
real and even	real and even
real and odd	imaginary and odd
imaginary	anti-Hermitian
imaginary and even	imaginary and even
imaginary and odd	real and odd

2 Establish the following properties of the continuous Fourier transformation

Spatial domain	Fourier domain	Property
$f(A \cdot \mathbf{x})$	$\hat{f}((A^{-1})^T \cdot \mathbf{u})/\|A\|$	Affine mapping
$f(U \cdot \mathbf{x})$	$\hat{f}(U \cdot \mathbf{u})$	Rotation
$f(\mathbf{x} - \mathbf{x}_o)$	$\exp(-j2\pi \mathbf{u}^T \mathbf{x}_o)\, \hat{f}(\mathbf{u})$	Spatial shift
$f(\mathbf{x} + \mathbf{x}_o) - f(\mathbf{x} - \mathbf{x}_o)$	$2j\, \sin(2\pi \mathbf{u}^T \mathbf{x}_o)\, \hat{f}(\mathbf{u})$	Finite difference
$\exp(j2\pi \mathbf{u}_o^T \mathbf{x})\, f(\mathbf{x})$	$\hat{f}(\mathbf{u} - \mathbf{u}_o)$	Fourier shift
$2\, \cos(2\pi \mathbf{u}_o^T \mathbf{x})\, f(\mathbf{x})$	$\hat{f}(\mathbf{u} + \mathbf{u}_o) + \hat{f}(\mathbf{u} - \mathbf{u}_o)$	Modulation
$\frac{\partial^m f(\mathbf{x})}{\partial x_p^m}$	$(j2\pi u_p)^m\, \hat{f}(\mathbf{u})$	Differentiation
$(-j2\pi x_p)^m\, f(\mathbf{x})$	$\frac{\partial^m \hat{f}(\mathbf{u})}{\partial u_p^m}$	
$\int_{\mathbb{R}^D} f(\mathbf{x})\, d\mathbf{x}$	$\hat{f}(\mathbf{0})$	Integration
$\int_{\mathbb{R}^D} x_p^m f(\mathbf{x})\, d\mathbf{x}$	$\left(-\frac{1}{j2\pi}\right)^m \left(\frac{\partial^m \hat{f}(\mathbf{u})}{\partial u_p^m}\right)_{\mathbf{u}=0}$	Moments

3 The straightforward calculation of the discrete Fourier series

$$\hat{f}(n) = \sum_{i=0}^{N-1} f(i) \cdot \exp\left(j2\pi \frac{n\,i}{N}\right)$$

for $n = 0, \ldots, N-1$ requires approximately N^2 operations (additions and multiplications, or accumulation operations). Prove that, if $N =$

$2 \cdot M$ is even, then

$$\begin{aligned}\hat{f}(n) &= \sum_{i=0}^{M-1} f(2i) \cdot \exp\left(j2\pi \frac{n\,i}{M}\right) \\ &+ \exp\left(j2\pi \frac{n}{N}\right) \cdot \sum_{i=0}^{M-1} f(2i+1) \cdot \exp\left(j2\pi \frac{n\,i}{M}\right)\end{aligned}$$

for $n = 0, \ldots, N-1$, which is equivalent to calculating two discrete Fourier series of length M, performed respectively on the even and odd samples, that require $C(M)$ operations each, followed by N operations to combine the intermediate results, i.e.,

$$C(N) = 2 \cdot C(N/2) + N,$$

where $C(N)$ denotes the complexity of calculating a discrete Fourier series of length N. If $N = 2^K$, then we can recursively apply this technique. Prove that the resulting, so-called **Fast Fourier Transformation (FFT)** algorithm requires $C(2^K) = K \cdot 2^K$ operations to derive the discrete Fourier transformation of a sequence of length 2^K. Compare the number of operations for a straightforward algorithm, i.e., 2^{2K}, with the number of operations required by this FFT algorithm, i.e., $K \cdot 2^K$, for $K = 1, \ldots, 10$, and verify that the FFT algorithm does indeed result in a substantial reduction in the number of operations.

4 Specify two different matrices for each of the (rectangular and quincunx) lattices Λ in Figure 6.3, hence illustrating that different matrices can indeed describe the same lattice. Find sublattices Λ_s with determinant $d(\Lambda_s) = 2 \cdot d(\Lambda)$ for each of the lattices Λ in Figure 6.3 and draw them.

5 Derive equations (6.40) and (6.87) for the relationship between the discrete Fourier transformation of a sampled signal and the continuous Fourier transformation of the original continuous signal.

Chapter 7

IMAGE SAMPLING AND INTERPOLATION

1. INTRODUCTION

In chapter 6, it was demonstrated how signals of finite extent can be mapped into periodic signals. This mapping is mostly of interest because it simplifies the description of signals in terms of a Fourier basis. Operations on periodic signals by linear time- and/or -space-invariant systems (i.e., convolutions) can for instance be described very concisely in the Fourier domain, because the harmonic signals that constitute Fourier bases are eigenvectors of such systems. Hence, periodic signals are adequate mathematical abstractions of signals of finite extent.

The treatment of discretization (or sampling) that was also presented in chapter 6 is, on the other hand, although mathematically correct, less adequate from a physical perspective. Sampling is modelled by assuming that the original continuous signal is known at the positions on a regular discretization lattice. This assumption that a signal value at a specific point can be measured with infinite precision is contrary to physical principles such as Heisenberg's uncertainty principle. In chapter 2, it was argued that physical measurements on a signal can be modelled more adequately by the inproduct $\langle \phi(\mathbf{x}), f(\mathbf{x}) \rangle$ between the input signal $f(\mathbf{x})$ and a test function $\phi(\mathbf{x})$. In section 2 we will therefore rephrase sampling for general test functions. We will show how repeated measurements at regular intervals with the same test function can be interpolated to create an approximation of the original signal.

In order to design linear systems for sampling and interpolation, a quantitative measure is needed for the distance between an original and an interpolated signal. It is discussed in section 3 how such a distance measure can be created. As part of this discussion, an expression is derived for the interpolated signal in the Fourier domain.

Although sections 2 and 3 describe in theory how the optimum interpolation function can be derived from the test function and the sampling lattice, direct application of this theory is often not advisable in practice, for instance because of the occurrence of unwanted signals, such as noise, within the system. Some relevant practical aspects concerning the derivation of interpolation functions are therefore discussed in section 4. Finally, in section 5, we describe a number of experiments that were conducted in order to obtain a better insight into the subjective appreciation of sampled and interpolated images.

2. SAMPLING OF PERIODIC SIGNALS

In this section, we discuss the theory of sampling and interpolation for general test functions. We assume that the input signals $f(\mathbf{x})$ into our sampling-and-interpolation system are periodic with periodicity lattice Λ_p. The unit cell $\wp_p$ of the periodicity lattice is a finite subset of the D-dimensional Euclidean space $\mathbb{R}^D$. We moreover assume that all bounded linear functionals, i.e., linear mappings from the test functions to the finite real/complex numbers, can be expressed as the inproduct $\langle \phi(\mathbf{x}), f(\mathbf{x}) \rangle$ between a test function $\phi(\mathbf{x})$ and a periodic input function $f(\mathbf{x})$. The space of test functions is denoted by V. Test functions are typically smooth localized non-periodic functions, such as Schwartz functions. The space of periodic input signals that result in finite measurements for all test functions in V is the dual space V' of bounded linear functionals on V.

2.1 SAMPLING WITH TEST FUNCTIONS

We define a **repetitive basis of test functions** as a set of *linearly-independent* test functions in V that are all translated copies of a fixed real-valued function $\phi(\mathbf{x})$ of V on some regular grid $\mathcal{P} = \Lambda \cap \wp_p$ in the unit cell $\wp_p$ of the periodicity lattice. We assume that the periodicity lattice Λ_p is a sublattice of the sampling lattice Λ, i.e., $\Lambda_p \subset \Lambda$, so that the number of test functions is $n = d(\Lambda_p)/d(\Lambda)$. The repetitive basis can be written as

$$B_\phi = \{\phi(\mathbf{x} - \mathbf{p}); \mathbf{p} \in \mathcal{P}\}, \tag{7.1}$$

for $\mathbf{x} \in \mathbb{R}^D$, and spans a subspace U of V of dimension n. Note that the number of test functions n only determines the density $d(\Lambda) = d(\Lambda_p)/n$ of the sampling lattice. The remaining freedom of choice for the sampling lattice Λ will be discussed in chapter 9.

Sampling with this repetitive basis is equivalent to performing the n measurements

$$[\langle\, \phi(\mathbf{x}-\mathbf{p}), f(\mathbf{x})\,\rangle, \mathbf{p} \in \mathcal{P}]^T, \tag{7.2}$$

where $\langle\, \cdot, \cdot\,\rangle$ denotes an inproduct with the first vector belonging to V and the second vector belonging to V'. The most frequently occurring inproduct is

$$\langle\, \phi(\mathbf{x}), f(\mathbf{x})\,\rangle = \int_{\mathbb{R}^D} \phi(\mathbf{x})\, f(\mathbf{x})\, d\mathbf{x}, \tag{7.3}$$

for $\phi(\mathbf{x}) \in V$ and $f(\mathbf{x}) \in V'$. Alternative inproducts are however also allowed.

The interest in repetitive bases of test functions can be motivated as follows. Signal processing systems are usually not allowed to make a priori assumptions about the input signals, such as where (in space and/or time) the interesting information occurs. This implies that, if $\phi(\mathbf{x})$, centered at the origin, is an interesting test function, then $\phi(\mathbf{x}-\mathbf{p})$, centered at any other position $\mathbf{p} \in \wp_p$, is a priori equally interesting. Also, if only a limited number of measurements can be made, a uniform distribution of the measurements across space seems an obvious choice that does not favor any specific region within the unit cell $\wp_p$. A lattice Λ is a convenient way of creating such a uniform sampling.

Using any set $\{\varphi'_{\mathbf{p}}(\mathbf{x}), \mathbf{p} \in \mathcal{P}\}$ of linearly independent **interpolation functions** in the space V' of periodic signals, we can define the linear vector operator

$$T_{\varphi\phi} : V' \mapsto U' : f(\mathbf{x}) \mapsto T_{\varphi\phi}\, f(\mathbf{x}) = \sum_{\mathbf{p}\in\mathcal{P}} \langle\, \phi(\mathbf{x}-\mathbf{p}), f(\mathbf{x})\,\rangle \cdot \varphi'_{\mathbf{p}}(\mathbf{x}). \tag{7.4}$$

Any signal of V' is mapped by this operator into a signal in the subspace U' spanned by the interpolation functions. There is a lot of freedom in the choice of the interpolation functions, so that many different interpolated signals can be derived from the same sample values. This freedom of choice for the interpolation functions was also observed in chapter 6.

Now suppose that V' is a Hilbert space with an inproduct $\langle\, \cdot, \cdot\,\rangle_p$; for example,

$$\langle\, f(\mathbf{x}), g(\mathbf{x})\,\rangle_p = \int_{\wp_p} f(\mathbf{x})\, g(\mathbf{x})\, d\mathbf{x}, \tag{7.5}$$

for $f(\mathbf{x}), g(\mathbf{x}) \in V'$. One obvious way of restricting the interpolation functions is to impose that the linear operator $T_{\varphi\phi}$ is a projection operator P onto a subspace U' of V' with a dimensionality equal to the number n of test functions. This assumption imposes the condition that sampling the interpolated signal $T_{\varphi\phi}\, f(\mathbf{x}) = Pf(\mathbf{x})$ returns the original sample values if $f(x) = Pf(x) \in U'$. This in turn implies that the biorthogonality condition between the test functions and the interpolation functions

$$\langle\, \phi(\mathbf{x}-\mathbf{p}), \varphi'_{\mathbf{q}}(\mathbf{x})\,\rangle = \delta_{\mathbf{pq}}, \tag{7.6}$$

must be satisfied for all $\mathbf{p}, \mathbf{q} \in \mathcal{P}$.

Now suppose that we additionally require that, if $\varphi'(\mathbf{x}) \in U'$, then also $\varphi'(\mathbf{x}-\mathbf{p}) \in U'$ for any displacement $\mathbf{p} \in \mathcal{P}$. This is another manifestation of the requirement that all sample positions should be equivalent. Since U' is a vector space of dimension n, it should be possible to find an element $\varphi'(\mathbf{x}) \in U'$ such that

$$B'_{\varphi} = \{\varphi'_{\mathbf{p}}(\mathbf{x}) = \varphi'(\mathbf{x}-\mathbf{p}); \mathbf{p} \in \mathcal{P}\} \tag{7.7}$$

is a repetitive basis of U'. For reasons of symmetry, the dual basis of B'_{φ} can then be assumed to also be repetitive, i.e.,

$$B'_{\phi} = \{\phi'_{\mathbf{p}}(\mathbf{x}) = \phi'(\mathbf{x}-\mathbf{p}); \mathbf{p} \in \mathcal{P}\} \tag{7.8}$$

for some function $\phi'(\mathbf{x}) \in U'$. In principle, the standard matrix algorithm for constructing a dual basis can be used to derive B'_{ϕ} from B'_{φ}, since U' is assumed to be finite-dimensional. This algorithm however requires the inversion of a Gramian matrix of size $n \times n$, which is often not practically feasible, since n may be very large[1]. We will however see shortly that the problem of finding the dual basis can be simplified, because of the specific nature of repetitive bases. This explicit algorithm for constructing the bases B'_{ϕ} and B'_{φ} from a specified test function $\phi(\mathbf{x})$ will also demonstrate that the assumptions that we make here, based on qualitative arguments, about the existence of such repetitive bases are indeed realistic.

If $T_{\varphi\phi} = P$ is the projection operator onto U', then, according to the general theory in chapter 2, this operator can be written as

$$P : V' \mapsto U' : f(\mathbf{x}) \mapsto Pf(\mathbf{x}) = \sum_{\mathbf{p}\in\mathcal{P}} \langle\, \phi'(\mathbf{x}-\mathbf{p}), f(\mathbf{x})\,\rangle_p \cdot \varphi'(\mathbf{x}-\mathbf{p}). \tag{7.9}$$

This latter expression for the projection operator is identical to the earlier expression, provided that

$$\langle\, \phi(\mathbf{x}-\mathbf{p}), f(\mathbf{x})\,\rangle = \langle\, \phi'(\mathbf{x}-\mathbf{p}), f(\mathbf{x})\,\rangle_p \tag{7.10}$$

[1]For a typical 512×512 image, $n = 262144$.

for all $f(\mathbf{x}) \in V'$. For instance, for the integral inproducts defined above, this can be satisfied with the choice

$$\phi'(\mathbf{x}) = \sum_{\mathbf{y} \in \Lambda_p} \phi(\mathbf{x} + \mathbf{y}), \tag{7.11}$$

for $\mathbf{x} \in \mathbb{R}^D$, which is the standard way of mapping a non-periodic function $\phi(\mathbf{x})$ into a periodic function $\phi'(\mathbf{x})$.

The following strategy for sampling and interpolation arises from the above reasoning. Given the sampling lattice Λ and the test function $\phi(\mathbf{x}) \in V$, we start by constructing a periodic function $\phi'(\mathbf{x}) \in V'$ that satisfies condition (7.10). Once $\phi'(\mathbf{x})$ is known, it can be used to construct a repetitive basis B'_ϕ. The dual basis B'_φ of B'_ϕ is also repetitive and contains the basis vector $\varphi'(\mathbf{x})$ that is needed in the interpolation formula

$$P : V' \mapsto U' : f(\mathbf{x}) \mapsto Pf(\mathbf{x}) = \sum_{\mathbf{p} \in \mathcal{P}} \langle\, \phi(\mathbf{x} - \mathbf{p}), f(\mathbf{x}) \,\rangle \cdot \varphi'(\mathbf{x} - \mathbf{p}). \tag{7.12}$$

We now return to the problem of finding the dual basis of a repetitive basis.

2.2 DUAL BASIS OF A REPETITIVE BASIS

Suppose that we want to construct the dual basis B'_φ of a given repetitive basis $B'_\phi = \{\phi'(\mathbf{x} - \mathbf{p}); \mathbf{p} \in \mathcal{P}\}$ of the subspace U'. For reasons of symmetry, we can assume that the dual basis is also repetitive and of the form $B'_\varphi = \{\varphi'(\mathbf{x} - \mathbf{p}); \mathbf{p} \in \mathcal{P}\}$. The basis vectors are related through the biorthogonality condition

$$\langle\, \phi'(\mathbf{x} - \mathbf{p}), \varphi'(\mathbf{x} - \mathbf{q}) \,\rangle_p = \delta_{\mathbf{p}-\mathbf{q}}, \tag{7.13}$$

for $\mathbf{p}, \mathbf{q} \in \mathcal{P}$, where $\delta_\mathbf{p}$ denotes Kronecker's delta. The dual basis vectors can be expressed as linear combinations of the given basis vectors, i.e.,

$$\varphi'(\mathbf{x}) = \sum_{\mathbf{r} \in \mathcal{P}} g_{\varphi\varphi}(\mathbf{r}) \cdot \phi'(\mathbf{x} - \mathbf{r}), \tag{7.14}$$

which, upon substitution in the biorthogonality condition, results in

$$\sum_{\mathbf{r} \in \mathcal{P}} g_{\varphi\varphi}(\mathbf{r}) \cdot g_{\phi\phi}(\mathbf{p}, \mathbf{q} + \mathbf{r}) = \delta_{\mathbf{p}-\mathbf{q}}, \tag{7.15}$$

where

$$g_{\phi\phi}(\mathbf{p}, \mathbf{q}) = \langle\, \phi'(\mathbf{x} - \mathbf{p}), \phi'(\mathbf{x} - \mathbf{q}) \,\rangle_p \tag{7.16}$$

for $\mathbf{p}, \mathbf{q} \in \mathcal{P}$, are the elements of the Gramian matrix $G_{\phi\phi}$.

We now show how the construction of a dual basis can be much simplified in case the Gramian matrix is a **circular matrix**. Applied to the case at hand, this corresponds to requesting that the elements of the Gramian matrix

$$g_{\phi\phi}(\mathbf{p}, \mathbf{q}) = g_{\phi\phi}(\mathbf{p} - \mathbf{q}) \tag{7.17}$$

for $\mathbf{p}, \mathbf{q} \in \mathcal{P}$, only depend on the relative displacement $\mathbf{p} - \mathbf{q}$, and not on the absolute positions $\mathbf{p}$ and $\mathbf{q}$. This condition is for instance satisfied for the integral inproduct defined above. It is again a condition that relates to our earlier demand that all lattice positions should be equivalent. The function $g_{\phi\phi}(\mathbf{p})$ is called the **kernel** of the circular Gramian matrix $G_{\phi\phi}$.

The biorthogonality condition simplifies to a discrete convolution

$$d(\Lambda) \sum_{\mathbf{r} \in \mathcal{P}} g_{\varphi\varphi}(\mathbf{r}) \cdot g_{\phi\phi}(\mathbf{p} - \mathbf{r}) = d(\Lambda)\, \delta_{\mathbf{p}}, \tag{7.18}$$

for $\mathbf{p} \in \mathcal{P}$, which has a simple interpretation. We are looking for the impulse response $g_{\varphi\varphi}(\mathbf{p})$, for $\mathbf{p} \in \mathcal{P}$, of a discrete linear space-invariant filter that is such that, if the signal $g_{\phi\phi}(\mathbf{p})$, for $\mathbf{p} \in \mathcal{P}$, is applied to the input of this filter, then the output signal will be the pulse signal $d(\Lambda)\,\delta_{\mathbf{p}}$, for $\mathbf{p} \in \mathcal{P}$. Specifying this linear space-invariant filter is much simpler if all signals are converted to their corresponding discrete Fourier series on the lattice Λ, i.e.,

$$G_{\phi\phi}(\mathbf{u}) = d(\Lambda) \sum_{\mathbf{p} \in \mathcal{P}} g_{\phi\phi}(\mathbf{p}) \exp(j2\pi \mathbf{u}^T \mathbf{p}) \tag{7.19}$$

and

$$G_{\varphi\varphi}(\mathbf{u}) = d(\Lambda) \sum_{\mathbf{p} \in \mathcal{P}} g_{\varphi\varphi}(\mathbf{p}) \exp(j2\pi \mathbf{u}^T \mathbf{p}), \tag{7.20}$$

for $\mathbf{u} \in \mathcal{P}^\star = \wp^\star \cap \Lambda_p^\star$, where $\wp^\star$ denotes the unit cell of the dual lattice $\Lambda^\star$ of Λ. Mapping the biorthogonality condition to the Fourier domain reveals that the discrete Fourier series of the Gramian kernels must be related through

$$\frac{G_{\varphi\varphi}(\mathbf{u})}{d(\Lambda)} = \frac{d(\Lambda)}{G_{\phi\phi}(\mathbf{u})}, \tag{7.21}$$

for $\mathbf{u} \in \mathcal{P}^\star$. Determining the inverse Fourier series

$$g_{\varphi\varphi}(\mathbf{p}) = d(\Lambda_p^\star) \sum_{\mathbf{u} \in \mathcal{P}^\star} G_{\varphi\varphi}(\mathbf{u}) \exp(-j2\pi \mathbf{u}^T \mathbf{p}), \tag{7.22}$$

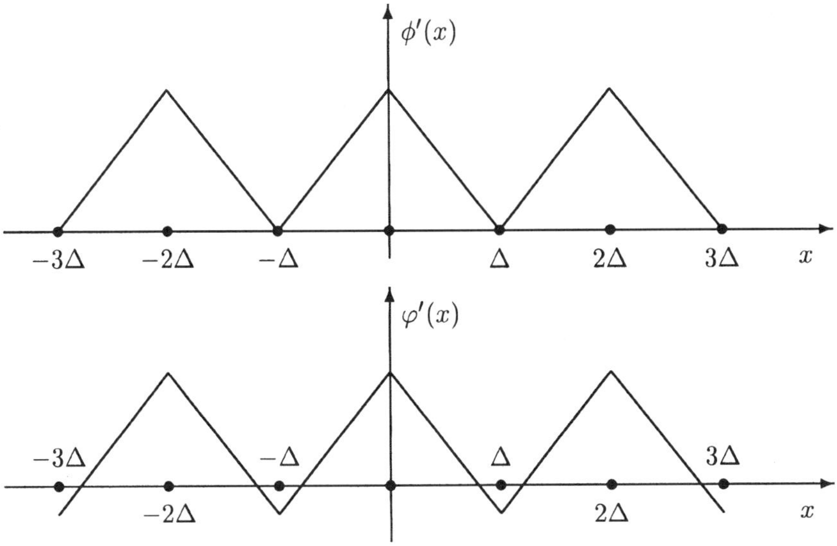

Figure 7.1. Piecewise linear sampling and interpolation function.

for $\mathbf{p} \in \mathcal{P}$, delivers the inverse Gramian kernel.

Constructing the dual basis using discrete Fourier series reduces the computational complexity from n^2 to $n \log n$. This is a very substantial reduction in the number of required operations if n is large. It is made possible by the availability of the Fast Fourier Transformation (FFT) algorithm for calculating discrete Fourier series of length N in $N \log N$, rather than N^2, operations (see exercise 3 in chapter 6).

2.2.1 EXAMPLE

We illustrate the above algorithm for a simple one-dimensional example. We assume the piecewise linear test function

$$\phi(x) = \begin{cases} 1 + \frac{x}{\Delta} & \text{if} \quad x \in [-\Delta, 0] \\ 1 - \frac{x}{\Delta} & \text{if} \quad x \in [0, \Delta] \\ 0 & \text{if} \quad x \notin [-\Delta, \Delta] \end{cases} \tag{7.23}$$

and a sampling distance equal to Δ for periodic signals with period 2Δ. The periodic signal $\phi'(x)$ that can be derived from this localized test function is plotted in the upper part of Figure 7.1.

Straightforward calculation shows that the Gramian matrix is equal to

$$G_{\phi\phi} = \begin{bmatrix} 2\Delta/3 & \Delta/6 \\ \Delta/6 & 2\Delta/3 \end{bmatrix} = \frac{\Delta}{6} \begin{bmatrix} 4 & 1 \\ 1 & 4 \end{bmatrix}. \qquad (7.24)$$

Of course, in this simple example, we can perform a straightforward inversion of this matrix to obtain the dual Gramian matrix

$$G_{\varphi\varphi} = \frac{2}{5\,\Delta} \begin{bmatrix} 4 & -1 \\ -1 & 4 \end{bmatrix}. \qquad (7.25)$$

Instead, we will use the Fourier technique described in this section to obtain the same result. The set of sampling coordinates is $\mathcal{P} = \{0, \Delta\}$, and we have already obtained that $g_{\phi\phi}(0) = 2\Delta/3$ and $g_{\phi\phi}(\Delta) = \Delta/6$. Since the sampling lattice Λ_p has periodicity 2Δ, the dual lattice $\Lambda_p^\star$ has periodicity $1/(2\Delta)$ and the index set for the Fourier series is $\mathcal{P}^\star = \{0, 1/(2\Delta)\}$. The Fourier series of length 2 can hence be expressed as

$$\begin{bmatrix} G_{\phi\phi}(0) \\ G_{\phi\phi}(1/(2\Delta)) \end{bmatrix} = \Delta \begin{bmatrix} 1 & 1 \\ 1 & -1 \end{bmatrix} \cdot \begin{bmatrix} g_{\phi\phi}(0) \\ g_{\phi\phi}(\Delta) \end{bmatrix} = \frac{\Delta^2}{6} \begin{bmatrix} 5 \\ 3 \end{bmatrix}, \qquad (7.26)$$

which in turn results in the values $G_{\varphi\varphi}(0) = 6/5$ and $G_{\varphi\varphi}(0) = 2$ for the Fourier series of the kernel $\{g_{\varphi\varphi}(0), g_{\varphi\varphi}(\Delta)\}$. Performing the inverse Fourier series transformation

$$\begin{bmatrix} g_{\varphi\varphi}(0) \\ g_{\varphi\varphi}(\Delta) \end{bmatrix} = \frac{1}{2\Delta} \begin{bmatrix} 1 & 1 \\ 1 & -1 \end{bmatrix} \cdot \begin{bmatrix} G_{\varphi\varphi}(0) \\ G_{\varphi\varphi}(1/(2\Delta)) \end{bmatrix} = \frac{2}{5\Delta} \begin{bmatrix} 4 \\ -1 \end{bmatrix} \qquad (7.27)$$

indeed delivers the first row of the circular Gramian matrix $G_{\varphi\varphi}$. The dual basis function is hence equal to

$$\varphi'(x) = \frac{2}{5\Delta} \left[4 \cdot \phi'(x) - \phi'(x - \Delta)\right] \qquad (7.28)$$

and is plotted in the lower part of Figure 7.1.

3. COMPARING ORIGINAL AND INTERPOLATED SIGNALS

3.1 QUADRATIC ERROR CRITERIUM

In the previous section, it was derived how sampling and interpolation can be implemented for any combination of a test function $\phi(\mathbf{x})$ and a lattice Λ (provided that $\Lambda_p \subset \Lambda$). This section concentrates on evaluating the result of such sampling and interpolation. This comparison can be performed between the original signal $f(\mathbf{x})$ and the signal

$Pf(\mathbf{x})$ after sampling and interpolation, or between two alternative processed signals, for instance obtained through different choices for the test function $\phi(\mathbf{x})$ and/or the sampling lattice Λ.

Comparing the interpolated signal $Pf(\mathbf{x})$ of equation (7.12) to the original signal $f(\mathbf{x})$ requires that both signals can be represented in a common basis. This can be accomplished in case the input signals $f(\mathbf{x})$ and the periodic test signal $\phi'(\mathbf{x})$ both belong to an intermediate space W' (i.e., $U' \subset W' \subset V'$) with a known basis $B_\psi = \{\psi_\mathbf{u}(\mathbf{x}); \mathbf{u} \in \mathcal{U}\}$ and dual basis $B_\chi = \{\chi_\mathbf{u}(\mathbf{x}); \mathbf{u} \in \mathcal{U}\}$. In such case, the quadratic error measure

$$\begin{aligned} &\| Pf(\mathbf{x}) - f(\mathbf{x}) \|_p^2 \\ &= \sum_{\mathbf{u}\in\mathcal{U}} \langle Pf(\mathbf{x}) - f(\mathbf{x}), \psi_\mathbf{u}(\mathbf{x}) \rangle_p \cdot \langle \chi_\mathbf{u}(\mathbf{x}), Pf(\mathbf{x}) - f(\mathbf{x}) \rangle_p \end{aligned} \tag{7.29}$$

can be used as a performance number, for instance to compare alternative interpolation schemes.

The comparison method is illustrated in case $W' = \mathcal{L}^2(\mathcal{U})$, with $\mathcal{U} = \wp_p \cap \Lambda_d$, is the Hilbert space of discrete periodic signals with periodicity lattice Λ_p and discretization lattice Λ_d, where $\Lambda_p \subset \Lambda_d$ and $d(\Lambda_p)/d(\Lambda_d) = N$. The Fourier basis

$$B_\psi = \left\{\psi_\mathbf{u}(\mathbf{x}) = d\left(\Lambda_p^\star\right) \cdot \exp(-j2\pi\mathbf{u}^T\mathbf{x}); \mathbf{u} \in \mathcal{U}^\star\right\}, \tag{7.30}$$

with index set $\mathcal{U}^\star = \wp_d^\star \cap \Lambda_p^\star$, is adopted to interpolate the signals (and hence to define the intermediate space W'). The unit cell of the dual lattice $\Lambda_d^\star$ is denoted by $\wp_d^\star$. The dual Fourier basis vectors are $\chi_\mathbf{u}(\mathbf{x}) = d(\Lambda_p) \cdot \psi_\mathbf{u}(\mathbf{x})$, for $\mathbf{u} \in \mathcal{U}^\star$ and $\mathbf{x} \in \mathbb{R}^D$. The inproduct between signals $f(\mathbf{x})$ and $g(\mathbf{x})$ in W' can be defined in both the spatial and frequency domain, i.e.,

$$\langle f(\mathbf{x}), g(\mathbf{x}) \rangle_p = d(\Lambda_d) \sum_{\mathbf{x}\in\mathcal{U}} f^\star(\mathbf{x})\, g(\mathbf{x}) = d(\Lambda_p^\star) \sum_{\mathbf{u}\in\mathcal{U}^\star} \hat{f}^\star(\mathbf{u})\, \hat{g}(\mathbf{u}), \tag{7.31}$$

where

$$\hat{f}(\mathbf{u}) = d(\Lambda_d) \sum_{\mathbf{x}\in\mathcal{U}} f(\mathbf{x}) \cdot \exp(j2\pi\mathbf{u}^T\mathbf{x}), \tag{7.32}$$

for $\mathbf{u} \in \mathcal{U}^\star$, denotes the discrete Fourier series on the lattice Λ_d of the signal $f(\mathbf{x})$. The squared error norm is then equal to

$$\begin{aligned} \| Pf(\mathbf{x}) - f(\mathbf{x}) \|_p^2 &= d(\Lambda_d) \sum_{\mathbf{x}\in\mathcal{U}} |Pf(\mathbf{x}) - f(\mathbf{x})|^2 \\ &= d(\Lambda_p^\star) \sum_{\mathbf{u}\in\mathcal{U}^\star} \left|\widehat{Pf}(\mathbf{u}) - \hat{f}(\mathbf{u})\right|^2, \end{aligned} \tag{7.33}$$

where $\widehat{Pf}(\mathbf{u})$ denotes the discrete Fourier series on the lattice Λ_d of the interpolated signal $Pf(\mathbf{x})$. This squared error norm is hence determined by comparing the signal values at the lattice positions $\mathbf{x} \in \mathcal{U}$.

3.2 FOURIER TRANSFORMATION OF AN INTERPOLATED SIGNAL

In order to make the above comparison between interpolated and original signal explicit, an expression is required for the discrete Fourier series of the interpolated signal in terms of the discrete Fourier series of the original signal. Readers not interested in this mathematical derivation may proceed directly to the result in equation (7.44). In order to perform the derivation, one additional assumption must be made for the sampling lattice Λ, i.e., that it is a sublattice of the discretization lattice Λ_d, i.e., $\Lambda_p \subset \Lambda \subset \Lambda_d$.

The interpolated signal $Pf(\mathbf{x})$ is a linear sum of displaced copies of the interpolation function $\varphi'(\mathbf{x})$. Upon Fourier transformation of the interpolated signal in equation (7.12) one obtains that

$$\widehat{Pf}(\mathbf{u}) = \frac{\hat{\varphi}(\mathbf{u}) \cdot \hat{F}_\varphi(\mathbf{u})}{d(\Lambda)}, \tag{7.34}$$

is a multiple of $\hat{\varphi}(\mathbf{u})$, the discrete Fourier series on the lattice Λ_d of the interpolation function $\varphi'(\mathbf{x})$. The factor

$$\hat{F}_\varphi(\mathbf{u}) = d(\Lambda) \sum_{\mathbf{p} \in \mathcal{P}} \langle\, \phi(\mathbf{x} - \mathbf{p}), f(\mathbf{x}) \,\rangle \cdot \exp\left(j2\pi \mathbf{u}^T \mathbf{p}\right) \tag{7.35}$$

for $\mathbf{u} \in \mathcal{U}$, can be interpreted as a discrete Fourier series on the lattice Λ. Since the original signal $f(\mathbf{x})$ is assumed to belong to W', its inverse Fourier series, i.e.,

$$f(\mathbf{x}) = d\left(\Lambda_p^\star\right) \sum_{\mathbf{v} \in \mathcal{U}^\star} \hat{f}(\mathbf{v}) \cdot \exp(-j2\pi \mathbf{v}^T \mathbf{x}), \tag{7.36}$$

for $\mathbf{x} \in \mathbb{R}^D$, can be substituted into this expression. Rewriting the inproduct, using equation (7.10), results in

$$\begin{aligned} \hat{F}_\varphi(\mathbf{u}) &= \sum_{\mathbf{v} \in \mathcal{U}^\star} \langle\, \phi'(\mathbf{x}), \exp(-j2\pi \mathbf{v}^T \mathbf{x} \,\rangle_p \, \hat{f}(\mathbf{v}) \\ &\cdot \; d(\Lambda)\, d(\Lambda_p^\star) \sum_{\mathbf{p} \in \mathcal{P}} \exp(-j2\pi (\mathbf{u} - \mathbf{v})^T \mathbf{p}). \end{aligned} \tag{7.37}$$

Since the periodic test signal $\phi'(\mathbf{x})$ is assumed to also belong to W', its inverse Fourier series, i.e.,

$$\phi'(\mathbf{x}) = d\left(\Lambda_p^\star\right) \sum_{\mathbf{u}\in\mathcal{U}^\star} \hat{\phi}(\mathbf{u}) \cdot \exp(-j2\pi\mathbf{u}^T\mathbf{x}), \tag{7.38}$$

for $\mathbf{x} \in \mathbb{R}^D$, can be substituted into the inproduct. Poisson's summation formula can be used to obtain the following expression for the inproduct

$$\langle\, \phi'(\mathbf{x}), \exp(-j2\pi\mathbf{v}^T\mathbf{x})\,\rangle_p = \hat{\phi}^\star(\mathbf{v}), \tag{7.39}$$

where $\hat{\phi}^\star(\mathbf{v})$, for $\mathbf{v} \in \mathcal{U}^\star$, denotes the complex conjugate of the discrete Fourier series on the lattice Λ_d of the function $\phi'(\mathbf{x})$. Poisson's summation formula can also be used to prove that the rightmost sum in equation (7.37) is zero except when $\mathbf{u} - \mathbf{v} \in \Lambda^\star$, so that equation (7.37) can be simplified to

$$\hat{F}_\varphi(\mathbf{u}) = \sum_{\mathbf{v}\in\mathcal{S}^\star} \hat{\phi}^\star(\mathbf{u}+\mathbf{v}) \cdot \hat{f}(\mathbf{u}+\mathbf{v}), \tag{7.40}$$

for $\mathbf{u} \in \mathcal{U}^\star$, where $\mathcal{S}^\star = \wp_d^\star \cap \Lambda^\star$.

In a similar way, it can be derived that the discrete Fourier series on the lattice Λ of the Gramian kernels $g_{\varphi\varphi}(\mathbf{p})$ and $g_{\phi\phi}(\mathbf{p})$, for $\mathbf{p} \in \mathcal{P}$, can be expressed as

$$G_{\varphi\varphi}(\mathbf{u}) = \sum_{\mathbf{v}\in\mathcal{S}^\star} \mid \hat{\varphi}(\mathbf{u}+\mathbf{v}) \mid^2, \tag{7.41}$$

and

$$G_{\phi\phi}(\mathbf{u}) = \sum_{\mathbf{v}\in\mathcal{S}^\star} \mid \hat{\phi}(\mathbf{u}+\mathbf{v}) \mid^2, \tag{7.42}$$

respectively, for $\mathbf{u} \in \mathcal{U}^\star$.

The linear relationship between the basis vectors, expressed in equation (7.14), can hence be transformed to the frequency domain, i.e.,

$$\hat{\varphi}(\mathbf{u}) = \frac{G_{\varphi\varphi}(\mathbf{u})\, \hat{\phi}(\mathbf{u})}{d(\Lambda)} = \frac{d(\Lambda)\, \hat{\phi}(\mathbf{u})}{G_{\phi\phi}(\mathbf{u})}, \tag{7.43}$$

for $\mathbf{u} \in \mathcal{U}^\star$, where we have used the relationship of equation (7.20) between the discrete Fourier series on the lattice Λ of the two Gramian kernels.

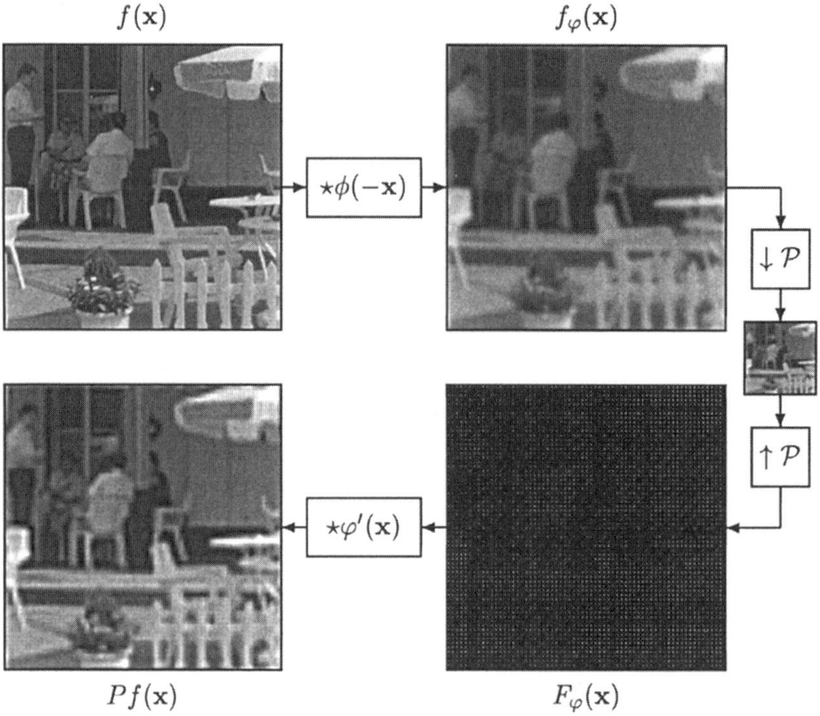

Figure 7.2. Sampling and interpolation in the spatial domain.

Combining the above expressions, it is obtained that the Fourier representation of the interpolated signal is related to the Fourier representation of the original signal by

$$\widehat{Pf}(\mathbf{u}) = \frac{\hat{\phi}(\mathbf{u})}{\sum_{\mathbf{v}\in\mathcal{S}^\star} |\, \hat{\phi}(\mathbf{u}+\mathbf{v}) \,|^2} \sum_{\mathbf{v}\in\mathcal{S}^\star} \hat{\phi}^\star(\mathbf{u}+\mathbf{v})\, \hat{f}(\mathbf{u}+\mathbf{v}), \tag{7.44}$$

for $\mathbf{u} \in \mathcal{U}^\star$. This expression for the Fourier transformation of an interpolated signal generalizes the expression for aliasing introduced in chapter 6, which corresponded to the case that $\hat{\phi}(\mathbf{u}) = 1$, for $\mathbf{u} \in \mathcal{P}^\star$, and zero elsewhere. This expression was first derived by Oackley & Cunningham [OCL90, OC90]. The interpretation of this result is that the sampling and interpolation process can be decomposed into three steps.

First, the periodic input signal $f(\mathbf{x})$ is passed though a convolution with impulse response $\phi'(-\mathbf{x})$, i.e., the periodic counterpart of the non-periodic test function $\phi(-\mathbf{x})$, with discrete Fourier series $\hat{\phi}^\star(\mathbf{u})$, for $\mathbf{u} \in \mathcal{U}^\star$. The signal after linear space-invariant filtering has Fourier series

$$\hat{f}_\varphi(\mathbf{u}) = \hat{\phi}^\star(\mathbf{u}) \cdot \hat{f}(\mathbf{u}), \tag{7.45}$$

for $\mathbf{u} \in \mathcal{U}^\star$.

Next, the filtered signal $f_\varphi(\mathbf{x})$ is pointwise sampled at the lattice points of Λ. This sampled signal $F_\varphi(\mathbf{x})$ is made equal to zero at positions $\mathbf{x} \notin \mathcal{P}$ and equal to $f_\varphi(\mathbf{p})$ for positions $\mathbf{p} \in \mathcal{P}$ that belong to the lattice Λ. As mentioned in the previous chapter, this implies that the Fourier series of the sampled signal is equal to

$$\hat{F}_\varphi(\mathbf{u}) = \sum_{\mathbf{v} \in \mathcal{S}^\star} \hat{f}_\varphi(\mathbf{u} + \mathbf{v}) \tag{7.46}$$

for $\mathbf{u} \in \mathcal{U}^\star$.

Last but not least, the sampled signal $F_\varphi(\mathbf{x})$ is passed though a convolution with impulse response $\varphi'(\mathbf{x})$ and discrete Fourier series

$$\hat{\varphi}(\mathbf{u}) = \frac{\hat{\phi}(\mathbf{u})}{\sum_{\mathbf{v} \in \mathcal{P}^\star} \mid \hat{\phi}(\mathbf{u} + \mathbf{v}) \mid^2}, \tag{7.47}$$

for $\mathbf{u} \in \mathcal{U}^\star$.

The different steps in this sampling and interpolation process[2] are illustrated in the spatial and frequency domain in Figures 7.2 and 7.3, respectively. All lattices are square and $d(\Lambda)/d(\Lambda_d) = 16$. The test function $\phi(\mathbf{x})$ is a two-dimensional Gaussian with standard deviation $\sigma = 2$ times the sampling distance of the square lattice Λ_d (see equation (7.54) for an exact expression). The corresponding interpolation function $\varphi'(\mathbf{x})$ is derived in the next section.

4. INVERSE OPERATORS

It may be obvious from the preceding discussion that the complicated part in the above sampling and interpolation process is inverting the

[2]If the input signal $f(\mathbf{x})$ and the output signal $Pf(\mathbf{x})$ are continuous, i.e., for $d(\Lambda_d) \to 0$, then the intermediate steps are physically impossible, due to Heisenberg's principle, as was already mentioned in the introduction. However, the combination of the first two steps (filtering+downsampling) is physically plausible, since it corresponds to performing a finite set of displaced measurements. Similarly, the combination of the last two steps (upsampling+interpolation) is also physically plausible, since it corresponds to taking a weighted sum of a finite number of displaced interpolation functions.

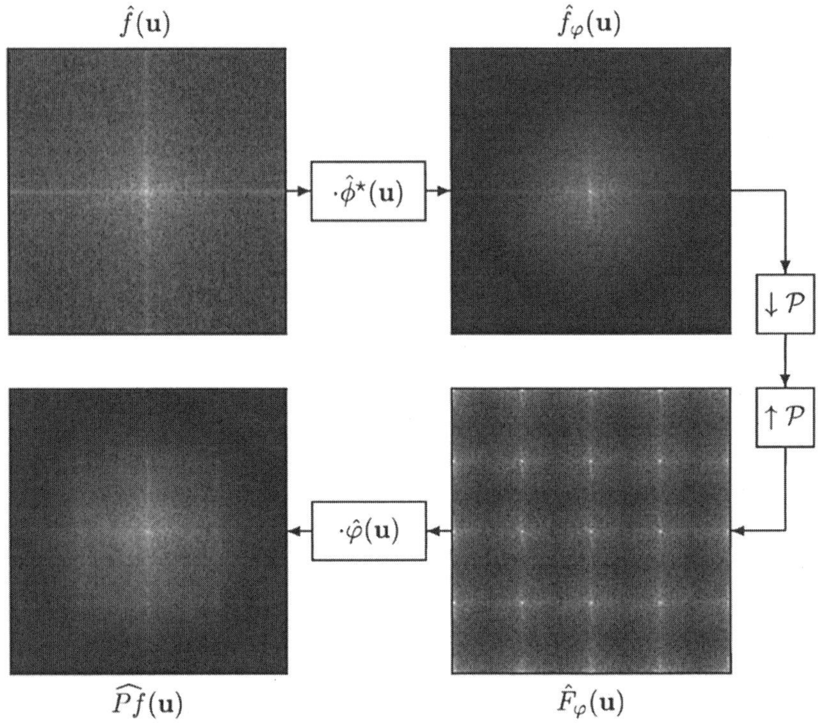

Figure 7.3. Sampling and interpolation in the frequency domain. The Fourier amplitudes are mapped by a power of 0.1 for display purposes.

Gramian operator $G_{\phi\phi}$. In classical treatments of the sampling theorem, this problem is circumvented by only considering orthogonal test functions $\phi(\mathbf{x})$ for which the Gramian operators $G_{\phi\phi}$ and $G_{\varphi\varphi}$ are proportional to the identity operator. However, this latter condition can only be satisfied with test functions such as the $\sin(x)/x$ function of the one-dimensional Shannon/Nyquist theorem. Such test functions are not Schwartz functions, and deviate significantly from the smooth localized test functions that are typically found in image processing practice. Two practical problems that can occur in implementing the theory for non-orthogonal test functions will now be discussed in more detail.

First, even if the test function $\phi(\mathbf{x})$ and the kernel of the Gramian operator $g_{\phi\phi}(\mathbf{p})$ have finite support, i.e., are zero outside a region of finite size, then the kernel of the dual Gramian operator $g_{\varphi\varphi}(\mathbf{p})$ and the corresponding interpolation function $\varphi'(\mathbf{x})$ usually extend over the entire unit cell $\wp_p$ of the periodicity lattice. Since such a large filter kernel is usually very unpractical, i.e., computationally expensive to implement, one is often interested in approximating the interpolation function $\varphi'(\mathbf{x})$ by a function that is localized in space and/or time. One possible approach for accomplishing this is discussed by means of an example in exercise 3.

Second, the sampling and interpolation process is usually implemented in three parts, i.e.,

$$\widehat{Pf}(\mathbf{u}) = \hat{\phi}(\mathbf{u}) \cdot \frac{1}{G_{\phi\phi}(\mathbf{u})} \cdot \sum_{\mathbf{v} \in \mathcal{S}^\star} \hat{\phi}^\star(\mathbf{u}+\mathbf{v})\, \hat{f}(\mathbf{u}+\mathbf{v}) \tag{7.48}$$

for $\mathbf{u} \in \mathcal{U}^\star$. An image sensor with impulse response $\phi(-\mathbf{x})$ and discrete Fourier series $\hat{\phi}^\star(\mathbf{u})$, in combination with a downsampling device maps the original signal $f(\mathbf{x})$ into the discrete representation $[Pf(\mathbf{x})]_\varphi = [\langle\, \phi(\mathbf{x}), f(\mathbf{x})\,\rangle; \mathbf{p} \in \mathcal{P}]$ in the basis B'_φ. A downsampling device with this functionality is usually referred to a an **analog-to-digital (A/D) convertor**. This discrete representation is subsequently mapped to the discrete representation $[Pf(\mathbf{x})]_\phi$ in the dual basis B'_ϕ by means of a discrete filter with impulse response $\tilde{g}_{\varphi\varphi}(\mathbf{p})$ and Fourier series on the lattice Λ equal to $\tilde{G}_{\varphi\varphi}(\mathbf{u}) = 1/G_{\phi\phi}(\mathbf{u})$. Finally, using an upsampling device in combination with a display element with impulse response $\phi'(\mathbf{x})$ and discrete Fourier series $\hat{\phi}(\mathbf{u})$, the interpolated signal is obtained. An upsampling device with this functionality is usually referred to as a **digital-to-analog (D/A) convertor**. The corresponding signal flow graph is depicted in Figure 7.4.

Due to practical limitations in the sensor, the input to the digital filter is not equal to $[Pf(\mathbf{x})]_\varphi$, but is modelled more accurately by $[\widehat{Pf}_\varphi(\mathbf{p}) + n(\mathbf{p}); \mathbf{p} \in \mathcal{P}]$. The unwanted (and unknown) contribution $n(\mathbf{p})$ may for instance be caused by quantization (see chapter 10) and/or noise in the image sensor. Hence, performing an inversion from $[Pf(\mathbf{x})]_\varphi$ to $[Pf(\mathbf{x})]_\phi$ is usually not the only requirement for the digital filter $\tilde{G}_{\varphi\varphi}(\mathbf{u})$. Indeed, if the filter $\tilde{G}_{\varphi\varphi}(\mathbf{u})$ magnifies the wanted component $[Pf(\mathbf{x})]_\varphi$ less than the unwanted noise component $[n(\mathbf{p})]$, then the relative contribution of this noise component will increase, possibly up to the point where it becomes disturbing. Finding a good balance between inverse filtering and noise suppression is a difficult problem for which there exists no general solution. This freedom of choice, which is caused by the fact that there usually exists only partial knowledge about the noise component and/or the signal being processed, has led to a large variety of inverse filter design techniques. Some of the more sophisticated

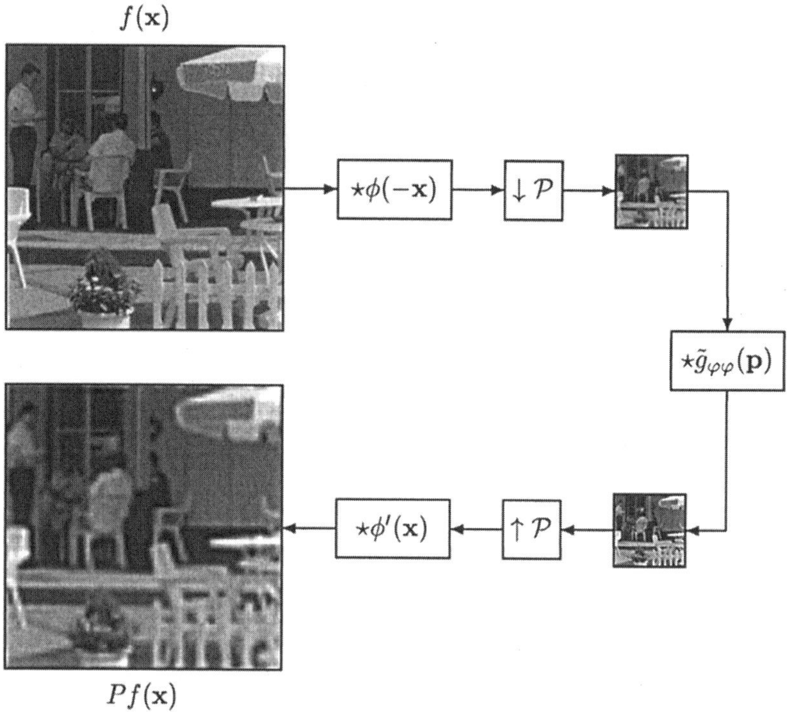

Figure 7.4. Sampling and interpolation with discrete filtering.

techniques adapt the digital filter locally, depending on estimated image features and/or noise characteristics [Mar90a].

A simple approach that gives reasonable results is to change the discrete Fourier series of the digital filter to

$$\tilde{G}_{\varphi\varphi}(\mathbf{u}) = \frac{1}{G(\mathbf{u}) + G_{\phi\phi}(\mathbf{u})}, \tag{7.49}$$

for $\mathbf{u} \in \mathcal{U}^\star$, where the term $G(\mathbf{u})$ is added to limit the magnification of the unwanted component. The filter $G(\mathbf{u})$ must be small for those frequencies that contribute more to the signal than to the noise (usually, the low frequencies), while $G(\mathbf{u})$ may be high for those frequencies that contribute fairly little to the signal. Note that an adequate choice for the filter $G(\mathbf{u})$ requires information about both the signal and the noise components.

We concentrate here on one, fairly simple, specific choice for the filter, i.e.,

$$G(\mathbf{u}) = (1-\alpha)\cdot[G_{\phi\phi}(\mathbf{0}) - G_{\phi\phi}(\mathbf{u})], \tag{7.50}$$

which results in a digital filter with a frequency characteristic equal to

$$\tilde{G}_{\varphi\varphi}(\mathbf{u}) = \frac{1}{(1-\alpha)\cdot G_{\phi\phi}(\mathbf{0}) + \alpha\cdot G_{\phi\phi}(\mathbf{u})}, \tag{7.51}$$

for $\mathbf{u} \in \mathcal{U}^\star$. By changing the **regularization parameter** α from 0 to 1, one can vary between no digital filtering and $\tilde{G}_{\varphi\varphi}(\mathbf{u}) = 1/G_{\phi\phi}(\mathbf{u})$, the theoretically optimal filter in a noise-free condition. The filter value is always correct for frequency $\mathbf{u} = \mathbf{0}$, so that input signals with constant amplitude undergo the theoretically optimum processing. Note that $G(\mathbf{u})$ is small for low frequencies, which usually have the highest contribution to the signal, and increases towards higher frequencies, for which the relative contribution of the noise is usually higher.

In chapter 2, the following general expression

$$\frac{1}{(1-\alpha)\cdot G_{\phi\phi}(\mathbf{0}) + \alpha\cdot G_{\phi\phi}(\mathbf{u})} = \sum_{n=0}^{\infty} \alpha^n \cdot (G_{\phi\phi}(\mathbf{0}) - G_{\phi\phi}(\mathbf{u}))^n, \tag{7.52}$$

has been mentioned for the inverse of a linear operator, provided that $\alpha\cdot \|G_{\phi\phi}(\mathbf{0}) - G_{\phi\phi}(\mathbf{u})\| < 1$. This condition can always be satisfied by making α sufficiently small. If $G_{\phi\phi}(\mathbf{u})$, for $\mathbf{u} \in \mathcal{P}^\star$, is a convolution operator with kernel $[g_{\phi\phi}(\mathbf{p}), \mathbf{p} \in \mathcal{P}]$, then the inverse operator is also a convolution operator with kernel

$$\begin{aligned} g_{\varphi\varphi}(\mathbf{p}) &= \delta(\mathbf{p}) + \alpha\cdot[G_{\phi\phi}(\mathbf{0})\cdot\delta(\mathbf{p}) - g_{\phi\phi}(\mathbf{p})] \\ &+ \alpha^2\cdot[G_{\phi\phi}(\mathbf{0})\cdot\delta(\mathbf{p}) - g_{\phi\phi}(\mathbf{p})] \star [G_{\phi\phi}(\mathbf{0})\cdot\delta(\mathbf{p}) - g_{\phi\phi}(\mathbf{p})] \\ &+ \ldots \end{aligned} \tag{7.53}$$

for $\mathbf{p} \in \mathcal{P}$. Successive approximations to the inverse kernel can hence be generated by ending the above series expansion after a selected number of terms. If the correlation kernel has finite support, then the support of the inverse kernel increases linearly with the number of terms considered in the series expansion. The parameter α must be chosen small enough to make this series expansion converge sufficiently fast.

4.1 GAUSSIAN FUNCTIONS

Consider the important special case that the test function is Gaussian with standard deviation σ, i.e.,

$$\phi(\mathbf{x}) \propto \exp\left(-\frac{\mathbf{x}^T\mathbf{x}}{2\sigma^2}\right), \tag{7.54}$$

for $\mathbf{x} \in \mathbb{R}^D$. The kernel of the Gramian matrix is then equal to

$$g_{\phi\phi}(\mathbf{p}) \propto \exp\left(-\frac{\mathbf{p}^T\mathbf{p}}{4\sigma^2}\right), \tag{7.55}$$

for $\mathbf{p} \in \mathcal{P}$. In Figure 7.5, we have drawn horizontal cross-sections through the spatial and the frequency representations of the test and interpolation functions for the case of sampling and interpolation with a Gaussian basis function on a square lattice. Two properties of the interpolation function are worth mentioning.

First, the interpolation filter has less attenuation at low frequencies than the test filter. In current image processing practice, where no intermediate discrete processing is usually applied, this property is reflected by the fact that the impulse response of the display filter, which can often also be approximated by a Gaussian function, usually has a smaller standard deviation than that of the sampling filter. In many textbooks, this is 'explained' by the somewhat cryptic remark that 'some aliasing in image sampling seems to be preferable'. Since the optimal interpolation function differs significantly from a Gaussian function, this is obviously a sub-optimal solution.

Second, the interpolation function has negative (or inhibitive) sidelobes in the spatial domain and amplification for intermediate frequencies in the frequency domain. This implies that, if discrete filtering is applied, it should involve negative coefficients. This was also realized heuristically, and has led to an image enhancement technique called **unsharp masking**. This technique was originally developed for photography, and later mimicked in digital image processing [ST85].

4.2 B-SPLINES

Gaussian functions have the practical disadvantage that they are infinitely long. In this section, we discuss a class of functions of finite extent with Gaussian-like shapes and properties, i.e., the B-splines [Sch69]. B-splines are piecewise polynomials that originate by recursive convolution of the sample-hold function

$$\beta^0(x) = \mu(x) - \mu(x-1), \tag{7.56}$$

where

$$\mu(x) = \begin{cases} 0 & \text{for } x < 0 \\ 1 & \text{for } x \geq 0 \end{cases} \tag{7.57}$$

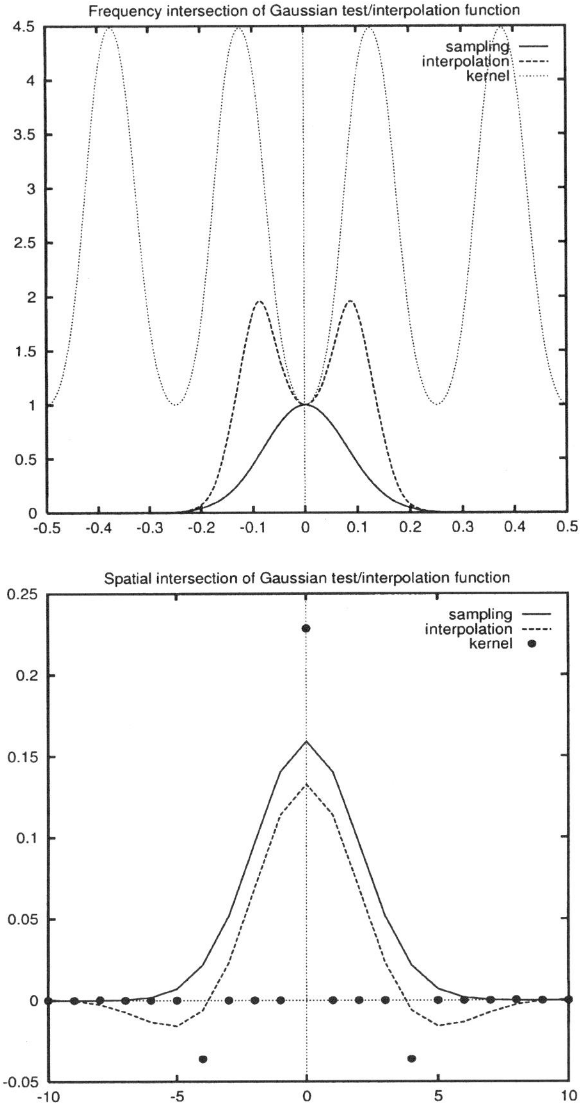

Figure 7.5. The drawn and dashed curves are horizontal cross-sections through the Gaussian test function (with $\sigma = 2$) and corresponding interpolation function in the spatial and frequency domain. The sampling lattice Λ is square with $d(\Lambda)/d(\Lambda_d) = 16$. The dotted line in the frequency domain is a horizontal section through the Fourier transformation $\tilde{G}_{\varphi\varphi}$ of the Gramian kernel $\tilde{g}_{\varphi\varphi}$ (regularization parameter $\alpha = 0.8$). The kernel values are indicated by the (nonzero) discrete dots in the spatial domain. The filters have been scaled differently in order to plot them in the same range.

denotes the unit step function. The B-spline of order n is defined by

$$\beta^n(x) = \sum_{j=0}^{n+1}(-1)^j \frac{n+1}{(n+1-j)!\,j!}(x-j)^n\,\mu(x-j), \tag{7.58}$$

for $x \in [0, n+1]$, and satisfies the recursive relationship

$$\beta^n(x) = \beta^{n-1}(x) * \beta^0(x), \tag{7.59}$$

for $n = 1, 2, \ldots$ The translated B-spline $\beta^n(x + [n+1]/2)$ is a symmetric function. The centered B-splines $\beta^n(x + [n+1]/2)$ of order $n = 0, 1, 2, 3$ are plotted in Figure 7.6. Note that their approximation to a Gaussian function improves with the order n. For $n \geq 5$, the centered B-spline of order n is approximately equal to a Gaussian function with standard deviation $\sigma = 0.433 \cdot \sqrt{n}$.

For applications with discrete signals, the discrete B-spline function of order n and sampling factor m can be defined, i.e.,

$$b_m^n(k) = \beta^n\left(\frac{k}{m}\right) \tag{7.60}$$

for $k = 0, 1, \ldots, m(n+1)$. The centered discrete B-spline function

$$\varphi_{m,n}(k) = b_m^n\left(k + \frac{m(n+1)}{2}\right) = \beta^n\left(\frac{k}{m} + \frac{n+1}{2}\right), \tag{7.61}$$

with $k \in \mathbb{Z}$, cn be used as interpolation function when subsampling discrete signals by a factor of m. The resulting repetitive basis is only orthogonal in case $n = 0$. If $n > 0$, then the dual signal $\phi_{m,n}(k)$ that is required as pre-filter before sampling the signals has to be obtained by means of the techniques discussed above (see also exercise 4). It will have infinite length and hence will have to be truncated. The sampling functions $\phi_{m,n}(k)$ corresponding to the interpolation functions $\varphi_{m,n}(k)$ of Figure 7.6 are plotted in Figure 7.7 for the case of sampling in one dimension with sampling distance $m = 4$. More details about B-spline sampling and interpolation can for instance be found in [UAE91, UAE92, UAE93].

5. PERCEPTUAL ASSESSMENT

Up to now, the minimum mean square error has been used as the criterium while developing sampling and interpolation. In reality, however, we are interested in optimizing the subjective quality of the images. Therefore, in this section, we discuss a number of experiments that have been conducted [Box93] in order to obtain a better insight into the subjective appreciation of sampled and interpolated images.

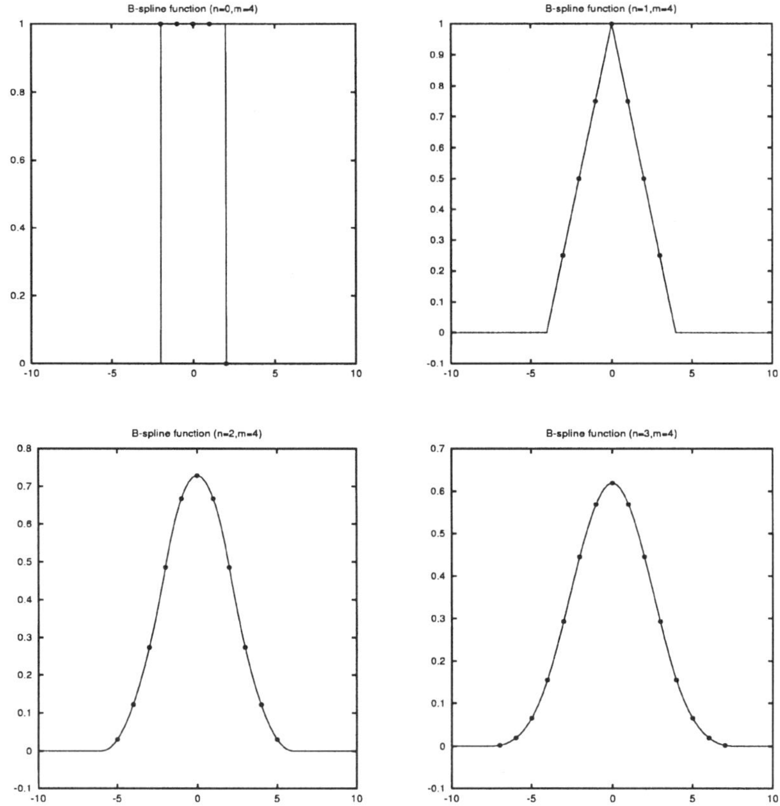

Figure 7.6. The draw curves represent the continuous B-splines $\beta^n(m[x+(n+1)/2])$ of order $n \leq 3$, centered at the origin and dilated by a factor m. The points on the curve correspond to the discrete B-splines with sampling factor $m = 4$. The B-spline of order $n = 0$ in the upper-left corner is also known as the sample-hold function or nearest-neighbor interpolation function. The B-spline of order $n = 1$ in the upper-right corner is also known as the linear interpolation function.

5.1 EXPERIMENTAL SET-UP

The two scenes that were used in the experiment will be referred to by their names "Stadhuis" and "Wanda", and are shown in Figure 7.8. They were selected because of their very different characteristics. Stadhuis is an outdoor picture of an old city hall and contains high contrasts, sharp edges and a clearly visible brick structure. Wanda is an indoor picture of a head-and-shoulder scene with less contrast. The original slides were digitized with a scanner to 2048x2048 pixels and 8 bits per

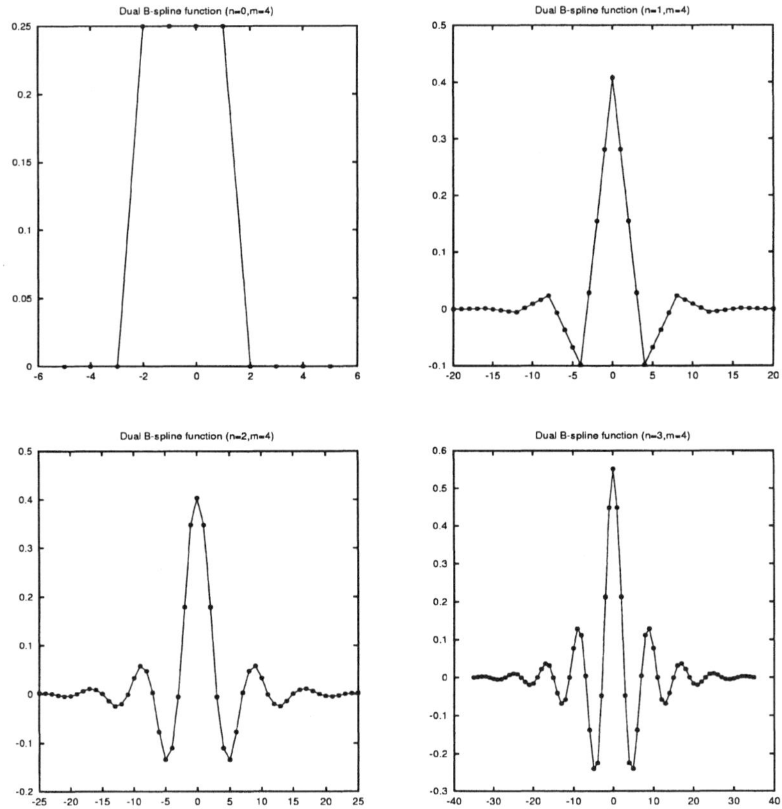

Figure 7.7. The points on the above curves represent the discrete sampling functions that correspond to the discrete B-spline interpolation functions of Figure 7.6. The sampling is assumed to be one-dimensional and the sampling distance is $m = 4$.

color. The color signals were subsequently mapped into a luminance signal (using the standard formula of PAL) and a 2x2 averaging filter was used to further reduce this luminance picture to 1024x1024 pixels. A region of 944x500 pixels was selected to be used in the actual experiments. This reduced size allowed to simultaneously display two images on the screen during the experiments. A high-resolution interlaced monitor, able to render 1024x1024 images, was used to display the images. The grey-value-to-luminance characteristic of this monitor was calibrated to have a gamma-characteristic with $\gamma = 1.5$. The monitor was placed in a dark room with a dimly lit white background. The viewing distance was equal to three times the image height, and was controlled by a head rest.

Figure 7.8. Scenes used in experiment with sampled images: Stadhuis (left) and Wanda (right).

At this viewing distance, the visual angle between successive pixels is approximately 1.1 arcmin, so that the line structure of the displayed image is just noticeable. All subjects that participated had experience with performing psychophysical scaling experiments. They all had normal or corrected-to-normal visual acuity.

5.2 EXPERIMENT 1 - EFFECT OF INTERPOLATION FILTER

Eight subjects took part in this first experiment that aimed at measuring the effect of the order of a B-spline interpolation filter on the image quality. There were eight stimuli per scene, i.e., the original image and 7 samped/interpolated images. Three different subsampling factors ($m = 2$, $m = 3$ and $m = 4$) were combined with separable B-spline interpolation filters $\varphi_{m,n}(k)$ of varying order ($n = 1$, $n = 5$ and $n = 9$). For each combination (m, n), the corresponding sampling filter $\phi_{m,n}(k)$ was determined in one dimension using the algorithm discussed in exercise 4. The two-dimensional separable sampling filter that was actually applied to the images was equal to the product of the one-dimensional filters in x- and y-direction. Since the images produced with $(m = 2, n = 5)$ and $(m = 2, n = 9)$ were hardly distinguishable from those produced with $(m = 2, n = 1)$, it was decided not to include them in the actual experiment.

The experiment itself consisted of four sub-sessions. In the first session, the subjects were asked to judge the dissimilarities between all 64 combinations of the 8 stimuli on an integer scale from 0 to 9. In the second session, the 7 sampled and interpolated images were paired with the original images and shown 6 times to the subjects. The processed images were shown 3 times on the left and 3 times on the right of the display. The subjects were asked to express the similarity between both images on a scale from 0 to 9. This experiment hence aimed at measuring the fidelity of the image reproduction. In these first two sessions, the stimulus pairs were displayed for approximately 20 seconds. In the final two sessions, the sampled and interpolated images were presented one at a time. Each stimulus was repeated 6 times and displayed for approximately 12 seconds. In the third and fourth session, the subjects were asked to scale the sharpness and "amount of ringing" in the images, respectively. An image with pronounced ringing, i.e., the "Stadhuis" image corresponding to $(m = 4, n = 9)$, was shown at the start of the fourth session to clarify the attribute. A trial session (with 8 stimuli in the first session, and 10 stimuli in the other sessions) always preceded the actual experiment, and allowed the subjects to get used to the experimental setting and to adjust the sensitivity of their rating scale. Stimuli were randomized and images corresponding to the two different scenes were alternated.

The experimental data were analyzed with the discrete version of the MDS program XGms (see chapter 5). The data were assumed to be nonmetric, and a non-linear quantization characteristic based on a monotonic spline function with 2 internal knot points was used in all cases to relate the geometrical properties of the MDS model (i.e., the interstimulus distances and the stimulus coordinates along attribute axes) to the scale values attributed by the subjects. All attributes (fidelity, blur and ringing) were modelled by a vector-product model. The subjects agreed very well on the attributes fidelity and sharpness, so that these attributes were modelled by a common vector for all subjects. In order to reflect the individual differences in the judgements of ringing, a separate "ringing" vector was adopted for each subject. Especially the diversity in these judgements of ringing necessitated a two-dimensional

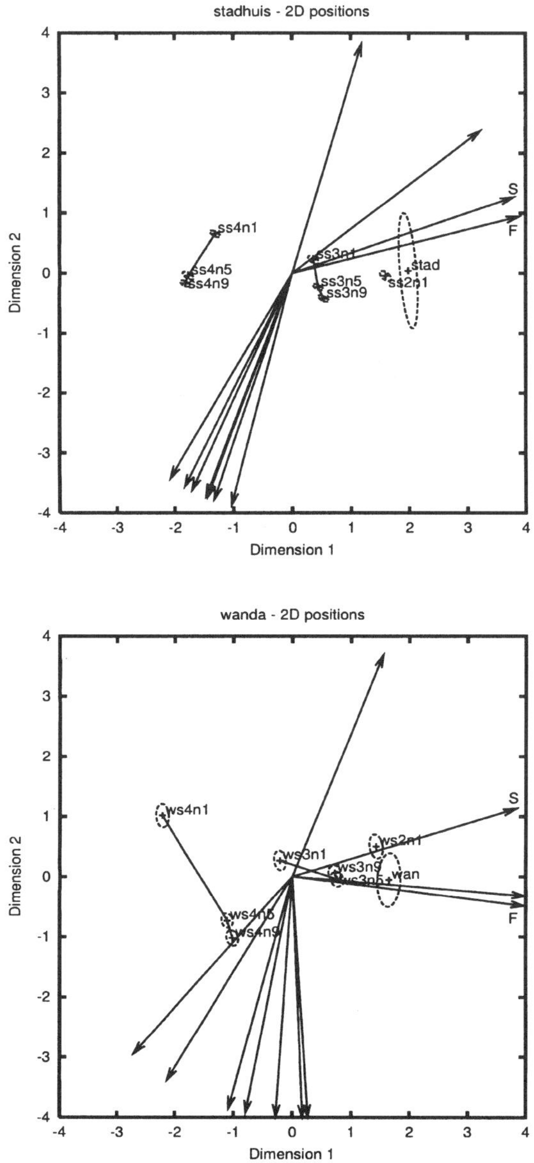

Figure 7.9. Stimulus configurations with 95 % uncertainty intervals derived from the first experiment on sampled and interpolated images. The attribute vectors for fidelity and sharpness are indicated by F and S, respectively, while the unnamed attribute vectors are the individual attribute vectors for ringing.

stimulus configuration. The 2D stimulus configurations for both scenes are presented in Figure 7.9 together with the estimated attribute vectors.

The conclusions that can be drawn from the experiment depend on the scene content. In case of the outdoor scene "Stadhuis", the attributes sharpness and fidelity coincide, so that ringing has little effect on the overall judgement. Only in case of a subsampling factor $m = 4$ does the order of the interpolation kernel have an effect on the judgement. For this outdoor scene, nearest-neighbor interpolation is preferred. In case of the indoor scene "Wanda", the attributes sharpness and fidelity make an angle, so that other attributes (such as ringing or the staircase effect) seem to have a larger influence on the fidelity judgement. The spline interpolations of order $n = 5$ and $n = 9$ differ very little, but are clearly preferred over the nearest-neighbor interpolation in this case. Two of the subjects whose "ringing" attribute vectors point in an opposite direction as the vectors for the majority of the subjects also seem to have confused the staircase effect, that occurs for images with nearest-neighbor interpolation, with ringing.

Even from this very small data set with two scenes, one can already conclude that the optimal sampling and interpolation depends on the scene contents. This scene-dependence is difficult to predict, since it is for instance not reflected in the mean-square error between original and processed image. This is illustrated in Figure 7.10, where the relationship between image fidelity, as derived from the experimental data of the second session, is plotted against signal-to-noise ratio. Note that, although the order n of the B-spline interpolation function can have a significant effect on the image fidelity, this is not always reflected in a variation in signal-to-noise ratio.

5.3 EXPERIMENT 2 - EFFECT OF SAMPLING FILTER

Six subjects participated in this second experiment that aimed at measuring the effect of the sampling filter on the image quality. Four combinations of two different subsampling factors m ($m = 2$ and $m = 3$) and two different orders n of the B-spline interpolation filter ($n = 1$ and $n = 5$) were combined with five different sampling filters, leading to a total of 20 images per scene. The first alternative, denoted by 'nf', consisted of subsampling the image without pre-filtering. In the second case, denoted by 'if', the interpolation filter was also used as sampling filter. The remaining three alternatives, denoted by 's(n+p)', consisted of combining the sampling filter $\phi_{m,n+p}(k)$ for a B-spline interpolation function $\varphi_{m,n+p}(k)$ of order $n+p$ with the interpolation function $\varphi_{m,n}(k)$, for $p = 0, 1, 2$.

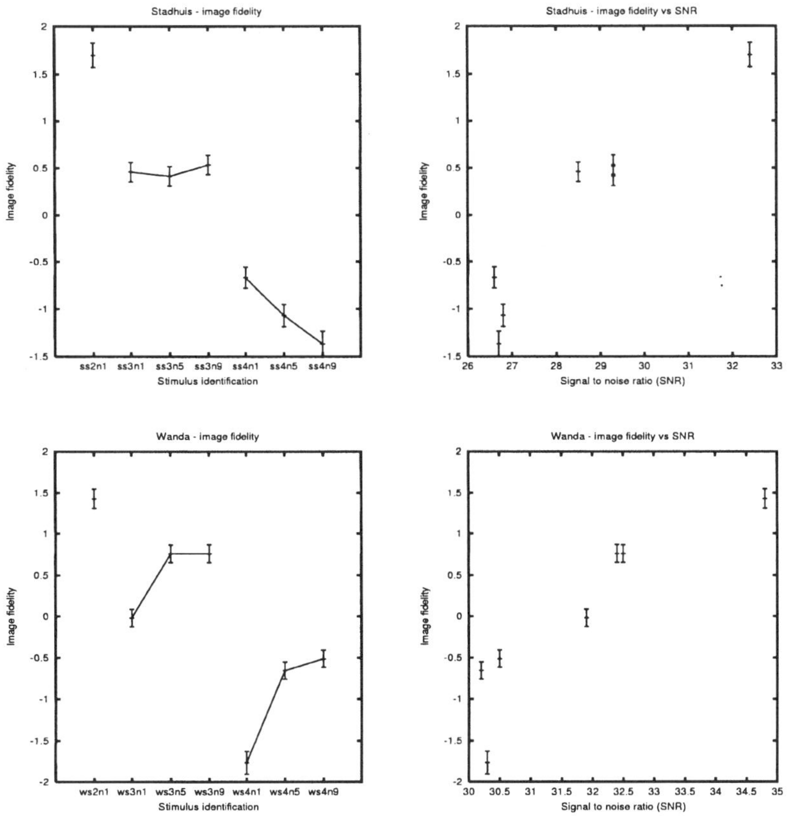

Figure 7.10. Image fidelity of the sampled and interpolated images in the first experiment as a function of stimulus identification (left) and signal to noise ratio.

Two experimental sessions were performed with the images. In the first session, the processed images were presented one at a time and the subjects were asked to rate their image quality. In the second session, the processed images were shown alongside the unprocessed image, and the subjects were asked to rate the similarity between both images. Hence, the procedure in this second session was similar to that in the second session of experiment 1, and aimed at measuring the image fidelity, rather than the image quality. Since three of the four images corresponding to condition 'sn', i.e., for (m, n) equal to $(2, 1)$, $(3, 1)$ and $(5, 1)$, occurred in both experiments, they can be used to cross-check the experimental results. In both sessions, each of the 40 stimuli was displayed 4 times for a duration of approximately 10 seconds. The subjects were asked to

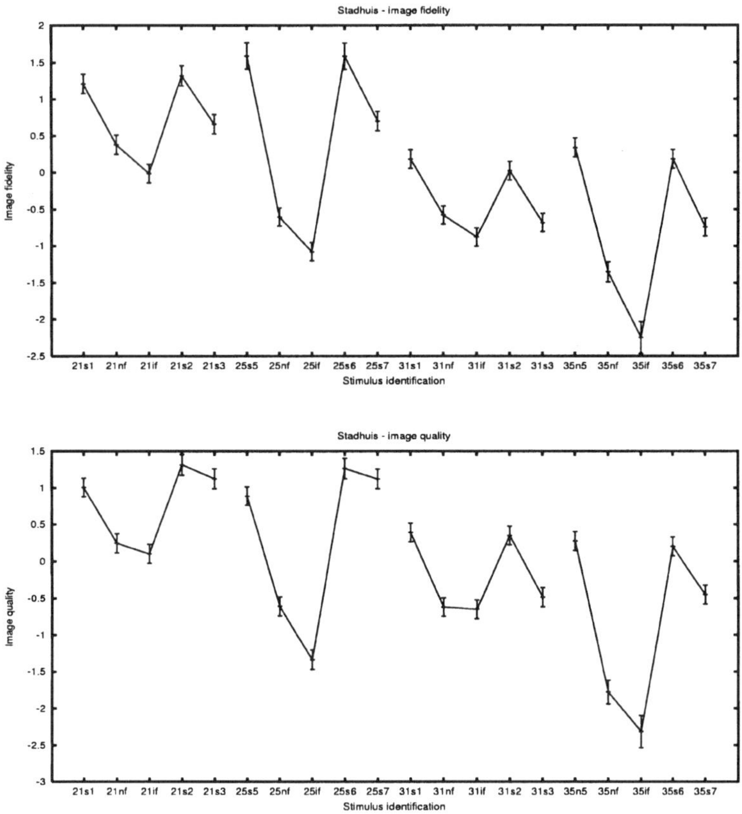

Figure 7.11. Attributes strengths for scene "Stadhuis" derived from the second experiment on sampled and interpolated images. The two different attributes, image fidelity and image quality, are presented in the upper and lower part, respectively.

rate them on a ten-point categorical scale from 0 to 9. A trial session containing 16 stimuli preceded the actual experiment in all cases.

The experimental data were again analyzed with the discrete version of the MDS program XGms. The data were assumed to be non-metric, and a non-linear quantization characteristic based on a monotonic spline function with 2 internal knot points was used in all cases to relate the stimulus coordinates to the scale values attributed by the subjects. The resulting one-dimensional stimulus configurations for both attributes, image fidelity and image quality, are presented in Figure 7.11 (scene "Stadhuis") and Figure 7.12 (scene "Wanda").

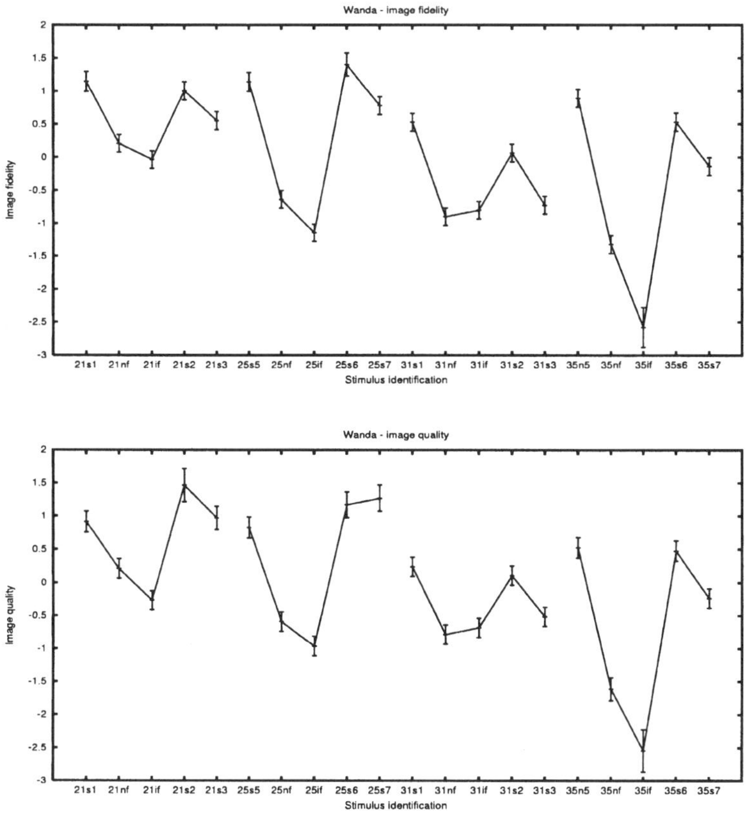

Figure 7.12. Attributes strengths for scene "Wanda" derived from the second experiment on sampled and interpolated images. The two different attributes, image fidelity and image quality, are presented in the upper and lower part, respectively.

It can be observed that the sampling filter has a important effect on both the image quality and fidelity. The images with a sampling filter are clearly preferred over the images without sampling filter (nf) or with a sampling filter equal to the interpolation function (if). Most of the images with a sampling filter and subsampling factor $m = 3$ are even preferred over images without sampling filter and subsampling factor $m = 2$. The effect also becomes more pronounced as the order n of the interpolation function increases. Using a sampling filter with an order higher than that of the interpolation filter leads to an increased quadratic error and a reduced image fidelity. However, in case of subsampling factor $m = 3$, the images produced with a sampling filter of order $n + 1$

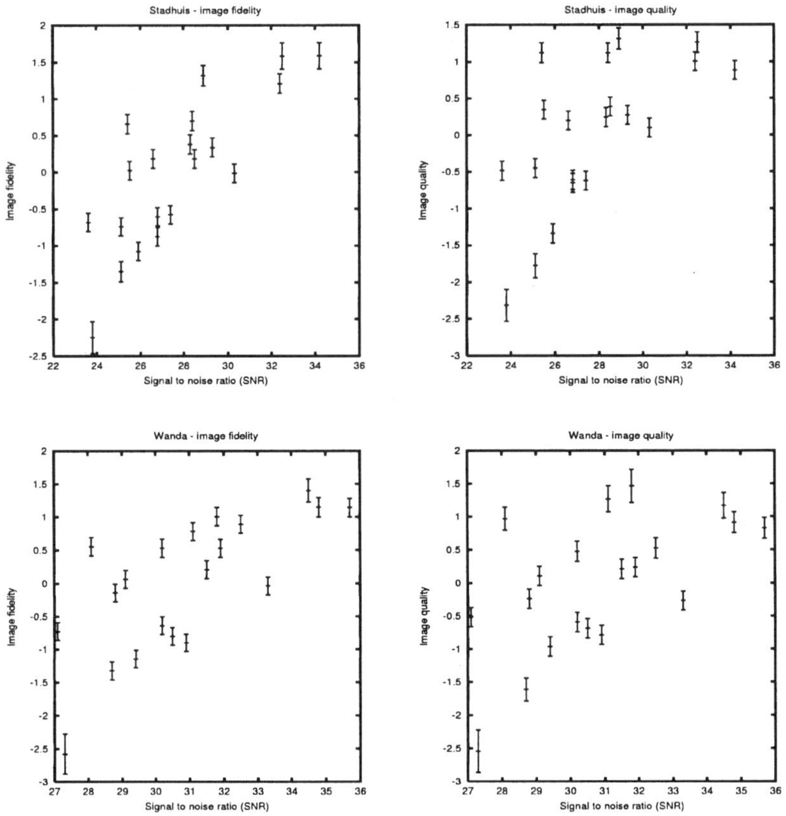

Figure 7.13. Image fidelity and image quality of the sampled and interpolated images in the second experiment as a function of the signal to noise ratio.

have approximately the same quality as the images produced with the optimum sampling filter. In case of subsampling factor $m = 2$, the images produced with the sampling filters of order $n+1$ and $n+2$ often even have a better quality than those produced with the sampling filter of order n. This is another instance of a phenomenon that was also observed in the case of color reproductions, i.e., that image fidelity and quality need not coincide. A human observer often prefers images that deviate (slightly) from the original images, despite the fact that (s)he can perceive the processing and is aware of the fact that the reproduction is not entirely faithful. In the current case, the increased sharpness of the images compensates for the increase in sampling artifacts, such as staircasing and ringing.

In Figure 7.13, the relationship between the image attributes quality and fidelity and the signal to noise ratio is shown. Note that the quadratic error is obviously not a good correlate for these attributes when compared across sampling distances and interpolation functions. Nevertheless, the optimization of the sampling filters, where the quadratic error criterium is used for a fixed sampling distance and interpolation filter, produces very good results.

EXERCISES

1 Consider the discrete periodic signals that are fully specified by $[f]_\psi = [f(i), i = 0, \ldots, 5]^T$. Assume for instance that the corresponding continuous periodic signals are piecewise constant, and obtained by interpreting the samples as the representation in the canonical basis

$$B_\psi = \left\{ \psi_i(x) = \sum_{k \in \mathcal{Z}} p_d(x - i - 6k); i = 0, \ldots, 5 \right\},$$

where

$$p_d(x) = \begin{cases} 1 & \text{if} \quad x \in [0,1[\\ 0 & \text{if} \quad x \notin [0,1[\end{cases}$$

denotes the unit pulse. The inner product in the resulting vector space V is defined by

$$\langle \sum_{i=0}^{5} f(i) \cdot \psi_i(x), \sum_{i=0}^{5} g(i) \cdot \psi_i(x) \rangle = \sum_{i=0}^{5} f(i) \cdot g(i).$$

Consider the subspace U of V spanned by the basis $B_\varphi = \{\varphi_i(x); i = 1, 2, 3\}$, with respective signal representations

$$\begin{aligned} [\varphi_1(x)]_\psi &= (1/2, 1/2, 1/2, 1/2, 0, 0)^T \\ [\varphi_2(x)]_\psi &= (0, 0, 1/2, 1/2, 1/2, 1/2)^T \\ [\varphi_3(x)]_\psi &= (1/2, 1/2, 0, 0, 1/2, 1/2)^T \end{aligned}$$

Solve the following questions:

- Motivate why B_φ satisfies the definition of a repetitive basis.
- Find the 3×3 Gramian matrix $G_{\varphi\varphi}$. What is the kernel of this matrix? Derive the dual Gramian matrix by using Fourier transformations (i.e., find three harmonic eigenvectors and the corresponding eigenvalues of the matrix).
- Find and draw the dual basis vectors $\{\phi_1(x), \phi_2(x), \phi_3(x)\}$ as a function of x.
- Find the 3×6 matrix $G_{\phi\psi}$ that maps the representation $[f]_\psi$ of a vector f in the basis B_ψ of V to the representation $[Pf]_\varphi$ of the orthogonal projection Pf (onto the subspace U) in the basis B_φ of U. Show that the mapping $G_{\phi\psi}$ can be implemented by a technical system that consists of a convolution, followed by a downsampling device (i.e., a device that retains only the

outputs of the convolution device at regular intervals). Hint: show how the 3×6 matrix can be decomposed into a 3×6 matrix containing only one nonzero element per row, and a 6×6 circular matrix. Draw the impulse response of the convolution filter (i.e., the response to the signal $\psi_0(x)$).

- Find the 6×3 matrix $G_{\chi\varphi}$ that maps the vector representation $[Pf]_\varphi$ in the basis B_φ of U to the representation $[Pf]_\psi$ in the basis B_ψ of V. Show that the mapping $G_{\chi\varphi}$ can be implemented by a technical system that consists of an upsampling device (i.e., a device that substitutes zeroes in between the known sample values), followed by a convolution. Hint: show how the 6×3 matrix can be decomposed into a 6×6 circular matrix and a 6×3 matrix containing only one nonzero element per column. Draw the impulse response of the convolution filter.
- Find and draw the projection Pf of the signal f with representation $[f]_\psi = [f(i) = i; i = 0, \dots, 5]^T$. Find the representation $[Pf]_\varphi$ (3 numbers) in the basis B_φ of U, as well as the representation $[Pf]_\psi$ (6 numbers) in the basis B_ψ of V.

2 Repeat the previous exercise assuming that the basis B_ψ is a Fourier basis. Which results would change, and which ones would stay the same?

3 Consider the one-dimensional sampling of (non-periodic) continuous signals at regular sampling positions $k\,\Delta, k \in \mathbb{Z}$. Assume for instance that the basis function is

$$\varphi(x) = \begin{cases} 1 + \frac{x}{\Delta} & \text{if } \; x \in [-\Delta, 0] \\ 1 - \frac{x}{\Delta} & \text{if } \; x \in [0, \Delta] \\ 0 & \text{if } \; x \notin [-\Delta, \Delta] \end{cases}$$

with

$$g_{\varphi\varphi}(p - q) = \langle\, \varphi(x - p\,\Delta), \varphi(x - q\,\Delta)\,\rangle = \begin{cases} \frac{2\,\Delta}{3} & \text{if } \; p = q \\ \frac{\Delta}{6} & \text{if } \; |p - q| = 1 \\ 0 & \text{if } \; |p - q| > 1 \end{cases}$$

This corresponds to **linear interpolation** between sample values. Do not try to explicitly determine the sampling function $\phi(x)$, since this function is infinitely long. A finite approximation to this sampling function can be of the form

$$\tilde{\phi}(x) = \tilde{g}_{\phi\phi}(-1) \cdot \varphi(x - \Delta) + \tilde{g}_{\phi\phi}(0) \cdot \varphi(x) + \tilde{g}_{\phi\phi}(1) \cdot \varphi(x + \Delta).$$

The weights of this approximate sampling function can be found by requiring that the combined action

$$\sum_{q=-1}^{1} g_{\varphi\varphi}(p-q) \cdot \tilde{g}_{\phi\phi}(q) = r_p$$

is in some way close to the identity operator (i.e., $r_p \approx \delta_p$). One possible way of expressing this is to demand that at least the low-order moments

$$\sum_{p \in Z} r_p \cdot (p\,\Delta)^n = \sum_{p \in Z} \delta_p \cdot (p\,\Delta)^n = \begin{cases} 1 & \text{if} \quad n = 0 \\ 0 & \text{if} \quad n > 0 \end{cases},$$

for n = 0, 1, ... are equal for the approximate and ideal case. Derive the resulting approximate sampling function $\tilde{\phi}(x)$ by solving for the coefficients $\tilde{g}_{\phi\phi}(-1)$, $\tilde{g}_{\phi\phi}(0)$ and $\tilde{g}_{\phi\phi}(1)$. Find numerical values for the resulting combined operator r_p. Make a drawing of $\tilde{\phi}(x)$.

Determine the approximate signals $\tilde{f}(x)$, after sampling with the approximate dual function $\tilde{\phi}(x)$ and interpolation with the function $\varphi(x)$, for the input signals $f(x) = 1, x$ and x^2. Plot the results on the interval $x \in [-2, 2]$, assuming $\Delta = 1$ and $\Delta = 2$.

The following intermediate results can be used in the above exercise

$$\begin{aligned} \int_{-\infty}^{+\infty} dx\, \varphi(x - p\Delta) &= \Delta \\ \int_{-\infty}^{+\infty} dx\, \varphi(x - p\Delta) \cdot x &= p \cdot \Delta^2 \\ \int_{-\infty}^{+\infty} dx\, \varphi(x - p\Delta) \cdot x^2 &= \left(p^2 + \frac{1}{6}\right) \cdot \Delta^3 \end{aligned}$$

4 In case of one-dimensional sampling, one can use the following (approximation) technique for inverting a Gramian kernel $g_{\varphi\varphi}(i)$, for $i \in \mathbb{Z}$. Provided that the values outside the interval $[-n, n]$ are negligibly small, then the **Z-transformation** of the Gramian kernel can be approximated by

$$G_{\varphi\varphi}(z) = \sum_{i=-\infty}^{\infty} g_{\varphi\varphi}(i)\, z^i \approx \sum_{i=-n}^{n} g_{\varphi\varphi}(i)\, z^i,$$

so that the Z-transformation of the inverse Gramian kernel can in turn be approximated by

$$G_{\phi\phi}(z) \approx \frac{1}{\sum_{i=-n}^{n} g_{\varphi\varphi}(i)\, z^i}$$

(the relationship between Z-transformation and Fourier transformation is established by substituting $z = \exp(j2\pi u)$). Derive how this inverse Gramian kernel can be obtained as a cascade of simple filters with impulse response

$$h(i; a) = a^{|i|},$$

for $i \in Z$, and Z-transformation

$$H(z; a) = \frac{1 - a^2}{(1 - a \cdot z)\,(1 - a \cdot z^{-1})}.$$

This method has been used to obtain the sampling functions in Figure 7.7.

Chapter 8

SPATIO-TEMPORAL CHARACTERISTICS OF THE HUMAN VISUAL SYSTEM

1. INTRODUCTION

The harmonic signals that have been introduced in chapter 6 as the eigensignals of linear space- and/or time-invariant systems have also been used extensively to study the spatial and temporal processing performed within the human visual system. These spatiotemporal characteristics of the visual system will be discussed in some detail in this chapter. They can for instance be applied to the optimization of sampling structures, as is demonstrated in chapter 9, or to the optimization of coding algorithms [Nad00].

The eye optics is the entry point for spatially and temporally varying visual information. As such, the limitations imposed by it also have an effect on successive stages in the visual system. It is therefore interesting to understand the characteristics of the eye optics in some more detail. Since the eye optics is approximately linear and space-invariant, it can be characterized by how it transforms spatially varying harmonic functions. This characteristic of the eye optics is usually referred to as the optical transfer function (OTF) of the eye, and existing mathematical models of the OTF are discussed in section 2.

The perception of harmonic patterns varying in space and/or time is not only determined by the optics of the eye, but also by the neural processing in the remainder of the visual system. The overall visual system cannot be described straightforwardly by a linear front-end system, because of a phenomenon called **luminance adaptation**. This adaptation means that the sensitivity of the visual system varies strongly as a function of the mean luminance in the scene. This property allows the human visual system to operate under a wide range of lighting circumstances (in dim, moonlight, as well as bright, sunshine, conditions). If images are described in terms of their variation (or **contrast**) with respect to a reference luminance, then the approximation by a linear front-end system is much more accurate, at least in cases where the contrast is relatively small. The contrasts required for detecting harmonic signals of different spatial and/or temporal frequencies are important characteristics of the visual system. These contrast detection thresholds are usually presented in the form of **contrast sensitivity functions** (CSF), and some existing CSF model descriptions are presented in section 3.

Although much remains unknown about the perception of colored patterns, spatial and temporal extensions of the CIELAB uniform color space have been proposed. The resulting models, which are discussed in section 4, can be used to compare images that differ in both color and spatial and/or temporal information.

2. EYE - OPTICAL TRANSFER FUNCTION (OTF)

2.1 OPTICS OF THE HUMAN EYE

In chapter 1, it was already mentioned that the Maxwell equations of electromagnetism imply that electromagnetic energy is transferred in a linear way in most media. It can therefore be assumed that the optical transfer in the human eye is a linear mapping from the object radiance $L_e(\lambda, x_d, y_d, t)$ into the retinal irradiation $E_e(\lambda, x_c, y_c, t)$. Corresponding positions (x_c, y_c) on the retina and (x_d, y_d) on the object[1] are usually designated by a direction vector $\boldsymbol{\rho} = (\rho, \theta)$ that is expressed in visual angles. In case of the perspective projection geometry of Figure 1.2, this direction vector $\boldsymbol{\rho}$ corresponds to a light ray that passes through the center of the lens and that makes a visual angle $\rho = |\boldsymbol{\rho}|$ with the optical axis. The visual angle ρ can vary between 0 degrees (the optical axis) and 90 degrees (orthogonal to the optical axis), and can be expressed in radians or degrees. All rays that make an angle ρ with the optical axis lie on a cone, and a unique ray may be identified by considering the intersection of this cone with a sphere of unit radius that is centered on the vertex of the cone. This intersection is a circle, so that a position on this circle can be specified uniquely by a planar angle θ between 0 and 360 degrees. In case of the flat camera target of Figure 1.2, the position on the target corresponding to the spherical coordinates (ρ, θ) is $(x_c, y_c) = (b\ \tan(\rho) \cdot \cos\theta, b\ \tan(\rho) \cdot \sin\theta)$. For small angles, the polar

[1]The object is very often an approximately flat display surface in case of image technology.

coordinate approximation $(x_c, y_c) \approx (b\,\rho \cdot \cos\theta, b\,\rho \cdot \sin\theta)$ can be used. This polar coordinate approximation is for instance sufficiently accurate for describing foveal vision (visual angles ρ smaller than 5 degrees).

Because of the high velocity of light, it can be assumed, for all practical purposes, that the transfer of electromagnetic energy from the display to the retina is instantaneous, so that the irradiance at time t is only determined by the radiance at time t. However, because of diffraction and aberration in the lens system, and due to light scattering in the eye media, a point on the display will be mapped into a spot of finite size on the retina. This response to a point light source in the direction $\boldsymbol{\rho}_s$ is called the **point spread function** $PSF(\boldsymbol{\rho}, \boldsymbol{\rho}_s)$ and will typically be centered on the direction $\boldsymbol{\rho}_s$. Since the optical system is linear, the relationship between radiance and irradiance can hence be expressed as

$$E_e(\lambda, \boldsymbol{\rho}, t) = \int PSF(\boldsymbol{\rho}, \boldsymbol{\rho}_s)\, L_e(\lambda, \boldsymbol{\rho}_s, t)\, d\boldsymbol{\rho}_s. \tag{8.1}$$

Usually, the PSF will not significantly change shape for directions $\boldsymbol{\rho}_s$ that are close to the optical axis, so that a single $PSF(\boldsymbol{\rho}, \boldsymbol{\rho}_s) = PSF(\boldsymbol{\rho} - \boldsymbol{\rho}_s)$ is usually assumed to describe foveal vision. The definition of convolution in chapter 6 indicates that this corresponds to assuming that the optical transfer for foveal vision can be described by a linear space-invariant system. It was shown that such a system can be alternatively described by the **optical transfer function** $OTF(\boldsymbol{\nu})$, which is the spatial Fourier transformation of the $PSF(\boldsymbol{\rho})$. If the spatial variations are measured in degrees of visual angle, than the corresponding frequencies are measured in number of **cycles per degree** (c/deg) of visual angle.

There are two widespread experimental techniques for measuring the OTF that give related but nevertheless slightly different results. In the double-pass technique [CG66, LNS93, WBMN94], point (or line) light sources are presented to the eye and the spatial distribution of the light that exits the eye after it has been reflected by the retina is measured. Since the reflected light has passed twice through the optics of the eye, the Fourier transform of this reflected pattern is not equal to $OTF(\boldsymbol{\nu})$, but to $OTF^2(\boldsymbol{\nu})$. Hence, the square root of the Fourier transform of the measured PSF is the required OTF. In the interference technique [CG65, WBMN94], sinusoidal interference fringes with known amplitude and frequency are created directly on the retina by means of interference of two coherent light sources. The OTF is then obtained by dividing the detection threshold for interference gratings by the detection threshold for externally presented gratings (see section 3 for more details on detection threshold measurements). Unlike the double-pass method, the interference technique also takes into account how the light that falls onto the retina is coupled into the photoreceptors[2].

There is no limitation on the spectral distribution of the light source in case of the double-pass technique. The interference technique, on the other hand, requires coherent light, so that these latter measurements are almost exclusively performed with monochromatic light sources (generated by a laser). Consequently, no chromatic aberration occurs in these latter measurements. A numerical model for chromatic aberration is hence required in order to relate results from both experimental techniques. Such a model will therefore also be discussed later in this section.

2.2 SMALL-ANGLE AND LARGE-ANGLE OTF

It is usually assumed that the OTF of the human eye can be described by a rotationally-symmetric function with two components, i.e.,

$$OTF(\nu) = w_{sa}\, OTF_{sa}(\nu) + (1 - w_{sa})\, OTF_{la}(\nu), \tag{8.2}$$

with ν in cycles per degree. The *small-angle* component $OTF_{sa}(\nu)$ is mostly determined by diffraction and aberration, while the *large-angle* component $OTF_{la}(\nu)$ is mostly determined by light scattering. Both components of the OTF are normalized such that $OTF_{sa}(0) = OTF_{la}(0) = 1$, which implies that a uniform field with frequency $\nu = 0$ passes unaltered through the OTF[3].

The $PSF(\rho)$ corresponding to this $OTF(\nu)$ is also rotationally symmetric in the spatial domain. Although it is known that deviations from rotational symmetry occur in the foveal PSFs of individuals [LNS93], the assumption seems realistic for an average OTF over a large number of subjects.

Light scattering in the eye typically increases with age, so that the relative weight w_{sa} on the small-angle component will decrease with age. This age-dependence has been modelled by

$$w_{sa}(age) = \frac{w_o}{w_o + (1 - w_o) * AF(age)}, \tag{8.3}$$

where

$$AF(age) = 1 + (age/70)^4 \tag{8.4}$$

[2]Since the photoreceptors are long elongated structures, light that enters from a direction that is aligned with the axis of the photoreceptors is more effective than light entering under an angle. This is called the ***Stiles-Crawford effect.*** Light rays that enter near the margin of the pupil have the largest aberration, but also make the largest angle with the optical axis, so that they are least effective in exciting the photoreceptors.

[3]The relevant factors in the relationship between radiance and irradiance for a uniform field were already discussed in chapter 1, and can therefore be neglected here.

is the age factor (with age expressed in years). The relative weight w_o for the young eye also depends on the pigmentation (i.e., blue or brown eyes). The value $w_{sa}(35) = 0.85$ (or equivalently, $w_o = 0.8576$) is a typical value for the blue Caucasian eye [IBS93].

Some representative OTF (and PSF) models that have been proposed for the case of white (halogen) light sources are now discussed in more detail. The large-angle OTF can for instance be approximated by a function that is the sum of (two) exponential functions (SOE), i.e.,

$$OTF_{la}(\nu) = c_{la}\ \exp(-\gamma_1 \cdot \nu) + (1 - c_{la})\ \exp(-\gamma_2 \cdot \nu), \tag{8.5}$$

where $\gamma_1 = 25.92$ and $\gamma_2 = 360$ degrees, and $c_{la} = 0.11$, are typical values for a 35-year-old blue-eyed Caucasian. The inverse Fourier transformation of $OTF(\nu) = \exp(-\gamma\, \nu)$ is expressed analytically by

$$PSF(\rho) = \frac{2\pi\, \gamma}{[(2\pi)^2 \sin^2 \rho + \gamma^2 \cos^2 \rho]^{3/2}}. \tag{8.6}$$

Because $\gamma_2 = 2\pi$, this latter result implies that the second term in the above expression for $OTF_{la}(\nu)$ corresponds to an irradiance offset that is spatially uniform. This constant offset is thought to be mostly caused by the light transmission through the ocular wall [IBS93].

In many practical applications, such as in display design and image coding, the images that have to be compared differ only in detail. These detailed differences in radiance will usually be observable in the small-angle component of the irradiance, but not in the large-angle component, so that the latter component of the OTF can often be neglected when comparing output images after optical filtering. Different mathematical models have been proposed for the small-angle OTF. In discussions on the small-angle OTF (for foveal vision), it can be assumed that the plane approximation is valid, so that the spatial coordinate $\boldsymbol{\rho} = (\rho, \theta)$ can be interpreted as a polar coordinate. The corresponding frequency coordinate $\boldsymbol{\nu} = (\nu, \phi)$ is then a polar coordinate in the Fourier plane.

A sum of (two) exponentials (SOE), i.e.,

$$OTF_{sa}(\nu) = c_{sa}(p)\ \exp[-\beta_1(p) \cdot \nu] + [1 - c_{sa}(p)]\ \exp[-\beta_2(p) \cdot \nu]. \tag{8.7}$$

is also the most-frequently adopted description for the small-angle OTF. The effect of one component $OTF(\nu) = \exp(-\gamma\, \nu)$ can be interpreted in terms of its response to a point source, i.e.,

$$PSF(\rho) = \frac{2\pi\, \gamma}{[(2\pi)^2 \sin^2 \rho + \gamma^2 \cos^2 \rho]^{3/2}} \approx \frac{2\pi\, \gamma}{[(2\pi\, \rho)^2 + \gamma^2]^{3/2}}, \tag{8.8}$$

or in terms of its response to a line source, i.e.,

$$LSF(\rho) = \frac{2\,\gamma}{[(2\pi)^2 \sin^2 \rho + \gamma^2 \cos^2 \rho]} \approx \frac{2\,\gamma}{[(2\pi\,\rho)^2 + \gamma^2]} \tag{8.9}$$

(see Ijspeert et al. [IBS93]). The rightmost results in the above equations hold under the (realistic) assumption that all visual angles of interest are small, so that the polar coordinate approximation can be adopted.

Since the $OTF_{sa}(\nu)$ is influenced by diffraction, its parameters must depend on the pupil diameter p (in mm). A good fit to the experimental data is obtained with

$$\begin{aligned} \beta_1(p) &= PF(p)\,\frac{1}{b\,p} \\ \beta_2(p) &= \frac{PF(p)}{PF(p)-1}\left(e - \frac{1}{b\,p}\right) \end{aligned} \tag{8.10}$$

where the parameters $b = 25 - 2.6\sqrt{AF}$ mm^{-1} and $e = 0.18\sqrt{AF}$ depend on the age factor AF (i.e., $b = 22.32$ mm^{-1} and $e = 0.1855$ for an age of 35 years), and

$$c_{sa}(p) = \frac{1}{PF(p)}, \tag{8.11}$$

depends on the pupil factor

$$PF(p) = 1 + (p/d)^2, \tag{8.12}$$

with $d = 3.2$ mm [IBS93]. For large pupil diameters, the relative weight on the sharp-peaked first term decreases, so that the broad-peaked second term becomes more manifest. The parameter values for the SOE model are listed in Table 8.1 as a function of pupil diameter p for a 35-year-old observer.

It should be noted that the SOE model has been introduced to describe OTF measurements for frequencies below 50 c/deg. These measurements probably contain some aliasing. If equation (8.8) is used to derive the small-angle PSF from the OTF in equation (8.7), then the SOE model is implicitly extrapolated to frequencies above 50 c/deg. Since the exponential functions in the SOE model do not decay very fast, the assumed Fourier amplitude at frequencies above 50 c/deg is most often higher than what can be expected for the actual OTF. Since this extrapolation by the SOE model to high frequencies is not very satisfactory, the PSF according to the SOE is usually narrower than the observed PSF [AN94]. Of course, one could limit the bandwidth

Table 8.1. Parameters of the SOE model of Ijspeert et al. for the small-angle OTF of the human eye as a function of pupil diameter p. The parameter values are for a 35-years-old observer.

p (mm)	β_1 (arcmin)	β_2 (arcmin)	c_{sa}
2	1.869	34.846	0.719
3	1.684	21.883	0.532
4	1.722	17.155	0.390
5	1.850	14.934	0.291
6	2.023	13.724	0.221
7	2.222	12.995	0.173
8	2.436	12.524	0.138

of the SOE model before calculating the PSF by inverse Fourier transformation. However, the way this bandwidth limitation is performed will have a significant influence on the shape of the PSF. In Figure 8.1, the OTF and PSF according to the SOE model are shown without and with bandwidth limitation to 50 c/deg. A raised cosine window that decreases from 1 to 0 for frequencies in the range from 25 to 75 c/deg has been used to accomplish the bandwidth limitation of the OTF.

An alternative often-used description of the small-angle OTF is by means of a sum of Gaussian functions (SOG), i.e.,

$$OTF_{sa}(\nu) = c_{sa}(p)\ \exp[-(\beta_1(p)\cdot\nu)^2] + [1-c_{sa}(p)]\ \exp[-(\beta_2(p)\cdot\nu)^2]. \tag{8.13}$$

The parameters $\beta_1(p)$, $\beta_2(p)$ and $c_{sa}(p)$ again depend on the pupil diameter p. Typical values for a $p = 2$ mm pupil diameter are $\beta_1 = 2$ arcmin, $\beta_2 = 9$ arcmin and $c_{sa} = 0.55$ [Gei84]. Because the OTF according to the SOG model decays much faster at high frequencies than the OTF according to the SOE model, the corresponding point response

$$\begin{aligned} PSF_{sa}(\rho) &= c_{sa}(p)\,\frac{\pi}{\beta_1^2(p)}\,\exp\left[-\left(\frac{\pi\,\rho}{\beta_1(p)}\right)^2\right] \\ &+ [1-c_{sa}(p)]\,\frac{\pi}{\beta_2^2(p)}\,\exp\left[-\left(\frac{\pi\,\rho}{\beta_2(p)}\right)^2\right] \end{aligned} \tag{8.14}$$

is usually a much better approximation to the observed PSF. The SOG-model is compared to the SOE-model for a pupil diameter of $p = 2$ mm in figure 8.1.

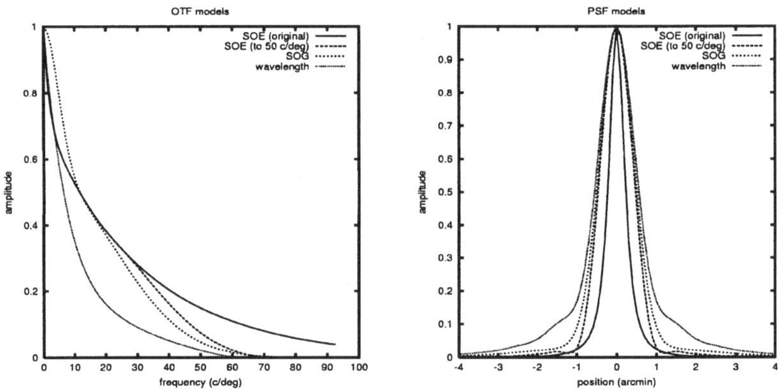

Figure 8.1. The left panel shows the amplitude of the small-angle component of the OTF of the human eye as a function of spatial frequency (in c/deg), while the right panel shows the corresponding PSF as a function of spatial position (in arcmin). The PSFs have been rescaled such that the maximum amplitude is one in all cases. The four curves relate to: the original SOE model, the SOE model with bandwidth limitation to 50 c/deg, the SOG model and the wavelength-dependent OTF model. The assumed pupil diameter is $p = 2$ mm in all cases.

2.3 WAVELENGTH-DEPENDENT OTF

The above models for the (small-angle) OTF are inadequate when color perception has to be modelled. Indeed, in chapter 1, it was discussed that the optical characteristic of the human eye varies significantly across wavelength due to chromatic aberration. This variation of the OTF with wavelength therefore also has to be taken into account in a realistic model[4].

The OTF of a defocused optical system has been derived by Hopkins [Hop55]. If an optical system is focused for a wavelength λ_o, and if $d(\lambda) = D(\lambda) - D_o$ denotes the difference in dioptic power $D(\lambda)$ at wavelength λ with the dioptic power D_o at wavelength λ_o, then

$$OTF_c(\nu, \lambda) = H(\tau, a) = \frac{4}{a\pi} \int_0^{\tau} \sin[a(\cos\theta - \cos\tau)] \cos\theta \, d\theta, \quad (8.15)$$

[4]It is usually assumed that a wavelength-independent description is sufficiently accurate for the large-angle OTF.

with

$$\begin{aligned} \tau &= \arccos\left(\frac{\lambda}{\lambda_o}\left|\frac{\nu}{\nu_o}\right|\right) \\ a &= 8\pi\ell(\lambda)\left|\frac{\nu}{\nu_o}\right|, \end{aligned} \tag{8.16}$$

expresses the OTF as a function of frequency ν and wavelength λ. If the optical path length error[5], expressed in multiples of the wavelength λ_o, i.e.,

$$\ell(\lambda) = \frac{p^2}{8\lambda_o} \frac{D_o\, d(\lambda)}{D_o + d(\lambda)} \tag{8.17}$$

is zero, then the above OTF reduces to

$$OTF_d(\nu, \lambda) = H(\tau, 0) = \frac{2}{\pi}(\tau - \cos\tau\ \sin\tau), \tag{8.18}$$

which is the OTF of an optical system that is only limited by diffraction (and hence has a perfect lens with no aberration). The frequency

$$\nu_o = \nu_d(\lambda_o) = \frac{p\, D(\lambda_o)}{c\, \lambda_o} \tag{8.19}$$

is the cut-off frequency at wavelength λ_o for a diffraction-limited imaging system with a circular aperture with diameter p. The factor $c =$ 3.43407 deg/μm is the ratio between visual angle and corresponding distance on the retina [WS82, MW94]. In Figure 8.2, the diffraction-limited $OTF_d(\nu, \lambda_o)$, with $\lambda_o = 580$ nm, is plotted for several values of the pupil diameter p. A realistic value for the power of the unaccommodated eye is $D_o = 59.9404$ diopters, which results in a cut-off frequency for diffraction at wavelength $\lambda_o = 580$ nm equal to $\nu_o = 30.094 \cdot p$ c/deg, where the pupil diameter p is expressed in mm.

The above result has been applied to model the chromatic aberration in the human eye [MW94]. The experimentally determined relative defocus of the human eye as a function of wavelength can be approximated by

$$d(\lambda) = d_f\, \frac{\lambda - \lambda_o}{\lambda - \lambda_\infty}. \tag{8.20}$$

[5]If a light source is focused at a distance $b(\lambda)$ from the lens for wavelength λ, then $\ell(\lambda) = b(\lambda) - b(\lambda_o)$ is the optical path length error.

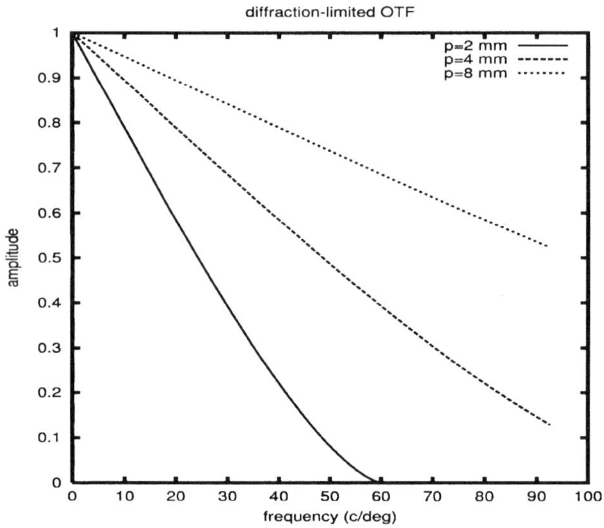

Figure 8.2. Diffraction limited $OTF_d(\nu, \lambda_o)$, with $\lambda_o = 580$ nm, for several values of the pupil diameter p.

where the following values for $\lambda_o = 580$ nm, $\lambda_\infty = 214.1$ nm and $d_f = 1.7312$ diopters can be adopted [MW94]. The resulting chromatic defocus $d(\lambda)$ was plotted as a function of wavelength in Figure 1.4.

Although axial chromatic aberration is the major source of optical aberration in the human eye, other wavelength-independent aberrations contribute as well. The overall small-angle wavelength-dependent OTF of the human eye is therefore modelled by

$$OTF_{sa}(\nu, \lambda) = OTF_b(\nu) \cdot OTF_c(\nu, \lambda), \tag{8.21}$$

where $OTF_b(\nu)$ models the wavelength-independent aberration. This factor can be estimated by measuring $OTF_{sa}(\nu, \lambda^*)$ at a single wavelength λ^*. Assuming that the eye is focused at this wavelength λ^*, we obtain that

$$OTF_{sa}(\nu, \lambda^*) = OTF_b(\nu) \cdot OTF_d(\nu, \lambda^*), \tag{8.22}$$

since the chromatic aberration reduces to diffraction.

Williams et al. [WBMN94] measured the OTF at wavelength $\lambda^* = 632.8$ nm and pupil diameter $p = 3$ mm using both an interference and a double-pass method. They derived the following estimate

$$OTF_b(\nu) = b + (1 - b)\ \exp(-\delta\, \nu), \tag{8.23}$$

Table 8.2. Parameter values for the SOE model of Artal and Navarro for the small-angle OTF of the human eye as a function of pupil diameter p. The model is based on double-pass measurements at $\lambda^* = 632$ nm and subjects with an average age of 28 years.

p (mm)	β_1^* (arcmin)	β_2^* (arcmin)	c_{sa}^*
2	3.356	9.462	0.406
3	2.960	9.697	0.369
4	2.937	11.180	0.332
5	3.109	13.749	0.295
6	3.428	17.613	0.258
7	3.888	23.208	0.221
8	4.501	31.216	0.184

for the wavelength-independent aberration, with $\delta = 7.272$ arcmin, $b = 0.3481$ for the interference method and $\delta = 8.238$ arcmin, $b = 0.1998$ for the double-pass method. The estimate from the interference method was adopted in the original paper by Marimont and Wandell [MW94].

An alternative estimate for $OTF_b(\nu)$ can be derived from the double-pass measurements of Artal and Navarro [AN94] at wavelength $\lambda^* = 632$ nm and a range of pupil diameters between 1 and 8 mm. The following analytical approximation to their experimental data, i.e.,

$$OTF_{sa}(\nu, \lambda^*) = c_{sa}^*(p)\ \exp[-\beta_1^*(p) \cdot \nu] + [1 - c_{sa}^*(p)]\ \exp[-\beta_2^*(p) \cdot \nu], \tag{8.24}$$

can be used to estimate $OTF_b(\nu) = OTF_{sa}(\nu, \lambda^*)/OTF_d(\nu, \lambda^*)$. The angles

$$\begin{aligned} \beta_1^*(p) &= \frac{1.69\ \exp(0.28\,p)}{\nu_o^*(p)} \\ \beta_2^*(p) &= \frac{3.53\ \exp(0.43\,p)}{\nu_o^*(p)} \end{aligned} \tag{8.25}$$

are modelled by exponential functions of the pupil diameter p. They are expressed in degrees if the cutoff frequency $\nu_o^*(p) = \nu_d(\lambda^*) = 26.45 \cdot p$ for the diffraction-limited system is expressed in c/deg. The relative weight

$$c_{sa}^*(p) = 0.48 - 0.037 \cdot p \tag{8.26}$$

is modelled by a linearly decreasing function of the pupil diameter p. The parameter values are listed in Table 8.2 as a function of pupil diameter p. In order to allow comparison with the model based on the data from

Table 8.3. Parameter values for the wavelength-independent aberration model of equation (8.23). These values have been derived from the SOE model of Artal and Navarro for the small-angle OTF of the human eye as a function of pupil diameter p.

p (mm)	δ (arcmin)	b
2	7.538	0.2653
3	8.088	0.2274
4	9.740	0.2073
5	12.412	0.1917
6	16.348	0.1759
7	22.003	0.1581
8	30.076	0.1375

Williams, we can continue to describe $OTF_b(\nu)$ by equation (8.23), but now with b and δ dependent on the pupil diameter p. By for instance equating the first two derivatives at frequency zero for both alternative descriptions, we can solve analytically for the parameters b and δ. The result is summarized in Table 8.3. Note that the parameter values for a pupil diameter of $p = 3$ mm agree reasonably well with those obtained by Williams et al.

The wavelength-dependent OTF model is rendered as a function of frequency and wavelength in Figure 8.3 for a pupil diameter of $p = 2$ mm. The expression for the wavelength-independent aberration $OTF_b(\nu)$ provided by Williams, with the parameter values obtained from his interference measurements, was used in the model. Figure 8.3 demonstrates the drastic reduction in spatial bandwidth at wavelengths below 500 nm that was already mentioned in chapter 1.

2.4 COLOR COORDINATE TRANSFORMATIONS

The wavelength-dependent description of the OTF can for instance be applied as follows in display analysis [MW94]. Suppose that the images of interest are created by an additive color mixture device with three primaries, i.e., all images can be expressed as

$$L_e(\boldsymbol{\rho}, \lambda) = \sum_{j=1}^{3} E_j(\boldsymbol{\rho}) \cdot P_j(\lambda), \qquad (8.27)$$

where $E_j(\boldsymbol{\rho})$ is the spatial image of the j-th primary and $P_j(\lambda)$ is the corresponding spectral power distribution. Spatial Fourier transformation

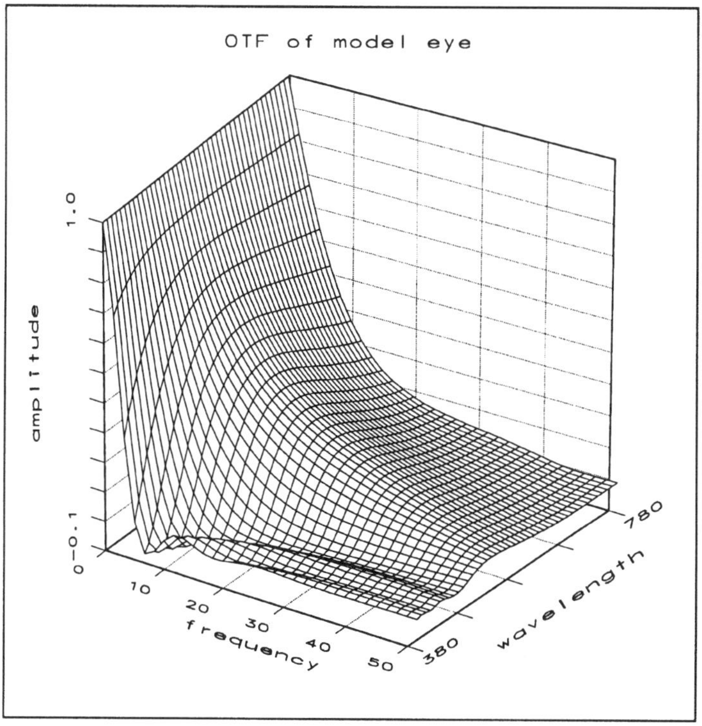

Figure 8.3. Optical transfer function of the model human eye as a function of spatial frequency (in c/deg) and wavelength (in nm) for a pupil diameter of $p = 2$ mm.

of this expression results in

$$\hat{L}_e(\boldsymbol{\nu}, \lambda) = \sum_{j=1}^{3} \hat{E}_j(\boldsymbol{\nu}) \cdot P_j(\lambda), \tag{8.28}$$

where $\hat{E}_j(\boldsymbol{\nu})$ denotes the Fourier transformation of $E_j(\boldsymbol{\rho})$. The cone absorptions, measured through the optics of the eye, depend on both the spatial and the wavelength variables through

$$R_i(\boldsymbol{\nu}, \lambda) = OTF(\boldsymbol{\nu}, \lambda) \cdot S_i(\lambda), \tag{8.29}$$

where $S_i(\lambda)$ describes the sensitivity of the i-th receptor family (i.e., L, M or S cones). The Fourier transformation $\hat{C}_i(\boldsymbol{\nu})$ of the spatial pattern $C_i(\boldsymbol{\rho})$ for the i-th receptor array arises by multiplying $R_i(\boldsymbol{\nu}, \lambda)$ by the Fourier transform $\hat{L}_e(\boldsymbol{\nu}, \lambda)$ of the image and integrating across wavelength. The result

$$\begin{bmatrix} \hat{C}_1(\nu) \\ \hat{C}_2(\nu) \\ \hat{C}_3(\nu) \end{bmatrix} = \begin{bmatrix} T_{11}(\nu) & T_{12}(\nu) & T_{13}(\nu) \\ T_{21}(\nu) & T_{22}(\nu) & T_{23}(\nu) \\ T_{31}(\nu) & T_{32}(\nu) & T_{33}(\nu) \end{bmatrix} \cdot \begin{bmatrix} \hat{E}_1(\nu) \\ \hat{E}_2(\nu) \\ \hat{E}_3(\nu) \end{bmatrix}, \tag{8.30}$$

with

$$T_{ij}(\nu) = \int S_i(\lambda)\, OTF(\nu, \lambda)\, P_j(\lambda)\, d\lambda, \tag{8.31}$$

is very similar to the matrix expressions that were obtained in section 4 for additive color mixture devices. The difference is that the vector elements have now been replaced by the Fourier transformations of the spatial patterns for the primaries (i.e, $\hat{E}_i(\nu)$, for $i = 1, 2, 3$) and the receptors (i.e, $\hat{C}_i(\nu)$, for $i = 1, 2, 3$), and that the matrix elements do not only depend on the spectral characteristics of the primaries and sensors, but also on the OTF of the eye. The result for frequency $\nu = 0$ is identical to that obtained earlier in section 6 of chapter 4, since $OTF(0, \lambda) = 1$.

The above derivation can also be applied to obtain an OTF model for white light conditions from the wavelength-dependent OTF. Indeed, if the spectral sensitivity of the sensor is $S(\lambda)$, then the small-angle OTF for a white light source with spectrum $W(\lambda)$ is equal to

$$OTF_{sa}(\nu) = \int S(\lambda)\, OTF(\nu, \lambda)\, W(\lambda)\, d\lambda. \tag{8.32}$$

The resulting OTF is also plotted in figure 8.1 for an equal energy light source $W(\lambda) = 1$, a receptor sensitivity $S(\lambda) = V(\lambda)$ equal to the photopic eye sensitivity curve, and a pupil diameter of $p = 2$ mm.

3. CONTRAST SENSITIVITY FUNCTION (CSF)

The obvious disadvantage of the OTF description in the previous section is that it only takes into account the spatial filtering introduced by the eye optics. The influence of any neural processing of spatial and temporal patterns in the rest of the brain is neglected. This is similar to what we encountered earlier in the study of color vision. The encoding performed by the wavelength sensitivities of the cones determines the mapping from physical spectra into fundamental color space. The neural mapping which is responsible for the interpretation of color coordinates (X, Y, Z values) as subjective color sensations (hue, saturation and brightness/lightness) is more difficult to study and understand. In order to obtain information on this mapping between linearly encoded color coordinates and subjective color sensations, experimental techniques such as detection, discrimination and scaling were used. We now discuss the results of detection and discrimination experiments that have been performed to obtain more insight into the neural mapping that is performed on spatially and/or temporally varying patterns.

3.1 DETECTION OF COUNTERPHASE GRATINGS

A large amount of psychophysical data is available on the detection of sinusoidal gratings. One pattern that is often used in such psychophysical experiments is the **counterphase grating** that can be expressed as follows

$$L(\boldsymbol{\rho}, t) = L_o\,[1 + m_c\,\cos(2\pi\,\boldsymbol{\nu}\cdot\boldsymbol{\rho})\cdot\cos(2\pi f\,t)]\,, \qquad (8.33)$$

in terms of the position $\boldsymbol{\rho}$ and time t[6]. The average luminance for the grating is L_o (expressed in cd/m^2), while f and $\boldsymbol{\nu}$ denote the **temporal frequency** (in cycles/second or Hertz, Hz) and **spatial frequency vector** of the grating, respectively (if $\boldsymbol{\nu} = (\nu, \phi)$ is expressed in polar coordinates, then spatial frequency ν is in c/deg, while orientation ϕ is in radians). We expect that this counterphase grating is visible if the **modulation depth** m_c exceeds a threshold value $m_{c,thr}$. This threshold will depend on the grating parameters L_o, $\boldsymbol{\nu}$ and f.

Detection thresholds are often measured using a **two-alternative-forced-choice** method (see also chapter 5). The observer is presented two stimuli: one containing the counterphase grating, the other containing a uniform field with the same average luminance L_o. The stimuli may be presented simultaneously or sequentially. The task of the subject is to select which one of the stimuli contains the grating. There is hence a 50 % chance of guessing correctly. The threshold is defined as the modulation depth for which the subject correctly selects the grating in a preset percentage[7] of the presentations.

---

[6]The inproduct between coordinate vectors is defined by $\boldsymbol{\nu}\cdot\boldsymbol{\rho} = \nu_x\,\rho_x + \nu_y\,\rho_y$ in case of cartesian representations $\boldsymbol{\rho} = (\rho_x, \rho_y)$ and $\boldsymbol{\nu} = (\nu_x, \nu_y)$, or equivalently, by $\boldsymbol{\nu}\cdot\boldsymbol{\rho} = \nu\rho\cos(\theta-\phi)$ in case of polar representations $\boldsymbol{\rho} = (\rho, \theta)$ and $\boldsymbol{\nu} = (\nu, \phi)$.

[7]In the up-down method [Lev71], the modulation depth is increased after every wrong answer, and decreased after n correct answers. At equilibrium, $p^n = 1 - p^n$, or $p = (1/2)^{1/n}$, where p is the probability of detection. For instance, $p = 0.707, 0.794, 0.841$ for $n = 2, 3, 4$.

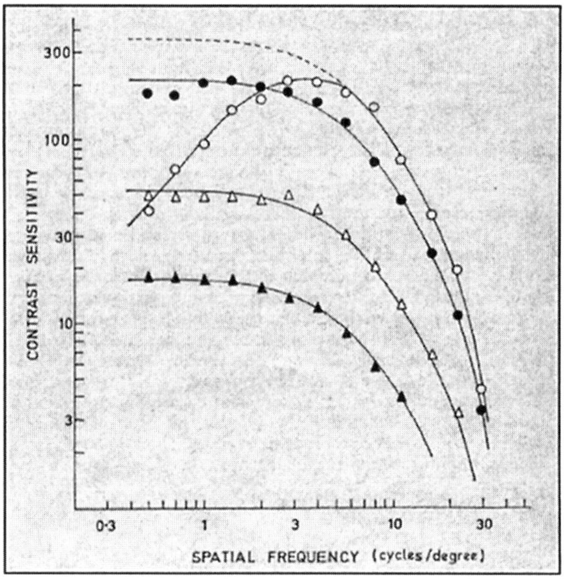

Figure 8.4. CSF of counterphase gratings as a function of spatial frequency (from [Rob66]). The curves for different temporal frequencies (one with a dashed low-frequency section) differ only in their position along the contrast-sensitivity scale (○ 1, • 6, △ 16 and △ 22 Hz). The contrast sensitivity decreases with the temporal frequency.

Instead of the threshold modulation depth $m_{c,thr}$, one usually plots the **contrast sensitivity function**[8] (CSF), defined as $CSF_c = 1/m_{c,thr}$. In Figure 8.4 [Rob66], experimentally determined CSF values[9] for vertical counterphase gratings are plotted as a function of the spatial frequency ν. The four curves correspond to different temporal frequencies f. The highest sensitivity ($CSF_{c,max} = 200$) is obtained for the lowest temporal frequency ($f = 1$ Hz) at a spatial frequency of approximately $\nu_{max} = 4$ c/deg. All spatial CSFs except the one for $f = 1$ Hz have a similar shape, and differ almost exclusively in their position along the log sensitivity axis.

In Figure 8.5 [Rob66], the contrast sensitivity for counterphase gratings is plotted as a function of the temporal frequency f. The four

[8]In case of the interference method of the previous section, where gratings with known modulation depth are created on the retina, the corresponding function is called the neural transfer function (NTF) [LNS93].

[9]The measurements of Robson were performed at an average luminance $L_o = 20$ cd/m^2, using a square display of size 2.5 degrees.

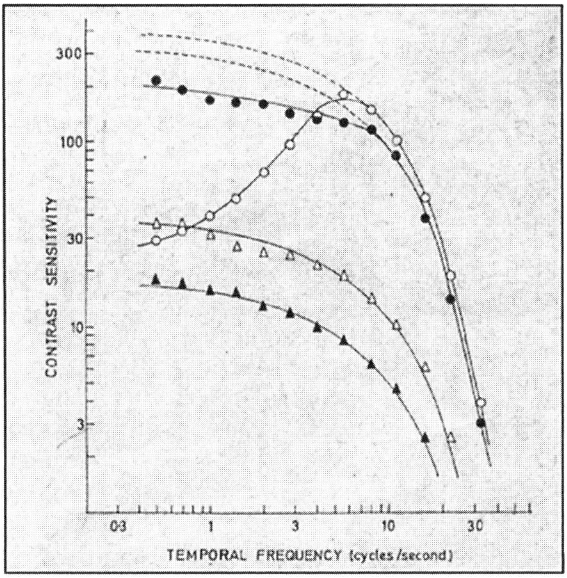

Figure 8.5. CSF of counterphase gratings as a function of temporal frequency (from [Rob66]). The curves for different spatial frequencies (two with dashed low-frequency sections) differ mostly in their position along the contrast-sensitivity scale (∘ 0.5, • 4, △ 16 and △ 22 cycles/degree). The CSF decreases with the spatial frequency.

curves correspond to different spatial frequencies ν. The highest sensitivity ($CSF_{c,max} = 200$) is obtained for the lowest spatial frequency ($\nu = 0.5$ c/deg) at a temporal frequency of approximately $f_{max} = 5$ Hz. The temporal CSFs for large spatial frequencies have a similar shape, and differ only in their position along the log sensitivity axis.

3.2 DETECTION OF DRIFTING GRATINGS

CSF measurements are influenced by the fact that the eyes are never completely stable. Although these involuntary eye movements are essential for normal vision, they create some uncertainty about the physical stimulus that is presented to the eye retina in an experiment. In order to avoid this uncertainty, CSF measurements have also been performed using images that were stabilized for eye movements. By moving the images on the display in synchrony with the subject's eye movements, a controlled stimulation of the retina can be realized [Kel79a]. In the experiments of Kelly [Kel79b], **drifting gratings** that can be mathematically expressed by

$$L(\rho, t) = L_o\,(1 + m_d \cos[2\pi\,(\boldsymbol{\nu}\cdot\boldsymbol{\rho} - f\,t)]) = L_o\left(1 + m_d \cos\left[2\pi\,\boldsymbol{\nu}\cdot\left(\boldsymbol{\rho} - \frac{\boldsymbol{\nu}}{\nu}\frac{f}{\nu}\,t\right)\right]\right) \quad (8.34)$$

are used to stimulate the eye retina, and detection thresholds $m_{d,thr}$ (or equivalently, sensitivities $CSF_d = 1/m_{d,thr}$) are measured[10].

The grating described in the previous equation is drifting in the direction $\boldsymbol{\nu}/\nu$ with velocity $v = f/\nu$ (in deg/s). The visual system is especially sensitive to such drifting gratings, in the sense that the detection of the previously introduced counterphase grating

$$L(\rho, t) = L_o\,[1 + m_c \cos(2\pi\,\boldsymbol{\nu}\cdot\boldsymbol{\rho})\cdot\cos(2\pi f\,t)] \quad (8.35)$$

$$= L_o\left(1 + m_c\,\frac{\cos[2\pi\,(\boldsymbol{\nu}\cdot\boldsymbol{\rho} - f\,t)] + \cos[2\pi\,(\boldsymbol{\nu}\cdot\boldsymbol{\rho} + f\,t)]}{2}\right)$$

is believed to be determined by the detection of the two constituting drifting gratings that drift in opposite directions. Experiments indeed confirm that

$$m_{c,thr} = 2\cdot m_{d,thr} \quad (8.36)$$

over a very broad range of conditions [Kel79b].

For velocities $v > 0.1$ deg/s), the spatial CSF for drifting gratings can be modelled by the function

$$CSF_d(\nu, v) = k(v)\,v\cdot\nu^2\,\exp\left[-2\,\frac{\nu}{\nu_{max}(v)}\right], \quad (8.37)$$

where the parameters depend on the velocity v. The spatial frequency

$$\nu_{max}(v) = \frac{45.9}{2\pi\,(v+2)} \quad (8.38)$$

at which the maximum sensitivity occurs is a monotonically decreasing function of the velocity v that is plotted in Figure 8.6. The scale factor

$$k(v) = (2\pi)^2\,\left[6.1 + 7.3\left|\log_{10}\left(\frac{v}{3}\right)\right|^3\right] \quad (8.39)$$

[10]The measurements of Kelly were performed at an average luminance $L_o = 70$ cd/m^2, using a circular display of 7.5 degrees in diameter.

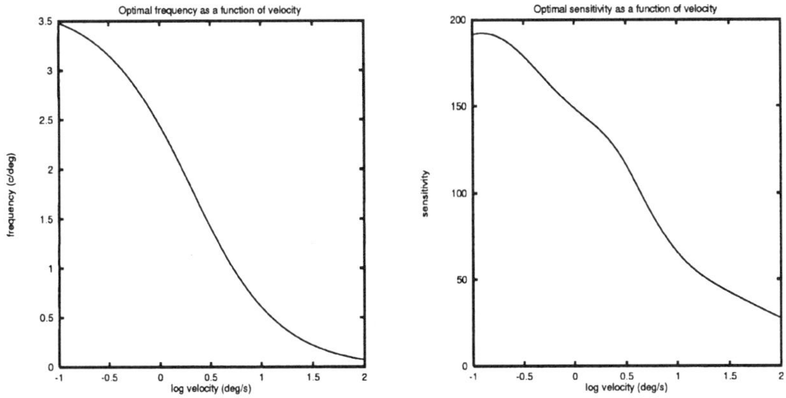

Figure 8.6. The spatial frequency ν_{max} at which the maximum sensitivity for a drifting grating occurs is plotted as a function of the logarithm of the grating velocity v in the left figure. The right figure plots the maximum sensitivity $CSF_{d,max}$ at this frequency ν_{max}.

determines the maximum contrast sensitivity

$$CSF_{d,max} = k(v)\, v \left[\frac{\nu_{max}(v)}{e}\right]^2, \tag{8.40}$$

which is also plotted as a function of velocity v in Figure 8.6. The spatial CSF at low velocity (i.e., $v = 0.15$ c/deg) approximates the spatial CSF under free-viewing conditions. However, the measurements of Kelly were restricted to relatively low spatial frequencies (i.e., $\nu < 10$ c/deg), so that it is dangerous to extrapolate the results to higher spatial frequencies. Instead, we advise to use the CSF model of the next section in case of static gratings.

The above spatial CSF as a function of velocity can be mapped to a spatio-temporal CSF for counterphase gratings, i.e.,

$$CSF_c(\nu, f) = \frac{1}{2}\, CSF_d\left(\nu, \frac{f}{\nu}\right), \tag{8.41}$$

since the contrast sensitivity for counterphase gratings is only halve of the contrast sensitivity for drifting gratings. It has been shown that this latter CSF_c, which is plotted in Figure 8.7, can be closely approximated by the linear difference $CSF_c(\nu, f) = S_e(\nu)\, T_e(f) - S_i(\nu)\, T_i(f)$ between two space-time separable filters (see [BK80, Kel83] and exercise 1). At high temporal and/or spatial frequencies, the contribution of the second term becomes negligible, so that this CSF model can be even

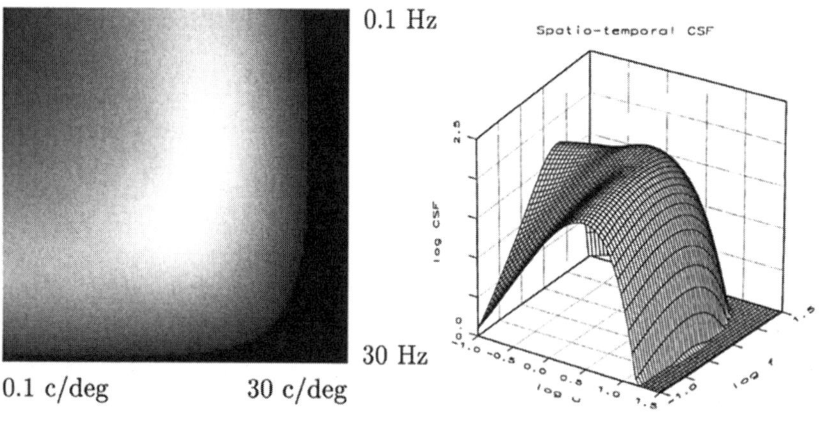

Figure 8.7. The spatio-temporal CSF for counterphase gratings is rendered as an image on the left, and as a mesh plot on the right. All axes are logarithmic.

further reduced to $CSF_c(\nu, f) \approx S_e(\nu)\, T_e(f)$ at these frequencies. This approximate spatio-temporal separability of the CSF for large spatial and temporal frequencies is exactly what is implied by the shift of the CSF curves in the Robson data.

3.3 DETECTION OF STATIC GRATINGS

Many alternative descriptions have been proposed for the CSF_s of static gratings in free-viewing conditions. Barten [Bar90] used experimental data from van Meeteren [Mee73] and van Nes [Nes68] to model this CSF as a function of the average luminance L_o, i.e.,

$$CSF_s(\nu; L_o) = a(L_o)\, \nu\, \exp[-b(L_o) \cdot \nu]\, \sqrt{1 + 0.06 \cdot \exp[b(L_o) \cdot \nu]}, \tag{8.42}$$

with parameter values

$$\begin{aligned} a(L_o) &= 540 \left(1 + \frac{0.7}{L_o}\right)^{-0.2} \\ b(L_o) &= 0.3 \left(1 + \frac{100}{L_o}\right)^{0.15} \end{aligned} \tag{8.43}$$

that depend on L_o.

Based on the experimental data from Carlson [Car82], the CSF description for static gratings was further extended to also include the

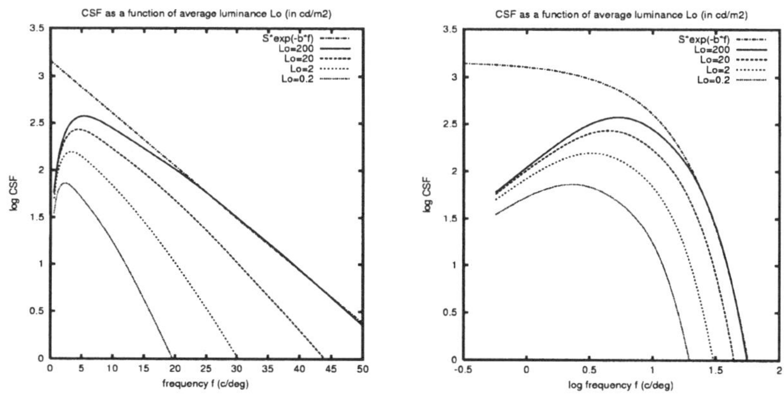

Figure 8.8. Relationships between the logarithm of the contrast sensitivity CSF_s and input frequency ν (left) or $\log \nu$ (right). The average luminance in cd/m^2 is $L_o = 0.2$, $L_o = 2$, $L_o = 20$ and $L_o = 200$ for the different curves, while the display height is 2.5 degrees in all cases. The excellent approximation of the CSF_s by an exponential function at large spatial frequencies is also illustrated for the case $L_o = 200$

effect of display size, i.e.,

$$CSF_s(\nu; L_o, h) = CSF_s(\nu; L_o) \cdot \left[1 + \frac{12}{h\,(1 + \nu/3)^2}\right]^{-1}, \tag{8.44}$$

where h is the height (or diameter) of the display in degrees of visual angle. The effect of average luminance L_o and display height h on the CSF_s are plotted in Figures 8.8 and 8.9, respectively.

The straight-line approximation to the CSF_s in Figure 8.8 illustrates that the contrast sensitivity is approximately an exponential function of spatial frequency, i.e.,

$$CSF_s(\nu) \approx S_o \cdot \exp[-\beta_o \cdot \nu], \tag{8.45}$$

at least for spatial frequencies ν above 20 c/deg. The required parameter values for the exponential approximation to the Barten CSF are provided in Table 8.4 for a number of average luminances L_o and display heights $h = 2.5$ and $h = 7.5$ degrees. Remember that the OTF of the human eye at large spatial frequencies was similarly described by an exponential function with power β_1. Since β_o is significantly larger than β_1 (see for instance Table 8.1), we must conclude that the reduction in sensitivity at large spatial frequencies is only partially caused by the OTF [LNS93].

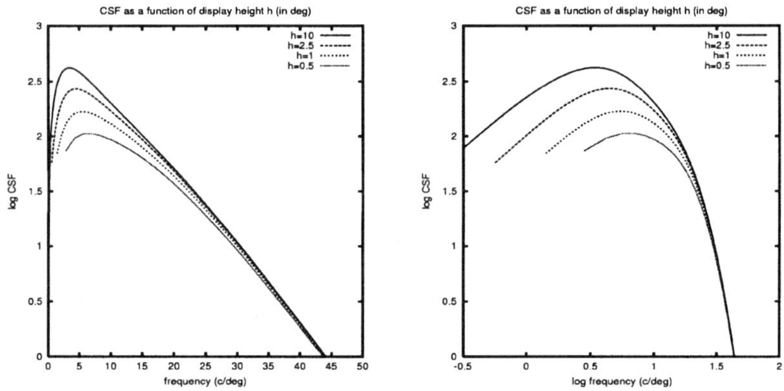

Figure 8.9. Relationships between the logarithm of the contrast sensitivity CSF_s and input frequency ν (left) or $\log \nu$ (right). The display height in degrees of visual angle is 10, 2.5, 1 and 0.5 for the different curves, while the average luminance is $L_o = 20$ cd/m^2 in all cases.

Table 8.4. Parameter values for the exponential approximation to the Barten CSF as a function of average luminance L_o and display height h.

	$h = 2.5$ degrees		$h = 7.5$ degrees	
L_o (cd/m^2)	S_o	β_o (arcmin)	S_o	β_o (arcmin)
200	1442	7.68	1543	7.75
150	1440	7.83	1541	7.91
100	1438	8.10	1538	8.17
50	1430	8.72	1441	8.67
20	1234	9.58	1336	9.68

Some of the reduced sensitivity at large spatial frequencies must hence be due to neural processing[11].

Figure 8.10 summarizes the results from an experiment by Campbell et al. [CKL66]. It illustrates that the CSF for static gratings also depends on the orientation of the pattern[12]. The highest sensitivity is obtained for horizontal and vertical gratings; the lowest sensitivity is obtained for gratings that are oriented in the oblique directions. This

[11]In case of the temporal CSF, it is obvious that the drop in sensitivity at high temporal frequencies has a neural origin.

[12]The measurements of Campbell et al. were performed at an average luminance L_o = 70 cd/m^2, using a circular display of size 7.5 degrees diameter.

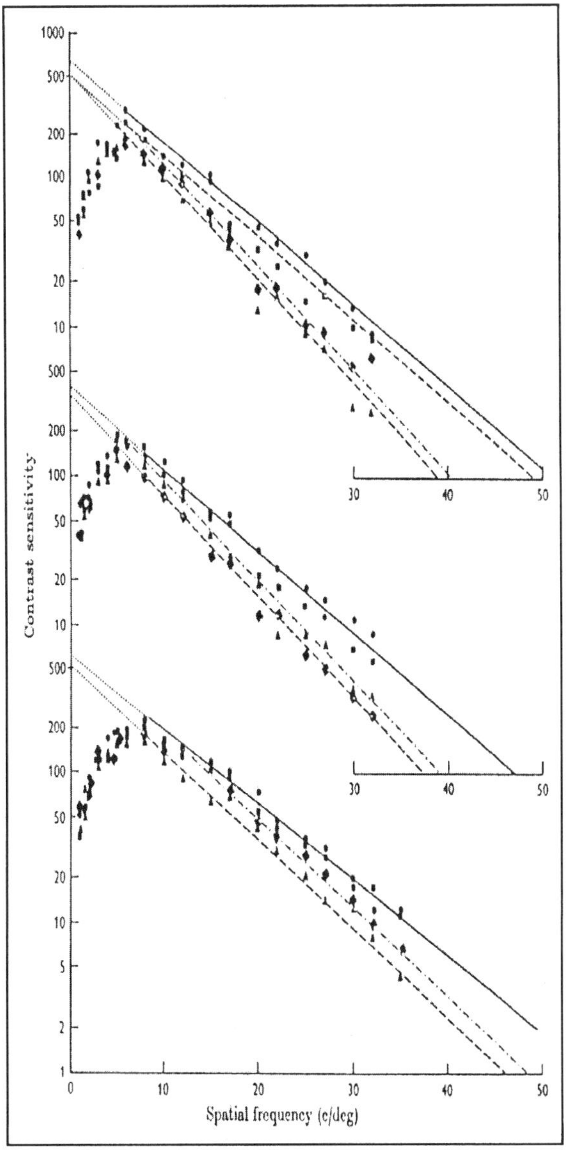

Figure 8.10. CSF of static gratings as a function of spatial frequency for three subjects (from [CKL66]). The contrast sensitivity is higher for the vertical and horizontal directions (upper curves) than for the oblique directions (45°, dashed curve, and 135°, dash-dotted curve).

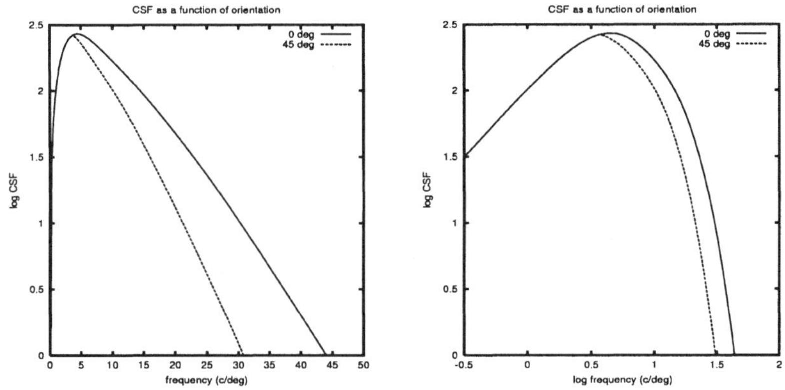

Figure 8.11. Relationships between the logarithm of the contrast sensitivity CSF_s and input frequency ν (left) or $\log \nu$ (right). The orientation is either 0 or 45 degrees, while the average luminance and display height are fixed at $L_o = 20$ cd/m^2 and $h = 2.5$ degrees of visual angle, respectively.

oblique effect has been modelled by Daly [Dal92, Dal93] through

$$CSF_s(\nu; L_o, h, \phi) = \min \left[CSF_s \left(\frac{\nu}{ob(\phi)}; L_o, h \right), CSF_s(\nu; L_o, h) \right], \tag{8.46}$$

where the oblique factor

$$ob(\phi) = \frac{1 + ob_f}{2} + \frac{1 - ob_f}{2} \cos(4\phi) \tag{8.47}$$

varies between $ob(0) = 1$ and $ob(\pi/4) = ob_f = 0.7$. The minimum function in the above expression guarantees that only the high-frequency part of the CSF is affected by the oblique factor. The straight-line approximations to the measurements in Figure 8.10 again illustrate that the contrast sensitivity is approximately an exponential function of spatial frequency[13]. Note that the above modification makes the CSF approximately exponential with exponent $\beta_o / ob(\phi)$ for orientation ϕ.

3.4 DETECTION OF COLORED GRATINGS

The previous sections have demonstrated that there are many sources of information for luminance CSFs. Most of these CSF measurements

[13] A rough estimate, based on the straight lines in Figure 8.10, leads to $\beta_o \approx 7$ arcmin for horizontal and vertical gratings, and $\beta_o \approx 9$ arcmin for oblique gratings.

have been performed on monochrome displays, so that the hue and chroma of the stimuli stayed fixed while the luminance was modulated sinusoidally. Many applications in the processing and coding of color images however also require information about the spatiotemporal processing in the chromatic channels. This information is supplied by measurements with isoluminant gratings, in which case the color is modulated (sinusoidally) along a line that is parallel to the chromaticity plane. Such experiments not only require much more expensive equipment (i.e., a color display with three high-precision D/A convertors), but are also complicated by the inter-subject variations in color perception [Mul90]. Some of the earliest color CSF measurements were performed by van der Horst and Bouman [HB69] and Granger and Heurtley [GH73]. They demonstrated that chrominance CSF's have a low-pass characteristic and that larger modulation depths are required for the detection of isoluminant gratings than for the detection of luminance-modulated gratings. These findings were confirmed in later measurements by Kelly [Kel83] and Mullen [Mul85].

Nadenau [Nad00] has argued that engineering applications require CSFs with the following characteristics:

1 the luminance and chrominance measurements should be performed under identical circumstances;

2 the display characteristics (amongst others, average luminance, display size and viewing distance) should be representative for a typical (i.e., TV and/or desk-top) viewing condition;

3 the color modulations should be performed along axes in fundamental color space that correspond to standardized color coordinate systems;

4 mathematical functions describing the CSFs, rather then plotted characteristics, are needed for most applications.

He concluded that CSF measurements with all the above characteristics were not available and therefore decided to conduct a new set of experiments. More specifically, he performed nine sets of threshold measurements for sinusoidal color modulations in fundamental color space along the principal axes of the following coordinate systems: the (ℓ, a, b) coordinate system corresponding to the Hering primaries (see section 5 of chapter 3), the (E_Y, E_B, E_R) coordinate system of ITU-R recommendation BT.709 (see section 4.2 of chapter 4) and the opponent color coordinate system (O_Y, O_R, O_B) of S-CIELAB (see next section). The horizontal gratings (with 1, 2, 3, 4, 6, 7, 8 and 10 c/deg) that he used in his experiments were contained within a circular area with a diameter of

Table 8.5. Parameter values of the Nadenau models in Equation (8.48) for the CSFs along the principal axes of the (ℓ, a, b) color coordinate system. The parameter values for the exponential approximations at large spatial frequencies are also included.

	Nadenau model			exponential	
	weight	size (arcmin)	exponent	weight	size (arcmin)
ℓ	$a_{11} = 58.7$	$b_{11} = 53.65$	$c_{11} = 1.995$	$S_1 = 1388$	$\beta_1 = 15.38$
	$a_{12} = 265.9$	$b_{12} = 57.64$	$c_{12} = 0.760$		
a	$a_2 = 38.00$	$b_2 = 10.08$	$c_2 = 0.873$	$S_2 = 25.73$	$\beta_2 = 7.33$
b	$a_3 = 59.24$	$b_3 = 10.34$	$c_3 = 0.905$	$S_3 = 43.36$	$\beta_3 = 8.14$

Table 8.6. Parameter values of the Nadenau models in Equation (8.48) for the CSFs along the principal axes of the (E_Y, E_B, E_R) color coordinate system. The parameter values for the exponential approximations at large spatial frequencies are also included.

	Nadenau model			exponential	
	weight	size (arcmin)	exponent	weight	size (arcmin)
E_Y	$a_{11} = 59.1$	$b_{11} = 53.67$	$c_{11} = 1.999$	$S_1 = 1382$	$\beta_1 = 15.21$
	$a_{12} = 266.7$	$b_{12} = 57.64$	$c_{12} = 0.758$		
E_R	$a_2 = 45.44$	$b_2 = 7.283$	$c_2 = 0.893$	$S_2 = 35.29$	$\beta_2 = 5.77$
E_B	$a_3 = 28.92$	$b_3 = 10.264$	$c_3 = 0.900$	$S_3 = 20.92$	$\beta_3 = 7.99$

Table 8.7. Parameter values of the Nadenau models in Equation (8.48) for the CSFs along the principal axes of the (O_Y, O_R, O_B) color coordinate system of S-CIELAB. The parameter values for the exponential approximations at large spatial frequencies are also included.

	Nadenau model			exponential	
filter	weight	size (arcmin)	exponent	weight	size (arcmin)
O_Y	$a_{11} = 40.9$	$b_{11} = 53.67$	$c_{11} = 2.000$	$S_1 = 1194$	$\beta_1 = 16.36$
	$a_{12} = 189.4$	$b_{12} = 54.15$	$c_{12} = 0.781$		
O_R	$a_2 = 35.89$	$b_2 = 7.206$	$c_2 = 1.094$	$S_2 = 45.32$	$\beta_2 = 7.49$
O_B	$a_3 = 35.76$	$b_3 = 7.337$	$c_3 = 1.100$	$S_3 = 48.19$	$\beta_3 = 9.04$

11 degrees of visual angle and were viewed from a distance of 0.72 meters. The average (grey) background had luminance $L_o = 50$ cd/m^2 and color coordinates $(x, y) = (0.28, 0.30)$.

Nadenau modelled his experimental data by the following exponential functions

$$\begin{aligned} D_1(\nu) &= a_{11}\,\exp\left[-(b_{11}\cdot\nu)^{c_{11}}\right] + a_{12}\,\nu^2\exp\left[-(b_{12}\cdot\nu)^{c_{12}}\right] \\ D_2(\nu) &= a_2\,\exp\left[-(b_2\cdot\nu)^{c_2}\right] \\ D_3(\nu) &= a_3\,\exp\left[-(b_3\cdot\nu)^{c_3}\right] \end{aligned} \tag{8.48}$$

for the luminance $(D_1 : \ell, E_Y, O_Y)$, red-green $(D_2 : a, E_R, O_R)$ and blue-yellow $(D_3 : b, E_B, O_B)$ channels, respectively. The parameter values for the (ℓ, a, b), (E_Y, E_B, E_R) and (O_Y, O_R, O_B) coordinate systems are listed in Tables 8.5, 8.6 and 8.7, respectively. In order to enable a comparison with the luminance CSFs presented above, we have also included the parameter values for exponential approximations $S_i \cdot \exp(-\beta_i \cdot \nu)$ to the high-frequency parts of these characteristics. The Nadenau models for the CSFs along the principal axes of the color coordinate systems (ℓ, a, b), (E_Y, E_B, E_R) and (O_Y, O_R, O_B) are plotted in Figure 8.12.

Note that the variations along the ℓ-axis and the E_Y-axis are both luminance variations for a neutral grey light source and therefore should give almost identical results, as indeed they do. The fact that the high-frequency part of the model of Nadenau for the luminance CSF ($S_1 = 1382$, $\beta_1 = 15.21$ arcmin) deviates from the approximation to the Barten model (see Table 8.4) is probably due to the fact that Nadenau's measurements were restricted to frequencies $\nu < 10$ c/deg. The different CSF characteristics for variations along the a-axis and the E_R-axis are however puzzling. According to Figure 4.12, both axes almost coincide in fundamental color space, so that the CSF characteristics should also be approximately equal, which is obviously not the case. Especially the slope of $\beta_2 = 5.77$ for the E_R characteristic seems unrealistically small. Since the b-axis and the E_B-axis do not coincide in fundamental color space, different CSFs are allowed for variations along these axes.

4. CIELAB EXTENSIONS

4.1 ASYMMETRIC COLOR MATCHING

The CIE(L^*, a^*, b^*) uniform color space that was introduced in chapter 3 is a very useful tool for assessing the accuracy of color reproductions. Because of the way it is derived, the color difference formula in CIE(L^*, a^*, b^*) is only expected to correlate well with perceived color differences if the two color patches being compared are relatively large (more than 1 degree of visual angle) and presented on the same uniform background. The CIE(L^*, a^*, b^*) metric can hence not be applied in a straightforward way for assessing the color accuracy in spatio-temporally varying patterns such as images. Rather then discarding the metric alto-

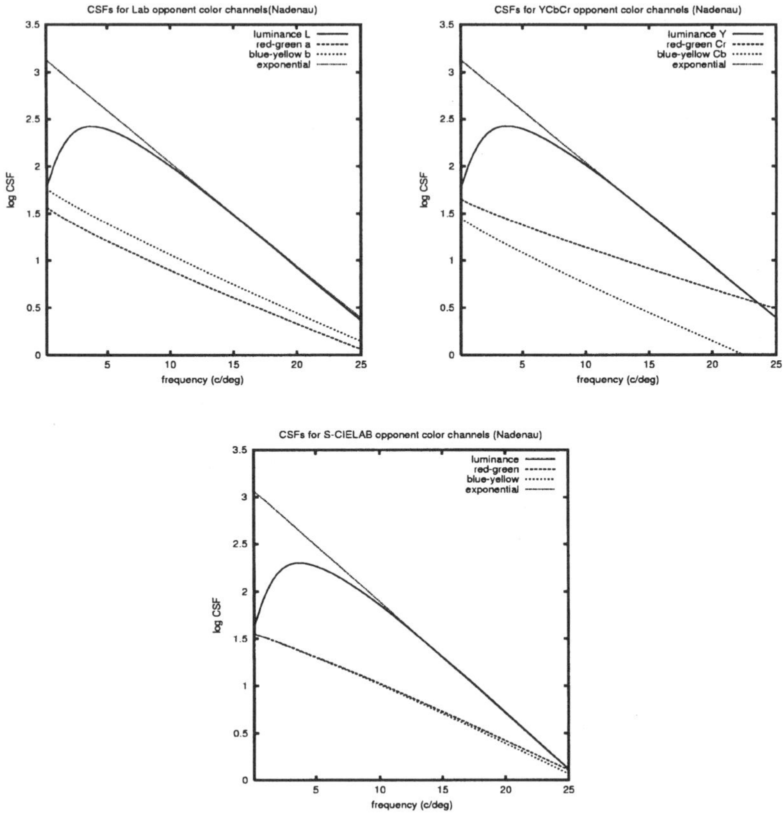

Figure 8.12. Relationships between the logarithm of the contrast sensitivity CSF and input frequency ν for sinusoidal variations along the principal axes of the (ℓ, a, b) (upper left), (E_Y, E_B, E_R) (upper right) and (O_Y, O_R, O_B) (lower middle) color coordinate systems. The straight lines depict the exponential approximations to the luminance CSFs at large spatial frequencies.

gether, it is worthwhile to consider if it cannot be modified to also obtain a better agreement with perceived color differences for more complex patterns.

Poirson and Wandell [PW93, PW96] used **asymmetric color matching** experiments to study the effect of spatial size on perceived color. Asymmetric color matching differs from the classical color matching described in chapter 1 by the fact that the color patches that have to be compared are allowed to have different shapes and/or sizes. More specifically, Poirson and Wandell used stimuli such as the one shown in

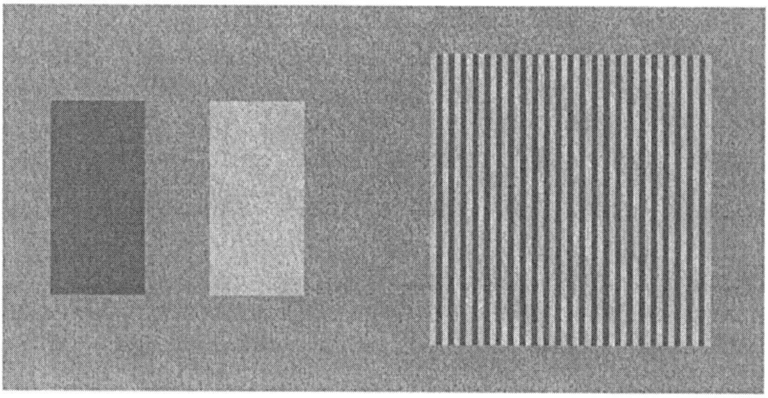

Figure 8.13. Stimulus used in asymmetric color matching. The task of the subject is to match the colors in the uniform rectangles to the colors of the stripes in the square-wave pattern.

Figure 8.13. The task of the subject is to match the colors of the stripes in the square-wave pattern on the right to the colors of the uniform rectangles on the left. The spatial frequency ν and the contrast of the square-wave stimulus are varied within the experiment, and the subjects have to adjust the colors of the uniform patches. Contrast is defined by a vector $\mathbf{m} = [\Delta C_L/C_L, \Delta C_M/C_M, \Delta C_S/C_S]^T$, where $[C_L, C_M, C_S]^T$ and $[C_L + \Delta C_L, C_M + \Delta C_M, C_S + \Delta C_S]^T$ denote the cone-receptor responses for the uniform background and the color patch of interest, respectively. The stripes in the square-wave pattern with frequency vector $\boldsymbol{\nu}$ have contrast $\mathbf{m}(\boldsymbol{\nu})$ and $-\mathbf{m}(\boldsymbol{\nu})$, or equivalently, the modulation depth of the square-wave pattern is $\mathbf{m}(\boldsymbol{\nu})$. Since the background is fixed throughout the experiment, the cone-receptor responses $[C_L, C_M, C_S]^T$ are constant. There is hence a one-to-one relationship between the contrast vector $\mathbf{m} = [\Delta C_L/C_L, \Delta C_M/C_M, \Delta C_S/C_S]^T$ and the response vector $\mathbf{C} = [\Delta C_L, \Delta C_M, \Delta C_S]^T$, i.e., a match in contrast also signifies a match in cone responses.

If the spatial frequency vector $\boldsymbol{\nu}$ of the square-wave pattern and the color of the background are fixed, then the asymmetric color matching can be described by a linear front-end system (LFS). This conclusion can be reached, based on the following experimental findings [PW93]:

1 If the contrast $\mathbf{m}(\boldsymbol{\nu})$ of a stripe in the square-wave pattern with frequency ν matches the contrast $\mathbf{m}_o$ of a uniform patch, then a stripe with contrast $\alpha\, \mathbf{m}(\boldsymbol{\nu})$ matches a uniform patch with contrast $\alpha\, \mathbf{m}_o$.

2 If the contrasts $\mathbf{m}(\boldsymbol{\nu})$ and $\mathbf{m}'(\boldsymbol{\nu})$ of the stripes match the contrasts $\mathbf{m}_o$ and $\mathbf{m}'_o$ of the uniform patches, respectively, then the stripe with contrast $\mathbf{m}(\boldsymbol{\nu})+\mathbf{m}'(\boldsymbol{\nu})$ matches the uniform patch with contrast $\mathbf{m}_o+\mathbf{m}'_o$.

The LFS must hence have a structure as shown in Figure 8.14. This figure is a modified version of Figure 1.13, in which the effect of spatial frequency $\boldsymbol{\nu}$ has been incorporated in the linear transformation $T(\nu)$. Color matching implies that

$$T(\nu)\, T_o \cdot \mathbf{m}(\boldsymbol{\nu}) = T_o \cdot \mathbf{m}_o. \tag{8.49}$$

The transformation matrix $T(\nu)$ depends on the frequency ν of the square-wave pattern[14] and is by definition equal to the identity matrix for uniform patches, i.e., $T(0) = I$. The matrix T_o is the known transformation matrix from cone coordinates to (ℓ, a, b) coordinates, i.e.,

$$T_o = \begin{bmatrix} 0 & \frac{1}{Y_n} & 0 \\ \frac{1}{X_n} & -\frac{1}{Y_n} & 0 \\ 0 & \frac{1}{Y_n} & -\frac{1}{Z_n} \end{bmatrix} \cdot C_{XYZ}^{-1}, \tag{8.50}$$

where (X_n, Y_n, Z_n) are the tristimulus coordinates of the uniform background and C_{XYZ}^{-1} is the matrix that maps the cone receptor curves $(\phi_L(\lambda), \phi_M(\lambda), \phi_S(\lambda))$ into the CIE tristimulus curves $(\overline{x}(\lambda), \overline{y}(\lambda), \overline{z}(\lambda))$ [SP75] (see also exercise 2 of chapter 3).

4.2 S-CIELAB

There is no a priori reason why the transformation matrices $T(\nu)$ for different spatial frequencies ν should be related. Poirson and Wandell however considered one such possible relation between the transformation matrices, i.e., that the effects of wavelength and spatial frequency can be separated. A mathematically equivalent condition for this so-called **color-pattern separability** is that the transformation matrices can be expressed as

$$T(\nu) = U\, S(\nu)\, V, \tag{8.51}$$

where U and V are 3×3 matrices of singular vectors that are independent of the spatial frequency ν, and $S(\nu)$ is a diagonal matrix of singular

[14] Only square-wave gratings in the horizontal direction were used in the experiment, but it is assumed that the system is approximately circularly symmetric, so that the color matching is the same for all spatial frequency vectors $\boldsymbol{\nu}$ with the same frequency amplitude $\nu = |\boldsymbol{\nu}|$.

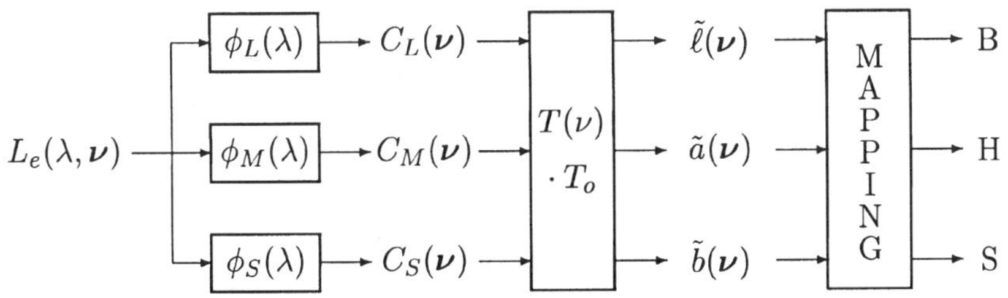

Figure 8.14. Three linear measurement functions, denoted by $\phi_L(\lambda)$, $\phi_M(\lambda)$ and $\phi_S(\lambda)$, representing the spectral sensitivity functions of the long, medium and short-wavelength cones, are applied to the square-wave pattern $L_e(\lambda, \boldsymbol{\nu})$ with frequency vector $\boldsymbol{\nu}$ to obtain three cone contrasts $\mathbf{m}(\boldsymbol{\nu}) = [C_L(\boldsymbol{\nu}), C_M(\boldsymbol{\nu}), C_S(\boldsymbol{\nu})]^T$ that encode the color information. For a (square-wave) pattern with spatial frequency ν these cone contrasts are mapped linearly by $T(\nu) \cdot T_o$ into the color coordinates $[\tilde{\ell}(\boldsymbol{\nu}), \tilde{a}(\boldsymbol{\nu}), \tilde{b}(\boldsymbol{\nu})]^T$. These color coordinates give rise to subjective brightness (B), hue (H) and saturation (S) by means of a mapping that is independent of the frequency vector $\boldsymbol{\nu}$.

values that captures the effect of spatial frequency on the transformation matrix $T(\nu)$. It is hence assumed that all matrices $T(\nu)$ share the same singular vectors. The physical interpretation of color-pattern separability is that the cone contrasts, summarized in the column vector $\mathbf{m}(\boldsymbol{\nu})$ are mapped to contrasts $V\,T_o \cdot \mathbf{m}(\boldsymbol{\nu}) = [O_Y(\boldsymbol{\nu}), O_R(\boldsymbol{\nu}), O_B(\boldsymbol{\nu})]^T$ in an intermediate coordinate system that is different from the (ℓ, a, b) coordinate system. The unique feature of this intermediate color coordinate system is that the spatial-frequency-dependent part of the mapping $S(\nu) V T_o \cdot \mathbf{m}(\boldsymbol{\nu})$ can be carried out independently for all three coordinates $O_Y(\boldsymbol{\nu})$, $O_R(\boldsymbol{\nu})$ and $O_B(\boldsymbol{\nu})$. The processed coordinates are subsequently mapped by the matrix U to the color coordinates $[\tilde{\ell}(\boldsymbol{\nu}), \tilde{a}(\boldsymbol{\nu}), \tilde{b}(\boldsymbol{\nu})]^T$ that determine the color perception. For uniform patches with frequency vector $\boldsymbol{\nu} = \mathbf{0}$ these latter coordinates $[\tilde{\ell}(\mathbf{0}), \tilde{a}(\mathbf{0}), \tilde{b}(\mathbf{0})]^T$ are equal to the $[\ell, a, b]^T$ coordinates that are used in chapter 3 to define the CIELAB uniform color space. It is assumed that the CIE(L^*, a^*, b^*) uniform color space transformations can also be applied to the coordinates $[\tilde{\ell}(\boldsymbol{\nu}), \tilde{a}(\boldsymbol{\nu}), \tilde{b}(\boldsymbol{\nu})]^T$ in order to derive psychometric lightness, hue and chroma in case $\nu \neq 0$. The overall transformation is referred to as the **spatial color fidelity metric S-CIELAB** and is depicted graphically in Figure 8.15.

The matrices $T(\nu)$ can be determined experimentally, based on Equation (8.49). Provided that $K \geq 3$ different color gratings with the same frequency $|\boldsymbol{\nu}| = \nu$ have been matched, these matches can be assembled

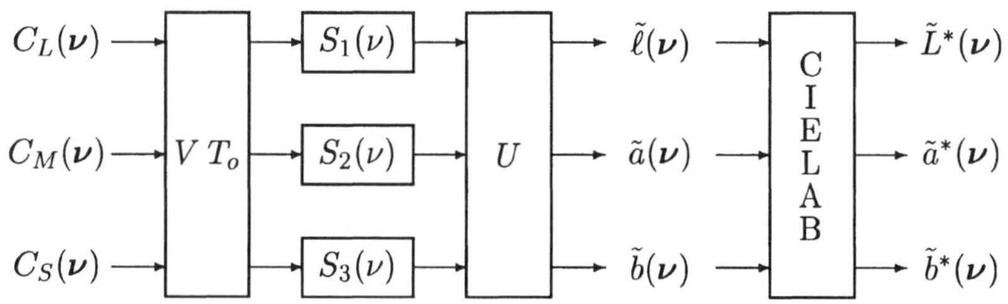

Figure 8.15. The separable linear mapping from cone coordinates $\mathbf{m}(\boldsymbol{\nu}) = [C_L(\boldsymbol{\nu}), C_M(\boldsymbol{\nu}), C_S(\boldsymbol{\nu})]^T$ to color coordinates $[\tilde{\ell}(\boldsymbol{\nu}), \tilde{a}(\boldsymbol{\nu}), \tilde{b}(\boldsymbol{\nu})]^T$, followed by the non-linear CIE(L^*, a^*, b^*) mapping to uniform color space defines the overall mapping to spatial-frequency dependent S-CIELAB coordinates $[\tilde{L}^*(\boldsymbol{\nu}), \tilde{a}^*(\boldsymbol{\nu}), \tilde{b}^*(\boldsymbol{\nu})]^T$.

in an equation that resembles Equation (8.49), i.e.,

$$T(\nu)\,T_o \cdot M_\nu = T_o \cdot M_o, \tag{8.52}$$

where M_ν and M_o are $3 \times K$ matrices of matching contrasts. The projection theorem of chapter 2 supplies us with the required solution for the transformation matrix

$$T(\nu) = T_o\,M_o\,M_\nu^T\,(M_\nu\,M_\nu^T)^{-1}\,T_o^{-1} \tag{8.53}$$

at spatial frequency ν. Poirson and Wandell found that the experimentally obtained transformation matrices $T(\nu)$ for different spatial frequencies ν confirm the assumption of color-pattern separability. They estimated the transformation matrix VT_o per subject from their experimental data, and proposed an average matrix $V\,T_o$ for use in the S-CIELAB metric [ZW98]. In S-CIELAB, this transformation from the cone coordinates (C_L, C_M, C_S) to the intermediate color coordinates (O_Y, O_R, O_B) used for spatial processing is equal to

$$\begin{bmatrix} O_Y \\ O_R \\ O_B \end{bmatrix} = \begin{bmatrix} 0.990 & -0.106 & -0.094 \\ -0.669 & 0.742 & -0.027 \\ -0.212 & -0.354 & 0.911 \end{bmatrix} \cdot \begin{bmatrix} C_L \\ C_M \\ C_S \end{bmatrix}. \tag{8.54}$$

Alternatively, the intermediate color coordinates (O_Y, O_R, O_B) can also be related to the 1951 CIE tristimulus values (X_{51}, Y_{51}, Z_{51}), i.e.,

$$\begin{bmatrix} O_Y \\ O_R \\ O_B \end{bmatrix} = \begin{bmatrix} 0.2829 & 0.7202 & -0.1185 \\ -0.4564 & 0.2943 & 0.0799 \\ 0.0884 & -0.5931 & 0.5484 \end{bmatrix} \cdot \begin{bmatrix} X_{51} \\ Y_{51} \\ Z_{51} \end{bmatrix}. \tag{8.55}$$

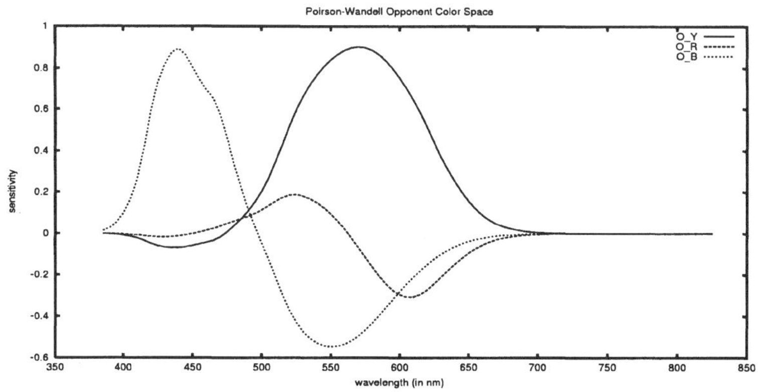

Figure 8.16. Spectral sensitivity curves for Poirson-Wandell opponent color space.

Table 8.8. Parameter values for the sum-of-Gaussian approximations to the S-CIELAB spatial filters.

filter	weight	size (arcmin)
D_1	$w_{11} = 0.921$	$s_{11} = 5.33$
	$w_{12} = 0.105$	$s_{12} = 25.07$
	$w_{13} = -0.108$	$s_{13} = 817.32$
D_2	$w_{21} = 0.531$	$s_{11} = 7.39$
	$w_{22} = 0.330$	$s_{12} = 93.12$
D_3	$w_{31} = 0.488$	$s_{31} = 10.10$
	$w_{32} = 0.371$	$s_{32} = 72.76$

The latter matrix can also be used to express the spectral sensitivity curves of the intermediate color coordinates in terms of the 1951 CIE tristimulus curves $(\overline{x}_{51}(\lambda), \overline{y}_{51}(\lambda), \overline{z}_{51}(\lambda))$. The resulting curves are plotted in Figure 8.16.

In the S-CIELAB proposal [ZW98], the spatial frequency characteristics of the three intermediate color channels are approximated by SOG characteristics, i.e.,

$$\begin{aligned} S_Y(\nu) &= w_{11}\,\exp[-(s_{11}\cdot\nu)^2] + w_{12}\,\exp[-(s_{12}\cdot\nu)^2] \\ &+ w_{13}\,\exp[-(s_{13}\cdot\nu)^2] \\ S_R(\nu) &= w_{21}\,\exp[-(s_{21}\cdot\nu)^2] + w_{22}\,\exp[-(s_{22}\cdot\nu)^2] \\ S_B(\nu) &= w_{31}\,\exp[-(s_{31}\cdot\nu)^2] + w_{32}\,\exp[-(s_{32}\cdot\nu)^2] \qquad (8.56) \end{aligned}$$

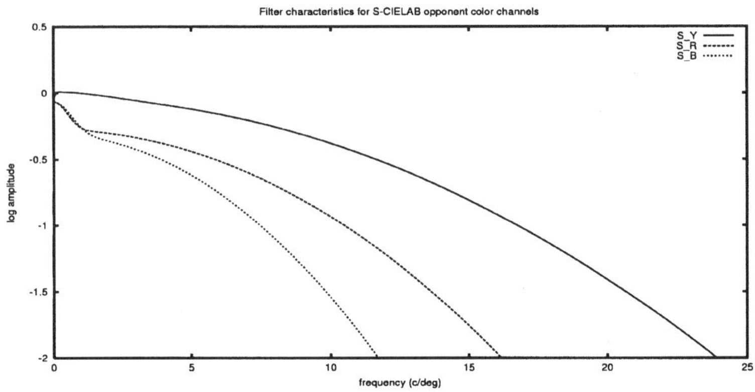

Figure 8.17. The logarithm of the S-CIELAB spatial filter characteristics as a function of the input frequency ν (in c/deg).

where the parameter values are listed in Table 8.8. The third term in $S_Y(\nu)$ is required to model the dip in sensitivity at low spatial frequencies. For many applications in image quality, where the images that have to be compared differ only in detail, the influence of this latter term to the S-CIELAB color difference can often be neglected. The resulting spatial filter characteristics are plotted in Figure 8.17. Note that these filter characteristics, which were derived from matching experiments, differ greatly from the filter characteristics in Figure 8.12 proposed by Nadenau, based on detection experiments. Further tests will be needed to establish which set of filters is most useful for applications.

The S-CIELAB mapping can for instance be used to define the dissimilarity between two processed or coded versions of an image. After mapping both images, a pointwise CIELAB difference ΔE^*_{ab} (see chapter 3) can be calculated. A summary statistic such as the mean CIELAB difference, can then be used as the dissimilarity measure between the images. How such dissimilarity measures can be used for quality prediction will be discussed in more detail in chapter 11.

The S-CIELAB mapping has also been extended to the spatiotemporal domain [THBL99], in which case spatiotemporal filters, based on the CSF measurements of Kelly (see also exercise 1) are applied to the (O_Y, O_R, O_B) signals of the Poirson and Wandell opponent color space. The resulting modified CIELAB metric is referred to as ST-CIELAB.

EXERCISES

1 Show that the spatio-temporal CSF for counterphase gratings derived from the experiments of Kelly, see equation (8.41), can be approximated by the linear difference

$$CSF_c(\nu, f) = K(\nu, f) = w_e\, S_e(\nu)\, T_e(f) - w_i\, S_i(\nu)\, T_i(f)$$

between two space-time separable filters that are called the excitatory and inhibitory mechanism [BK80, THBL99]. The derivation is based on two assumptions. First, the inhibitory mechanism is assumed to be zero for spatial frequencies above $\nu_1 = 10$ c/deg and/or temporal frequencies above $f_1 = 19$ Hz. Second, the shape of the CSF at large spatial (or temporal) frequencies (i.e., $\nu \geq \nu_1$ and/or $f \geq f_1$) does not change with temporal (or spatial) frequency, as can also be observed in the experimental data of Robson. These CSF shapes as a function of spatial and temporal frequency can hence be determined by sampling the CSF at a single low temporal frequency $f_2 = 1$ Hz and at a single low spatial frequency $\nu_2 = 0.5$ c/deg, respectively. This assumption leads to the following expressions for the spatial filter

$$S_e(\nu) = \begin{cases} \tilde{S}_e(\nu) & \nu \leq \nu_1 \\ K(\nu, f_2) & \nu \geq \nu_1 \end{cases}$$

and the temporal filter

$$T_e(f) = \begin{cases} \tilde{T}_e(f) & f \leq f_1 \\ K(\nu_2, f) & f \geq f_1 \end{cases}$$

in the excitatory mechanism. Derive the following expressions for the unknown filters

$$\tilde{S}_e(\nu) = K(\nu, f_1)\, \frac{K(\nu_1, f_2)}{K(\nu_1, f_1)},\ \tilde{T}_e(f) = K(\nu_1, f)\, \frac{K(\nu_2, f_1)}{K(\nu_1, f_1)}$$

by verifying the CSF at spatial frequency ν_1 (or temporal frequency f_1) as a function of temporal frequency f (or spatial frequency ν). Verify the CSF at spatial frequency ν_2 and temporal frequency f_2 in order to derive that

$$\tilde{S}_e(\nu_2) = \tilde{T}_e(f_2) = \frac{1}{w_e} = \frac{K(\nu_1, f_2) \cdot K(\nu_2, f_1)}{K(\nu_1, f_1)},$$

and to show that the filters of the inhibitory mechanism can be expressed as $S_i(\nu) = \tilde{S}_e(\nu) - K(\nu, f_2)$ and $T_i(f) = \tilde{T}_e(f) - K(\nu_2, f)$, respectively. Determine the weighting factor w_i for the inhibitory mechanism.

Kelly [Kel83] also used the same filters to model the CSF for achromatic gratings. He observed that adding, rather than subtracting the filters, i.e.,

$$CSF_{cc}(\nu, f) = \frac{1}{w_c} \left[w_e \, S_e(\nu) \, T_e(f) + w_i \, S_i(\nu) \, T_i(f) \right],$$

provided a good fit to his experimental data for isoluminant red-green gratings, provided that a factor $w_c = 30$ was included to describe the lower sensitivity for isoluminant gratings. Both CSF_c and CSF_{cc} are used in the ST-CIELAB color difference metric [THBL99].

2 Let us model the detection behavior of the visual system by a linear space and/or time-invariant filter with impulse response $h(\mathbf{x})$, followed by a peak detector, i.e., we assume that a signal is detected if the maximum value at the output of the filter exceeds a fixed threshold value. Suppose that the filter is such that the output to a constant input signal is zero. We can experimentally determine the detection thresholds ΔL_1 and ΔL_2 for input signals with one impulse, $L_o + \Delta L_1 \cdot \delta(x)$, and with two impulses, $L_o + \Delta L_2 \cdot [\delta(\mathbf{x}) + \epsilon \cdot \delta(\mathbf{x} - \mathbf{y})]$. If the ratio ϵ between the impulse amplitudes is small, then we can assume that the peak in the filter output, which determines the detection, will be at the same position for both input signals. Show how this so-called *perturbation technique* [RB81, Blo87] can be used to measure the shape of the impulse response, i.e., show how $h(\mathbf{y})$ is related to the measured thresholds ΔL_1 and $\Delta L_2(\mathbf{y})$.

Chapter 9

OPTIMIZING SAMPLING STRUCTURES

1. INTRODUCTION

In this chapter, it is discussed how the theoretical results of chapter 7 on sampling and interpolation and the knowledge about human spatial perception of chapter 8 can be combined in the design of a sampling-and-interpolation system. Such a system contains an image sensor, such as a camera, that converts analog input images into discrete images, and a display that converts discrete images into analog output images that are intended for a human observer. The observer characteristics play a role in the design, since it is the combined characteristic of the display and the observer that determines the interpolation function.

The starting point in designing a display with a given technology is usually that a specified number of display elements can be placed in one unit area of the display, i.e., the determinant $d(\Lambda)$ of the sampling lattice is known. Some of the important design issues that remain to be resolved are the following. How can the sampling lattice Λ be optimized? For which viewing distance should the display be designed? How can the impulse response $\varphi_d(x)$ of the display element be optimized? What is the optimal sampling function between the continuous input image and the discrete weights of the display elements (in the specified viewing situation)? How can the sampling function be implemented by combining the impulse response of the camera with a discrete convolution? We successively discuss these design issues.

2. SAMPLING LATTICE OPTIMIZATION

2.1 PERIODIC STRUCTURES

We start by determining how the sampling lattice can be optimized. The examples in chapter 6 and chapter 7 have illustrated that the sampled and interpolated images contain two distinguishable components: the required image and some regular (i.e., quasi-periodic) artifacts that are due to the sampling process. These artifacts that do not belong to the image are referred to by the common name of aliasing and include **Moiré**[1], **staircases**[2] and **periodic structures**. Periodic structure is the most prominent artefact in most natural scenes and is especially noticeable in uniform regions of the image. This can be easily understood. The interpolated output of a sampled system with display function $\varphi_d(\mathbf{x})$ can be expressed as

$$Pf(\mathbf{x}) = \sum_{\mathbf{p}\in\mathcal{P}} \widehat{Pf}_{\varphi}(\mathbf{p}) \cdot \varphi_d(\mathbf{x}-\mathbf{p}), \tag{9.1}$$

for $\mathbf{x} \in \mathbb{R}^D$, where $[Pf]_{\varphi} = [\widehat{Pf}_{\varphi}(\mathbf{p}), \mathbf{p} \in \mathcal{P}]$ is the discrete signal representation that is applied to the display. For a constant input signal, the output is proportional to

$$\sum_{\mathbf{p}\in\mathcal{P}} \varphi_d(\mathbf{x}-\mathbf{p}), \tag{9.2}$$

which has a Fourier transformation equal to

$$\hat{\varphi}_d(\mathbf{u}) \sum_{\mathbf{v}\in\Lambda^\star} \delta_{\mathbf{u}-\mathbf{v}} = \sum_{\mathbf{v}\in\Lambda^\star} \hat{\varphi}_d(\mathbf{v}) \cdot \delta_{\mathbf{u}-\mathbf{v}}, \tag{9.3}$$

for $\mathbf{u} \in \mathbb{R}^D$. This output signal is periodic with lattice Λ and only contains harmonic components at the frequencies $\mathbf{v} \in \Lambda^\star$ of the reciprocal lattice $\Lambda^\star$. These components correspond to the peaks in the Fourier amplitudes in Figure 6.8 and Figure 7.3, for example. The amplitude of the harmonic component at frequency $\mathbf{v} \in \Lambda^\star$ is proportional to $|\hat{\varphi}_d(\mathbf{v})|$.

Since spatial characteristics of the human visual system are specified in degrees of visual angle, we need to know the viewing distance in order

[1]Moiré is caused by the multiplicative interaction of the periodic sampling structure with frequency vector $\mathbf{u}_1$ with a periodic structure in the image with frequency vector $\mathbf{u}_2$, where $\mathbf{u}_2$ is only slightly different from $\mathbf{u}_1$. This interaction creates periodic patterns at the frequency vectors $(\mathbf{u}_1 + \mathbf{u}_2)/2$ and $(\mathbf{u}_1 - \mathbf{u}_2)/2$ that are not present in the original image. Especially the interaction at the low frequency $(\mathbf{u}_1 - \mathbf{u}_2)/2$ is usually easily observed.

[2]The discrete sampling can convert an edge or line that is perfectly straight into one that wiggles around the average direction.

to relate these characteristics to the display. A distance of Δ on the display corresponds to $\delta = (360/\pi) \cdot \arctan(1/(2a)) \approx 180/(\pi \cdot a)$ degrees of visual angle, when viewed from a distance of $a \cdot \Delta$ from the display. For instance, the standardized viewing distance for broadcast TV is 6 times the image height. Since there are 575 visible horizontal lines on a TV-screen, we obtain that the lines on a TV-screen are $\delta = (180 \cdot 60)/(6 \cdot 575 \cdot \pi) \approx 1$ arcmin apart. This corresponds to a line periodicity of $\nu = 1/\delta = 60$ c/deg. More generally, the following relationships exist

$$\frac{x}{\Delta} = \frac{\rho}{\delta}, \; u \cdot \Delta = \nu \cdot \delta \tag{9.4}$$

between spatial distance x and frequency u on the display, and the corresponding units ρ and ν in visual angles and cycles per degree.

2.2 STILL IMAGES

We first consider lattice optimization for static patterns. We assume that the first lattice vector points in the horizontal direction[3], i.e., $\mathbf{v}_1 = (x_1, 0)$. Without loss of generality, the second lattice vector can be assumed to be of the form $\mathbf{v}_2 = (a \cdot x_1, x_2)$, with $-1/2 \leq a \leq 1/2$. The following matrices

$$V = \begin{bmatrix} x_1 & a \cdot x_1 \\ 0 & x_2 \end{bmatrix}, \; U = \begin{bmatrix} 1/x_1 & 0 \\ -a/x_2 & 1/x_2 \end{bmatrix} \tag{9.5}$$

hence specify the lattice and dual lattice (since $U^T V = I$). The columns $\mathbf{u}_1$ and $\mathbf{u}_2$ of the matrix U are the **fundamental frequency vectors** of the lattice, and have frequencies

$$u_1 = \sqrt{\frac{1}{x_1^2} + \frac{a^2}{x_2^2}} \text{ and } u_2 = \frac{1}{x_2}. \tag{9.6}$$

When distances are measured in degrees of visual angle, then the corresponding fundamental frequency vectors are denoted by $\boldsymbol{\nu}_1 = \mathbf{u}_1 \cdot \Delta/\delta$ and $\boldsymbol{\nu}_2 = \mathbf{u}_2 \cdot \Delta/\delta$, with frequencies $\nu_1 = |\boldsymbol{\nu}_1| = u_1 \cdot \Delta/\delta$ and $\nu_2 = |\boldsymbol{\nu}_2| = u_2 \cdot \Delta/\delta$ (in c/deg), respectively. The orientation of the first fundamental component satisfies $\theta_1 = \arctan(-(a x_1)/x_2)$, while the second fundamental component is oriented vertically (i.e., $\theta_2 = 90°$).

---

[3]This is equivalent to assuming that the display has horizontal scanlines.

If the modulation depths[4] of two sinusoidal gratings with frequency vectors $\mathbf{u}_1$ and $\mathbf{u}_2$ are $m_1(\mathbf{u}_1)$ and $m_2(\mathbf{u}_2)$, with respective detection thresholds $m_{thr}(\mathbf{u}_1)$ and $m_{thr}(\mathbf{u}_2)$, then it is assumed that these gratings are equally perceivable if their **normalized modulation depths** are equal, i.e., if

$$\tilde{m}(\mathbf{u}_1) = \frac{m_1(\mathbf{u}_1)}{m_{thr}(\mathbf{u}_1)} = \frac{m_2(\mathbf{u}_2)}{m_{thr}(\mathbf{u}_2)} = \tilde{m}(\mathbf{u}_2), \tag{9.7}$$

provided of course that this ratio exceeds one. By definition, this assumption is valid for modulation depths $m_1(\mathbf{u}_1)$ and $m_2(\mathbf{u}_2)$ that are close to their threshold.

The detection of periodic structures is mostly determined by the variation in luminance, especially when the spatial frequency of the periodic pattern exceeds 10 c/deg. As was discussed in chapter 8, a large amount of psychophysical data is available on the detectability of periodic (luminance) gratings on a uniform background. The CSFs of chapter 8 indicate that the modulation depth required for detection of static luminance gratings can be approximately described by

$$m_{thr}(\nu) \approx m_o \cdot \exp[\beta_o(\theta) \cdot \nu], \tag{9.8}$$

at least for spatial frequencies ν above 10 c/deg. Typical values for the parameters are $m_o = 1/S_o = 1/1500$, $\beta_o(0) = 7.5$ arcmin for horizontal and vertical gratings and $\beta_o(45) = 9$ arcmin for gratings in the oblique directions (plus and minus 45 degrees). These functions are plotted in Figure 9.1.

Due to the exponential decrease in contrast sensitivity with spatial frequency, it can be assumed that the visibility of non-sinusoidal periodic patterns of sufficiently high frequency is mainly determined by the modulation depths of the fundamental frequency vectors. If the modulation detection threshold is modelled by the exponential function of equation (9.8), then equation (9.7) implies that the sinusoidal patterns corresponding to the fundamental frequency vectors are equally visible if

$$\log m_1(\nu_1) - \beta_o(\theta_1) \cdot \nu_1 = \log m_2(\nu_2) - \beta_o(\theta_2) \cdot \nu_2. \tag{9.9}$$

[4]Because the images and sampling structures that we consider are real-valued, frequency vectors occur in pairs $(\mathbf{u}, -\mathbf{u})$ with Fourier coefficients $A(\mathbf{u}) \cdot \exp(j\theta(\mathbf{u}))$ and $A(\mathbf{u}) \cdot \exp(-j\theta(\mathbf{u}))$, respectively. We will however behave as if the modulation depth $m(\mathbf{u}) = 2 \cdot A(\mathbf{u})/A(\mathbf{0})$ of a grating with frequency vector $\mathbf{u}$ is assigned to the frequency vector $\mathbf{u}$, while in actuality it is distributed evenly over the pair of frequency vectors.

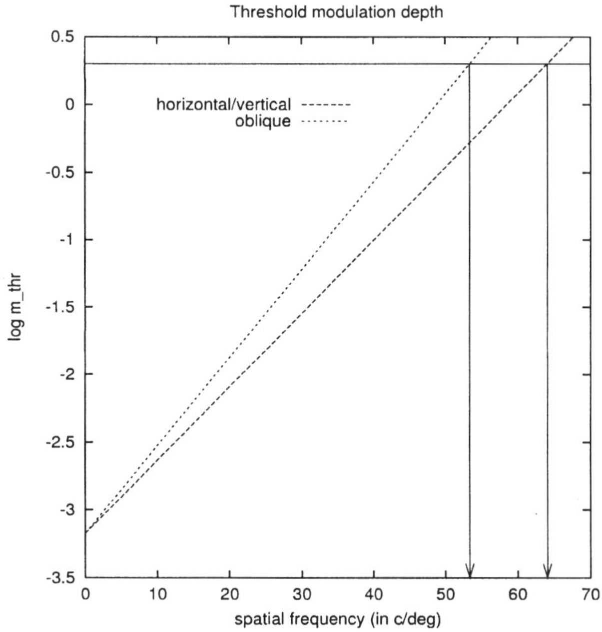

Figure 9.1. Approximate functions describing the logarithm (base 10) of the modulation depth required for detection of a sinewave luminance grating as a function of spatial frequency ν (in c/deg). One curve is for gratings in the horizontal and vertical directions, while the other curve is for gratings in the oblique directions. The largest possible modulation depth of 2 ($\log(2) = 0.3$) occurs for unfiltered sampling structures. Fundamental frequencies above 64 c/deg and 53 c/deg are hence invisible in case of horizontal/vertical and oblique patterns, respectively.

If $m_1(\nu_1) = m_2(\nu_2)(= 2)$, i.e., in case no filtering is applied to the lattice and all frequency vectors in the periodic pattern have equal amplitude, then the ratio of the fundamental frequencies

$$\frac{u_2}{u_1} = \frac{\nu_2}{\nu_1} = \frac{\beta_o(\theta_1)}{\beta_o(\theta_2)} = b(\theta_1) \tag{9.10}$$

should be equal to the exponent ratio. This exponent ratio $b(\theta_1)$ depends only on the orientation θ_1, since $\theta_2 = 90°$ is fixed.

The a priori condition $x_1 \cdot x_2 = \Delta^2 = d(\Lambda)$ on the sampling density can be satisfied by choosing

$$x_1 = \frac{\Delta}{c(\theta_1)}\,, \; x_2 = c(\theta_1)\,\Delta\,, \tag{9.11}$$

which allows to express the fundamental frequencies as

$$\nu_1 = \frac{c(\theta_1)}{\delta} \sqrt{1 + \tan^2 \theta_1}, \ \nu_2 = \frac{1}{c(\theta_1)} \frac{1}{\delta}. \tag{9.12}$$

The above ratio condition on the fundamental frequencies results in the following expression for the unknown factor

$$c(\theta_1) = \left[b^2(\theta_1)\,(1 + \tan^2 \theta_1)\right]^{-1/4}. \tag{9.13}$$

The relationship between the horizontal displacement a of the second sampling vector and the orientation θ_1 can also be expressed in this notation, i.e.,

$$a(\theta_1) = -c^2(\theta_1)\, \tan \theta_1. \tag{9.14}$$

According to our assumption, the visibility of the optimized lattice is determined by the normalized modulation depth

$$\tilde{m}(\nu_1) = \tilde{m}(\nu_2) = 2\, S_o\, \exp\left[-\frac{\beta_o}{\delta}\, \frac{1}{c(\theta_1)}\right], \tag{9.15}$$

with $\beta_o = \beta(90°)$, since all unfiltered components have a modulation depth of 2. Consequently, the visibility of this sampling lattice can be decreased maximally by minimizing $c(\theta_1)$, or $c^2(\theta_1)$, as a function of θ_1.

If the visual system were equally sensitive in all directions, i.e., if $b(\theta_1) = 1$, then the goal of lattice optimization would be to move the fundamental frequency vectors as far away as possible from the origin, i.e., to minimize $c = (1 - a^2)^{1/4}$. The optimum lattice in this isotropic case is a hexagonal lattice with a equal to its extreme value $a = \pm 1/2$ and $x_2 = x_1 \cdot \sqrt{3/4}$. It is well known that such a hexagonal lattice allows for the most dense packing of (circularly symmetric) disks in a plane.

In order to discuss the effect of the anisotropic nature of the visual system on the lattice optimization, a mathematical expression is needed for the relationship between the exponent β_o in equation (9.8) and the orientation θ_1 of the fundamental grating. As discussed in chapter 8, the experimental data suggest that the description

$$\beta_o(\theta_1) = \beta - \Delta\beta\, \cos(4\theta_1), \tag{9.16}$$

where the exponent varies sinusoidally from $\beta - \Delta\beta$ (=7.5 arcmin) for horizontal and vertical gratings to $\beta + \Delta\beta$ (=9 arcmin) for diagonal gratings, is a good first-order approximation. This leads to the following expression for the exponent ratio

$$b(\theta_1) = \frac{\beta - \Delta\beta\, \cos[4\theta_1]}{\beta - \Delta\beta} = \frac{1 - m_\beta \cos[4\theta_1]}{1 - m_\beta}, \tag{9.17}$$

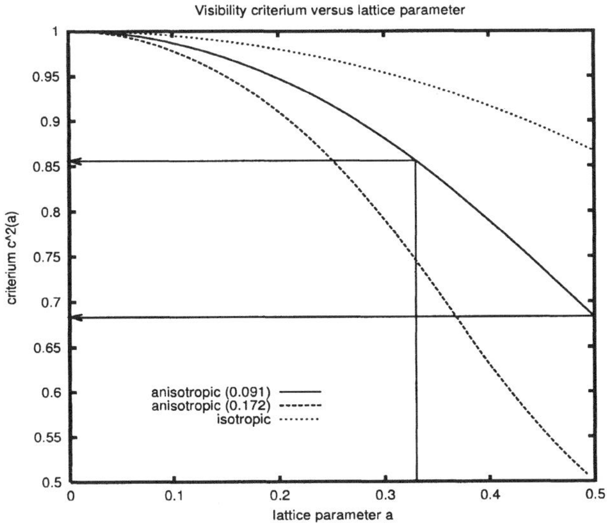

Figure 9.2. The optimization criterium $c^2(a)$ as a function of the lattice parameter a for three values of the exponent modulation depth m_β, i.e, $m_\beta = 0$ (isotropic), $m_\beta = 0.091$ (visual system) and $m_\beta = 0.172$.

where $m_\beta = \Delta\beta/\beta$ denotes the modulation depth of the exponent. The resulting relationship between the optimization criterium $c^2(\theta_1)$ of Equation (9.13) and $a(\theta_1)$ is plotted in Figure 9.2 for three values of m_β, including the isotropic case ($m_\beta = 0$). In all cases, it is observed that c^2 is minimal when a reaches its extreme value $a = 1/2$. This corresponds to the case of a quincunx lattice. The relationship between the optimum criterium value $c^2(1/2)$ and m_β is plotted in Figure 9.3. It can be observed that the optimum lattice would be a diagonal lattice with $x_2 = x_1/2$ if $m_\beta = (1 - 1/\sqrt{2})/(1 + 1/\sqrt{2}) \approx 0.172$. This lattice allows for the most dense packing of (tilted) squares in a plane. In case of the visual system $m_\beta = 0.091$ is a realistic choice, which results in $x_2 \approx 0.68 \cdot x_1$. This optimum quincunx lattice from the point of view of the visual system is hence in between a hexagonal and a diagonal lattice.

It is shown in Figure 9.4(a) how a quincunx lattice, with $x_2/x_1 = c^2(1/2)$, can be realized in a color display. Since this structure is fairly difficult to implement, because of the non-alignment of display elements in successive rows, it is often replaced by the suboptimum lattice in Figure 9.4(b), with lattice parameter $a = 1/3$ and $x_2/x_1 = c^2(1/3)$, or by a square lattice, with $a = 0$ and $x_2 = x_1$, in which alike-colored display elements are organized in columns. Equation (9.15) implies that

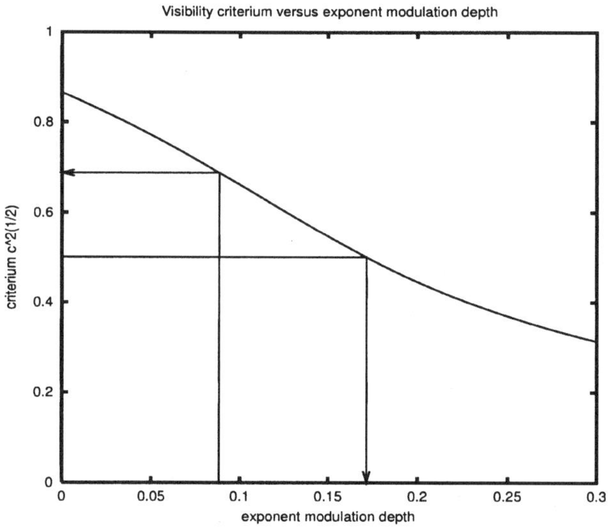

Figure 9.3. The optimization criterium $c^2(a)$ as a function of the exponent modulation depth m_β in case of a quincunx lattice with $a = 1/2$. Two points on the graph, i.e., for the actual visual system ($c^2 \approx 0.68$) and the diagonal lattice ($c^2 = 1/2$), are highlighted.

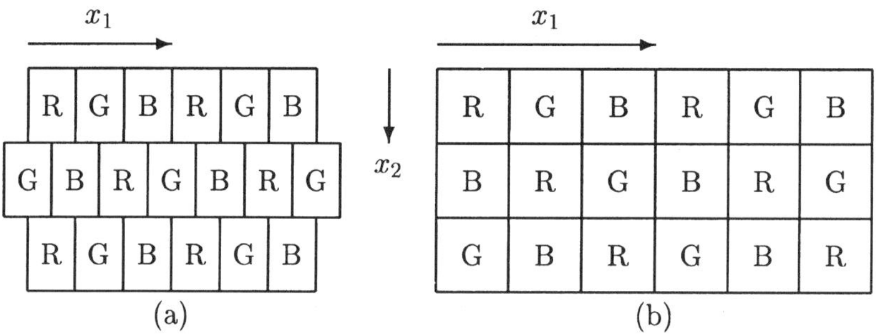

Figure 9.4. Sampling structures for color displays.

two optimized lattices with different lattice parameter values a_1 and a_2 are equally visible if the sampling densities δ_1^2 and δ_2^2 satisfy

$$\frac{\delta_2^2}{\delta_1^2} = \frac{c^2(a_1)}{c^2(a_2)}. \tag{9.18}$$

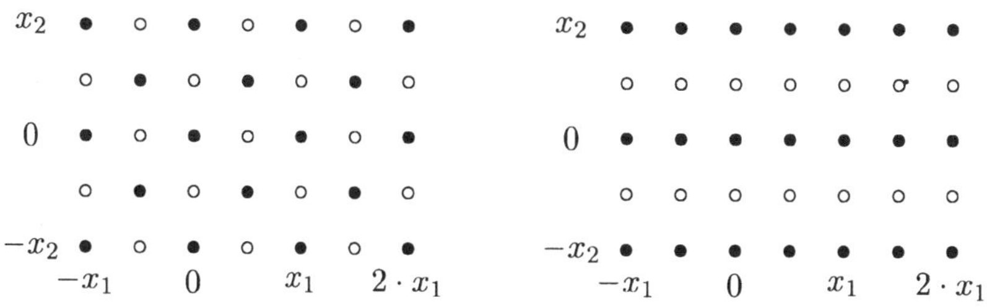

Figure 9.5. Field-quincunx (left) and field-rectangular (right) lattices: the filled and open circles represent the spatial sampling points in the even and odd fields of the time sequence.

Because $c^2(0) = 1$, this implies that Figure 9.2 can also be used to establish the reduction in sampling density of an optimized lattice with parameter a as compared to a square lattice with equal visibility. For example, we obtain a decrease of approximately 15 and 32 percent in display elements for the case $a = 1/3$ and $a = 1/2$, respectively.

2.3 IMAGE SEQUENCES

The quincunx lattice is also used in our current **interlace** TV system. The 2D sampling in this case is along the combined vertical spatial dimension and the time dimension. The motivation for this interlace structure is similar to the motivation for the quincunx lattice in the spatial domain.

In digital TV standards, where sampling is used along all three dimensions, the two previous sampling structures (quincunx in the spatial dimensions, together with interlace) can be combined into a 3D sampling lattice called **field-quincunx** [AAH+86]. This 3D lattice is specified by the matrices

$$V = \begin{bmatrix} x_1 & x_1/2 & x_1/2 \\ 0 & x_2/2 & 0 \\ 0 & 0 & x_3 \end{bmatrix}, U = \begin{bmatrix} 2/x_1 & 0 & 0 \\ -1/x_2 & 2/x_2 & 0 \\ -1/(2\,x_3) & 0 & 1/x_3 \end{bmatrix}, \quad (9.19)$$

with $x_1 = x_2$. The spatial samples for the odd and even time indices are indicated by the filled and open circles in Figure 9.5, respectively. They form the odd and even **fields** of the sequence. A combination of successive odd and even fields, as is shown in Figure 9.5, is called a **frame**. A still image extracted from an image sequence will usually correspond to one such frame.

Many display systems at this time use a square sampling lattice in the spatial dimensions, combined with interlace, which can be specified by the matrices

$$V = \begin{bmatrix} x_1/2 & 0 & 0 \\ 0 & x_2 & x_2/2 \\ 0 & 0 & x_3 \end{bmatrix}, U = \begin{bmatrix} 2/x_1 & 0 & 0 \\ 0 & 1/x_2 & 0 \\ 0 & -1/(2\,x_3) & 1/x_3 \end{bmatrix}, \quad (9.20)$$

with $x_1 = x_2$. A comparison of fundamental frequencies reveals that this **field-rectangular** lattice, which is shown on the right hand side of Figure 9.5, is suboptimal.

3. DISPLAY FUNCTION OPTIMIZATION

3.1 DISPLAY FUNCTIONS

The demands on the display function can be analyzed once a sampling lattice has been selected. The modulation depth of the frequency vector $\mathbf{u} \in \Lambda^\star$ is equal to $2 \cdot |\hat{\varphi}_d(\mathbf{u})|/\hat{\varphi}_d(\mathbf{0})$ after filtering with the display function. The maximum modulation depth that can occur for a positive-valued display function φ_d (a reasonable physical limitation) is two, and is obtained if the support of the display function φ_d is much smaller than the size of the unit cell of the sampling lattice. This case is closely approximated in many existing color displays, where three non-overlapping primaries have to be placed within one sampling interval (see Figure 9.4). Since the modulation depth is always smaller than 2, it can be derived from Figure 9.1 that luminance patterns with a spatial frequency above 64 cycles/degree are imperceivable. This for instance implies that the standardized viewing distance in TV is such that the visibility of the sampling pattern is close to the threshold. Similarly, luminance gratings with a temporal frequency above 70 Hz are not perceived either. Harmonic components that fall above these frequencies need therefore not be taken into account when discussing sampling artifacts [WAJF86].

Display elements have a low-pass spatial characteristic that is most often modelled by either a rectangular (or square) point spread function, such as in the case of Liquid Crystal Displays, or a Gaussian profile

$$\varphi_d(\boldsymbol{\rho}) \propto \exp\left(-\frac{\boldsymbol{\rho}^T\boldsymbol{\rho}}{2\sigma_d^2}\right), \quad (9.21)$$

such as in the case of Cathode Ray Tubes and/or optical filtering. The temporal low-pass behavior of the display is most often modelled by a first-order characteristic

$$\varphi_d(t) \propto \exp\left(-\frac{t}{\tau_d}\right), \quad (9.22)$$

for $t \in \mathbb{R}$, where τ_d is called the integration time of the display. Most displays allow fairly little integration, i.e., $\tau_d < 1$ ms, in order to avoid motion smearing.

3.2 THRESHOLD VISIBILITY

The low-pass behavior of the display reduces the visibility of the sampling structure, since it reduces the modulation depth for all nonzero frequency components. For instance, for a Gaussian profile with a standard deviation of σ_d, in degrees of visual angle, the modulation depth at frequency ν c/deg is

$$m(\nu) = 2 \cdot \exp\left[-2\;(\pi\,\sigma_d\,\nu)^2\right]. \tag{9.23}$$

Note that a modulation depth $m = 2$ is indeed obtained for $\sigma_d \to 0$. If the CSF can be described by the exponential function of Equation (9.8), then the normalized modulation depth is equal to

$$\tilde{m}(\nu) = 2\,S_o\,\exp\left[-\beta_o\,\nu - 2\,(\pi\,\sigma_d\,\nu)^2\right]. \tag{9.24}$$

If the normalized modulation depths of all fundamental frequencies ν of the lattice Λ are smaller than 1, then the sampling structure is assumed to be invisible. Further increase of σ_d beyond this value will hence have no influence on the visibility of the sampling structure.

The Gaussian filter spread σ_d that corresponds to a normalized modulation depth $\tilde{m}(\nu) = 1$ is plotted in Figure 9.6 as a function of the average sampling density δ (i.e., lattice density $d(\Lambda) = \delta^2$). The drawn curve corresponds to a square lattice with fundamental frequencies $\nu_1 = \nu_2 = 60/\delta$ c/deg in the horizontal and vertical directions, and assumes parameter values $m_o = 1/S_o = 1/1500$ and $\beta_o = 7.5$ arcmin for the CSF in equation (9.8). The dashed and dotted curves are for the visually optimized quincunx lattice of the previous section. The fundamental frequencies for this quincunx lattice are at $\nu_1 = 0.943 \cdot 60/\delta$ c/deg in the direction $\theta_1 = 36°$, with modulation detection threshold exponent $\beta_o(\theta_1) = 8.86$ arcmin, and at $\nu_2 = 1.11 \cdot 60/\delta$ c/deg in the vertical direction. The unfiltered sampling structure is invisible if $\delta < \beta_o/(\log(2) - \log(m_o)) = 0.94$ arcmin, for the square lattice, and $\delta < 1.11 * 0.94 = 1.04$ arcmin, for the quincunx lattice.

Note that the quincunx lattice is no longer optimal if filtering by the display element is included. With increasing standard deviation σ_d of the Gaussian filter, the oblique component in the quincunx lattice will become easier to detect than the horizontal/vertical components in the square lattice, as can be observed in Figure 9.6. The underlying reason

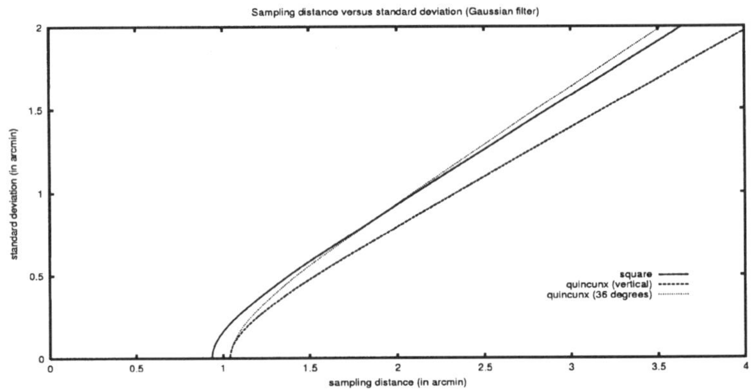

Figure 9.6. The curves show the required standard deviation σ_d (in arcmin) of the Gaussian filter needed to bring the visibility of the sampling structure below threshold, as a function of the square root δ (in arcmin) of the sampling density. The drawn curve is for a square lattice, while the dashed and dotted curves are for the two fundamental components in a visually optimized quincunx lattice.

is that, with increased filtering, the two fundamental frequency vectors of the quincunx lattice, which have unequal frequency, no longer retain equal modulation depth, so that the general equation (9.9), rather than the simplified equation (9.10), needs to be applied to optimize the lattice. More specifically, in this more general case, the visibility of the lattice is reduced optimally by minimizing the normalized modulation depth

$$\tilde{m}(\nu_1) = \tilde{m}(\nu_2) = 2\, S_o\, \exp\left[-\frac{\beta_o(\theta_2)}{\delta\, c(\theta_1)} - 2\left(\frac{\pi\, \sigma_d}{\delta\, c(\theta_1)}\right)^2\right], \qquad (9.25)$$

with $\theta_2 = 90°$, under the following nonlinear equation

$$\frac{\beta_o(\theta_1)\, c(\theta_1)}{\delta}\sqrt{1+\tan^2\theta_1} + 2\left(\frac{\pi\, \sigma_d\, c(\theta_1)}{\delta}\right)^2 (1+\tan^2\theta_1) = \frac{\beta_o(\theta_2)}{\delta\, c(\theta_1)} + 2\left(\frac{\pi\, \sigma_d}{\delta\, c(\theta_1)}\right)^2 \qquad (9.26)$$

for the factor $c(\theta_1)$, where $\beta_o(\theta_1)$ may be expressed by equation (9.16).

3.3 IMPAIRMENT MODEL

The low-pass behavior of the display, although beneficial for reducing the visibility of the sampling structure, also has negative consequences, since the input image $f(\mathbf{x})$ is also filtered by the display characteristic. For the spatial information, it is known from psychophysical experiments that the quality of images is unaffected by filtering with a Gaussian, provided that the standard deviation of this Gaussian is limited to below 0.5 arcmin [Wes91]. From Figure 9.6, it can be derived that filtering can hence be applied without affecting the image quality for sampling structures with a sampling distance δ below 1.5 arcmin. Increasing the standard deviation of the Gaussian display function beyond 0.5 arcmin further decreases the visibility of the sampling lattice, however at the cost of affecting the sharpness of the original image. Only psychophysical experiments can at this point help in determining how these two degrading factors can be optimally balanced. Nijenhuis [Nij93, NB96, NB97a, NB97b] performed such experiments and showed that they were in good agreement with the impairment model that is described below.

Nijenhuis first developed models for the individual impairments "Gaussian blur" and "visibility of sinusoidal patterns" using an approach that is similar to the one used for deriving the relationship between luminance (ratio) and psychometric lightness (see section 4.2). More specifically, he assumed that the perceptual strength S_b of Gaussian blur with standard deviation σ_d is a differentiable function of the normalized standard deviation $\tilde{\sigma}_d = \sigma_d/\sigma_o$, i.e., $S_b = S_b[\tilde{\sigma}_d]$, where σ_o denotes a fixed standard deviation that is a characteristic of human blur perception. Similarly, he assumed that the perceptual strength S_p expressing the "visibility of sinusoidal patterns" is a differentiable function of the normalized modulation depth $\tilde{m}(\nu)$ defined in Equation (9.7), i.e., $S_p = S_p[\tilde{m}(\nu)]$. Variations in these perceptual strengths S_b and S_p can then be related to variations in the corresponding physical parameters $\tilde{\sigma}_d$ and $\tilde{m}(\nu)$, i.e.,

$$\Delta S_b \approx \frac{dS_b}{d\tilde{\sigma}_d} \cdot \Delta\tilde{\sigma}_d[\tilde{\sigma}_d],\ \Delta S_p \approx \frac{dS_p}{d\tilde{m}(\nu)} \cdot \Delta\tilde{m}(\nu)[\tilde{m}(\nu)], \qquad (9.27)$$

where ΔS_b and ΔS_p denote the increase in sensation strength of S_b and S_p for an increase in the physical parameters from $\tilde{\sigma}_d$ to $\tilde{\sigma}_d + \Delta\tilde{\sigma}_d[\tilde{\sigma}_d]$ and from $\tilde{m}(\nu)$ to $\tilde{m}(\nu) + \Delta\tilde{m}(\nu)[\tilde{m}(\nu)]$, respectively. If we adopt Fechner's assumption that the just noticeable differences (JNDs) that can be measured in discrimination experiments with perceptual attributes are independent of the attribute strengths themselves[5], then ΔS_b and ΔS_p

[5]This is equivalent to assuming that sensation strengths are internally represented on an interval scale.

are constant, and we obtain the following relationships

$$\frac{dS_b}{d\tilde{\sigma}_d} = \frac{k_b}{\Delta\tilde{\sigma}_d[\tilde{\sigma}_d]}, \; \frac{dS_p}{d\tilde{m}(\nu)} = \frac{k_p}{\Delta\tilde{m}(\nu)[\tilde{m}(\nu)]}, \tag{9.28}$$

with k_b and k_p unknown constants, between the derivatives to the sensation strengths and the measurable discrimination thresholds. Integration of these expressions leads to

$$S_b[\tilde{\sigma}_d] = k_b \cdot \int_0^{\tilde{\sigma}_d} \frac{1}{\Delta\tilde{\sigma}_d[\hat{\sigma}_d]} \, d\hat{\sigma}_d, \; S_p[\tilde{m}(\nu)] = k_p \cdot \int_0^{\tilde{m}(\nu)} \frac{1}{\Delta\tilde{m}(\nu)[\hat{m}]} \, d\hat{m}, \tag{9.29}$$

where the constants k_b and k_p can be selected such that the sensation strengths vary between 0 and 1 (for variations within the range $\sigma_d \in [0, \infty]$ and $m(\nu) \in [0, 2]$).

Blur discrimination experiments by Watt and Morgan [WM83] can be modelled by the following relationship

$$\Delta\tilde{\sigma}_d[\tilde{\sigma}_d] = \left[1 + \tilde{\sigma}_d^2\right]^{3/4}, \tag{9.30}$$

provided that a value of $\sigma_o = 1.2$ arcmin is adopted for normalizing the standard deviation σ_d of the blur. This implies that the discrimination threshold is proportional to $\tilde{\sigma}_d^{3/2}$ for values of $\tilde{\sigma}_d \gg 1$. The corresponding sensation strength for blur is equal to

$$S_b[\tilde{\sigma}_d] = 1 - \left[1 + \tilde{\sigma}_d^2\right]^{-1/4} \tag{9.31}$$

and is plotted in the left part of Figure 9.7. It has been verified experimentally that the impairment I_b caused by Gaussian blur in complex images, which can for instance be measured using the scaling techniques of chapter 5, is linearly related to the sensation strength S_b, which is based on discrimination thresholds for simple patterns, so that we can adopt $I_b = \lambda_b \cdot S_b$.

Experimental results from Legge [Leg83] and Carlson&Cohen [CC78] indicate that $\Delta\tilde{m}(\nu)$ is proportional to $\tilde{m}(\nu)^{\alpha(\nu)}$, where the exponent $\alpha(\nu)$ depends linearly on spatial frequency, i.e.,

$$\alpha(\nu) = \alpha_o \left(1 + \frac{\nu - \nu_o}{\nu_f}\right), \tag{9.32}$$

with $\alpha_o = 0.7$, $\nu_o = 12.7$ c/deg and $\nu_f = 98.4$ c/deg [Nij93]. Integration leads to the following relationship

$$S_p[\tilde{m}(\nu)] = \left[\frac{\tilde{m}(\nu)}{\tilde{m}_2(\nu)}\right]^{1-\alpha(\nu)}, \tag{9.33}$$

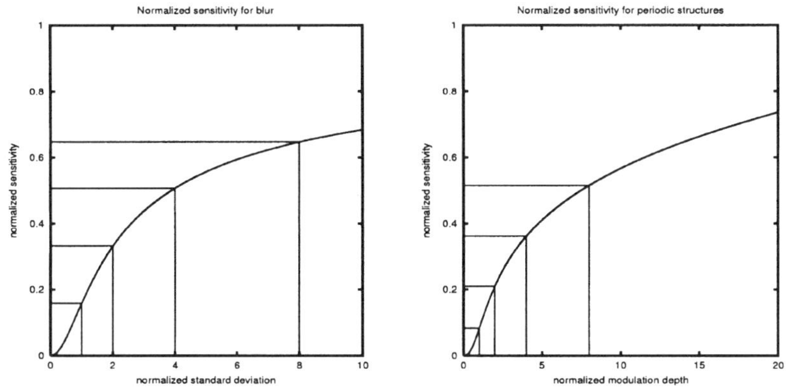

Figure 9.7. Left figure: sensation strength S_b for blur as a function of the normalized standard deviation $\tilde{\sigma}_d$ of a Gaussian blurring kernel. Right figure: sensation strength S_p for sinusoidal patterns as a function of the normalized modulation depth $\tilde{m}(\nu)$ in case $\alpha(\nu) = 0.7$ and the maximum normalized modulation depth is $\tilde{m}_2(\nu) = 50$.

where $\tilde{m}_2(\nu) = 2/m_{thr}(\nu)$ is the normalized modulation depth that corresponds to the maximum physically realizable modulation depth of $m(\nu) = 2$. This expression for the sensation strength is however not sufficiently accurate for modulation depths close to threshold, so that the modified relationship proposed by Wilson [Wil80],

$$S_p[\tilde{m}(\nu)] = \frac{P[\tilde{m}(\nu)]}{P[\tilde{m}_2(\nu)]}, \tag{9.34}$$

with

$$P(\tilde{m}) = \left[(1 + \tilde{m}^3)^{1/3} - 1\right] \tilde{m}^{-\alpha(\nu)}, \tag{9.35}$$

is adopted instead. Note that $P(\tilde{m}) \approx \tilde{m}^{1-\alpha(\nu)}$ for $\tilde{m} \gg 1$. The function $S_p[\tilde{m}(\nu)]$ is plotted in the right part of Figure 9.7.

For a two-dimensional periodic sampling pattern with a sufficiently high density, only the two fundamental frequencies ν_1 and ν_2 usually play a role in the visibility of this pattern. It was established experimentally that a Minkowski metric with power 4 provides an adequate description for the total impairment strength

$$I_p = \left(S_p^4[\tilde{m}(\nu_1)] + S_p^4[\tilde{m}(\nu_2)]\right)^{1/4} \tag{9.36}$$

of a periodic pattern as a function of the sensation strengths for the fundamental frequencies.

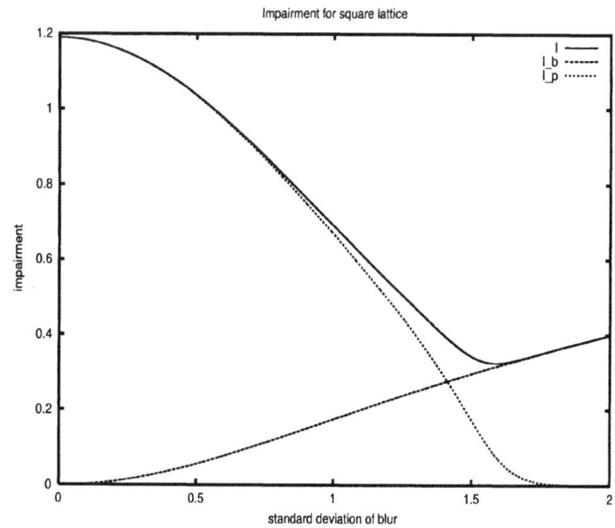

Figure 9.8. Impairment strengths I_b, I_p and I as a function of the standard deviation σ_d of the Gaussian interpolation function for a square lattice with sampling distance equal to $\delta = 3$ arcmin.

The total impairment I in a sampled image is a combination of blur and the visibility of the sampling pattern. It has been observed that a Minkowski metric with power 2 is often a good combination rule for such independent distortions [Rid92]. For the case at hand, the overall impairment is expressed as

$$I = \left[I_b^2 + I_p^2\right]^{1/2}, \tag{9.37}$$

where the relative weight of $\lambda_b = \sqrt{2}$ on the blur impairment was established by comparing model predictions with experimental data on the impairment of sampled images.

The implications of the model are demonstrated for a square sampling pattern with fundamental frequencies $\nu_1 = \nu_2 = 1/\delta$ in the horizontal and vertical directions. If the display function is Gaussian with standard deviation σ_d, then the normalized modulation depths satisfy

$$\tilde{m}(\nu_1) = \tilde{m}(\nu_2) = 2\, S_o\, \exp\left[-\frac{\beta_o}{\delta} - 2\left(\frac{\pi\,\sigma_d}{\delta}\right)^2\right] \tag{9.38}$$

if the CSF can be described by the exponential function of Equation (9.8). In Figure 9.8, we show how the individual impairments I_b, I_p and the total impairment I vary as a function of σ_d in case the sampling distance is $\delta = 3$ arcmin. The same values as above are adopted for $S_o = 1500$ and $\beta_o = 7.5$ arcmin. The simulation indicates that the impairment due to the sampling pattern dominates over the impairment due to blur, so that the optimal amount of blur is very close to the amount of blur required to bring the visibility of the sampling structure below threshold (see Figure 9.6).

4. SAMPLING FUNCTION OPTIMIZATION

Once the sampling lattice Λ and display function $\varphi_d(\mathbf{x})$ are known, then the basis function $\varphi(\mathbf{x})$, together with the corresponding sampling function $\phi(\mathbf{x})$, can be determined using the theory of chapter 7. Since a display is intended to be viewed by a human observer, these functions should include characteristics of both the display and the viewing condition. We discuss one possible approach that is based on a very simple linear model for human perception.

We hypothesize that the human visual system can be closely modelled by a linear system with an impulse response $\varphi_e(\mathbf{x})$, at least for the comparison of images that differ only in their high-frequency components, such as densely sampled images [Blo87]. As shown in Figure 9.9, this implies that our sampling system, instead of minimizing the distance between the original signal $f(\mathbf{x})$ and the interpolated discrete signal

$$Pf(\mathbf{x}) = \sum_{\mathbf{p}\in\Lambda} \widehat{Pf}_{\varphi}(\mathbf{p}) \cdot \varphi_d(\mathbf{x}-\mathbf{p}), \tag{9.39}$$

for $\mathbf{x} \in \mathbb{R}^D$, should rather be designed to minimize the difference between the filtered input $f_e(\mathbf{x}) = f(\mathbf{x}) \star \varphi_e(\mathbf{x})$ and the filtered output

$$Pf_e(\mathbf{x}) = Pf(\mathbf{x}) \star \varphi_e(\mathbf{x}) = \sum_{\mathbf{p}\in\Lambda} \widehat{Pf}_{\varphi}(\mathbf{p}) \cdot \varphi(\mathbf{x}-\mathbf{p}), \tag{9.40}$$

where the basis function $\varphi(\mathbf{x}) = \varphi_d(\mathbf{x}) \star \varphi_e(\mathbf{x})$, for $\mathbf{x} \in \mathbb{R}^D$, combines the display function and the visual filter. The Fourier transformation of this basis function is equal to $\hat{\varphi}(\mathbf{u}) = \hat{\varphi}_d(\mathbf{u}) \cdot \hat{\varphi}_e(\mathbf{u})$, for $\mathbf{u} \in \mathbb{R}^D$.

If the correlation kernel $[g_{\varphi\varphi}(\mathbf{p}) = \{\varphi(\mathbf{x}) \star \varphi(\mathbf{x})\}_{\mathbf{x}=\mathbf{p}}, \mathbf{p} \in \Lambda]$ has an inverse correlation kernel $[g_{\phi\phi}(\mathbf{p}), \mathbf{p} \in \Lambda]$, then the following optimal sampling function

$$\phi_d(\mathbf{x}) = \sum_{\mathbf{p}\in\Lambda} g_{\phi\phi}(\mathbf{p}) \cdot \varphi(\mathbf{x}-\mathbf{p}), \tag{9.41}$$

$\mathbf{x} \in \mathbb{R}^D$, is obtained for operating on the filtered input image $f_e(\mathbf{x})$. This sampling function $\phi_d(-\mathbf{x})$ that is applied to the filtered image $f_e(\mathbf{x})$ must

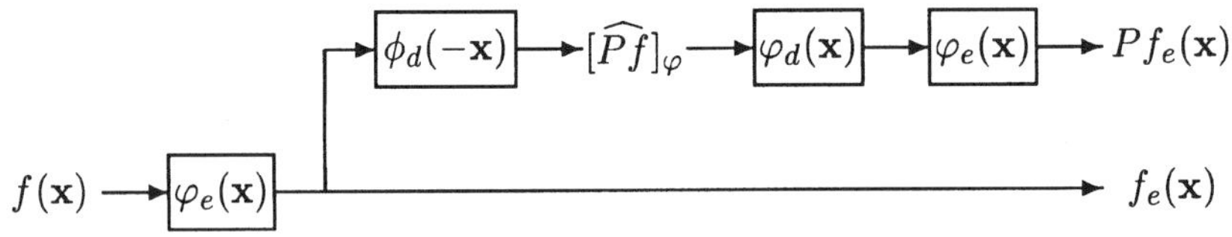

Figure 9.9. Design of the sampling function based on a minimum-square-error criterium between the filtered original image and the filtered displayed image, where $\varphi_e(\mathbf{x})$ approximates the filtering performed by the visual system.

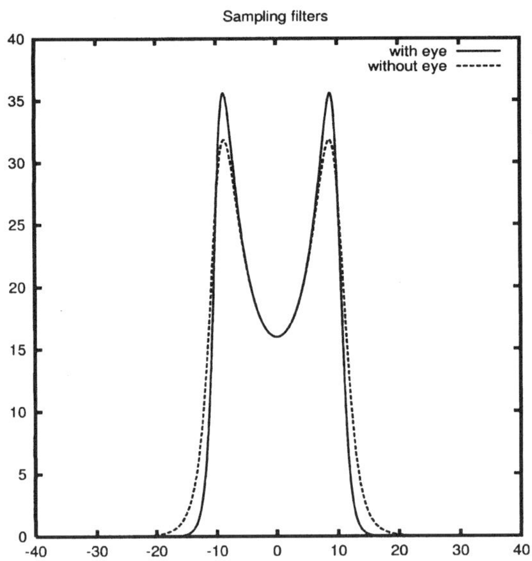

Figure 9.10. Optimum filter characteristic without and with inclusion of a visual filter. The curves are actually horizontal cross-sections through the 2-D filter characteristics.

have a frequency characteristic equal to

$$\hat{\phi}_d^\star(\mathbf{u}) = \frac{\hat{\varphi}^\star(\mathbf{u})}{\sum_{v\in\Lambda^\star} |\hat{\varphi}(\mathbf{u}+\mathbf{v})|^2} = \frac{\hat{\varphi}_d^\star(\mathbf{u}) \cdot \hat{\varphi}_e^\star(\mathbf{u})}{\sum_{v\in\Lambda^\star} |\hat{\varphi}_d(\mathbf{u}+\mathbf{v}) \cdot \hat{\varphi}_e(\mathbf{u}+\mathbf{v})|^2}, \quad (9.42)$$

for $\mathbf{u} \in \mathbb{R}^D$, as can be derived by applying the theory on optimal sampling and interpolation to the upper signal path in Figure 9.9. This implies that the sampling function $\phi(-\mathbf{x}) = \phi_d(-\mathbf{x}) \star \varphi_e(\mathbf{x})$ that is applied to the original image $f(\mathbf{x})$ must have a frequency characteristic equal to

$$\hat{\phi}^\star(\mathbf{u}) = \frac{\hat{\varphi}_d^\star(\mathbf{u}) \cdot |\hat{\varphi}_e(\mathbf{u})|^2}{\sum_{v\in\Lambda^\star} |\hat{\varphi}_d(\mathbf{u}+\mathbf{v}) \cdot \hat{\varphi}_e(\mathbf{u}+\mathbf{v})|^2}, \tag{9.43}$$

for $\mathbf{u} \in \mathbb{R}^D$. The sampling filter $\hat{\phi}^\star(\mathbf{u})$ can hence be implemented as the cascade of an analog sensor with filter characteristic $\hat{\varphi}_d^\star(\mathbf{u}) \cdot |\hat{\varphi}_e(\mathbf{u})|^2$, and a discrete filter with characteristic

$$G_{\phi\phi}(\mathbf{u}) = \frac{1}{G_{\varphi\varphi}(\mathbf{u})} = \frac{1}{\sum_{v\in\Lambda^\star} |\hat{\varphi}_d(\mathbf{u}+\mathbf{v}) \cdot \hat{\varphi}_e(\mathbf{u}+\mathbf{v})|^2}, \tag{9.44}$$

for $\mathbf{u} \in \mathbb{R}^D$. Approximations to this discrete filter $G_{\phi\phi}(\mathbf{u})$ can be realized in a similar way as before (see section 4 of chapter 7).

The effect of the visual filter on the sampling filter is demonstrated by means of an example. We assume a square sampling lattice with sampling distance $\delta = 3$ arcmin and a visual filter with an exponential low-pass characteristic $\hat{\varphi}_e(\nu) = \exp(-\beta \cdot \nu)$, with $\beta = 7.5$ arcmin. The filter characteristics[6] with and without the visual filter are shown in Figure 9.10. Note that the effect of the visual filter is to increase the emphasis on intermediate frequencies, while reducing the contribution of high frequencies.

[6]The simulations were performed on discrete images with a sampling distance of 0.75 arcmin, so that the obtainable frequencies occupy the range from -40 to 40 c/deg. By subsampling by a factor of 4, we obtained the desired sampling distance $\delta = 4 \cdot 0.75 = 3$ arcmin between the display elements. This also explains why the filters are equal to $16 = 4^2$ at zero frequency.

EXERCISES

1 Verify that equation (9.23) expresses the modulation depth of the first harmonic of a one-dimensional periodic signal, with period δ, and Gaussian display function, with spread σ_d.

2 Calculate the modulation depth of the first harmonic of the one-dimensional periodic signal

$$f(x) = \begin{cases} 1 & \text{if} \quad x \in [k\delta, k\delta + d[, \quad \text{for } k \in \mathcal{Z} \\ 0 & \text{if} \quad x \notin [k\delta, k\delta + d[, \quad \text{for } k \in \mathcal{Z} \end{cases}$$

This signal is a fair description of a Liquid Crystal Display with a rectangular display element.

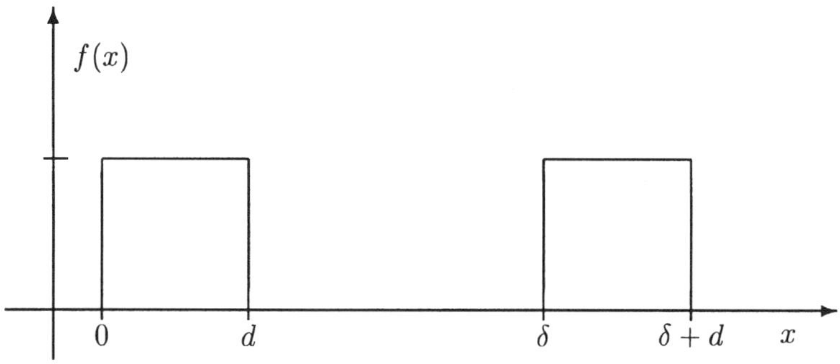

Plot the modulation depth as a function of the ratio d/δ. Use this modulation depth and the visibility threshold of equation (9.8) to derive the minimum required block width d, as a function of the sampling distance δ, that is required to make the modulation of this signal invisible, where d and δ are expressed in visual angles (similar to Figure 9.6).

Chapter 10

IMAGE QUANTIZATION

1. INTRODUCTION

The signals that are of interest in image technology were shown to have continuous variations along a number of dimensions (wavelength, space, time and amplitude). By applying insights into human color perception, it could be derived that the relevant information contained in the continuous variation in wavelength can be represented by three color signals. Subsequently, based on the limited resolution of the visual system in space and time, it was argued that adequate discrete approximations to the continuous variations in space and time can be obtained by representing these color signals on a repetitive basis. Since only images of finite spatial and temporal extent are considered, a repetitive basis with a finite number of basis signals suffices in practice. Consequently, each of the three color signals can be described by a (discrete) approximation on a finite repetitive basis.

Let $[f(\mathbf{p}), \mathbf{p} \in \mathcal{P}]$ denote the discrete representation of a color signal (of dimension D). Because of the continuous variation in the amplitude, this signal still contains an infinite amount of information, even if the amplitude is limited to a finite range (as will be the case for any practical system). The continuous variation in amplitude can be removed by applying **amplitude quantization**. In this procedure, the amplitude $f(\mathbf{p})$ at position $\mathbf{p}$ is mapped into an amplitude $g(\mathbf{p})$ that is taken from a finite set of possible amplitudes.

In sections 2 and 3, we concentrate on independent memoryless quantization of the color signals. This means that the different color signals are quantized separately, and that the quantization at position $\mathbf{p}$ is not affected by the quantization at neighboring positions. Such a memoryless quantization scheme of scalar signals is based on the assumption that the distortion between an input image $[f(\mathbf{p}), \mathbf{p} \in \mathcal{P}]$ and an output image $[g(\mathbf{p}), \mathbf{p} \in \mathcal{P}]$ can be made sufficiently small by imposing some fixed pointwise **distortion criterium** $d[f(\mathbf{p}), g(\mathbf{p})] < d_Q$ for all positions $\mathbf{p}$. The choice for a **distortion measure** $d(f, g)$ that is pointwise and fixed, i.e., that cannot be influenced by the signal amplitudes in neighboring positions, does not allow to optimally exploit the limitations of the visual system. However, this choice is often made since it leads to very simple quantization schemes that are easy to implement. More complex signal-approximation techniques, that are based on more complex distortion measures that do take into account neighboring image pixels, are usually referred to as coding techniques, and are not discussed in this book. For an excellent discussion on how CSF-based distortion measures can be applied in image coding, we refer to the PhD thesis of Nadenau [Nad00].

If the distortion measure and the **distortion bound** or **distortion threshold** d_Q are fixed for the entire image, then the required threshold value for the distortion bound d_Q is determined by the worst situation that occurs in an image. In order to get more insight into the distortions introduced by quantization, the error introduced by uniform quantization is studied in some detail in section 4. It is shown that the worst-case image features from the point of view of quantization are slowly-varying amplitudes over space and/or time. In such a case, the quantization error correlates highly with the input signal and cannot be considered as noise. It is derived in section 5 how the quality of quantized images can be improved by dithering. In this technique, one alternates, for successive positions, between a number of distinct quantizers that all satisfy the same pointwise distortion criterium. Although the pointwise distortion $d[f(\mathbf{p}), g(\mathbf{p})]$ is not influenced by these techniques, the differences between local averages in the input and output images can be decreased with these methods. This turns out to be especially beneficial, from a perceptual point of view, for those image features, such as the slowly-varying amplitudes, that typically determine the distortion bound for memoryless quantization. As a result, the distortion threshold d_Q can be raised for these worst-case features, and hence also for the entire image.

Some specific methods of dithering will be discussed in more detail. The least expensive method of dithering is called ordered dithering, and is discussed in section 6. In this case, a regular (periodic) pattern determines how the quantizer is varied across position. Section 7 discusses noise dithering, in which case a random (noise) source with specified statistics is used to vary the quantizer. Although noise dithering requires more computations and/or memory than ordered dithering, it does not lead to a better quality for the quantized image, so that this method is considered only marginally interesting for image quantization[1]. A better technique for improving the quality of quantized images is discussed in section 8. In this error diffusion technique, the quantization error, i.e., the difference between the original amplitude $f(\mathbf{p})$ and the quantized amplitude $g(\mathbf{p})$, is spread (diffused) over a number of neighboring sample positions in order to guarantee that the average quantization error is zero. This corresponds to a variation of the quantizer over position in a way that is signal-dependent. A perceptual evaluation of the different dithering techniques is presented in section 9.

In section 10, we summarize some guidelines on how the theory of quantization of scalar signals can be extended to color signals.

2. QUANTIZATION WITH BOUNDED DISTORTION

2.1 DISTORTION MEASURES

In order to evaluate the effect of amplitude quantization, some means of assessing the error between the original signal amplitudes $f(\mathbf{p})$ and the quantized amplitudes $g(\mathbf{p})$ is needed. In the case of memoryless quantization, it is assumed that this error can be evaluated on a point-by-point basis, i.e., it is assumed that the quantization at position $\mathbf{p}$ can be evaluated using only the distortion $d[f(\mathbf{p}), g(\mathbf{p})]$ at position $\mathbf{p}$. Since we are only interested at this point in memoryless quantization, we delete the discrete position index $\mathbf{p}$ for the time being.

It is usually very instructive to draw a 2D diagram with coordinates f and g in which curves of constant distortion $d(f, g)$ are shown. For instance, in figure 10.1, the case of a distortion measure

$$d(f, g) = |f - g|^{\alpha} \tag{10.1}$$

that only depends on the difference between the signal amplitudes is illustrated. This is the most frequently used distortion metric in engineering. It is however not in agreement with known adaptation phenomena in human vision. For instance, the classical experiment by Weber, in which a disc of luminance $L + \Delta L$ has to be detected on a background with luminance L, can be modelled, over a large range of luminance

[1]Noise dithering seems to perform much better when applied to the quantization of audio signals.

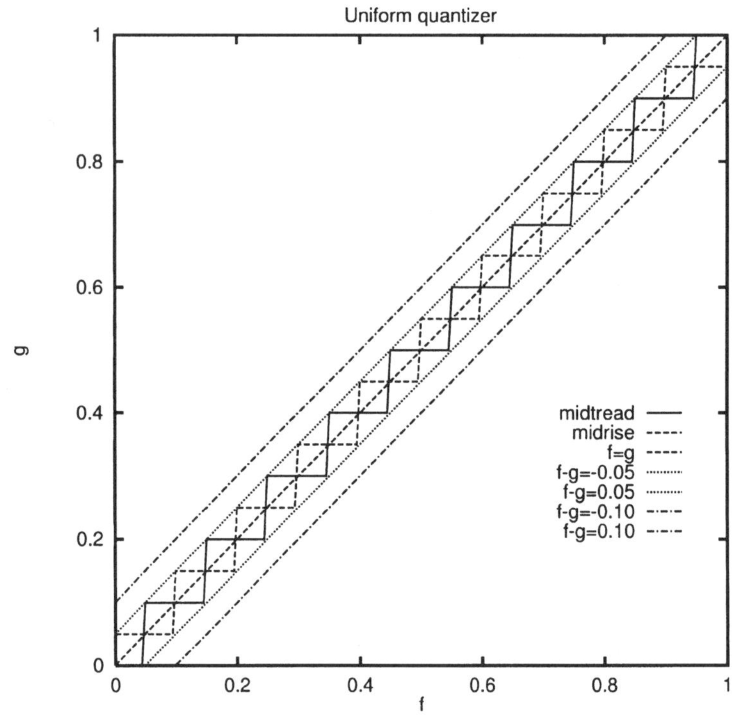

Figure 10.1. Amplitude distortion measure of equation (10.1) with $\alpha = 1$. The outer parallel curves connect points of constant distortion (with $d = 0.05$ and $d = 0.1$). The drawn curve shows a midtread quantizer characteristic, with $f_Q = 0$, for which the distortion is limited to $d_Q = 0.05$. The corresponding midrise quantizer is plotted as the dashed curve.

values L, by requiring that the distortion

$$d(L, L + \Delta L) = \frac{\Delta L}{L} \tag{10.2}$$

exceeds a fixed threshold value. Similarly, Lubin and Pica [Lub93] have measured the detectability of a vertical step edge halfway across a $17° \times 5.25°$ rectangle as a function of the base luminance L. Their experimental data can be described by

$$d(L, L + \Delta L) = \frac{\Delta L}{L^\beta} > d_o, \tag{10.3}$$

with $\beta = 0.85$ and $d_o = 0.01667$ (if L is expressed in cd/m^2).

Adaptation implies that the perceived distortion for a fixed difference $|f - g|$ decreases with increasing amplitude $|f|$ of the signal. This

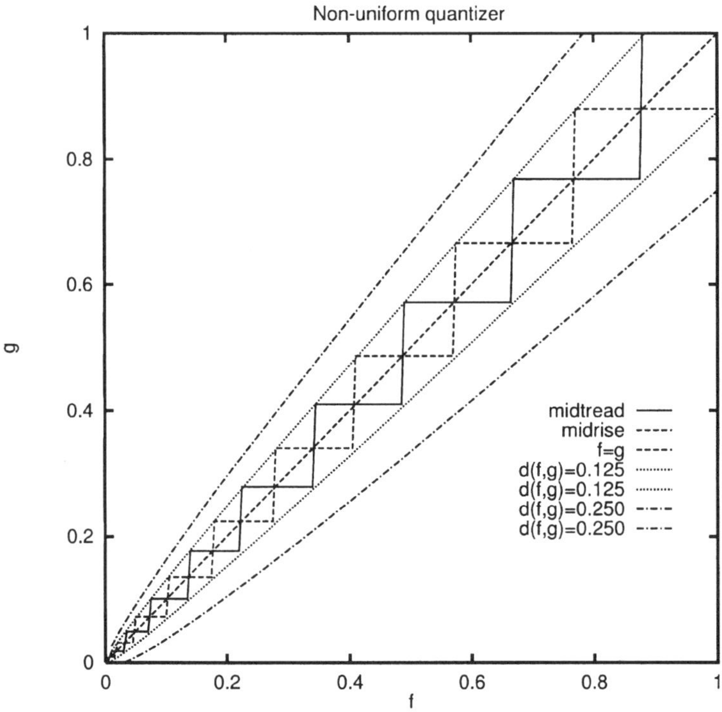

Figure 10.2. Amplitude distortion measure of equation (10.4) with $\alpha = 1, \beta = 0.6$ and $\delta = 0$. The outer curves connect points of constant distortion (for $d = 0.125$ and $d = 0.25$). The drawn curve shows a midtread quantizer characteristic, with $f_Q = 0$, for which the distortion is limited to $d_Q = 0.125$. The corresponding midrise quantizer is plotted as the dashed curve.

phenomenon can for instance be described by the distortion measure

$$d(f,g) = \frac{|f-g|^\alpha}{\max\left[\delta^\beta, |f|^\beta\right]} = \left(\frac{|f-g|}{\max\left[\delta^{\beta/\alpha}, |f|^{\beta/\alpha}\right]}\right)^\alpha . \qquad (10.4)$$

For small signal amplitudes ($|f| < \delta$), this distortion depends only on the amplitude difference $|f - g|$, while for larger signal amplitudes, the distortion is inversely proportional to $|f|^\beta$. The case of perfect adaptation ($\beta/\alpha = 1$) is unusual, and partial adaptation ($\beta/\alpha < 1$) is most frequently encountered. In figure 10.2, we show curves of constant distortion for the measure of equation (10.4).

2.2 QUANTIZER DESIGN

Designing a quantizer consists of specifying a finite set $\{g_1, \ldots, g_K\}$ of possible output values, called the **representation levels** or **reconstruction values**. If the input amplitude f falls within the interval $\Im_k = \{f_{k-1} \leq f < f_k\}$, then the output level g_k is produced, for $k = 1, \ldots, K$. The input amplitudes f_k, for $k = 0, 1, \ldots, K$, are called the **decision levels** or **decision thresholds**. The range of possible inputs is hence $[f_0, f_K[$. Both f_0 and f_K may be infinite. The overall mapping $g = Q(f)$ is called the **quantizer characteristic**, and is by definition a staircase function.

A straightforward way of designing a quantizer characteristic is to require that the distortion is bounded, i.e.,

$$d[f, Q(f)] \leq d_Q \tag{10.5}$$

for all possible input amplitudes f. The quantizer characteristic is obtained by constructing a staircase function that passes through a specified point (f_Q, f_Q) on the diagonal, and that stays within the distortion bounds for d_Q. This design technique is graphically illustrated in figures 10.1 and 10.2. The quantizer characteristics of figure 10.1 are called **uniform**, since there is a constant difference between the successive representation levels, as well as between the successive decision levels. The more general quantizer characteristics in figure 10.2 are called **nonuniform**. Note that the parameter α in the distortion measures of equations (10.1) and (10.4) can, without loss of generality, be equal to 1; it suffices to replace the distortion bound by $d_Q^{1/\alpha}$ if $\alpha \neq 1$.

Two types of quantizers are often distinguished. A quantizer for which f_Q is a representation level is called a **midtread** quantizer. If f_Q is not a representation level, but a decision level, then the quantizer is called **midrise**. For instance, if the distortion measure of equation (10.1) is used, with $\alpha = 1$, then the uniform midtread quantizer with $f_Q = 0$ has decision levels

$$f_0 = 0;\ f_k = (2k-1) \cdot d_Q,\ k \geq 1, \tag{10.6}$$

and representation levels

$$g_k = 2(k-1) \cdot d_Q,\ k \geq 1, \tag{10.7}$$

while the uniform midrise quantizer with $f_Q = 0$ has decision levels

$$f_k = 2k \cdot d_Q,\ k \geq 0, \tag{10.8}$$

and representation levels

$$g_k = (2k-1) \cdot d_Q,\ k \geq 1. \tag{10.9}$$

The **quantization stepsize** of these uniform quantizers, i.e., the difference between successive representation and decision levels, is $\Delta = 2 \cdot d_Q$.

2.3 COMPANDING

Any memoryless nonuniform quantizer characteristic can be realized using a **companding** technique [JN84]. This implies that the original signal amplitude f is *com*pressed to $c_1(f)$ by means of a nonlinear transfer characteristic, subsequently quantized by a uniform quantizer, and finally ex*pand*ed using the transfer characteristic $c_2(\cdot)$ to obtain the non-uniformly quantized output $Q(f)$. For instance, if the intermediate uniform quantizer is midtread with $f_Q = 0$ and quantization step $\Delta = 2 \cdot d_Q$, then the nonlinear function $c_1(\cdot)$ must be such that the decision levels f_k of the nonuniform quantizer are mapped into the decision levels

$$c_1(f_0) = 0,\ c_1(f_k) = (2k-1) \cdot d_Q\ (k \geq 1), \tag{10.10}$$

of the uniform quantizer. Similarly, the nonlinear function $c_2(\cdot)$ must map the representation levels of the uniform quantizer into the representation levels

$$c_2(2k \cdot d_Q) = g_k\ (k \geq 1), \tag{10.11}$$

of the nonuniform quantizer. Since the overall mapping from input to output amplitude must be linear, it is required that $c_2^{-1}(f) = c_1(f) = c(f)$, so that the common function $c(f)$ must satisfy the following conditions

$$c(f_0) = 0,\ c(f_k) = (2k-1) \cdot d_Q\ (k \geq 1),\ c(g_k) = 2k \cdot d_Q\ (k \geq 1), \tag{10.12}$$

in order to realize the nonuniform quantizer with the adopted uniform midtread quantizer. How the monotonically increasing function $c(f)$ varies between these specified points can be freely decided. The different steps in this compander design are illustrated graphically in figure 10.3.

If the companding function $c(f)$ is the (unique) piecewise linear function that satisfies equations (10.12), then the following interesting interpretation of the companding technique can be derived. If the uniform quantizer is designed such that

$$|c(f) - c_Q(f)| < d_Q, \tag{10.13}$$

i.e., using the simple distortion measure of equation (10.1), then the nonuniform quantizer of the companding technique satisfies

$$|c'(f)| \cdot |f - Q(f)| < d_Q, \tag{10.14}$$

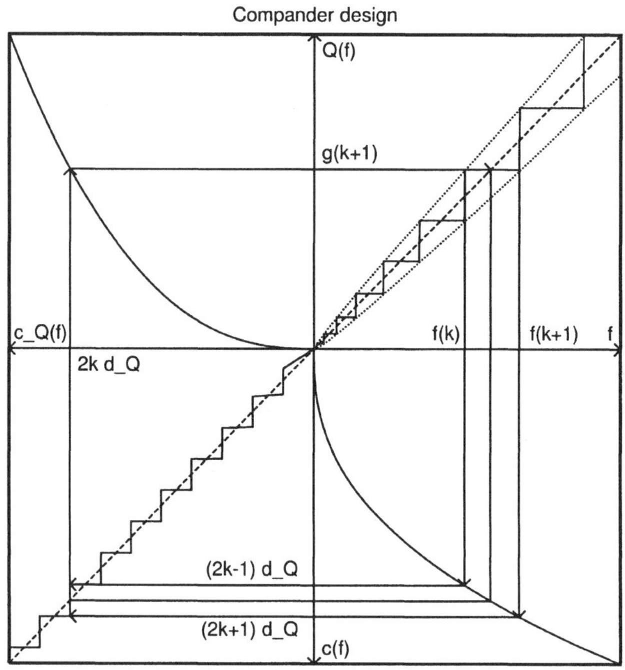

Figure 10.3. Companding technique for implementing nonuniform quantization. First quadrant: nonuniform quantizer characteristic. Fourth quadrant: the input amplitude f is mapped into the compressed amplitude $c(f)$. Third quadrant: the compressed amplitude $c(f)$ is uniformly quantized to $c_Q(f)$ by a midtread quantizer with $f_Q = 0$. Second quadrant: the compressed and quantized amplitude $c_Q(f)$ is mapped to $Q(f) = c^{-1}(c_Q(f))$.

where $c'(f)$ denotes the derivative of $c(f)$. If the quantization step is sufficiently small, then the piecewise constant function $c'(f)$ can be approximated by a continuous function.

In many imaging situations, displays such as CRTs are used. They have a nonlinear characteristic that can be approximated by $Q(L) = c^{-1}(V) \propto V^{\gamma}$ (or equivalently, $c(V) \propto V^{1/\gamma}$), where V is the grey value, $Q(L)$ is the displayed (i.e., quantized) luminance L, and $\gamma = 2.2$ is a typical value (see section 4 of chapter 4). Substituting this function in the previous expression results in the following distortion condition

$$\frac{|L - Q(L)|}{\gamma \cdot |L|^{1-\frac{1}{\gamma}}} < d_Q \qquad (10.15)$$

for the luminance in case the grey value V is quantized uniformly using a quantizer with distortion bound d_Q (or quantization stepsize $\Delta = 2 \cdot d_Q$). Hence, for $\gamma > 1$, the nonuniform companding quantizer is bounded for a distortion measure that is similar to equation (10.4), i.e., the distortion decreases with an increasing luminance L if the difference $|L - Q(L)|$ is fixed. As mentioned earlier, such a distortion measure with partial adaptation (with power $\beta = 1 - 1/\gamma$) approximates the distortion impression of a human observer.

3. STATISTICAL QUANTIZATION

In communication theory, a different approach towards designing quantizer characteristics is adopted. It is assumed that the probability density function[2] (PDF) $p_f(f)$ of the input f is known. The optimum quantizer characteristic is defined as the one that minimizes the expected value of the distortion, i.e.,

$$\min_Q E[d] = \min_Q \int d[f, Q(f)] \cdot p_f(f)\, df, \tag{10.16}$$

subject to known boundary conditions on the quantizer characteristic $Q(f)$ (such as a prescribed number of quantization steps). This is a non-linear optimization problem that can only be solved analytically for certain distortion measures $d(f, g)$ and boundary conditions. It is discussed at length in most textbooks on image coding [JN84, Jai89]. For instance, if the distortion measure is $d(f, g) = |f - g|^\alpha$, and the number of quantization steps is sufficiently high, then the optimum quantizer is equivalent to a compander design with

$$c'(f) \propto p_f(f)^{1/(1+\alpha)}. \tag{10.17}$$

Hence, a statistical quantization approach can be made equivalent to a quantization design based on bounded distortion, by selecting an adequate distortion measure. For most of the PDFs that have been considered (uniform, Gaussian, Laplacian), the equivalent distortion measure, obtained by substituting equation (10.17) into equation (10.14), deviates substantially from the psychophysical distortion measure of equation (10.4). More specifically, large signal amplitudes, that often have low probability (if a Gaussian or Laplacian PDF is adopted, for instance), are usually too coarsely quantized by a statistical quantizer. Consequently, very large errors can occur in few positions in the image. These large errors are easily perceived and have a strong negative impact on the image quality.

[2]The PDF expresses that amplitudes in the interval $[f, f + df]$ occur with probability $p_f(f)\,df$.

There are additional reasons why the approach of section 2 is thought to be superior to the statistical design approach. First, in the former approach, the distortion measure is made explicit, and the quantizer design is very simple. Second, the assumption that a fixed PDF $p_f(f)$ can accurately model the amplitude variations in an image is an oversimplification. The range of amplitudes that occur in a local image region typically varies within an image and across images. Third, the distortion measure in the statistical approach is supposed to depend solely on the statistical variations of the amplitude within the image. It is however more likely that the distortion measure is primarily determined by the observer and not by the image. Because the statistical quantizer optimization is usually performed on grey value images, some observer characteristics are implicitly taken into account (i.e., in an uncontrolled way) by the non-linearities in the A/D and D/A conversion. Last but not least, selecting a simple distortion measure such as $|f-g|^{\alpha}$ to determine the optimum quantizer characteristic is clearly a practical choice, made in order to enable a solution to the optimization problem, and is not supported by any insight into the visual perception of distortions. Indeed, these insights into perception have led to distortion measures such as equation (10.4) that are hard to handle in the statistical quantization approach.

In view of the above arguments, only quantization characteristics that are realized by means of a companding technique will be considered in the remainder of this chapter. The companding function $c(\cdot)$ is assumed to be determined by the distortion measure, so that one can concentrate on the design of the uniform quantizer. For the design of such a uniform quantizer one can, without loss of generality, assume the simple distortion measure of equation (10.1), with $\alpha = 1$.

4. UNIFORM QUANTIZATION

4.1 QUANTIZATION ERROR

The difference between the original amplitude f and the quantized amplitude $Q(f)$ is called the **quantization error**

$$q = g - f = Q(f) - f. \tag{10.18}$$

If $Q(f)$ is a uniform quantizer characteristic, then the difference between successive decision levels is equal to the quantization stepsize, i.e., $\Delta = f_k - f_{k-1} = 2 \cdot d_Q$, for $k = 1, \dots, K$. Since the representation levels

$$g_k = \frac{f_{k-1} + f_k}{2} \tag{10.19}$$

are located halfway in the decision intervals $[f_{k-1}, f_k[$, for $k = 1, \ldots, K$, the quantization error is restricted to $-d_Q < q \leq d_Q$.

We are interested in modelling the quantization error and its effect on the observer. In order to do this, it is necessary to make simplifying assumptions about the image and the observer. It was already remarked in the introduction that quantization effects are most probably not judged pointwise, but are more likely averaged over small areas within the image. We assume that the PDF of the input image within such a small area is given by $p_f(f)$. This PDF $p_f(f)$ will usually vary over the image. Note that we do not assume that the amplitude distribution over the entire image can be modelled by a single function $p_f(f)$.

From the definition it is easily derived that the PDF of the quantizer error q (within the small area) is equal to

$$p_q(q) = \sum_{k=1}^{K} p_f(g_k - q), \tag{10.20}$$

for $-d_Q < q \leq d_Q$. This PDF can hence be obtained by subdividing the original PDF $p_f(f)$ into slices of width $\Delta = 2 \cdot d_Q$ and adding all slices together [JN84]. If the quantization step d_Q is sufficiently small and/or the original PDF is sufficiently smooth and broad, then the quantization error will have a uniform PDF. The mean value of this uniform distribution is of course equal to zero, while the variance is equal to $d_Q^2/3$. One specific case where the quantization error will not have a uniform distribution is when all input values f are concentrated around a single value, i.e., if $p_f(f)$ has a peaked distribution with a peak width that is smaller than the quantization step. This situation for instance occurs in a uniform, or slowly varying, region of the image.

The above intuitive reasoning can also be made mathematically precise, i.e., it is possible to derive the necessary and sufficient condition on the PDF of the input signal in order for the PDF of the quantization error to be uniform [LWV92]. This condition is expressed most concisely in terms of characteristic functions. The **characteristic function** of a random variable is defined as the Fourier transformation of its PDF, i.e.,

$$P_q(u) = \int_{-\infty}^{+\infty} p_q(q)\, e^{j2\pi uq}\, dq, \tag{10.21}$$

where $u \in \mathbb{R}$ denotes frequency.

By properly choosing the companding function, it can always be accomplished that the uniform quantizer is midtread with representation levels $g_k = k \cdot \Delta$, for k integer. The PDF of the quantization error can then be expressed as

$$p_q(q) = \pi_\Delta(q) \cdot \sum_k p_f(k\,\Delta - q), \tag{10.22}$$

where

$$\pi_\Delta(q) = \begin{cases} 1, & -d_Q < q \leq d_Q \\ 0, & \text{otherwise} \end{cases} \tag{10.23}$$

denotes a rectangular window of width $\Delta = 2 \cdot d_Q$. The multiplication by this rectangular windows guarantees that the quantization error is zero outside the interval $]-d_Q, d_Q]$. The Fourier transformation of this rectangular window is equal to a **sinc function**

$$\Pi_\Delta(u) = \frac{\sin(\pi \Delta u)}{\pi \Delta u}, \tag{10.24}$$

for $u \in \mathbb{R}$. The rightmost sum in equation (10.22) can be interpreted as the interpolation of a constant signal (with amplitude equal to one) on a repetitive basis with basis function $p_f(-q)$. This basis function has Fourier transformation $P_f(-u)$, if $P_f(u)$ denotes the characteristic function of the input f. Such an interpolation can be decomposed in an upsampling operation, followed by a convolution with the interpolation function. In the current case this leads to

$$p_q(q) = \pi_\Delta(q) \cdot [p_f(-q) \star W_\Delta(q)], \tag{10.25}$$

where $\star$ denotes convolution, and

$$W_\Delta(q) = \sum_k \delta(q - k\,\Delta) \tag{10.26}$$

is a train of Dirac impulses separated by intervals of width Δ (i.e., the result of the upsampling operation on a constant discrete input signal). It was derived in equation (9.3) of chapter 9 that the Fourier transformation of such an interpolated signal is equal to $P_f(-u) \cdot W_{1/\Delta}(u)$.

It was observed in chapter 6 that a Fourier transformation maps a convolution in the space/time domain into a multiplication in the frequency domain. Because of the symmetry in the forward and inverse Fourier transformation, it is easily verified that a multiplication in the space/time domain must conversely be mapped into a convolution in the frequency domain. Applied to the case at hand, we obtain that the characteristic function of the quantization error is equal to

$$\begin{aligned} P_q(u) &= \Pi_\Delta(u) \star \left[P_f(-u) \cdot W_{1/\Delta}(u)\right] \\ &= \sum_k P_f\left(-\frac{k}{\Delta}\right) \frac{\sin[\pi\Delta(u - k/\Delta)]}{\pi\Delta(u - k/\Delta)}. \end{aligned} \tag{10.27}$$

Figure 10.4. Amplitude quantization as a sampling operator on the PDF. The PDF p_f of the input signal f (with continuous amplitude) is sampled, after filtering with a rectangular window π_Δ of width Δ, to obtain the PDF p_g of the quantized output signal g (with discrete amplitude).

If the quantization noise is to be uniformly distributed, then the characteristic function should reduce to a single sinc function centered at the origin. A necessary and sufficient condition is that

$$P_f\left(\frac{k}{\Delta}\right) = 0 \tag{10.28}$$

for $k = \pm 1, \pm 2, \ldots$. If the characteristic function of f is sufficiently smooth, i.e., bandlimited to the frequency range $]-1/\Delta, 1/\Delta[$, then this condition is automatically satisfied.

4.2 QUANTIZER OUTPUT

The output of the quantizer can only assume discrete values, so that its PDF is given by

$$p_g(g) = \sum_{k=1}^{K} \delta\left(g - g_k\right) \cdot \int_{f_{k-1}}^{f_k} p_f(x)\, dx, \tag{10.29}$$

i.e., the probability of the output g_k is equal to the probability that the input falls in the range $[f_{k-1}, f_k[$. Note that quantization hence corresponds to sampling the PDF of the input, as is illustrated in figure 10.4. The PDF of the input is first passed through a sampling filter π_Δ with a rectangular impulse response, and the discrete PDF of the output is obtained by sampling the filtered output on a 1D lattice with period Δ.

A similar derivation as for the quantization error reveals that the PDF of the output signal can be rewritten as

$$p_g(g) = [\Pi_\Delta \star p_f](g) \cdot W_\Delta(g), \tag{10.30}$$

with characteristic function

$$P_g(u) = \sum_k P_f\left(u - \frac{k}{\Delta}\right) \frac{\sin[\pi\Delta(u - k/\Delta)]}{\pi\Delta(u - k/\Delta)}. \tag{10.31}$$

This implies that the PDF of the input can be recovered from the PDF of the output if the former PDF is bandlimited, i.e., if $P_f(u) = 0$ for $|u| \geq 1/(2\Delta)$. Obviously, this result resembles the classical sampling theorem, which states that a band-limited signal can be recovered from discrete time samples. The notable difference is of course that the quantization theorem pertains not to time sampling, but to amplitude quantization of a signal.

In practice, knowing the PDF of a signal is often unnecessary. It is usually sufficient to recover the low-order moments of the PDF, such as the average value and the standard deviation. The m-th order moment of the output signal can be expressed as

$$E[g^m] = \int_{-\infty}^{+\infty} g^m \cdot p_g(g)\, dg = \left(\frac{1}{j2\pi}\right)^m \left.\frac{d^m P_g(u)}{du^m}\right|_{u=0}, \tag{10.32}$$

where $E[\cdot]$ denotes expected value. Based on the above expression for the characteristic function $P_g(u)$, it can be derived that the m-th order moment of the output signal can be expressed as

$$E[g^m] = \sum_{k=0}^{\lfloor m/2 \rfloor} \frac{m!}{(2k)!\,(m-2k)!} \frac{d_Q^{2k}}{2k+1} E[f^{m-2k}] \tag{10.33}$$

if

$$\frac{d^m}{du^m}\left[\frac{\sin(\pi\Delta u)}{\pi\Delta u} P_f(u)\right]\Bigg|_{u=k/\Delta} = 0 \tag{10.34}$$

for $k = \pm 1, \pm 2, \ldots$ [LWV92]. If $P_f(u)$ is bandlimited, i.e., $P_f(u) = 0$ for $|u| \geq 1/\Delta$, then this latter condition is obviously satisfied for arbitrary m.

4.3 CORRELATED QUANTIZATION ERRORS

The m-th order moment of a random variable with a rectangular PDF in the interval $[-d_Q, d_Q]$ is 0 if m is odd, and $d_Q^m/(m+1)$ if m is even. Equation (10.33) hence implies that the m-th order moment of $g = f + q$ is identical to the m-th order moment of $f + w$, where w is a random variable with a rectangular PDF that is statistically independent of f. This does however not imply that the quantization error q is statistically independent of the input signal f. Indeed, if the input signal f is known, then the quantization error q is even uniquely determined (i.e., if $f \in [f_{k-1}, f_k[$, then $q = f - g_k$).

Since the quantization error q is not statistically independent of the input signal f, it is not allowed to model the quantization error as an additive input-independent noise source. An approximation that is sometimes made is to model the output of a quantizer by

$$g = A_Q \cdot f + w, \tag{10.35}$$

where A_Q is a gain factor, and the noise signal w, with zero mean and variance σ_w^2, is uncorrelated with the input f (i.e., $E[f \cdot w] = 0$). This **gain-plus-additive-noise model** assumes that the quantization error consists of two contributions [JN84]. Part of the error is correlated with the input and can therefore be viewed as an error in reproduction, i.e., the output value $A_Q \cdot f$ differs from the input value f. The remaining error is uncorrelated with the input and can be characterized as noise. By equating the variances of the output and error amplitudes

$$\begin{aligned} E[g^2] &= A_Q^2 \cdot E[f^2] + E[w^2] = E[f^2] + 2\,E[f \cdot q] + E[q^2] \\ E[(g-f)^2] &= (A_Q - 1)^2 \cdot E[f^2] + E[w^2] = E[q^2] \end{aligned} \tag{10.36}$$

for the above model and the original model ($g = f + q$), the following expressions

$$\begin{aligned} A_Q &= 1 + \frac{E[q \cdot f]}{E[f^2]}, \\ E[w^2] &= E[q^2] - \frac{E[q \cdot f]}{E[f^2]} \cdot E[q \cdot f] \end{aligned} \tag{10.37}$$

are obtained for the gain factor A_Q and the noise variance $\sigma_w^2 = E[w^2]$.

The mathematical conditions under which quantization errors at different positions are statistically independent have also been determined using similar techniques as used above for deriving the first-order statistics [LWV92]. They confirm the following intuitive argument. If a pointwise amplitude quantization is applied to a discrete signal representation, then the quantization errors at distinct coordinates are approximately independent if the amplitude difference between neighboring discrete coordinates is much larger than the quantization step. Indeed, if several quantization levels are crossed to go from one amplitude level to the next, then it is highly improbable that the positions within the quantization intervals are correlated.

Obviously, the quantization errors are not statistically independent in image regions where the amplitude is varying slowly. The false contours that occur in quantized images illustrate this very well (see figure 10.5). These correlated quantization errors are very easily perceived, and determine the maximum distortion d_Q that can be allowed for the uniform quantizer. The fact that these correlated quantization errors are so annoying is yet another indication of the fact that the visual system does not judge distortions on a pointwise basis.

Figure 10.5. Image after quantization with a midrise quantizer with $f_Q = 128$ and $d_Q = 16$. The outer border of the original image increases linearly from 0 to 255. The contouring introduced by the pointwise quantization is clearly visible.

5. DITHERING

Because the quantization error is correlated with the input value, this error cannot be considered as noise in the usual sense. In order to get more insight into the consequences (and possible remedies) of this correlation, the mapping from input values f to output values $g = f + r$, where $E[f \cdot r] \neq 0$, is analyzed in a more general context. If the joint probability density function for the input and output values (always within the area over which the observer is assumed to integrate) is $p(f, g)$, then the error variance can be decomposed into

$$\sigma_r^2 = \int \int df \, dg \, p(f,g) \cdot (f-g)^2 = I_1 + I_2, \tag{10.38}$$

with

$$
\begin{aligned}
I_1 &= \int df \, p_f(f) \cdot (f - E[g|f])^2 \\
I_2 &= \int df \, p_f(f) \int dg \, p_{g|f}(g|f) \cdot (g - E[g|f])^2 ,
\end{aligned}
\tag{10.39}
$$

where

$$E[g|f] = \int dg \, p_{g|f}(g|f) \cdot g \tag{10.40}$$

is the conditional mean of the output g for a given input f [JN84]. The quantity I_2 is termed the **apparent noise** and represents the variation in the output with respect to the conditional mean output value. For the quantizers that we have considered up to now, $I_2 = 0$, since there is no variation in the output value g for a given input value f. The quantity I_1 is called the **tonal impairment** and expresses the deviation of the conditional mean output value from the true input. Note that $I_1 = 0$ implies that $E[g|f] = f$ for all possible input values f. Moreover, $I_1 = 0$ is equivalent to the demand $E[r \cdot f] = 0$ that the input and error signal are statistically uncorrelated.

Experimental evidence (see section 9) confirms that the degradation represented by I_1 is perceptually more annoying than the one described by I_2. Modifications to the fixed quantizer characteristic $g = Q(f) = f + q$ that reduce the term I_1, at the cost of an increase in I_2, are therefore worth considering. A simple modification that has proven very effective is to add a pattern p to the input f before quantization. By subtracting this pattern after quantization, the following output

$$g = Q(f + p) - p = f + \tilde{q}, \tag{10.41}$$

is obtained, where $\tilde{q}$ denotes the quantization error on the signal $f + p$. This technique is called **subtractive dithering**. A consequence of dithering is that the output value g for a given input value f is no longer fixed, but has some variation due to the varying pattern p, so that I_2 becomes larger than zero. The variance of the quantization error is not influenced by the dithering procedure, so that the sum $I_1 + I_2$ remains constant. This implies that I_1 must decrease, or equivalently, that the correlation $E[f \cdot \tilde{q}]$ between the input value and the quantization error is lower than the correlation $E[f \cdot q]$ in the case without dithering. This is because the quantization error $\tilde{q}$ not only correlates with the input signal, but also with the pattern p.

Suppose that the original quantizer Q has decision levels $\{f_k; k = 0, \ldots, K\}$ and representation levels $\{g_k; k = 1, \ldots, K\}$, then the relation between the input and output values $g = Q(f + p) - p = Q_p(f)$

defines a quantizer Q_p with decision levels $\{f_k - p; k = 0, \ldots, K\}$ and representation levels $\{g_k - p; k = 1, \ldots, K\}$. Hence, the function of the dithering pattern p is to select one out of a number of possible quantizers Q_p. Note that all the quantizers Q_p satisfy the same distortion bound d_Q. Referring to Figure 10.1, quantizer Q_p is obtained by shifting quantizer Q_0 along the diagonal, such that the horizontal and vertical shift are equal to p. Shifting over a distance that is larger than the quantization stepsize $\Delta = 2 \cdot d_Q$ does not produce new quantizer characteristics, so that $-d_Q < p \leq d_Q$ is an obvious demand for the dithering pattern in case of a uniform quantizer with distortion bound d_Q.

In many display situations, the output amplitude g is necessarily quantized. This implies that the dithering pattern p cannot be subtracted at the output, and that the quantized signal $g = Q(f + p)$ is retained as output. This corresponds to a horizontal shift of the quantizer characteristic, rather than a shift along the diagonal. Consequently, if $-d_Q < p \leq d_Q$, then the maximum distortion increases from d_Q to $2\, d_Q$. Even in this case of **non-subtractive dithering**, the quality of the dithered image is usually better then in the case without dithering. The decrease in the tonal impairment I_1 is the same as before, but the sum $I_1 + I_2$ increases. The fact that the quality of the output image increases nevertheless, as will be verified in section 9 for a number of dithering schemes, substantiates the earlier claim that the tonal impairment is more important then the apparent noise. In the remainder of the chapter, we will mostly consider non-subtractive dithering. Especially important is the case of **halftoning**, which corresponds to non-subtractive dithering with a binary quantizer (i.e., a quantizer with only two representation levels).

Designing a dithering pattern involves a number of conflicting demands. First, the pattern must have sufficient variation to decrease the correlation $E[f \cdot \tilde{q}]$ between the input value and the quantization error, or equivalently, to guarantee that the expected output is close to the original input for all possible input values. Second, especially in the case of non-subtractive dithering, the dithering pattern is an unwanted part of the output, and hence its visibility must be as low as possible. Third, in order to keep the cost of the quantization system as low as possible, the dithering pattern must be easy to generate, with a minimum required number of operations and/or memory.

6. ORDERED DITHERING

The simplest, and therefore least-expensive, method of dithering is called **ordered dithering**. In this case, a pre-computed pattern is used for dithering. The pattern is chosen periodic in order to reduce the amount of memory needed to store it.

The signal before quantization is assumed discrete, with sampling lattice Λ and unit cell $\wp$, and periodic, with periodicity lattice Λ_p and unit cell $\wp_p$. A small dither matrix is repeated periodically on a lattice Λ_o that is intermediate between Λ and Λ_p (i.e., $\Lambda_p \subset \Lambda_o \subset \Lambda$), i.e.,

$$p(\mathbf{x}) = p(\mathbf{x} + \mathbf{y}), \tag{10.42}$$

for $\mathbf{y} \in \Lambda_o$ and $\mathbf{x} \in \Lambda$. The dithering pattern is completely specified by the values $p(\mathbf{x}), \mathbf{x} \in (\wp_o \cap \Lambda)$, within one unit cell $\wp_o \cap \Lambda$ of the dither lattice Λ_o. Usually the pattern is pre-computed and stored in a memory of size $d(\Lambda_o)/d(\Lambda)$, where $d(\Lambda_o)$ and $d(\Lambda)$ are the determinants of the lattices Λ_o and Λ, respectively. Because the dithering pattern is position-variant, we re-introduce the position index $\mathbf{x}$ at this point of the discussion.

Suppose that the original quantizer has K different representation levels. The mean output value over one unit cell $\wp_o \cap \Lambda$ of the dithering pattern can then assume maximally $K \cdot d(\Lambda_o)/d(\Lambda)$ different values. Hence, provided that the visual system of the observer integrates over an area that is larger than the unit cell, then the number of possible output values increases proportional to $d(\Lambda_o)$ with the help of ordered dither. The difference between the input f and the expected output $E[g|f]$ obviously decreases with an increase in the number of representation levels. Hence, large dithering matrices are advantageous to reduce the tonal impairment.

In order to study the visibility of periodic patterns such as the dithering pattern, it is advantageous to use a spatial frequency representation. A large body of psychophysical data is available on the detectability and visibility of such periodic patterns. The detection thresholds can be specified as a function of the modulation depths of the frequencies that constitute the pattern, while pattern visibility can be expressed in terms of the normalized modulation depths (see chapter 9). Because the dithering pattern $p(\mathbf{x})$ is periodic with lattice Λ_o, its discrete Fourier series is equal to

$$\begin{aligned} P(\mathbf{u}) &= d(\Lambda) \sum_{\mathbf{x} \in (\wp_p \cap \Lambda)} p(\mathbf{x}) \exp\left(j 2\pi \mathbf{u}^T \mathbf{x}\right) \\ &= \begin{cases} \frac{d(\Lambda_p)\, d(\Lambda)}{d(\Lambda_o)} \sum_{\mathbf{x} \in (\wp_o \cap \Lambda)} p(\mathbf{x}) \exp\left(j 2\pi \mathbf{u}^T \mathbf{x}\right) & \text{if } \mathbf{u} \in (\wp^\star \cap \Lambda_o^\star) \\ 0 & \text{if } \mathbf{u} \notin (\wp^\star \cap \Lambda_o^\star) \end{cases} \end{aligned} \tag{10.43}$$

for $\mathbf{u} \in (\wp \cap \Lambda_p^\star)$. The modulation depth at frequency $\mathbf{u}$ is hence equal to $2 \cdot P(\mathbf{u})/f$ if the dithering pattern $p(\mathbf{x})$ is superimposed on a uniform background with value f. The dithering pattern contains only frequencies that belong to the reciprocal lattice $\Lambda_o^\star$ of the dither lattice Λ_o.

Since the human visual system is more sensitive to low-frequency components than to high-frequency components, the dithering pattern is usually designed to make the frequency components $P(\mathbf{u}), \mathbf{u} \in \wp^\star \cap \Lambda_o^\star$, that are closest to the origin as small as possible [RAA90]. For instance, the condition

$$\sum_{\mathbf{x} \in (\wp_o \cap \Lambda)} p(\mathbf{x}) = 0 \tag{10.44}$$

is equivalent to the demand that $P(\mathbf{0}) = 0$.

Some of the design procedures lead to recursive algorithms for generating the dither matrices [RA88]. For example, dither matrices of size $d(\Lambda_o)/d(\Lambda) = 2^k \times 2^k$ can be generated by starting with a well-designed 2×2 matrix such as

$$M_1 = \begin{bmatrix} m_{00} & m_{01} \\ m_{10} & m_{11} \end{bmatrix} = \begin{bmatrix} 0 & 2 \\ 3 & 1 \end{bmatrix}, \tag{10.45}$$

followed by a recursion relationship of the form

$$M_k = \begin{bmatrix} 4 \cdot M_{k-1} + m_{00} \cdot E_{k-1} & 4 \cdot M_{k-1} + m_{01} \cdot E_{k-1} \\ 4 \cdot M_{k-1} + m_{10} \cdot E_{k-1} & 4 \cdot M_{k-1} + m_{11} \cdot E_{k-1} \end{bmatrix}, \tag{10.46}$$

where E_k is a matrix of size $2^k \times 2^k$ with all elements equal to one. For instance, the first recursion produces the 4×4 matrix

$$M_2 = \begin{bmatrix} 0 & 8 & 2 & 10 \\ 12 & 4 & 14 & 6 \\ 3 & 11 & 1 & 9 \\ 15 & 7 & 13 & 5 \end{bmatrix} \tag{10.47}$$

[JN84]. The actual dither matrix of size $2^k \times 2^k$ has elements

$$p(\mathbf{x}) = \frac{2 \cdot A_p}{2^{2k} - 1} \left(M_k(\mathbf{x}) - \frac{2^{2k} - 1}{2} \right), \tag{10.48}$$

for $\mathbf{x} \in \wp_o \cap \Lambda$, that are obtained by subtracting an offset (in order to guarantee that the mean value of the dither elements is zero) and scaling. The resulting dither pattern has amplitudes in the range $[-A_p, A_p]$. The amplification factor is usually such that the maximum dither amplitude is slightly smaller than the distortion bound d_Q of the quantizer, i.e., $A_p \approx d_Q$. The dithering patterns of sizes 4×4 and 8×8, together with their Fourier transformations, are shown in Figure 10.6. Extensive

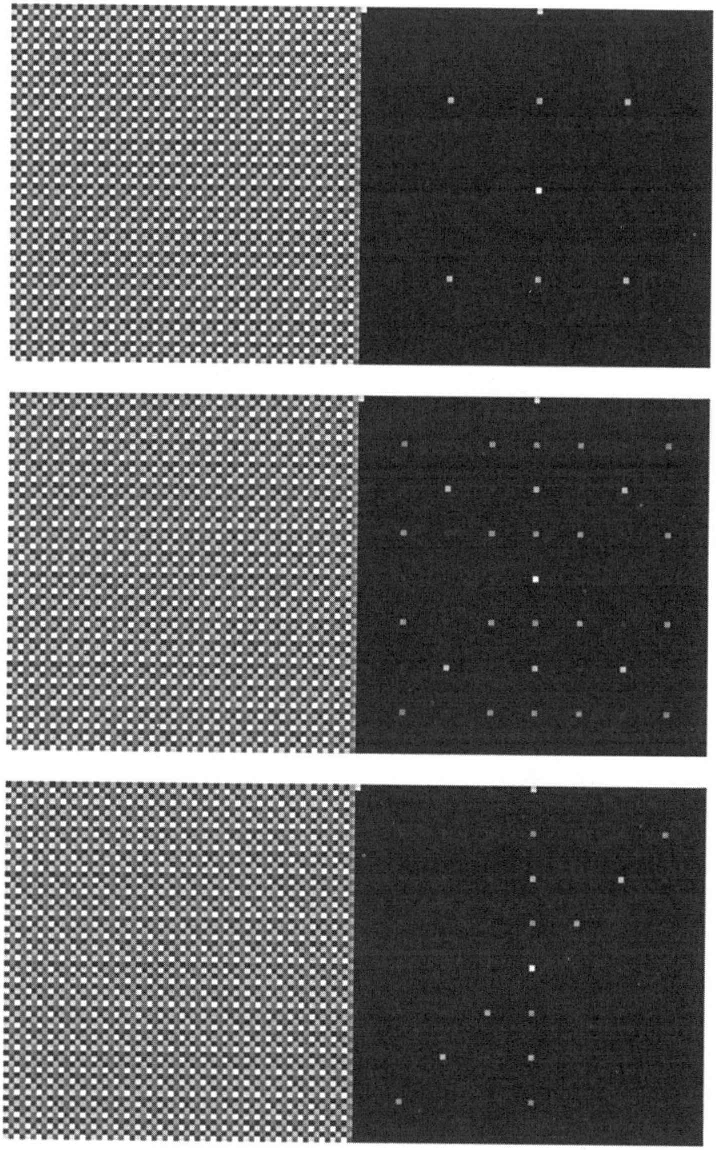

Figure 10.6. Ordered dithering patterns of sizes 4×4 (top) and 8×8 (middle and bottom) on a uniform background, together with their Fourier transformations. The Fourier amplitude has been mapped to $|P(u)|^{0.2}$ for display purposes.

details on the properties of specific ordered dithering patterns can be found in some of the references (e.g., [RA88]). An alternative dithering pattern of size 8×8 from the above reference is also shown in the bottom part of Figure 10.6.

In case of non-subtractive dithering on a binary display that can only generate two levels (on/off), the integers in the above matrix M_k indicate the order in which the pixels are turned on for a constant input signal. Depending on the (constant) input value, one out of $2^k \times 2^k + 1$ different patterns is generated. Pattern 0 has all pixels turned off, while pattern i, for $i = 1, \ldots, 2^{2k}$, has pixels with integer code 0 through $i - 1$ turned on. A well-known example of such a binary display is an ink-jet printer that can either eject ink or leave the page blank at the position of the printer head. A complicating factor with such printing devices is that not all dithering patterns can be equally well reproduced. Because the size and shape of a single ink dot is usually not defined very precisely, patterns with alternating zeros and ones can for instance not be produced accurately, because the ink on the "on" positions will also partly cover the blank paper at the "off" positions. Therefore, most printers use ordered dither patterns that have been specifically designed for them by the manufacturer.

Increasing the size $d(\Lambda_o)$ of the unit cell $\wp_o$ introduces harmonics of decreasing frequencies in the (repeated) dither pattern, see Figure 10.6, which implies that the pattern will be easier to perceive. Therefore, selecting the value of $d(\Lambda_o)$ is a compromise between a sufficient number of representation levels, requiring a large value for $d(\Lambda_o)$, and a low visibility of the dithering pattern, requiring a small value for $d(\Lambda_o)$. At this point there does not seem to exist a systematic procedure to determine how these two degrading distortions can be balanced. The best strategy is to generate different alternatives and compare the results subjectively. The current consensus is that, in typical display situations, where the sampling density $d(\Lambda)$ is close to the acuity of the visual system, dithering matrices with a unit cell of size 4×4 to 8×8 are close to optimum for a binary display. Smaller dither matrices, such as 2×2, do not provide sufficient gray values, while larger dither matrices, such as 16×16, provide gray values that are not needed, at the cost of a more conspicuous dither pattern. Figure 10.7 shows an image that was generated on a binary display using an 8×8 ordered dither matrix.

The ordered dither method has two important advantages. First, the simple, low-cost implementation makes it ideally suited for use in systems that have limited computational resources, such as printers. Second, since the dither pattern is selected a priori, it can be optimized for a specific output device. This device-specific optimization is especially

Figure 10.7. Binary reproduction of an image, using an 8 × 8 dither matrix on a square lattice.

important if the output device can no longer be described by the simple pointwise nonlinearity that was assumed in the compander design, but also involves spatial interactions between neighboring pixels. If the output is a linear display device that can be characterized by a single display response function $\varphi_d(\mathbf{x})$, such as a CRT or LCD, then this display function usually has little influence on the results discussed so far, since it only influences the modulation depths of the harmonic components in the dithering pattern (in a similar way as in the case of the display optimization discussed in chapter 8). Device-specific solutions will however be required in cases where the output device is clearly non-linear (such as most printing devices).

The quality of ordered-dithered images is limited by two important observations regarding the perception of regular patterns. First, regular (periodic) patterns are much easier to perceive than irregular patterns of the same amplitude [Cow85]. Second, the regular pattern, or texture, in the quantized output changes with the amplitude. Hence, although the difference between the mean values of two connecting patches may be small, the patches may have very distinct textures. Small, insignificant differences in amplitude can thus give rise to large, obvious changes in texture [FS86]. Alternative dithering methods that can potentially reduce these problems have therefore been studied.

7. NOISE DITHERING

Since periodic patterns are more easily visible than unstructured patterns, it is straightforward to try to improve the ordered dither method of the previous section by using uncorrelated noise[3] as the dithering pattern. In comparison to ordered dithering, noise dithering requires more calculations, needed for implementing the noise generator, and/or more memory, needed for storing the noise pattern. Noise dithering is therefore only interesting if a substantial quality improvement in comparison to ordered dithering can be obtained. The topic of noise dithering is discussed in depth in the excellent review paper by Lipshitz et al. [LWV92], and we therefore restrict ourselves to summarizing the main results.

We first discuss the case of subtractive dithering. The input to the quantizer is $f + n$, where the noise n, with PDF $p_n(n)$, is statistically independent of the input f. The output is $g = Q(f+n) - n$. In analogy with section 4.1, it can be derived that the PDF of the error $q = g - f$ can be expressed as

$$p_q(q) = \pi_\Delta(q) \cdot [p_f(-q) \star p_n(-q) \star W_\Delta(q)] , \tag{10.49}$$

where $p_f(f)$ denotes the PDF of the input signal f. The characteristic function of the error q is hence equal to

$$P_q(u) = \frac{\sin(\pi \Delta u)}{\pi \Delta u} \star \left[W_{1/\Delta}(u) \cdot P_f(-u) \cdot P_n(-u)\right] , \tag{10.50}$$

which implies that the condition

$$P_n\left(\frac{k}{\Delta}\right) = 0, \tag{10.51}$$

for $k = \pm 1, \pm 2, \ldots$ is sufficient to guarantee that the error is uniformly distributed. It has been demonstrated [LWV92] that this condition is also necessary and sufficient to guarantee that the error q is statistically

[3] A short introduction to stochastic processes, noise, autocorrelation functions, power spectra, etc. is provided in Appendix 10.A.

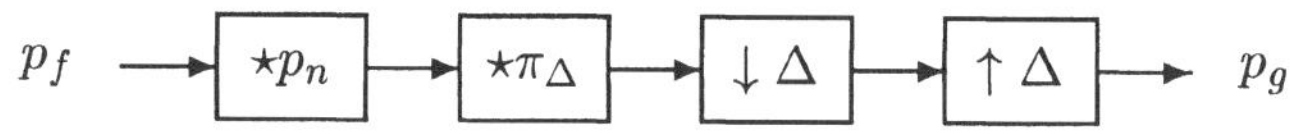

Figure 10.8. Non-subtractive noise dithering as a sampling operator on the PDF. The PDF p_f of the input signal f is convolved with the PDF p_n of the additive noise n and with a rectangular window π_Δ of width Δ, and subsequently sampled to obtain the PDF p_g of the output signal g.

independent of the input f. Hence, from a theoretical point of view, subtractive noise dithering seems interesting.

If the dither noise satisfies the above condition, then the output $g = f + q$ is the sum of two statistically independent random variables, and its PDF

$$p_g(g) = [\Pi_\Delta \star p_f](g) \tag{10.52}$$

is obtained by convolving the input PDF with a rectangular PDF.

An interesting special case worth mentioning is the case that the dither noise n is obtained by adding N statistically independent random variables with rectangular PDF in the interval $[-d_Q, d_Q]$. The PDF of n, with $n \in [-N \cdot d_Q, N \cdot d_Q]$, is then an $(N-1)$-th order spline function with characteristic function

$$P_n(u) = \left[\frac{\sin(\pi\Delta u)}{\pi\Delta u}\right]^N, \tag{10.53}$$

that obviously satisfies the above condition. The noise has variance $N \cdot d_Q^2/3$, so that the variance of the output is given by

$$E[g^2] = E[f^2] + N \cdot \frac{d_Q^2}{3}. \tag{10.54}$$

Uniform noise, corresponding to $N = 1$, results in the minimum noise variance and is therefore most often used in subtractive dithering.

The case of non-subtractive dithering is very similar to the case of no dithering. The input to the quantizer is $f + n$, instead of f, where the noise n is statistically independent of the input signal f. The PDF of this sum of two statistically independent signals is equal to the convolution of the individual PDFs, i.e., $p_{f+n} = p_f \star p_n$. The output is $g = Q(f + n)$. Figure 10.8 illustrates how non-subtractive noise dithering can be described as a sampling operation. This result can be derived in a straightforward way by analogy with section 4.1.

The output error $q = g - f$ is no longer statistically independent of the input f in the case of non-subtractive noise dithering[4], but some of the low-order (conditional) error moments $E[q^m|f]$ can be made independent of f if and only if

$$\frac{d^m}{du^m}\left[\frac{\sin(\pi\Delta u)}{\pi\Delta u}\cdot P_n(u)\right]\Bigg|_{u=k/\Delta} = 0 \tag{10.55}$$

for $k = \pm1, \pm2, \ldots$. If this condition is satisfied, then the m-th order error moment can be expressed as

$$E[q^m] = \sum_{k=0}^{\lfloor m/2\rfloor} \frac{m!}{(2k)!\,(m-2k)!}\,\frac{d_Q^{2k}}{2k+1}\,E[n^{m-2k}]. \tag{10.56}$$

If the above condition is satisfied for all derivatives up to order M, then the m-th order moment of the output is given by

$$E[g^m] = \sum_{k=0}^{m} \frac{m!}{k!\,(m-k)!}\,E[q^k]\,E[f^{m-k}], \tag{10.57}$$

for $m \leq M$ [LWV92].

Specifically, if the dither noise n is obtained by adding N (with $N \geq 2$) statistically independent random variables with rectangular PDF in the interval $[-d_Q, d_Q]$, then the variance of the output is given by

$$E[g^2] = E[f^2] + (N+1)\cdot\frac{d_Q^2}{3}. \tag{10.58}$$

Noise with a triangular PDF, corresponding to $N = 2$, is therefore very often used in the non-subtractive dithering of audio signals.

Figure 10.9 shows an image that was reproduced on a binary display using non-subtractive dithering with noise with a triangular PDF. Note that, although the average tonal impairment is better than in the case of ordered dithering, the apparent noise degrades the image up to the point where relevant image details can no longer be distinguished. Hence, noise dithering is not very useful for binary displays. If a quantizer with a smaller stepsize is used, then the apparent noise will be smaller, and noise dithering can help to improve the quality of quantized images. All in all, it can however be concluded that noise dithering seems much more useful for audio quantization, for which it was originally developed, than for image quantization. This will also be confirmed by the experimental data in section 9.

[4]This implies that $p(q|f)$ cannot be made independent of f by properly choosing the PDF $p_n(n)$ of the dither noise n.

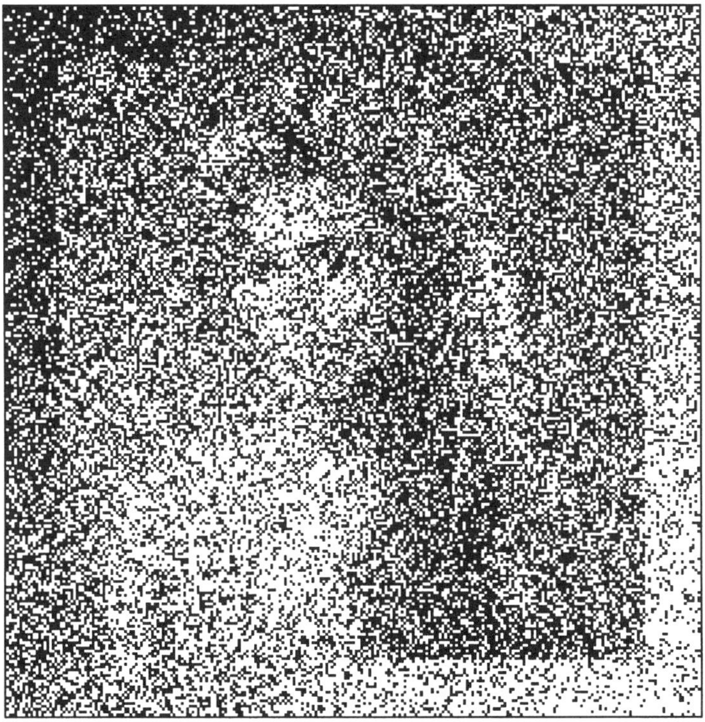

Figure 10.9. Binary reproduction of an image, using non-subtractive dithering with noise with a triangular PDF.

8. ERROR DIFFUSION

The purpose of dithering is to guarantee that, on the average, all input amplitudes are reproduced correctly. Up to now it was assumed that the observer averages the quantization error over a sufficiently large area in the image, so that this average error is expected (on statistical arguments) to be zero. The possibility that quantization errors of the same sign aggregate locally can however not be excluded. Significant tonal impairments will result at those positions.

Instead of relying on statistical averaging of the quantization error, one could also adjust the dithering pattern to the signal such that the averaged quantization error is forced to zero. This has been realized in a method called **error diffusion**, or **error correction**. It is assumed that the input samples are processed in a scanning order from left to

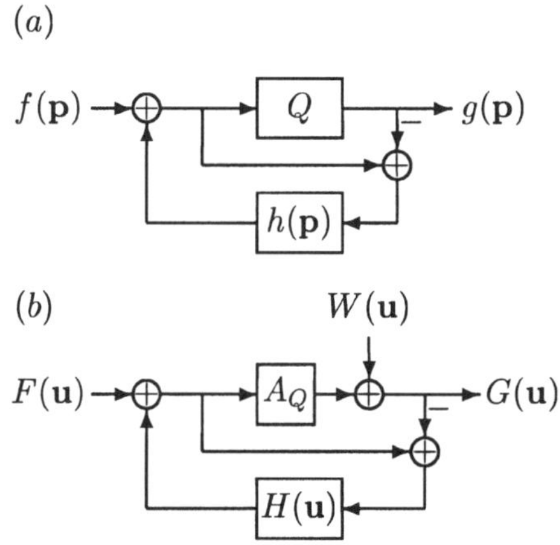

Figure 10.10. (a) Error diffusion algorithm: the dithering pattern is obtained by low-pass filtering the quantization error with the filter $h(\mathbf{p})$ with Fourier transformation equal to $H(\mathbf{u})$; (b) Using a gain-plus-additive-noise model for the quantizer allows to derive an expression between the discrete Fourier transformations $F(\mathbf{u})$ and $G(\mathbf{u})$ of the input and output image, respectively.

right and top to bottom[5]. The quantization error of the current sample is not lost, but distributed among the input samples that have not yet been processed. The resulting algorithm is illustrated in figure 10.10.

The most frequently used error-diffusion filters are the four-neighbor filter

$$\begin{matrix} & \cdot & 7 \\ 3 & 5 & 1 \end{matrix} \left(\times\frac{1}{16}\right) \tag{10.59}$$

and the twelve-neighbor filter

$$\begin{matrix} & & \cdot & 7 & 5 \\ 3 & 5 & 7 & 5 & 3 \\ 1 & 3 & 5 & 3 & 1 \end{matrix} \left(\times\frac{1}{48}\right) \tag{10.60}$$

[Uli88], where $\cdot$ indicates the current sample. This implies for instance, for the four-neighbor filter, that 7/16-th of the current quantization error is added to the next sample on the same line, before this next sample is quantized. Figure 10.11 shows an image that was generated on a binary display using error diffusion with the four-neighbor filter. Figure 10.12

[5]Serpentine scanning in which horizontal lines are alternatively scanned from left to right and right to left has also been considered, see [Uli88]

Figure 10.11. Binary reproduction of an image, using error diffusion with a four-neighbor filter.

shows the same image, using error-diffusion with the twelve-neighbor filter. Note that the latter image appears to be sharper, but also has a more easily perceivable dithering pattern.

By using the gain-plus-additive-noise model for the quantizer, we can obtain the following approximate relationship between the discrete Fourier transformations of the input and output images, i.e.,

$$G(\mathbf{u}) \approx \frac{1}{1 - H(\mathbf{u}) \cdot (1 - A_Q)} \left(A_Q \cdot F(\mathbf{u}) - [1 - H(\mathbf{u})] \cdot W(\mathbf{u}) \right), \tag{10.61}$$

for $\mathbf{u} \in \wp$, where $W(\mathbf{u})$ denotes the Fourier transformation of the additive noise. The low-pass filter $H(\mathbf{u})$ is such that $H(\mathbf{0}) = 1$ and $H(\mathbf{u}) \to 0$ for large frequencies $\mathbf{u}$. This implies that the filter $1/[1 - H(\mathbf{u}) \cdot (1 - A_Q)]$ varies from $1/A_Q$, for $\mathbf{u} = \mathbf{0}$, to 1, for large $\mathbf{u}$. Since A_Q is usually close to 1, the effect of this filter is fairly limited.

Figure 10.12. Binary reproduction of an image, using error diffusion with a twelve-neighbor filter.

A number of conclusions can be drawn from the above expression. First, $G(\mathbf{0}) = F(\mathbf{0})$, i.e., on average all constant inputs are reproduced correctly. Second, the main effect of error diffusion is that the power spectrum of the noise is filtered by $|1 - H(\mathbf{u})|^2$. This is called **noise shaping**. Low-frequency components in the quantization noise are effectively removed by error diffusion. The resulting high-frequency noise is sometimes called **blue noise**, and is perceptually less annoying than white noise with a uniform spectrum [Uli88]. The quantization error signal with error diffusion is usually not a regular pattern, so that it is more difficult to perceive than the regular ordered dither error patterns. Regular patterns may however occur for constant signal inputs with specific values. If necessary, these remaining patterns can be suppressed by adding a small ordered dither pattern if the input is close to a value for which regular quantization patterns are expected [FS86].

9. EXPERIMENTAL COMPARISON

Partly because of the nonlinearities involved in quantization, it is very difficult to get a good insight into the pros and cons of different quantization techniques solely on the basis of theoretical arguments. This is especially true if the quantization is very coarse, such as in the case of halftoning. In this section, we present the results of a psychophysical experiment that was conducted in order to compare the different dithering techniques that were presented in the previous sections.

Two different scenes, Terrace and Wanda, were used in the experiments. The original images contained 240×470 pixels, and 8 bits per pixel, and are displayed in Figure 5.1. They were quantized using the following techniques:

1 non-subtractive ordered dithering with a 4×4 Rao dither matrix;

2 subtractive ordered dithering with a 4×4 Rao dither matrix;

3 non-subtractive ordered dithering with a 8×8 Rao dither matrix;

4 subtractive ordered dithering with a 8×8 Rao dither matrix;

5 non-subtractive ordered dithering with noise with a uniform PDF;

6 subtractive ordered dithering with noise with a uniform PDF;

7 non-subtractive ordered dithering with noise with a triangular PDF;

8 subtractive ordered dithering with noise with a triangular PDF;

9 error diffusion with the four-neighbor filter;

10 error diffusion with the eight-neighbor filter;

(the above numbering will be used to refer to the different techniques). The amplitude range for the ordered dither patterns and the uniform noise was $[-d_Q, d_Q]$, while this range was $[-2\,d_Q, 2\,d_Q]$ for the triangular noise. Four different values of the distortion bound d_Q (=128, 32, 16, 8) were used in order to create uniform midrise quantizers, centered at $f_Q = 128$, with two, four, eight and sixteen levels. All quantized images were limited to the range [0,255] before being displayed. The calibrated display characteristic and the viewing conditions were identical to those mentioned in section 2 of chapter 5.

Including the original image, there were 41 images per scene. These images were divided into four subsets denoted by S_2, S_4, S_8 and S_{16}. Images processed with a uniform quantizer with n levels were contained in the subset S_n (the original image was added to subset S_{16}). Three additional subsets $S_{n,2n}$ of six images per scene were created by combining three images from subset S_n with three images from subset S_{2n}. All pairwise combinations of images in a subset were judged with respect to preference. Preference could be expressed by means of integer numbers in the range $[-9, 9]$. Three or four subjects participated in each of the seven experimental sessions.

The combined experimental data was analyzed using the discrete version of the MDS program XGms. A one-dimensional vector-product model was assumed to describe the preference scores, and the non-metric nature of the data was modelled by means of power-like functions (see chapter 5). The resulting stimulus configurations are shown in Figure 10.13, and reflect the quality sensations attributed to the different images. The following conclusions can be drawn from these experimental results:

- Images quantized with non-subtractive (1,3) and subtractive (2,4) ordered dithering have the same quality at all levels of quantization.
- Subtractive noise dithering (6,8) delivers images that are substantially better in quality then those obtained with non-subtractive noise dithering (5,7).
- Noise dithering is clearly inferior to the other dithering schemes.
- The quality of error-diffused images with a four-neighbor filter (9) is slightly better than that of images quantized with ordered dithering (1,2,3,4).
- Error-diffusion with a twelve-neighbor filter (10) is less appreciated than error-diffusion with a four-neighbor filter (9).
- The quality of error-diffused images with a four-neighbor filter (9) is about equal for a 2-level and a 4-level quantizer.
- Error-diffusion with a four-neighbor filter (9) and a quantizer with 16 levels provides almost the same quality as memoryless quantization to 256 levels.
- The portrait image Wanda is more sensitive to quantization errors than the outdoor scene Terrace.

10. QUANTIZATION OF COLOR IMAGES

Suppose that color images are produced by additive mixing, so that three color signals (f_1, f_2, f_3), representing the weights of the primaries, need to be quantized. A straightforward extension of the 1D quantization technique of this chapter consists of assigning a triplet (g_1, g_2, g_3),

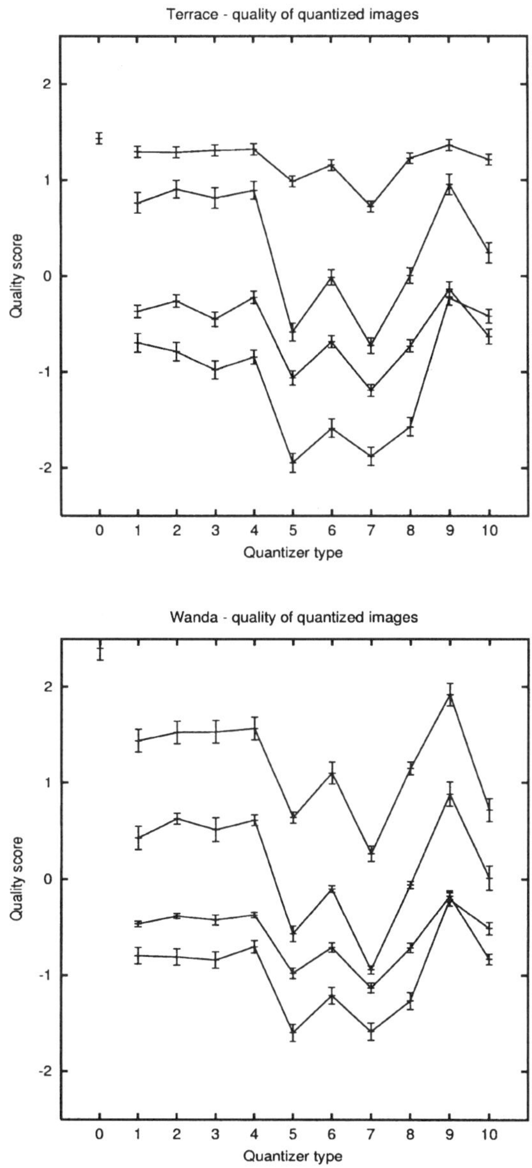

Figure 10.13. Quality scores for the scenes Terrace (top) and Wanda (bottom), as derived from the experiments with quantized images. The curves from top to bottom connect the images quantized with 16, 8, 4, and 2 levels, respectively. The isolated points corresponds to the original 8 bit images. See the text for a description of the 10 quantizer types.

from a finite set of possible triplets, to each input combination (f_1, f_2, f_3). This corresponds to partitioning the 3D space, spanned by (f_1, f_2, f_3), into decision regions, and assigning the same output (g_1, g_2, g_3) to all input combinations falling within one decision region. A color distortion measure and a maximum allowable distortion are required to design such a quantizer.

There are a number of reasons why the above general color quantization scheme is very seldom applied. First, it is difficult to find a reliable color distortion measure. The existing color distortion formulas, such as the CIELAB color difference formula, are intended for describing the perceived color differences of fairly large color patches[6]. In natural images, where color patches of different sizes occur, this measure is not necessarily applicable, especially for large color differences. Indeed, in natural images, the preservation of image structure, which is largely mediated by the luminance of the image, is more important than an accurate reproduction of the hue and saturation. Therefore, for natural images, the importance of luminance errors is probably underestimated in the CIELAB measure [Sta90]. A quantizer design should therefore put more emphasis on minimizing the luminance errors than is indicated by the CIELAB measure. Second, designing and implementing a general color quantizer is usually much more complex than quantizing each of the color signals separately, so that the latter approach is usually preferred from a practical point of view. There are also indications that separate quantization of color signals might perform about equally well as combined (3D) quantization, especially if the quantization is applied in combination with error diffusion [GWA90].

Suppose that the three color signals are quantized separately, using a distortion measure $d(f, g) = |f - g|$. For each of the color signals, a maximum allowable distortion must be selected. It was argued [Cow85], based on a photon-detector model of human vision, that the distortion bound for the i-th color should be proportional to $1/\sqrt{L_i}$, where L_i is the luminance of the i-th primary. This rule emphasizes the need for an accurate luminance reproduction, and is also supported by a posteriori subjective evaluation. Indeed, the number of bits needed for the accurate reproduction of the red, green and blue components of an image on a color monitor (using error diffusion) is predicted fairly well by this rule.

An accurate luminance reproduction becomes even more important if dithering is applied. Dithering is based on the ability of the visual system to average over the visual information within a local area of the image.

[6]The new S-CIELAB metric, discussed in chapter 8, may help to remedy this problem.

This area over which the local averaging is performed is about four times larger for color-difference variations (hue and saturation) than for luminance variations. Therefore, if dithering is applied, then the dithering patterns for the three color signals should not be independent. Instead, the dithering patterns should be related in such a way that the luminance distortion of the combined dithering patterns is minimized. The computer-aided design of ordered dithering patterns for color printing, based on an explicit model of human color perception, is for instance discussed in a recent paper [BA02].

EXERCISES

1 Derive numerical values for the decision and representation levels of the nonuniform midtread quantizer of figure 10.2.

2 Prove equation (10.14) for the distortion of the nonuniform quantizer, implemented using a compander technique.

3 Verify that the distortion measure of equation (10.4) also applies in case uniform quantization is combined with the linearized display proposed in equation (3.27) of chapter 3 (i.e., signal values f are proportional to psychometric lightness). Derive an expression for this distortion measure and the distortion bound, assuming that the error in psychometric lightness should be smaller than 1.

4 Prove that the m-th order moment of a quantized output $g = Q(f)$ is indeed given by equation (10.33) if condition (10.34) is satisfied. Make use of the fact that

$$\frac{d^k}{du^k}\left[\frac{\sin(\pi\Delta u)}{\pi\Delta u}\right]\Bigg|_{u=0} = \begin{cases} 0, & k \text{ odd} \\ (j2\pi)^k \frac{(\Delta/2)^k}{k+1}, & k \text{ even} \end{cases} .$$

5 Determine the Fourier series expansion of the dither pattern of equation (10.48) for the specific case of $k = 1$. Determine the nonzero Fourier component(s) of the lowest frequency for the case $k = 2$. The sampling lattice Λ is square with determinant $d(\Lambda) = 1$, the dither lattice Λ_o is square with determinant $d(\Lambda_o) = 2^{2k}$, and the periodicity lattice Λ_p is square with determinant $d(\Lambda_p) = 2^{2N}$, with $N > 2$.

6 Implement the error-diffusion with the four-neighbor filter of equation (10.59). Assume that the input image has values in the range [0,255], and that the display can only produce output values 0 and 255.

APPENDIX 10.A: STOCHASTIC PROCESSES

The signal representations and manipulations that were considered in chapter 2 are aimed at deterministic signals. However, in many applications, the signals that have to be dealt with are not known exactly, either because of incomplete information, or because a known input signal is perturbated by an unknown 'noise' source. Often, only certain properties of the signal are assumed to be known. Such a situation is most often modelled by assuming that the signal at hand belongs to a **stochastic process**. We view a stochastic process as consisting of a, usually infinitely large, set of signals that share common properties, together with a probabilistic rule for selecting a signal from this set. We provide a short introduction to stochastic processes for the case of one-dimensional signals. The extension to multiple dimensions is fairly trivial.

Consider a set of real-valued signals $\{g_s(x); s \in \mathcal{S}\}$ with coordinate x on the real axis $\mathbb{R}$. The index set $\mathcal{S}$ can be discrete or continuous. Assume that these signals share some common properties. More specifically, assume that the **mean value**

$$\mu = \lim_{T \to \infty} \frac{1}{2T} \int_{-T}^{+T} g_s(x)\, dx,$$

as well as the **autocorrelation function**

$$R(x) = \lim_{T \to \infty} \frac{1}{2T} \int_{-T}^{+T} g_s(x+y)\, g_s(y)\, dy$$

are the same for all signals g_s in the set. Some well-known properties of the autocorrelation function are

1 The *power* (energy per unit interval) of the signal is $R(0) \geq 0$.

2 $R(x) = R(-x), \forall x$ (symmetric)

3 $R(0) \geq |R(x)|, \forall x$

4 $R(0) \geq \mu^2$

Since the autocorrelation function $R(x)$ is a signal on the real axis, it possesses a Fourier transformation

$$S(u) = \int_{-\infty}^{+\infty} R(x)\, e^{j2\pi ux}\, dx$$

that is called the **power spectrum**. It can be shown that $S(u)$ is a real, positive-valued, even function of u.

Note that the signals that are introduced here are essentially different from the signals that were considered in chapter 2. They do not tend to zero for large values of $|x|$, and hence have infinite energy. They do however have finite power. These signals are called **permanent signals**. It is often assumed that observed signals are pieces of permanent signals. A good agreement with the theoretical results will already be obtained if the signals are known in the interval $(-T, +T)$, for T sufficiently large to reach the limit values of μ and $R(x)$.

The reason for the interest in permanent signals is that, for the important case of a position-invariant linear system, permanent signals with a common mean value μ and autocorrelation function $R(x)$ are transformed into permanent signals that still share a common mean and autocorrelation function. Assuming that the position-invariant linear system is described by its Fourier transformation $\hat{h}(u)$, then $\mu \cdot \hat{h}(0)$ and $|\hat{h}(u)|^2 \cdot S(u)$ are the mean value and power spectrum of the output signals.

It often happens that one cannot predict the output signal $g(x)$ that is generated by a system, but that it is known (or at least, assumed) that it is a permanent signal with given mean value μ and correlation function $R(x)$. The remaining uncertainty is then modelled by saying that the signal is selected from the class of permanent signals $\{g_s(x); s \in \mathcal{S}\}$ in a way that $g(x)$ is equal to $g_s(x)$ with a probability p_s. The signal output at a given coordinate x is hence considered as a stochastic variable with **statistical mean value**

$$\mu_g(x) = \int_S p_s\, g_s(x)\, ds$$

and **variance**

$$\sigma_g^2(x) = \int_S p_s\, g_s^2(x)\, ds - \mu_g^2(x).$$

In a similar way, all higher-order moments can be calculated, so that the probability distribution $p[g(x)]$ of the amplitude $g(x)$ of the signal at coordinate x can be determined.

Higher-order statistics of the output signal, such as the second-order statistic $p[g(x_1), g(x_2)]$, can also be calculated. Of particular interest will be the **statistical autocorrelation function**, defined as the expected value of the signal amplitude $g(x_1)$ at coordinate x_1 times the signal amplitude $g(x_2)$ at coordinate x_2, i.e.,

$$E[g(x_1) \cdot g(x_2)] = R_g(x_1, x_2) = \int_S p_s\, g_s(x_1)\, g_s(x_2)\, ds.$$

In order to measure the probability distribution p_s, we would need an infinite number of copies of the system that we want to model. Since this is clearly impossible, we usually assume a **stationary process**, which implies that μ_g is independent of the position and $R_g(x_1, x_2) = R_g(x_1 - x_2)$ only depends on the distance between the coordinates. Stationarity implies that $\mu_g = \mu$ and $R_g(x) = R(x)$, i.e., that the statistical mean and autocorrelation function are equal to the signal mean and autocorrelation function. The latter can be derived by observing one signal over a sufficiently long interval, so that explicit knowledge of the probability distribution p_s is not needed. If $R_g(x) = \mu_g^2$, then the signal is *uncorrelated* for coordinate values that are a distance x apart.

It should be realized that the **stochastic process** that is described here is only a special case, called a stationary ergodic process, of a much more general definition. It is however believed to be the most interesting case from the point of view of applications. In many applications, we even restrict ourselves to **Gaussian** (or **normal**) stationary stochastic processes for which

$$p[g(x_1), \ldots, g(x_N)] = \frac{1}{(2\pi)^{N/2}\sqrt{|K|}} \exp\left(-\frac{1}{2}\underline{g}^T K^{-1} \underline{g}\right),$$

with $\underline{g} = [g(x_1) - \mu \;\ldots\; g(x_N) - \mu]^T$, is the joint probability of the amplitudes $g(x_i)$ at a number of coordinates x_i, $i = 1, \ldots, N$. This distribution if fully determined by the mean value μ and the **variance-covariance matrix**

$$K = [R(x_i - x_j) - \mu^2]_{i,j=1,\ldots,N}.$$

For instance, noise in physical systems is often modelled by a Gaussian process with mean $\mu = 0$. The term **white noise** refers to a Gaussian process with a covariance matrix that is proportional to the identity matrix, i.e., $K = \sigma^2 \cdot I$, where σ^2 is the noise variance.

An interesting property of Gaussian stationary stochastic processes with zero mean is that higher-order moments can be expressed in terms of the statistical autocorrelation function, i.e.,

$$E[g(x_1) \cdot g(x_2) \cdot \ldots \cdot g(x_k)] = \sum \prod E[g(x_i) \cdot g(x_j)] = \sum \prod R(x_i - x_j),$$

for k even (and $= 0$ for k odd), where the $\sum \prod$ ranges over all distinct ways in which the coordinates $x_1, x_2, \ldots, x_k$ can be sorted into pairs. For example, in case $k = 4$, this results in

$$\begin{aligned} E[g(x_1) \cdot g(x_2) \cdot g(x_3) \cdot g(x_4)] &= E[g(x_1) \cdot g(x_2)] \cdot E[g(x_3) \cdot g(x_4)] \\ &+ E[g(x_1) \cdot g(x_3)] \cdot E[g(x_2) \cdot g(x_4)] \\ &+ E[g(x_1) \cdot g(x_4)] \cdot E[g(x_2) \cdot g(x_3)], \end{aligned}$$

with special case

$$E[g^4(x)] = 3 \cdot \left(E[g^2(x)]\right)^2 = 3 \cdot R^2(0).$$

Chapter 11

DOUBLE-ENDED INSTRUMENTAL MODELS OF IMAGE QUALITY

1. INTRODUCTION

Over the past few years, we have witnessed a growing academic and industrial interest into understanding and modelling the quality discrimination process that underlies the human judgement of images. The ability to model this process is of interest for many applications, including designing new image coding or processing systems, and monitoring the performance of existing systems. We may wonder how realistic such an enterprise is, and how far we have progressed towards the set goal over the past few years[1]. The two concluding chapters of this book are therefore devoted to this subject of image quality modelling. They intend to illustrate alternative approaches and to identify some of the major issues still to be resolved.

Quality is not a physical property of an image, but a sensation that can only be experienced by humans. We nevertheless assume that quality can be quantified in a consistent and meaningful way. In chapter 5, we have described how quantitative information about image quality sensations can be obtained by means of psychophysical experiments. Understanding the limitations of these psychophysical procedures is essential when modelling a sensation like image quality. In a psychophysical experiment on image quality, the subject is asked to convert his/her internal quality sensation into an external response, such as a number or a category, that can be observed by the experimenter. Recovering internal quality sensations from these external responses assumes a model for the response mechanism of the subject [Fal86, LK88]. This model determines

[1]This chapter is based on two earlier publications, i.e., [MM98] and [MM01].

the necessary data processing for retrieving the strengths of the quality sensations from the subject responses.

We often assume that the strengths of the quality sensations that are triggered by different images can be represented by positions on a **one-dimensional** axis [Thu27, Tor58], and that the subject responses are related to these positions in a simple way [PW86]. For instance, in the case of categorical scaling, the response categories (such as excellent, good, fair, poor, bad) are assumed to correspond to non-overlapping intervals on the one-dimensional quality axis [AP81, Agr89, Agr90]. Since perception processes are inherently noisy, the stimulus positions on this axis on repeated presentations of the same image vary stochastically. This models the fact that subjects do not necessarily give identical quality responses for the same image on repeated trials. The average strengths of the quality sensations can consequently only be estimated with a limited precision from the subject responses. A possible approach towards deriving sensation strengths from experimental data is offered by the MDS models of chapter 5.

The goal of the research on image quality modelling is to construct algorithms that take as input a number of images (usually, representing the same scene, but processed in different ways) and output quality scores for these images. Hence, similarly as in the case of psychophysical procedures, it is assumed that images can be positioned on a one-dimensional quality axis in a consistent way. Those positions are required to correlate highly with the averaged positions derived for these images in a psychophysical quality experiment. Although such an algorithm is often referred to as an objective measure for image quality, we prefer to use the term **instrumental measure**. We believe that the latter term more accurately reflects the fact that such a measure is not necessarily more reliable (i.e., more objective) than human judgement. On the contrary, experimental data on human quality judgements are typically used to tune and test these instrumental measures. In chapter 5, we have also demonstrated how an instrumentally generated stimulus configuration can be compared against experimental data through regression. We will present more examples in this chapter.

In most cases, experimental validation is only feasible in the design phase of a system, and even then only with a limited number of images and subjects. Therefore, instrumental measures that correlate well with experimental data on image quality are needed once systems become operational or when systems have to be tested on a large set of images. Many different computational image quality metrics have therefore been proposed, especially over the past few years. The most obvious way of approaching the problem of image quality modelling is to try and mimic, as closely as possible, the image processing as performed by the human visual system [AJ93, EB98]. The advantage of this so-called **mechanistic approach** [Lub93] is that it can be based on a large amount of available psychophysical data about detection and discrimination of simple patterns by the early visual system.

Most of the early vision models have aimed at predicting when two images are indistinguishable. Obviously, images that cannot be discriminated by the early visual system will be indistinguishable at later stages, and hence have equal quality. A plethora of such discrimination models are now available. At first, most metrics were limited to monochrome still images [ZH89, Dal92, Dal93, Lub93, TH94, Lub95, EB98]. More recently, extensions towards color and image sequences have been proposed [BL96b, BL96a, LBL96, Fib97]. Simplified models are also (re)gaining interest, since the complexity of some models makes them unsuited for most applications, especially applications where quality has to be monitored in real-time.

Many authors of image discrimination models have claimed that these models can also be used for above-threshold quality prediction. Mostly, the application has been towards predicting the quality loss (or impairment) of coded versions of an input image. The output of the discrimination model defines the distance between the input image and the coded image, and is claimed to be monotonically related to the perceived image quality. Although this may well be the case if the images are obtained by controlling the bit-rate in a fixed coder (such as baseline JPEG [PM93] or MPEG-2), we will show that this is not true in general, for at least two reasons. First, the distance measure incorporated in the discrimination model may not agree with human dissimilarity perception. In section 3, we will for instance demonstrate that existing models cannot adequately predict the image discrimination as performed by subjects if the images contain multiple, independent impairments, such as noise and blur [MM98]. The relative weight attributed to the different impairments is different for the models and the human observers. Second, image fidelity and image quality often do not coincide, as was observed in earlier chapters for color reproductions and sampling artifacts, as well as in a recent experimental study with printed images [SF96]. Therefore, the image that most closely approximates te original image does not necessarily posses the best quality.

Although many image discrimination models have been proposed, it remains unclear exactly what has been accomplished by these models. The parameters in the models are tuned such that the models give good predictions for the detection and discrimination of simple patterns such as sinewaves. Their accuracy in predicting discrimination (or quality) for more complex scenes is largely unknown, partly because extensive experimental reference data were not available until quite recently [CKT+99].

Undertaking a comparison study between alternative models is nontrivial for at least two reasons. First, because of the commercial interest in these models, the required implementation details are often not available. Second, large resources in both human and computer time are required for implementing these models, running them on a large data base of image sequences, compiling a representative data base of image sequences, and obtaining subjective quality judgements for these sequences. Last but not least, the test results and the experimental conditions (monitor characteristics, viewing distance, experimental procedure, etc.) have to be documented. Despite such problems, VQEG (Video Quality Evaluation Group), a consortium of interested users of image quality models, has recently undertaken such a comparative study. The most important conclusion of this study is that none of the proposed instrumental quality measures can outperform a simple measure such as signal-to-noise ratio. As demonstrated in section 3, this agrees with our own findings when comparing existing instrumental measures with experimental data.

There is hence a growing consensus that many of the proposed models may have very similar performance. In view of the fact that the computational complexity of these models can be very high, simplified models are also required to make them amenable for most practical (real-time) applications. A comparative study between a representative subset of models will therefore be performed in this chapter, in order to reveal what are the most important model features that are worth preserving in simplified models.

The above models are referred to as **double-ended image quality models**, since they need both the input image and the processed (coded) image to make a quality prediction. This is an important limitation, because in many applications, the reference image may not be available as an input to the instrumental quality measure. Moreover, double-ended measures cannot account for the fact that humans are very well capable of making quality judgements without having access to the original image. Subjects can also compare image quality across different scenes, something which is also clearly beyond the scope of discrimination measures. A growing interest in **single-ended quality models** that can operate without a reference image has therefore been witnessed over recent years. Such models will be discussed in the next chapter.

2. INSTRUMENTAL DISSIMILARITY MEASURES

Most image quality metrics that have been proposed up to now use as inputs two images, where most often one of the images is the original. The images are assumed to be perfectly aligned. Usually, a model of the early visual pathway is used to map the two separate image inputs I_1 and I_2 into two sets of visual system outputs $P(I_1)$ and $P(I_2)$ [AJ93]. These models typically include aspects of luminance adaptation, decomposition into visual channels with different frequency/orientation tuning, and masking (both within and across channels). The two sets of visual system outputs are subsequently integrated into a single number $Q[P(I_1), P(I_2)] = D(I_1, I_2)$ that is assumed to be an instrumental measure for the perceived distance between the two images. It is moreover often assumed that the distance from the original image is a measure for the impairment, and that the image quality decreases linearly with the distance from the original.

Given the amount of assumptions and the number of parameter choices that have to be made in the mapping $I \rightarrow P(I)$, and given the limited experimental evidence on which most of these assumptions and choices are based[2], it is at least surprising that these 'image-quality metrics' have been so widely adopted, albeit in many varieties. If nothing else, it indicates the need for reliable instrumental measures of image quality.

The assumption that image quality decreases linearly with the distance from the original image is not essential for the above distance measures to be used for image quality modelling [MM98]. Indeed, similarly as in the case of the MDS approach of chapter 5, one can use the results of an instrumental distance measure to position images in a multidimensional space. This requires not only determining the distances between the original image and all processed images, but also the distances between (a sufficiently large subset of) all image pairs. The stimulus configuration resulting from such an instrumental measure can then be compared against the stimulus configuration resulting from dissimilarity, preference and attribute judgements by subjects. Moreover, it can be investigated if such a multidimensional stimulus configuration is better suited for predicting image quality than the original one-dimensional configuration where only the distance from the original image is used.

In order to test the above ideas against experimental data, a representative subset of models has to be selected. We therefore start with a

[2]Most models are tuned based on experimental data for simple patterns such as sinewaves and their quality predictions for natural scenes have only marginally been tested against experimental data.

detailed description of the double-ended instrumental models that have been implemented as part of such a study.

2.1 SARNOFF MODEL

The Sarnoff model is regarded by many as the current de facto standard for an instrumental image quality metric, and many alternative models may be viewed as simplified versions of it [AJ93]. The model for monochrome still images is fairly well documented [Lub93][Lub95], although some parameters still have to be tuned for the specific viewing conditions. The model version for color image sequences is available as a commercial product [Fib97], which explains why little or no detailed information is published about this latter model. We now describe the model for monochrome still images that we have implemented in our study, including all model parameters. The model follows as closely as possible the original model description, although some small but inconsequential modifications (in the optical and pooling filter) have been made to simplify the implementation.

First, a gray-value image is converted into a luminance image. If the grey-value-to-luminance relationship is not known, then one of the standardized characteristics discussed in chapter 4 may for instance be adopted. The optical filtering by the human eye is simulated by a Gaussian filter with a standard deviation of 0.5 arcmin. This filtering will typically have to be applied to sampled images, so that the sampling distance, expressed in arcmin, has to be known. If the exact viewing conditions (i.e., display height and viewing distance) are unknown, then a standardized viewing condition may be adopted. For instance, a viewing distance equal to 6 or 3 times the display height is adequate for standard TV and high-resolution TV, respectively. In both cases, this results in a distance of approximately 1 arcmin between successive pixels.

The filtered luminance image is converted into a Gaussian pyramid with 7 levels using a separable filter with coefficients

$$\phi = (0.05, 0.25, 0.4, 0.25, 0.05)$$

[BA83]. The result is illustrated on an example image in the left part of Figure 11.1. The large image at the top is pyramid level G_0 and is equal to the original image (after optical filtering), while the smaller images are obtained in a recursive way. In each recursion step k, for $k = 0, 1, \ldots, 6$, the last obtained image G_k if filtered by the separable filter and subsampled by a factor of 2 in both the horizontal and vertical direction to obtain the next level G_{k+1}. Reflection of the images at the boundaries is applied during the filter operation. The Gaussian pyramid is subsequently mapped to a contrast pyramid with 5 levels, where the contrast at level k is as defined by Peli [Pel90], i.e.,

$$C_k = \frac{G_k - G^i_{k+1}}{G^{ii}_{k+2} + \Phi/4^k}, \tag{11.1}$$

for $k = 0, \ldots, 4$. The offset $\Phi = 0.1$ avoids divisions by zero. The image G^i_{k+1} arises by interpolating Gaussian pyramid level G_{k+1} with a separable interpolation filter with coefficients

$$\varphi = (0.1, 0.5, 0.8, 0.5, 0.1).$$

Applying this interpolation twice to Gaussian pyramid level G_{k+2} results in G^{ii}_{k+2}. The images G_k, G^i_{k+1} and G^{ii}_{k+2} all have the same size. The contrast pyramid for the example image is shown in the right part of Figure 11.1

In the next step, the contrast images C_k are mapped into channel responses. For this purpose the contrast images are convolved with four pairs of spatially oriented filters. The filters used are second derivatives of a Gaussian and their Hilbert transform, for four different orientations. The standard deviation of the Gaussian is approximately equal to the sampling distance, and the actual filter tabs are specified in tables IV and VI of the Freeman & Adelson paper on steerable filters [FA91]. The four oriented responses r_{kl}, for pyramid level index $k = 0, \ldots, 4$ and orientation index $l = 0, \ldots, 3$, are obtained by taking the square root of the sum of the squared filtered images for the Hilbert pairs with the same orientation.

The gains $g_{kl} = g_k$ for the different response channels have to be calibrated such that the peak sensitivities follow the contrast sensitivity function (CSF) of the visual system. This is accomplished by determining the peak response in the different channels to sinewave gratings with varying spatial frequency and with modulation depth equal to the inverse of the CSF for that frequency. The channel gain g_{kl} has to be adjusted such that the maximum response in channel (k, l) is equal to one for the spatial frequency and orientation for which that channel is most sensitive. We used the formula proposed by Barten [Bar87] for an average luminance of 20 cd/m^2 as a mathematical description of the CSF.

Although the above procedure seems straightforward, it runs into practical difficulties for spatial frequencies close to the sampling frequency. Hence, we have only used it to set the channel gains for pyramid levels 2, 3 and 4. The resulting gains were in agreement with the relative sensitivities of frequency channels with different peak frequencies reported in the contrast perception model of Cannon & Fullenkamp

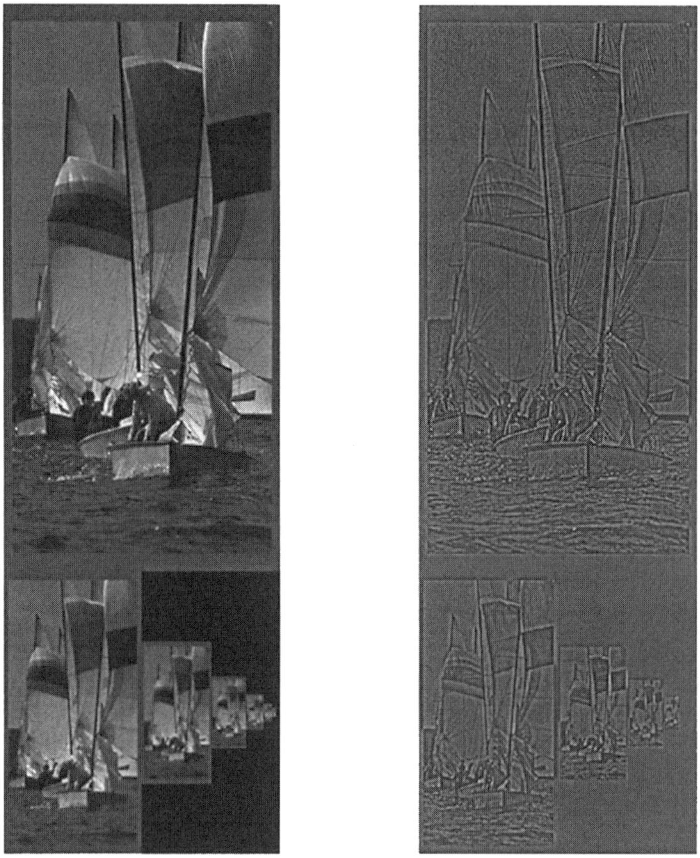

Figure 11.1. Gaussian (left) and contrast (right) pyramids for the image 'Boats'. The contrast has been mapped by $f(x) = \text{sgn}(x) \cdot \sqrt{x}$ for display purposes.

[CF91]. They also propose relative sensitivities for higher-frequency channels that seem more realistic than those resulting from the above calibration procedure, and which we have therefore adopted to set the channel gains for pyramid levels 0 and 1. The resulting channel gains depend on the viewing distance (because the CSF is specified in cycles per degree of visual angle), and therefore we specify these channel gains in table 1 for sampling distances of 1 and 2 arcmin. Note that the peak sensitivity shifts to a channel with a smaller channel number when the sampling distance increases from 1 to 2 arcmin.

Within-channel visual masking is taken into account by mapping the (amplified) channel responses $a_{kl} = g_{kl} \cdot r_{kl}$, for $k = 0, \ldots, 4$ and $l =$

Table 11.1. The two leftmost columns specify the channel gains g_k for the original Sarnoff model and sampling distances of 1 and 2 arcmin, respectively. The two rightmost columns give similar information for the simplified Sarnoff model.

channel	original Sarnoff		simplified Sarnoff	
number	1 arcmin	2 arcmin	1 arcmin	2 arcmin
0	25	60	170	420
1	85	133	450	960
2	125	111	845	885
3	90	68	670	535
4	50	37	385	295

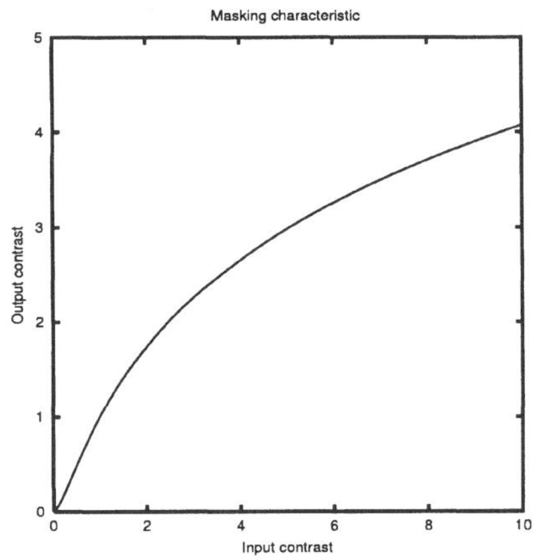

Figure 11.2. The within-channel masking characteristic used in the Sarnoff model.

$0, \ldots, 3$, through a sigmoid non-linearity of the form

$$T(a) = \frac{(2+c)\, a^s}{1 + a^{s-l} + c\, a^{s-w}}, \tag{11.2}$$

where $s = 1.5$, $l = 0.4$, $w = 0.068$ and $c = 0.1$ are typical values [Lub93]. The function that corresponds to these parameter values is plotted in Figure 11.2. We illustrate the masked responses for all four orientations in Figure 11.3.

The 20 masked response images $T(a_{kl})$ are filtered using a pooling filter equal to a uniform filter of size 5x5, and subsequently interpolated to the input image size using the interpolation filter mentioned in the contrast pyramid definition. The resulting response images $R_{kl}^{(1)}(x, y)$

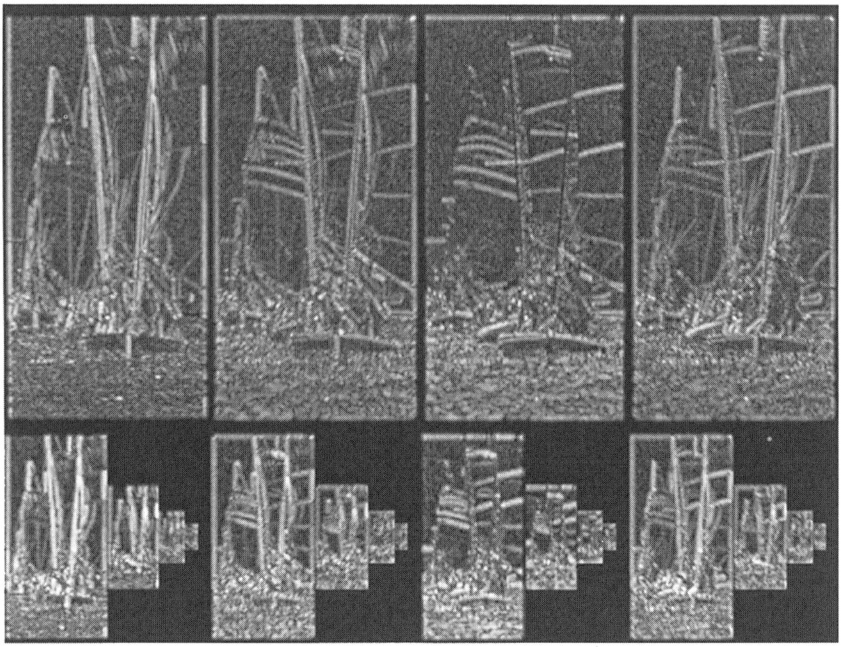

Figure 11.3. Masked responses $T(a_{kl})$ in the Sarnoff model for different orientations. The horizontal and vertical direction correspond to orientation index $l = 0$ and $l = 2$, respectively, while the oblique directions have orientation index $l = 1$ and $l = 3$.

and $R_{kl}^{(2)}(x, y)$ for the first and second input image I_1 and I_2, respectively, are pointwise combined into one distortion image using the Minkowski metric

$$D_{12}(x, y) = \left\{ \sum_{k=0}^{4} \sum_{l=0}^{3} \left| R_{kl}^{(1)}(x, y) - R_{kl}^{(2)}(x, y) \right|^{Q} \right\}^{1/Q} \tag{11.3}$$

with exponent $Q = 2.4$. If the model is properly tuned, then a value of $D_{12}(x, y) = 1$ indicates 1 JND (Just Noticeable Difference), i.e., a probability of 75 % to see a difference between both images at the corresponding location (x, y). The distortion image resulting from a comparison between the original "boats" image and a JPEG-coded version of it is illustrated in Figure 11.4. It agrees with our intuition that JPEG coding artifacts, such as blocking, are most easily observed in uniform regions of the image. In this sense, they behave similarly as the quantization artefact discussed in chapter 10.

Figure 11.4. The original and coded image are shown on the left and right, respectively. The distortion image, according to the Sarnoff model, is shown in the middle.

In order to obtain a single dissimilarity number for a pair of images, we can perform a spatial Minkowski integration with exponent P across the distortion image $D_{12}(x, y)$, i.e.,

$$D_S(I_1, I_2) = \left\{ \frac{1}{N} \sum_{x,y} |D_{12}(x, y)|^P \right\}^{1/P}, \tag{11.4}$$

where N is the number of pixels at pyramid level 0. By varying P from one to infinity, we can vary from taking the average to taking the (scaled) maximum of the distortion values as the relevant distance measure. The 'Sarnoff' instrumental measure $D_S(I_1, I_2)$ that we use in the rest of the paper is calculated with $P = 1$.

2.2 SIMPLIFIED SARNOFF MODEL

The original Sarnoff model described in the previous section requires a very large number of computations, and therefore we have also developed a simplified version of it. Especially the orientation and pooling filtering and the full-size interpolation at the end determine the total number of operations. We were interested to know if removing these components would seriously affect the predictions made by the model.

The directional filtering can be avoided by using an alternative method for mapping the contrast images C_k into channel responses. The purpose of using a Hilbert pair of oriented filters is to realize a phase independence for the channel responses, e.g., to make the responses insensitive to the exact position of a feature (with respect to the sampling lattice). Oriented filters are only required in case orientation masking is applied [TH94], which is not (currently) the case in the Sarnoff model. An alternative, and simpler, way of realizing phase-independent responses is by taking the residue amplitudes[3] (or local standard deviations) [Mar95], i.e.,

$$r_k(x,y) = \left\{ w(x,y) \star C_k^2(x,y) - [w(x,y) \star C_k(x,y)]^2 \right\}^{1/2}, \qquad (11.5)$$

where a Gaussian window

$$w(x,y;\sigma) = \frac{1}{\pi\sigma^2} \exp\left(-\frac{x^2+y^2}{\sigma^2} \right) \qquad (11.6)$$

with approximately the same standard deviation $\sigma/\sqrt{2}$ as in case of directional filtering is used. More specifically, this filter (with $\sigma = 1.5$ times the sampling distance) is implemented by a separable 9-tap filter with coefficients

$$\begin{aligned} w(4) &= w(-4) = 0.00048 \\ w(3) &= w(-3) = 0.00880 \\ w(2) &= w(-2) = 0.06965 \\ w(1) &= w(-1) = 0.23997 \\ w(0) &= 0.36217 \end{aligned}$$

Unlike in the original Sarnoff model, the number of channels hence remains the same as the number of pyramid levels.

A consequence of the alternative channel response mechanism is that the channel gains g_k have to be calibrated accordingly. These alternative channel gains are also given in table 1 for viewing distances of 1 and 2 arcmin. The amplified channel responses are denoted by $a_k = g_k \cdot r_k$, for $k = 0, \ldots, 4$. Similarly as in the case of the original Sarnoff model, the masked channel responses $T_k = T(a_k)$ are obtained by applying the sigmoid non-linearity in equation (11.2). A pooling filter is also not included in the simplified model. The standard deviation of the Gaussian

[3] Residue amplitudes will be discussed in more detail in the next chapter.

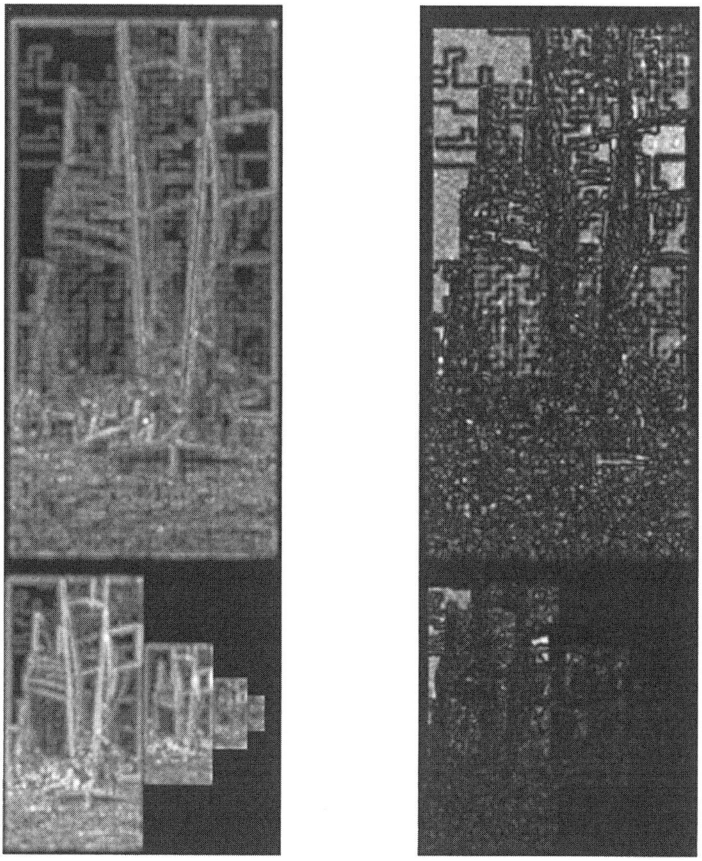

Figure 11.5. The left figure shows masked responses $T(a_k)$ for all five pyramid levels in the simplified Sarnoff model for the original image. The right figure shows the absolute values of the differences between the masked channel responses for the original and the coded image of Figure 11.4.

residue amplitude filter $w(x, y)$ can be increased if response averaging over a larger area is required.

The differences between the masked channel responses $T_k^{(1)}(x, y)$ and $T_k^{(2)}(x, y)$ for both input images can be spatially integrated at each pyramid level k. More specifically, let

$$D_{12}(k) = \left\{ \frac{1}{N_k} \sum_{x,y} \left| T_k^{(1)}(x, y) - T_k^{(2)}(x, y) \right|^P \right\}^{1/P}, \tag{11.7}$$

where N_k is the number of pixels at pyramid level k, for $k = 0, \ldots, 4$, denote the result of this spatial averaging with a Minkowski metric with exponent P. The overall instrumental measure for the perceived distance between two images I_1 and I_2 is obtained by Minkowski integration across pyramid levels

$$D_{SS}(I_1, I_2) = \left\{ \sum_{k=0}^{4} |D_{12}(k)|^Q \right\}^{1/Q} \tag{11.8}$$

with an exponent equal to Q. Correspondence with the original Sarnoff model implies the choice $P = 1$ and $Q = 2.4$. The need for full-size interpolation is avoided in the simplified model by reversing the order of across-channel integration and spatial integration. The consequence is that no single distortion image is available anymore. However, most of the time, we are only interested in an overall measure for dissimilarity.

The simplified model retains four basic properties of the original model: contrast analysis at multiple spatial scales, contrast calibration against human CSF, within-channel contrast masking, and Minkowski integration (with different exponents across space and across channels).

2.3 SIMPLE DISTANCE MEASURES

The original and simplified Sarnoff models are obviously complex distortion metrics that require a substantial amount of computations to implement. It is therefore worthwhile to study how they compare to simple measures, such as the root-mean-square error. Many of these simple measures that have been proposed can be discussed in a common framework, since they contain the same three stages: front-end processing, image filtering and integration, as shown in Figure 11.6. We will consider a number of alternative measures that all fit within this framework.

2.3.1 FRONT-END PROCESSING

One of the aspects that is often neglected when developing instrumental measures for image quality is deciding what is the best choice for the input signal into the measure. We discuss this aspect in some more detail for monochrome still images and indicate possible extensions for colored image sequences.

The choice for the input signal can be guided by both practical and perceptual considerations. It may be a practical requirement that the measure has to operate directly on the grey-value image, for instance because it is not possible to specify unique display and viewing conditions, or in order to limit the computational complexity. Perceptual

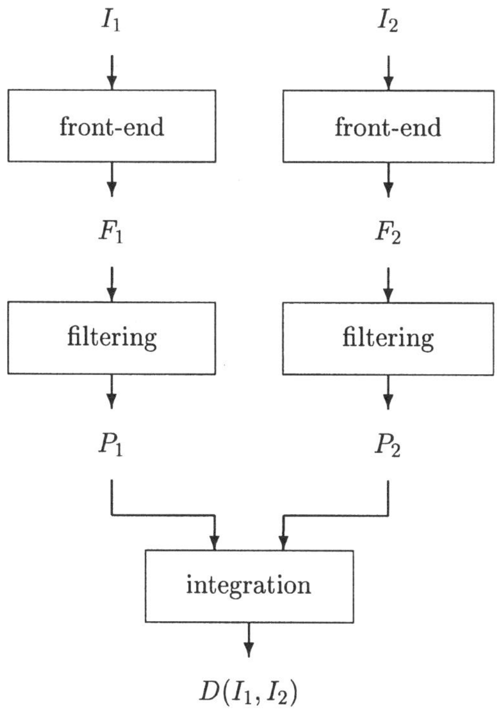

Figure 11.6. Many double-ended instrumental measures for image dissimilarity can be subdivided into three stages. A front-end stage converts the grey values in the input image I to an appropriate unit F for further processing, such as psychometric lightness. The image filtering stage maps the image F to a processed image P. In the integration stage, the processed images P_1 and P_2 are combined into a single number $D(I_1, I_2)$ that expresses the dissimilarity between both images.

considerations can be used in case the display and viewing conditions are known.

The display converts the grey values in the digital image I to luminances L on the screen. As was shown in chapter 1, luminance is the proper physical unit for describing the optical transfer of the image from the screen to the visual system. A commonly used approximation in case the exact grey-value-to-luminance characteristic of the display is unknown is $L \propto I^{\gamma}$. A value of $\gamma = 2.5$ is often a reasonable guess if the display is a CRT (cathode ray tube) [Poy93]. The spatial distance between the pixels is converted into a visual angle between pixels from the observer's point of view. For a fixed number of image lines, this angle is determined by the ratio of the display height over the viewing distance. For certain applications, such as TV broadcasting, this ratio is standardized, resulting in a viewing angle of approximately 1 arcmin between pixels. Both the display and the visual system apply some spatial filtering to the image. The filtering by the visual system is often the dominant one. It can be approximated by a Gaussian filter with a standard deviation of approximately 0.5 arcmin.

A useful characteristic for the input image F into an instrumental measure is that its amplitude has interval properties. This means that equal differences ΔF in input value, from F to $F + \Delta F$, are equally significant, independent of the base value F. This interval property greatly simplifies implementations, since processing need not depend on the base value. Luminance does however not have this interval property. Psychophysical experiments have demonstrated that the just perceivable luminance increment or decrement ΔL needed to detect a pattern depends on the base luminance L. Very often this relationship can be expressed as

$$\Delta L = a \cdot L^b. \tag{11.9}$$

Unfortunately, the numerical values for a and b depend on the pattern that has to be detected. For instance, for the detection of extended vertical step edges, values of $a = 0.017$ and $b = 0.85$ have been found experimentally [Lub93].

If we assume that F is a differentiable function of the luminance L, then approximately $\Delta F = [dF(L)/dL] \cdot \Delta L$. If we require that the increment ΔF needed for detection is constant, and equal to one, then $dF(L)/dL = 1/\Delta L$. Substitution of the experimental relationship between the detection threshold ΔL and the base luminance L, followed by integration, results in

$$F = \frac{1}{a \cdot (1-b)} L^{1-b}. \tag{11.10}$$

This signal has the required interval property, at least when the detection task corresponding to the a and b values used can be considered as the relevant sensitivity measure. An additional advantage of this input signal is that is has a unit value of 1 JND (just noticeable difference). Hence, differences in values smaller than 1 should be imperceivable.

If the grey-value-to-luminance relationship is not known exactly, then the approximate power-law relationship $F \propto I^{\gamma \cdot (1-b)}$ can be used to convert from grey values to input values. If $\gamma \cdot (1 - b)$ can be assumed to be close to 1, then this is an argument for applying the instrumental measure directly on the grey values of the digital image.

In many of the experiments that we have performed in our laboratory, we have used 8-bit input images and a calibrated display with a characteristic equal to

$$L = \max\left[L_{min}, L_{max}(I/I_{max})^{\gamma}\right], \tag{11.11}$$

with $I_{max} = 255$, $L_{max} = 60$ cd/m^2, $L_{min} = 0.2$ cd/m^2, and $\gamma = 2.5$. We have often used this display characteristic, together with the CIE 1976 relation for psychometric lightness [Hun78, ZW98],

$$F(L) = \begin{cases} 116\left(\frac{L}{L_{max}}\right)^{1/3} - 16, & \frac{L}{L_{max}} > 0.008856 \\ 903.3\left(\frac{L}{L_{max}}\right), & \frac{L}{L_{max}} \leq 0.008856 \end{cases} \tag{11.12}$$

with $0 \leq F \leq 100$, to convert the grey values to input values for an instrumental measure. Strictly speaking, the CIE relation given above expresses psychometric lightness and is only valid for secondary light sources (reflecting bodies) of fairly large size (2 degrees of visual angle). It is often adopted because it is one of the few standardized formulas that expresses the nonlinear transformation in the visual system. It is part of the CIELAB and CIELUV color difference metrics, so that extensions for color images are straightforward with this choice for the front-end processing. As discussed in chapter 8, the CIELAB color difference metric has recently also been extended with spatial and temporal characteristics of the visual system in S-CIELAB [ZW98] and ST-CIELAB [THBL99]. These more elaborate front-ends can hence be used when deemed necessary. It is however our experience that the performance of most instrumental measures is fairly robust against modest variations in the front-end processing. This is another reason why we have most often settled on the CIE relationship.

2.3.2 FILTERING

It is well-known that humans are much more sensitive to changes in light level than to absolute light levels. Amongst others, this is manifested by the low-frequency drop-off in the CSF. It has therefore been proposed that some sort of (nonlinear) filtering should be performed on images before they are being compared. The Sarnoff model is of course a very complex way of implementing this filtering. In case of the simple alternative measures discussed here, we will consider the following three alternatives for the filtering stage: no filtering (F_N), residue amplitude (F_R) and gradient amplitude (F_G).

In order to obtain the residue amplitude for a Gaussian window of size σ, we determine

$$P(x,y) = \left\{ w(x,y;\sigma) \star F^2(x,y) - \left[w(x,y;\sigma) \star F(x,y) \right]^2 \right\}^{1/2} , \quad (11.13)$$

in a similar way as was done in case of the simplified Sarnoff measure (see equation (11.6) for a definition of the Gaussian window $w(x, y; \sigma)$). The size parameter σ is again chosen equal to 1.5 times the sampling distance in our implementation. This is a suitable value to judge the high-frequency errors that are mostly of interest in image processing and coding. If a single window size is not sufficient to detect all distortions of interest, then an extension to a pyramid-like approach, similar to the one presented for the Sarnoff model, can be created in a fairly straightforward way.

In case of the gradient amplitude, we apply two linear filters with impulse responses

$$\begin{aligned} d_x(x,y;\sigma) &= -\frac{x\sqrt{2}}{\sigma}\, w(x,y;\sigma) \\ d_y(x,y;\sigma) &= -\frac{y\sqrt{2}}{\sigma}\, w(x,y;\sigma) \end{aligned} \quad (11.14)$$

to the input image $F(x, y)$, in order to obtain the gradient images along the x- and y-direction, i.e.,

$$\begin{aligned} F_x(x,y) &= d_x(x,y;\sigma) \star F(x,y) \\ F_y(x,y) &= d_y(x,y;\sigma) \star F(x,y) \end{aligned} \quad (11.15)$$

These gradient images are subsequently combined into a gradient amplitude image

$$P(x,y) = \left[F_x^2(x,y) + F_y^2(x,y) \right]^{1/2} . \quad (11.16)$$

The main difference between residue and gradient filtering is that the former filter reacts to all variations in the image, while the latter filter reacts mostly to patterns with odd symmetry (such as edges).

2.3.3 INTEGRATION

The purpose of the integration stage is to compare the filtered images $P_1(x, y)$ and $P_2(x, y)$ that are derived from the corresponding input images $I_1(x, y)$ and $I_2(x, y)$ in order to deliver a single number D that expresses the dissimilarity between both input images. No firm theoretical or experimental grounds are currently available for designing such integration rules. Therefore, it is worthwhile to explore different alternatives and to compare their performance against experimental data.

First-order statistics : Many simple image dissimilarity measures are based on the scaled difference image

$$D_{12}(x,y) = g_p\,[P_1(x,y) - P_2(x,y)], \tag{11.17}$$

where the choice of the scaling factor g_p will be discussed later. Since it is our aim to develop simple measures, we first consider integration rules that are based on the amplitude distribution of this difference image. Such a simplification obviously neglects all possible effects of spatial integration, but is a logical starting point for our exploration.

In line with the Sarnoff model, we include Minkowski measures

$$D_\alpha = \left[\sum_{x,y} |D_{12}(x,y)|^\alpha\right]^{1/\alpha}, \tag{11.18}$$

for exponents $\alpha = 1, 2, 3, 4$. Increasing values for α imply more emphasis on large differences. It is well-known from statistics that such an emphasis on large differences makes a measure more sensitive to outliers, and hence less robust. Alternative "robust" integration rules that do not suffer this drawback have therefore been proposed [BM98b, Mee02] and we have also included some of them in the current study. More specifically, integration rules according to Huber

$$D_H = \sum_{x,y} H[D_{12}(x,y)],\ H[D] = \left\{ \begin{array}{cl} D^2/2 & \text{if}\,|D| \le 1 \\ |D| - 1/2 & \text{if}\,|D| \ge 1 \end{array} \right., \tag{11.19}$$

Lorentz

$$D_L = \sum_{x,y} L[D_{12}(x,y)],\ L[D] = \log(1 + D^2), \tag{11.20}$$

and Tukey

$$\begin{aligned} D_T &= \sum_{x,y} T[D_{12}(x,y)], \\ T[D] &= \left\{ \begin{array}{cl} (75\,D^2 - 15\,D^4 + D^6)/96 & \text{if}\,|D| \le \sqrt{5} \\ 125/96 & \text{if}\,|D| \ge \sqrt{5} \end{array} \right. \end{aligned} \tag{11.21}$$

are considered. These "robust" difference measures, which are plotted in Figure 11.7, change their shape around the value $D = 1$, so that an

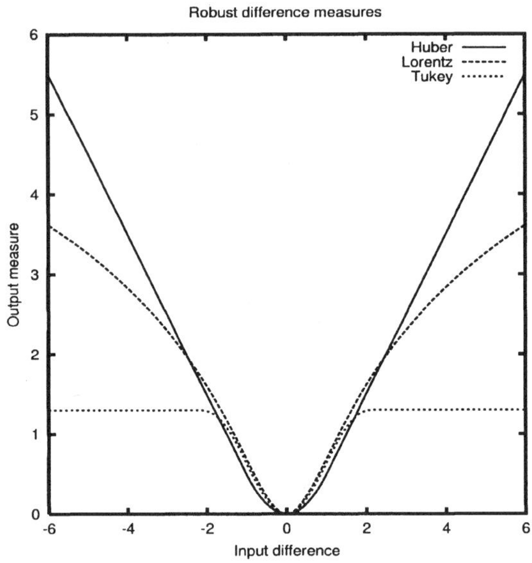

Figure 11.7. Robust difference measures according to Huber, Lorentz and Tukey.

adequate scaling of the difference image is essential when using these integration rules.

Some procedure is hence required to select a proper scaling factor g_p for the difference image. Moreover, this scaling factor will also have to depend on the processing that precedes the integration stage. In case psychometric lightness is used as input, then a value of $g_p = 1/2.5 = 0.4$ puts the changeover point for the robust statistics at a difference of $\Delta L^\star = 2.5$ psychometric units, which is known to be a noticeable difference. We will derive in the next chapter that an stepwise change in psychometric lightness with amplitude $\Delta L^\star$ gives rise to a residue amplitude equal to $\Delta L^\star/2$ and a gradient amplitude equal to $\Delta L^\star/\sqrt{2\pi}$. We therefore select $g_p = 0.4 \cdot 2 = 0.8$ and $g_p = 0.4 \cdot \sqrt{2\pi} = 1$ in case the filtering stage determines the residue and gradient amplitude, respectively.

Second-order statistics : It has been argued recently [WB01] that integration rules that are based on a difference signal are inherently weak, since, unlike the visual system, they treat all errors as equivalent. They propose an alternative "universal" quality index that is supposed to better capture the structural errors that are primarily of interest, i.e.,

$$Q_U = \frac{1}{N} \sum_{x,y} Q_U(x,y), \tag{11.22}$$

where N is the number of pixels in the images to be compared, and

$$Q_U(x,y) = \frac{\sigma_{12}(x,y)}{\sigma_1(x,y)\,\sigma_2(x,y)} \cdot \frac{2\,m_1(x,y)\,m_2(x,y)}{m_1^2(x,y)+m_2^2(x,y)} \cdot \frac{2\,\sigma_1(x,y)\,\sigma_2(x,y)}{\sigma_1^2(x,y)+\sigma_2^2(x,y)}, \tag{11.23}$$

is an image fidelity metric that is derived from the local averages and variances, i.e.,

$$\begin{aligned} m_i(x,y) &= w(x,y) \star P_i(x,y), \\ \sigma_i^2(x,y) &= w(x,y) \star P_i^2(x,y) - m_i^2(x,y), \end{aligned} \tag{11.24}$$

for $i = 1, 2$, and the local covariance

$$\sigma_{12}(x,y) = w(x,y) \star P_1(x,y) \cdot P_2(x,y) - m_1(x,y) \cdot m_2(x,y), \tag{11.25}$$

within a normalized window $w(x,y)$. A uniform 8×8 window is proposed in the original reference, but alternative windows (such as a Gaussian window) could clearly also be considered. The first term in the image fidelity metric $Q_U(x,y)$ is the weighted linear correlation coefficient $Q_C(x,y)$ between both filtered images, while the second and third term are included to penalize differences in average values and standard deviations, respectively. In order to judge the effect of these latter terms on the quality index, we will also consider a quality index

$$Q_C = \frac{1}{N} \sum_{x,y} Q_C(x,y) \tag{11.26}$$

that is based solely on the linear correlation. The quality measures Q_W and Q_C can vary from -1 to +1, and can be mapped to distance measures $D_W = 1 - Q_W$ and $D_C = 1 - Q_C$ that are strictly positive.

An alternative simple method of judging differences between signals is to use the gain-plus-additive-noise method that was also used to evaluate quantization errors. We can define a local vector product between images by

$$\langle f, g \rangle = \int_{-\infty}^{+\infty} \int_{-\infty}^{+\infty} w(x-p, y-q)\, f(x,y)\, g(x,y)\, dx\, dy. \tag{11.27}$$

Note that this vector product is based on the image values around position (p, q). Using this vector product we can express any image $g(x, y)$ as

$$g(x,y) = A_{gf} \cdot f(x,y) + n_{gf}(x,y), \tag{11.28}$$

where the image $n_{gf}(x, y)$ is locally uncorrelated[4] with $f(x, y)$, i.e., $\langle n_{gf}, f \rangle = 0$, and

$$A_{gf} = \frac{\langle g, f \rangle}{\langle f, f \rangle}. \tag{11.29}$$

The weighted squared distance between both images can be decomposed as follows

$$D_w^2 = \langle f - g, f - g \rangle = (1 - A_{gf})^2 \cdot \langle f, f \rangle + \langle n_{gf}, n_{gf} \rangle \tag{11.30}$$

into a correlated and an uncorrelated error term. We can obviously exchange the roles of $f(x, y)$ and $g(x, y)$ to obtain the following alternative expression

$$D_w^2 = \langle f - g, f - g \rangle = (1 - A_{fg})^2 \cdot \langle g, g \rangle + \langle n_{fg}, n_{fg} \rangle, \tag{11.31}$$

where $n_{fg}(x, y)$ is uncorrelated with $g(x, y)$, and

$$A_{fg} = \frac{\langle f, g \rangle}{\langle g, g \rangle}. \tag{11.32}$$

Averaging both expressions for the weighted squared distance D_w^2 leads to the following decomposition $D_w^2 = D_{wc}^2 + D_{wu}^2$ into a "correlated" squared distance

$$\begin{aligned} D_{wc}^2 &= \frac{(1 - A_{gf})^2 \cdot \langle f, f \rangle + (1 - A_{fg})^2 \cdot \langle g, g \rangle}{2} \\ &= \frac{[\langle f, f \rangle - \langle f, g \rangle]^2}{2 \langle f, f \rangle} + \frac{[\langle g, g \rangle - \langle f, g \rangle]^2}{2 \langle g, g \rangle} \end{aligned} \tag{11.33}$$

and an "uncorrelated" squared distance

$$\begin{aligned} D_{wu}^2 &= \frac{\langle n_{gf}, n_{gf} \rangle + \langle n_{fg}, n_{fg} \rangle}{2} \\ &= \frac{\langle f, f \rangle \cdot \langle g, g \rangle - \langle f, g \rangle^2}{2} \left[\frac{1}{\langle f, f \rangle} + \frac{1}{\langle g, g \rangle} \right]. \end{aligned} \tag{11.34}$$

[4]In the degenerate case that $\langle f, f \rangle = 0$, we adopt $A_{gf} = 0$ and $n_{gf}(x, y) = g(x, y)$.

We can apply this local decomposition of the squared distance to images $f(x,y) = P_1(x,y) - m_1(p,q)$ and $g(x,y) = P_2(x,y) - m_2(p,q)$ that are equal to the difference between a filtered image $P_i(x,y)$ and its local average $m_i(p,q)$ at position (p,q). The following spatially-variant expressions for the "correlated" squared distance

$$D^2_{wc}(p,q) = \frac{\left[\sigma_1^2(p,q) - \sigma_{12}(p,q)\right]^2}{2\,\sigma_1^2(p,q)} + \frac{\left[\sigma_2^2(p,q) - \sigma_{12}(p,q)\right]^2}{2\,\sigma_2^2(p,q)}, \tag{11.35}$$

the "uncorrelated" squared distance

$$D^2_{wu}(p,q) = \frac{\sigma_1^2(p,q)\,\sigma_2^2(p,q) - \sigma_{12}^2(p,q)}{2}\left[\frac{1}{\sigma_1^2(p,q)} + \frac{1}{\sigma_2^2(p,q)}\right] \tag{11.36}$$

and "total" squared distance

$$D^2_w(p,q) = D^2_{wc}(p,q) + D^2_{wu}(p,q) = \sigma_1^2(p,q) + \sigma_2^2(p,q) - 2\,\sigma_{12}(p,q) \tag{11.37}$$

arise. A single distance measure D_w, D_{wc} and D_{wu} can be obtained for a pair of images by averaging $D_w(p,q)$, $D_{wc}(p,q)$ and $D_{wu}(p,q)$ across space.

The different second-order distortion measures D_U, D_C, D_w, D_{wc} and D_{wu} that were presented above will be included in the comparison between alternative image quality metrics that is presented in the next section. The Gaussian window $w(x,y;\sigma)$ of equation (11.6) with a size parameter σ equal to 1.5 times the sampling distance will be adopted as the integration window in all cases.

2.3.4 POSSIBLE COMBINATIONS

We will consider the 36 possible dissimilarity measures that arise by combining the three filter options (F_N, F_R, F_G) with the twelve integration rules (D_1, D_2, D_3, D_4, D_H, D_L, D_T, D_U, D_C, D_w, D_{wc} and D_{wu}). The different models will be denoted by a two-component code, e.g., $D_{R,2}$ to indicate a Minkowski norm with exponent $\alpha = 2$, applied to the differences in residue amplitude. The mapping from grey value to psychometric lightness will be used in all cases for the front-end. Although this mapping is almost linear for the well-calibrated set-up used in our experiments, it seemed preferable to base the measures on the well-standardized unit of psychometric lightness rather than on the ill-specified grey values.

3. PERFORMANCE OF DOUBLE-ENDED MEASURES

3.1 EXPERIMENTAL DATA

3.1.1 IMAGES WITH NOISE AND BLUR

In chapter 5, we have discussed psychophysical experiments with images that were degraded by noise and blur. These experimental data will be used to test the performance of the proposed instrumental measures in a case where several image distortions have been varied independently over a fairly wide range. The display that was used in these experiments was calibrated to have the gray-value-to-luminance characteristic of equation (11.11), while the viewing distance was equal to six times the image height, corresponding to a separation between pixels of approximately 1 arcmin of visual angle.

3.1.2 JPEG-CODED IMAGES

Another data set that will be used to compare between experimental data and instrumental measures comes from a study on JPEG-coded images [MM02]. In this case, different kinds of distortions such as blur, ringing and blockiness arise by a variation in a single coding parameter. The range over which image distortions vary is also more limited than in the previous experiment.

The four natural scenes that were used in this experiment will be referred to as Boats, Child, Girls and Lighthouse, and were taken from a Kodak PhotoCD demonstration disc. The original color images were converted into monochrome images. A public-domain software package for JPEG encoding and decoding (developed by the Independent JPEG Software Group, http://www.jig.org/) was used to code these images at quality levels 60, 40, 30, 25 and 20. Including the original, we obtained six images per scene. A region of interest of 480 rows by 240 columns was selected from the images in order to allow simultaneous display of two images. The uncoded images are shown in Figure 11.8. The display was again calibrated to have the gray-value-to-luminance characteristic of equation (11.11). In order to allow subjects to adequately judge the small coding differences, the viewing distance was reduced to 80cm, corresponding to a separation between pixels of approximately 2 arcmin of visual angle.

All 15 distinct combinations of the six stimuli were judged by 10 subjects. The subjects were asked to rate dissimilarity, as well as differences in overall quality. Quality differences (or preferences) were rated on a category scale from -5 to 5, while dissimilarity was judged on a scale

Figure 11.8. Images used in the JPEG experiment: 'Boats' (upper left), 'Child' (upper right), 'Girls' (lower left) and 'Lighthouse' (lower right).

from 0 to 10. More details about the experimental set-up can be found in the original paper [MM02].

3.2 COMPARISON METHODS

We will present two alternative ways of comparing instrumental dissimilarity measures amongst themselves and with experimental data. We suppose that there are N stimuli in an experimental data set (i.e., $N = 16$ in case of the images with noise and blur, and $N = 6$ in case of the JPEG-coded images) and M dissimilarity measures to be evaluated.

In the first comparison method, we use the M instrumental measures to create $N \times N$ dissimilarity matrices for the images used in the experiment. We subsequently use the XGms program for continuous data to derive a 2-D stimulus configuration from each of these M dissimilarity matrices. The input dissimilarities are interpreted as non-metric, in the sense that the stimulus distances in the resulting 2-D configuration are only required to be monotonically related to the interstimulus distances provided by the instrumental measure. More precisely, a monotonically increasing spline function with three internal knot points is used as the non-linear relationship to be optimized. Even in cases where the input dissimilarities can be interpreted as metric, and such a non-linearity has little or no effect, it is included nevertheless in order to obtain that all measures are analyzed in a similar way. This first stage results in M two-dimensional stimulus configurations that are subsequently evaluated with respect to their ability to predict the experimental data. Therefore, in a second stage, the non-linear regressions between the M two-dimensional stimulus configurations and the experimental data are determined in separate XGms sessions. The experimental data are interpreted as discrete and non-metric. A monotonic spline function with two internal knot points is used in all cases to model the non-linear relationship between the model predictions and external responses. As discussed in chapter 5, the optimum likelihood function value L expresses the goodness of fit, and this measure is therefore used to compare the alternative measures.

A second comparison is performed between a selected number M_S of instrumental measures. In this case, XGms is first used to obtain stimulus configurations from the experimental data. The data are again interpreted as discrete and non-metric. Both one and two-dimensional stimulus configurations are evaluated in case of the JPEG-coded images, while a two-dimensional stimulus configuration is required to model the images with noise and blur. The distances between all pairwise combinations of stimulus positions can be used to create a dissimilarity matrix for the experimental data. Since this matrix is symmetric and has zero values on the diagonal, it is specified by its $N_D = N \times (N-1)/2$ values above (or below) the diagonal. In the second stage, this experimentally-based dissimilarity matrix is compared against the M_S dissimilarity matrices for the instrumental measures. We propose that two dissimilarity matrices $D_p = [d_p(i,j)]$ and $D_q = [d_q(i,j)]$ are equivalent if there exists a monotonic relationship between corresponding distances. This strength of monotonic association can be tested by means of the **Spearman rank order correlation coefficient**

$$R_{pq} = 1 - 6\,\frac{\sum_{i=2}^{N}\sum_{j=1}^{i-1}\left\{Rank[d_p(i,j)] - Rank[d_q(i,j)]\right\}^2}{N_D \cdot (N_D^2 - 1)}, \qquad (11.38)$$

where $Rank[d_p(i,j)]$ expresses the rank order of dissimilarity $d_p(i,j)$ in the matrix D_p. This rank order is a number between 1 and N_D. The Spearman coefficient, which varies between -1 and 1, can be used to define the dissimilarity

$$D_{pq} = \sqrt{1 - R_{pq}^2} \qquad (11.39)$$

between dissimilarity matrices D_p and D_q. The output of the second stage of this comparison method is hence a $(M_S+1) \times (M_S+1)$ matrix of dissimilarities between instrumental measures and/or experimental data. In the third stage, this dissimilarity matrix is approximated and depicted by a two-dimensional configuration in order to better appreciate the mutual relationship between the predictions made by the different measures. Again, XGms for continuous non-metric data is used for this purpose.

This second method of comparison is only performed on a subset of the available instrumental measures (i.e., $D_{N,2}$, $D_{G,2}$, $D_{R,2}$, $D_{N,wu}$, $D_{G,wu}$, $D_{R,wu}$, $D_{N,wc}$, $D_{G,wc}$, $D_{R,wc}$, D_S and D_{SS}) in order not to unduly complicate the resulting figure. Note that this second comparison method not only allows to depict the distances between instrumental measures and experimental data, but also the mutual distances between alternative instrumental measures. Alternative instrumental measures that are very close together are, for all practical purposes, equivalent (at least for the stimulus set under consideration). Even in the absence of experimental data, this second comparison method may be used for selecting a representative subset from a large set of available measures [Mee02]. Indeed, if such a subset of measures covers the same area as the total set of measures, then little is lost by retaining only this subset. The method may also be used to evaluate if new measures do indeed give fundamentally different predictions from existing measures, and are therefore worth exploring in more depth.

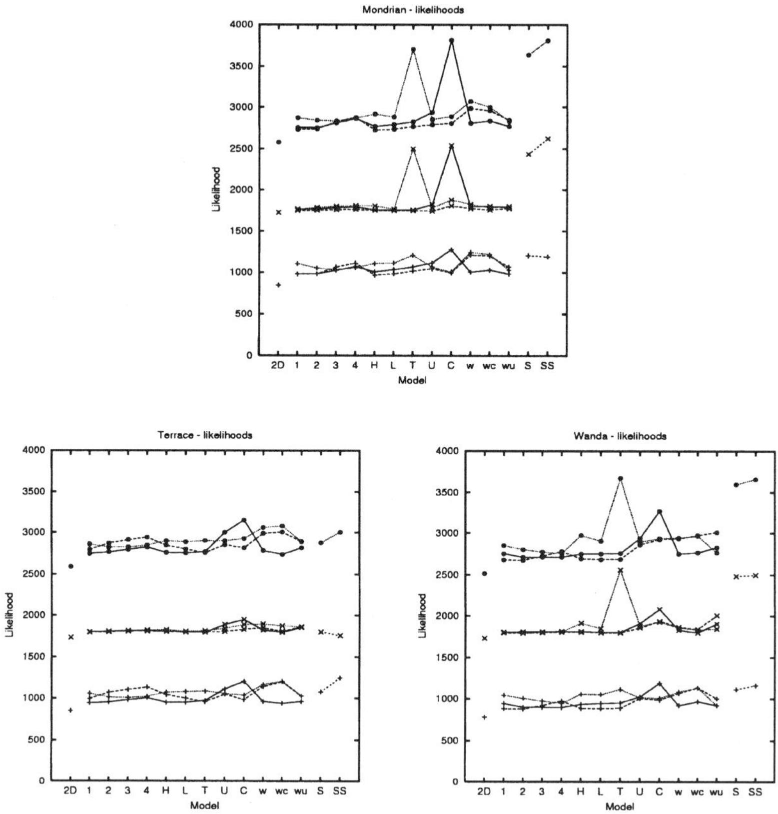

Figure 11.9. The optimized likelihood function value L expresses how well different 2-D stimulus configurations, derived from instrumental measures, can describe the experimental data for images with noise and blur. The total likelihood (denoted by the filled dots) is the sum of the likelihood for attribute data (denoted by the crosses) and the likelihood for dissimilarity data (denoted by the pluses). The optimum 2D-model, derived using MDS, is denoted by "2D", while the human-vision based measures are denoted by S and SS. The three alternative filter stages in the simple measures are indicated by the drawn lines (no filtering), dashed lines (gradient filtering) and dotted lines (residue filtering).

3.3 INSTRUMENTAL MEASURES VERSUS EXPERIMENTAL DATA

3.3.1 IMAGES WITH NOISE AND BLUR

In Figure 11.9, the optimized likelihood function value L is plotted for all considered instrumental measures. The first thing to notice is that the dissimilarity data discriminate more between the alternative measures than the attribute data, and hence provide a more critical test for instrumental measures [MM98]. Second, there is a significant difference between the likelihood value for the optimized 2-D model, as obtained using MDS, and the value for the instrumental measures. This implies that none of the considered instrumental measures is able to capture all the effects that are present in the experimental data. Third, the human-vision-based measures (S and SS) perform worse than the simple measures, which indicates that these former models, which are based on threshold experiments, are not able to adequately handle the wide range of distortions in the stimulus set. Fourth, simple measures that operate directly on the psychometric lightness most of the time perform better than measures that operate on filtered versions of this lightness. Gradient filtering is also preferred over residue filtering. Fifth, the robust integration rules (H, L and T) and the correlation measures (U and C) can give very unreliable results. The former methods probably depend too critically on the scaling factor that is applied to the difference image. The first-order Minkowski metrics (1, 2, 3 and 4) and the second-order weighted squared distances (w, wc and wu) are much better behaved in this respect and demonstrate the best overall performance.

In order to gain more insight into the differences between the stimulus configurations predicted by instrumental measures and the experimentally-based stimulus configurations, we present a direct comparison for the $D_{N,2}$-model, i.e., the root-mean-square difference between the lightness images, in Figure 11.10. The stimulus configurations derived from the instrumental measure have been transformed, using the optimal scaling, translation and unitary transformation described in Appendix 5.C, before performing the comparison with the experimentally-based configurations. Note that the instrumental measure is not able to make an adequate prediction for the experimental data, since the arrows that connect corresponding points in both configurations are typically much larger than the sizes of the uncertainty ellipses. The average distance between the image points in both configurations, as well as the inner-product correlations are also reported in Figure 11.10. The configurations derived from the instrumental measure are much more rectangular (and regular), and hence fail to model the interactions between noise and blur that are present in the experimental data. Also, for the images Mondrian and Wanda, the variations due to blur are compressed with respect to the variations due to noise. Hence, the relative importance assigned to both kinds of distortions by the instrumental measure is different from the relative importance adopted by subjects.

The results for the second comparison method between instrumental measures are shown in Figure 11.11. In this case, the points represent

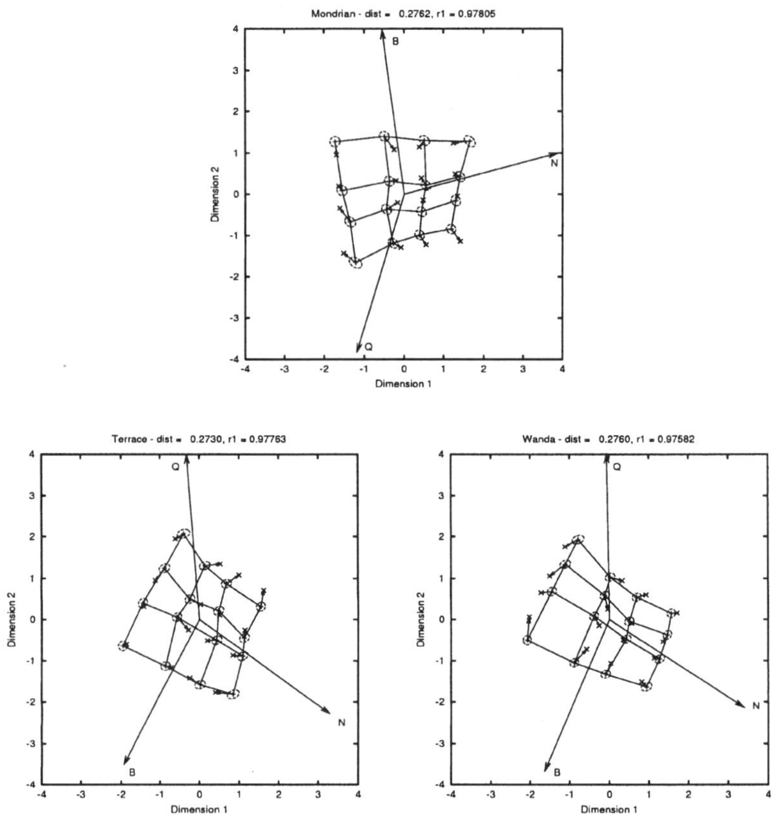

Figure 11.10. Experimental stimulus configurations in the blur/noise experiment versus stimulus configurations from the instrumental model $D_{N,2}$. The experimental data correspond to the centers of the 95% confidence ellipses, while the transformed instrumental model data are the end points of the arrows. The average distance and the inner-product correlation between the two stimulus configurations are also indicated.

different measures. The distances between these points are monotonically related to the differences between the model predictions, as expressed by the rank order correlation coefficient between dissimilarity predictions. Note that the distances of the instrumental measures to the experimentally-based configuration (denoted by "sub") agree well with the rank ordering derived in the first comparison method. The integration rule tends to have a bigger impact on the model predictions than the nature of the filtering stage. The correlated distance is an improvement over the uncorrelated distance and the root-mean-square error when ap-

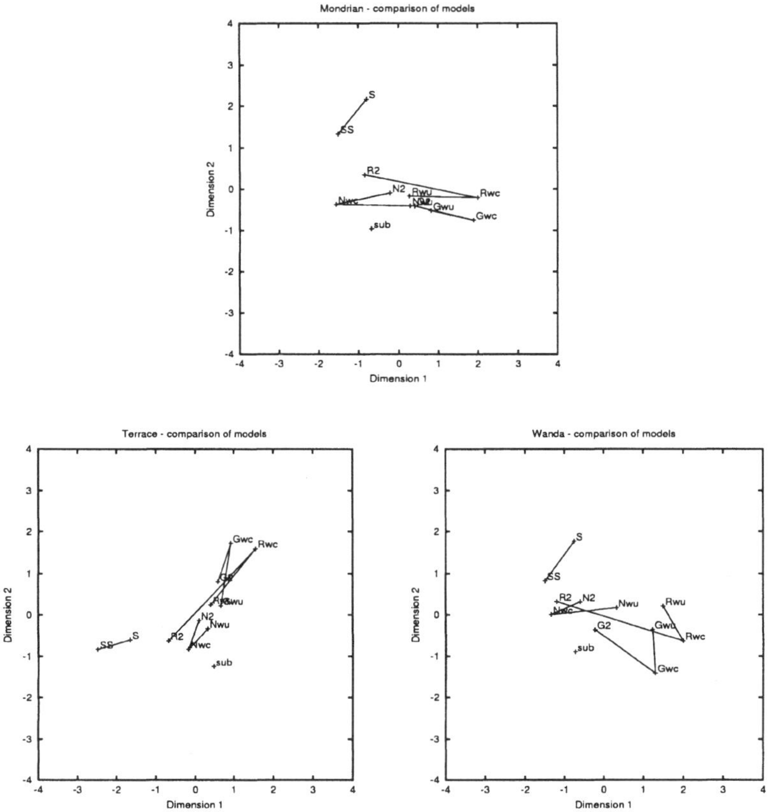

Figure 11.11. Two-dimensional configurations representing different instrumental measures and a 2-D experimentally-based model (sub) for images with noise and blur. The simple models with the same filtering stage (N: no filtering, G: gradient filtering, R: residue filtering) and different integration rule (2, wc and wu), as well as the two vision-based models (S and SS), are connected by lines.

plied to psychometric lightness (Nwc versus NWu and N2), but has a clear negative effect when applied to the residue- or gradient filtered versions (Rwc and Gwc).

3.3.2 JPEG-CODED IMAGES

In Figure 11.12, the optimized likelihood function value L is plotted for all considered instrumental measures. The dissimilarity and preference data provide about equally critical tests for the models in this case. There is again a significant difference between the likelihood value for the

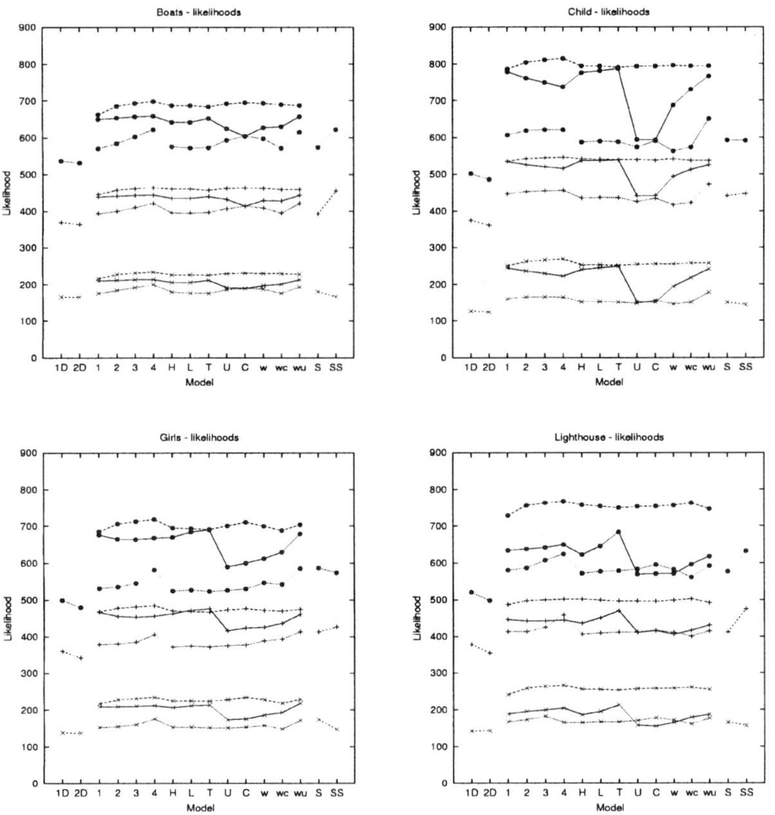

Figure 11.12. The optimized likelihood function value L expresses how well different 2-D stimulus configurations, derived from instrumental measures, can describe the experimental data for JPEG-coded images. The total likelihood (denoted by the filled dots) is the sum of the likelihood for preference data (denoted by the crosses) and the likelihood for dissimilarity data (denoted by the pluses). The optimum 1D- and 2D-models, derived using MDS, are denoted by "1D" and "2D", while the human-vision based measures are denoted by S and SS. The three alternative filter stages in the simple measures are indicated by the drawn lines (no filtering), dashed lines (gradient filtering) and dotted lines (residue filtering).

optimized 1-D and 2-D models, as obtained using MDS, and the value for the instrumental measures, implying that none of the considered instrumental measures is able to capture all the effects that are present in the experimental data. The human-vision-based measures (S and SS) perform reasonably well in this case and have a performance that is similar to that of the best simple measures. This confirms that these measures

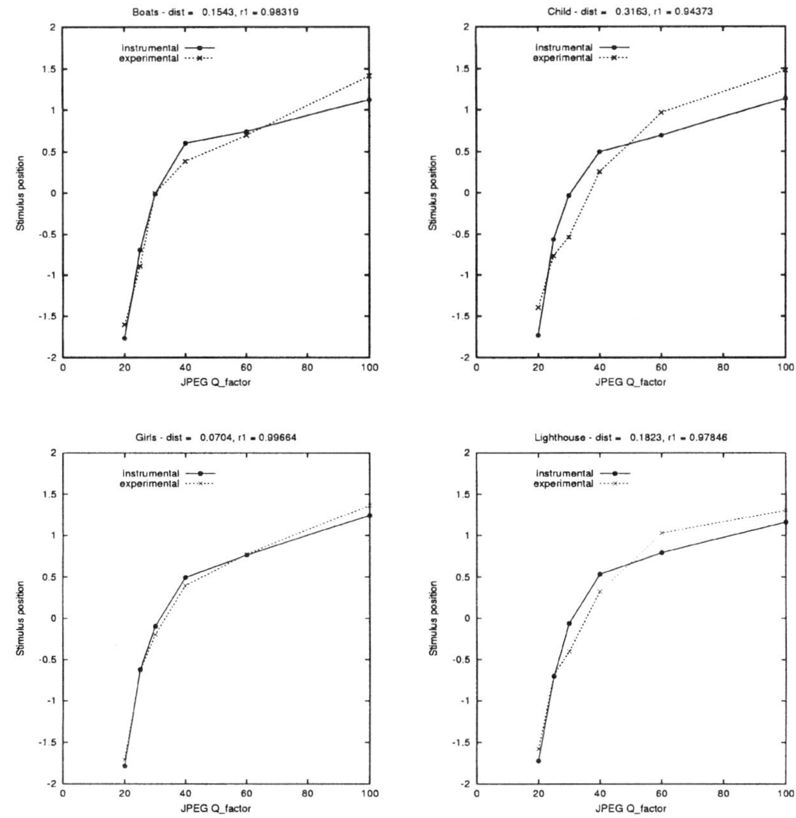

Figure 11.13. Experimental stimulus configurations for JPEG-coded images versus stimulus configurations from the instrumental model $D_{R,2}$. The experimental data are represented by the crosses, while the transformed instrumental model data are represented by the filled dots. The average distance and the inner-product correlation between the two stimulus configurations are also indicated.

are mostly suited for evaluating relatively small distortions. Simple measures that operate on the residue-filtered psychometric lightness most of the time outperform measures that operate directly on the psychometric lightness or on the gradient-filtered psychometric lightness. No filtering is also to be preferred over gradient filtering. This indicates that focusing on errors in the reproduction of edges is not sufficient. Also note that the rank order of the instrumental measures is very different from the case of images with noise and blur. No distortion measure hence wins out over the others across both experiments.

The stimulus configurations resulting from instrumental models can be compared directly with the experimentally-based stimulus configurations. We present this comparison in Figure 11.13 for one of the best models, i.e., the root-mean-square difference $D_{R,2}$ between residue-filtered lightness images. As in the previous section, comparisons are performed after optimal transformation of the instrumental model configurations. For the 1-D configurations that are considered in Figure 11.13, this reduces to an arbitrary linear transformation between stimulus positions. The average distance between the image points in both configurations, as well as the inner-product correlations, are also reported in Figure 11.13. The main reason why the instrumental measure cannot adequately predict the experimental configurations is that the instrumental measure saturates for values of the JPEG Q-factor above 40, while the subjective quality keeps increasing.

The results for the second comparison method between instrumental measures are shown in Figure 11.14. Note that the distances of the instrumental measures to the experimentally-based configurations (denoted by "sub1" and "sub2" for the 1-D and 2-D case, respectively) agree well with the rank ordering derived in the first comparison method. Since the quality variations in the JPEG-coded images are accomplished by the systematic variation of one coder parameter (i.e., a scaling factor for the quantization matrix), we a priori expected that a 1-D prediction model would perform much better than in the previous case of two simultaneous but independent distortions. Therefore we have included both 1-D and 2-D models for the experimental data in Figure 11.14.

4. CONCLUSIONS

Many existing double-ended models for perceived image quality are based on a distance metric between the original image and processed or coded versions of it. In this chapter we have shown how such instrumental measures can be compared mutually and with respect to their ability to predict experimental data. In the comparisons presented in this chapter, we have used the predicted distances between all pairwise combinations of images present in an experiment, rather than only the distances between processed and original image. In this way, a more critical test for the instrumental measures could be established, since the predicted configurations could also be compared against experimentally obtained stimulus configurations.

In none of the examined cases could a clear advantage of complex vision-based distance measures such as the Sarnoff model be demonstrated over simple measures. For two-out-of-the-three images degraded by noise and blur, the vision-based models performed even significantly

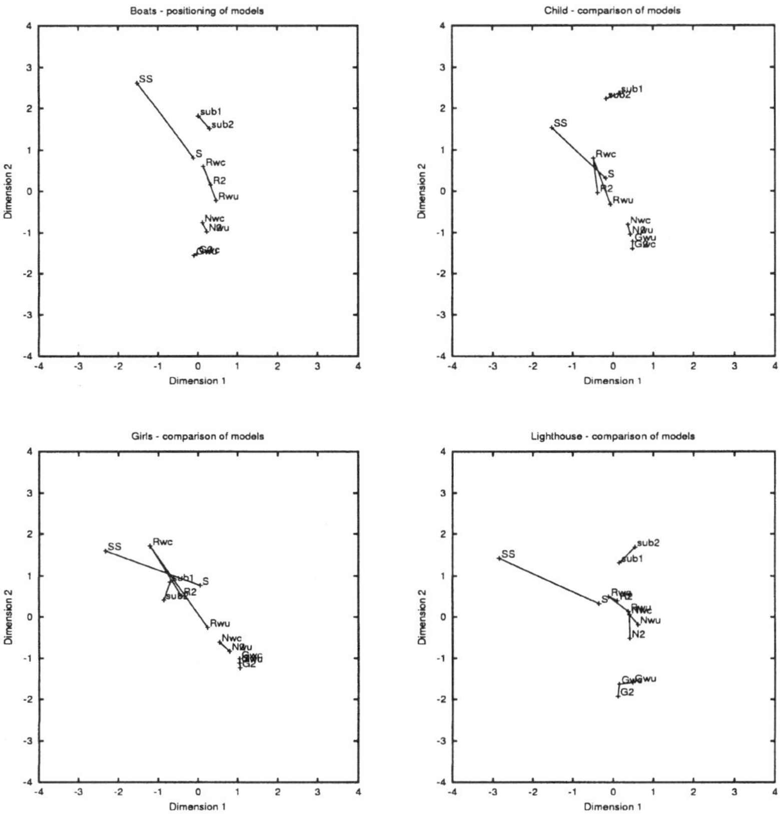

Figure 11.14. Two-dimensional configurations representing different instrumental measures and 1-D (sub1) and 2-D (sub2) experimentally-based models for JPEG-coded images. The simple models with the same filtering stage (N: no filtering, G: gradient filtering, R: residue filtering) and different integration rule (2, wc and wu), as well as the two vision-based models (S and SS), are connected by lines.

worse than the simple measures. In the case of JPEG-coded images, the vision-based models performed similarly to the best simple measures.

If more than one dimension is needed to model the stimulus configurations, then more than one independent psychological dimension is probably involved. The data for combined noise and blur demonstrate that the existing models are not able to balance the different kinds of distortions and to predict interactions between distortions. Consequently, the stimulus configurations that are derived from instrumental measures do not correlate very well with the stimulus configurations derived from experimental data.

An important distinction between both experimental data sets in this chapter is that the blurred and noisy images differ much more from the original image than the JPEG-coded images. In case of the JPEG-coded images, the non-metric analyses of the dissimilarities provided by the instrumental measures indicated that a non-linear threshold function was required for most measures to relate these dissimilarities to distances between stimulus positions. In the case of images with blur and noise, no such non-linearity was required, since all reported dissimilarities were well above threshold. The MDS approach promoted in this chapter hence provides an automated means for establishing such nonlinearities.

The goodness of fit between the instrumental measures and the experimental data, for instance expressed in terms of the likelihood of a regression model, depends to a large extent on the kind of distortions that are present in the image. It can be observed that the measures that performed best in case of the JPEG-coded images (using residue filtering) performed worst in case of images with noise and blur. This seems to indicate that different measures capture different distortions, which in turn points the way to an alternative approach towards image quality modelling. In this approach, which is discussed in more detail in the next chapter, distortion metrics that are specifically tailored to certain kinds of distortions are combined into a multi-dimensional model.

Chapter 12

SINGLE-ENDED INSTRUMENTAL MODELS OF IMAGE QUALITY

1. INTRODUCTION

In the previous chapter, we have studied double-ended instrumental measures for image quality. The two most common characteristics of existing double-ended measures are that they apply image filtering that is inspired by known perceptual mechanisms in the visual system, and that they integrate the differences between the filtered original and processed or coded image to arrive at a distortion measure. In this integration step, all differences between the filtered images are treated as being equivalent. In this chapter, an alternative perspective on image quality modelling is proposed that is in better agreement with the experimental approach towards measuring image quality that was discussed in chapter 5. This more cognitively-based perspective is inspired by two observations, i.e.,

1 that people are able to analyze and justify their image quality judgements, and

2 that people are able to make such judgements in the absence of an "original" (or undistorted) image.

The first observation implies that people can distinguish different kinds of distortions in coded [YW98] or processed [Ove96, FRYB97] images, and are aware of the fact that their overall quality judgement is determined by the relative weight that they attribute to these individual impairments. They are able to report their sensation strengths for these individual impairments in a similar way as they can report their overall quality sensations [ERMR95, KM96c]. All differences between images are hence not treated in the same way by human observers. In view of this, the assumption underlying many double-ended dissimilarity metrics, i.e., that image quality is a one-dimensional percept, should be reconsidered. The multidimensional approach that is proposed here as an alternative considers the impairments as **image quality attributes** that together determine overall quality [KM96d]. According to this philosophy, instrumental measures are not only required to predict overall quality, but also to indicate the presence and express the strength of underlying impairments [KK95, MM02]. From a practical point of view, such an approach has the added value that it may help to identify problems, i.e., to understand why the image quality does not meet the expectation or norm.

The second of the above observations motivates the interest in single-ended instrumental measures that can make attribute predictions without requiring the original image. Unlike in the case of double-ended image quality metrics, where the differences between the original and processed image are easily obtained, single-ended measures need general assumptions about what can be expected in (original) images in order to distinguish unwanted distortions from image features. This could take the form of an inventory of distortions that are most likely to occur [YW98, BHN+96]. This inventory obviously depends on the applications that are being considered (i.e., coding, processing or display). Instrumental measures that can detect and quantify (some of) these distortions need to be developed [NHT+97, MM02]. Similarly as in the case of double-ended models, the algorithms used to measure these distortions will often be inspired by discriminating processes that are known (or assumed) to exist within the visual system.

In summary, a single-ended quality measure that is based on a multi-dimensional analysis of image quality will need to address the following problems:

1 decide which impairments have to be modelled;

2 develop instrumental measures for these impairments, using only the impaired image;

3 find out how individual impairments combine into overall quality.

This chapter concentrates on the second problem, and addresses the following aspects in developing instrumental measures for specific impairments:

1 physical characterization of the impairment (i.e., identifying the prototypical distortion pattern);

2 development of an algorithm for detecting this distortion pattern and estimating its relevant parameters from the distorted image;

3 use of the estimated parameters to obtain a local measure for the impairment strength (this may involve psychophysical knowledge about the human sensitivity for detecting and discriminating the prototypical distortion pattern);

4 integrating the local impairment strengths into a global measure for the impairment strength (i.e., for the entire image);

5 comparison of the impairment measure against subjective judgements of this impairment (in complex scenes).

Note that, in order to test such measures, we not only require experimental data on overall image quality, but also ratings on image dissimilarity and/or specific image quality attributes such as noisiness, perceived blur, blockiness, etc.

When concentrating on specific applications (i.e., codecs), several impairments may vary in a highly correlated way, as was for instance observed in a study on JPEG-coded images [MM98]. In such a case, it makes sense to only develop measures for a limited number of impairments. The multidimensional characterization of (subjective) image quality in terms of underlying attributes, as discussed in chapter 5, can for instance provide information on which impairments have the highest correlation with overall quality. These impairments are the most interesting ones to model first. The choice between a number of highly correlated impairments may also be motivated by other aspects, such as the estimated effort needed to develop or implement the measure, the expected stability of the measure, etc. This expectation that many impairments will be highly correlated for a given application, so that only a limited number of distortion models are needed to construct an accurate quality model, is an important factor in thinking that the proposed approach is practically feasible. It is probably also the reason why existing double-ended methods often have reasonable performance when compared against subjective judgements for a limited data base of images.

We have been studying the feasibility of the above multidimensional approach to image quality modelling for a number of years. This chapter describes some of the results that have been obtained. Up to now, we have only considered the luminance component of still images with image-independent distortions, such as noise and blur, and simple (baseline JPEG) image coding. We are aware of the fact that many possible extensions to the approach remain to be explored. For instance, the extension towards predicting the quality of (colored) image sequences after coding with a number of distinct image coders remains to be done.

2. COMPUTATIONAL APPROACH

The multidimensional approach towards image quality modelling requires that one can differentiate between unwanted image distortions and actual image information. Unlike the human observer, an instrumental measure has no understanding of what the image represents, and therefore has to base this distinction on low-level signal features. Moreover, in order to limit the complexity of the measures, the required processing is restricted to localized areas within the image. This is also partly motivated by the fact that most of the distortions introduced by systems with the highest practical interest, such as image coders, tend to be localized.

The above limitations have led us to pursue a so-called computational approach [Mar82] in which all quality judgements are based on the analysis of uniform regions and edges [Kay95, KM96d]. This is an obvious but essential simplification of reality, made in order to keep the problem tractable. In the proposed approach, assumptions are made about how uniform regions and edges are affected by typical distortions. If such distorted uniform regions or edges are found in an image, they are considered as evidence for the presence of the corresponding impairment. Some examples of computational assumptions that have been used to guide the development of instrumental measures may help to clarify the approach:

- noisiness: original scenes are assumed to contain perfectly uniform regions, corresponding to a constant or slowly varying coloring of (part of) the objects in the scene; if such regions do not occur in the image, then this is very likely due to noise in the imaging process [KM96b, Mar95];

- blur and ringing: the original scene is assumed to contain ideal edge transitions, corresponding to sharp boundaries between objects or cast shadows; if such edges do not occur in the image, then this is very likely due to blur and/or ringing introduced in the imaging process [KM94, KM96a];

- quantization and blockiness: low-amplitude edge transitions in otherwise uniform, or slowly varying, regions of the image are artifacts of the imaging process that are very likely caused by quantization; horizontal/vertical edges are especially introduced by block coding algorithms [MM02].

It is important that the assumptions on which the instrumental measures are based are made explicit. This allows to discuss these assumptions separately from their implementation in a specific instrumental measure.

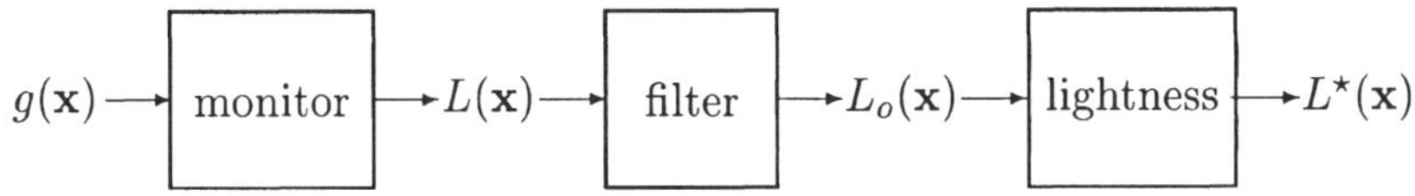

Figure 12.1. Common front-end processing for all single-ended quality measures. The grey-value image $g(\mathbf{x})$ is mapped into the luminance image $L(\mathbf{x})$ by the monitor characteristic. A Gaussian low-pass filter $w_g^2(\mathbf{x}; b_{eye})$ models the optical filtering. The filtered luminance image $L_o(\mathbf{x})$ is subsequently mapped into a psychometric lightness image $L^\star(\mathbf{x}) = f(\mathbf{x})$ that is used as input into the single-ended measures.

The computational approach also leads to a number of subtasks that need to be performed by the instrumental measures. Since edges and uniform regions are the basic features used in the measures, they first have to be detected. Second, once they have been detected, they need to be analyzed in more detail in order to estimate the parameters that will be used as indicators of the above distortions. Finally, the parameter estimates need to be integrated across the image in order to arrive at a single number for the degree of distortion. For each of these subtasks, many alternative implementations (algorithms) can be imagined. In the following sections, we will present some of the algorithms that have been proposed up to now for performing these subtasks.

Note that the discussion on a suitable front-end for double-ended instrumental measures, as presented in the previous chapter, is equally relevant for single-ended measures, and hence need not be repeated here. In casu, all the measures that will be presented in this chapter use the front-end depicted in Figure 12.1. The low-pass filtering performed by the eye optics is modelled by a simple Gaussian characteristic

$$w_g^2(\mathbf{x}; b_{eye}) = \frac{1}{\pi b_{eye}^2} \exp\left(-\frac{\|\mathbf{x}\|^2}{b_{eye}^2}\right), \tag{12.1}$$

with b_{eye} equal to 0.65 times the sampling distance. This is a reasonable model in case the viewing distance is such that the sampling distance corresponds to approximately 1 arcmin of visual angle, in which case the standard deviation of the Gaussian filter is $0.65/\sqrt{2} = 0.46$ arcmin.

3. FEATURE DETECTION

An instrumental measure for image quality has no a priori knowledge about the image. It is therefore important to derive in an early stage where the important image features, i.e., the edges and uniform regions, are most likely located. A simple algorithm is used to accomplish a fast pre-selection, while more advanced processing is only performed at the positions that are selected in the first stage. This approach also helps to keep the computational complexity of the measures low.

3.1 RESIDUE ENERGY

The first processing step aims at distinguishing uniform from non-uniform regions in an image. This is accomplished by means of a simple localized image processing technique. Let $f(\mathbf{x})$, for $\mathbf{x} \in \mathcal{F}$, denote the input image that is defined over a finite region $\mathcal{F}$ of the Euclidean plane. The localized processing is described by a normalized window function $w(\mathbf{x})$ with a limited support. This implies that $w(\mathbf{x}-\mathbf{p})$ is zero except in the neighborhood of position $\mathbf{p} \in \mathcal{F}$, and $\int_{\mathcal{F}} w^2(\mathbf{x})\, d\mathbf{x} = 1$. Any image $f(\mathbf{x})$ can be locally decomposed, i.e.,

$$w(\mathbf{x}-\mathbf{p}) \cdot f(\mathbf{x}) = w(\mathbf{x}-\mathbf{p}) \cdot [\overline{f}(\mathbf{p}) + (f(\mathbf{x}) - \overline{f}(\mathbf{p}))], \tag{12.2}$$

into its weighted mean value

$$\overline{f}(\mathbf{p}) = \int_{\mathcal{F}} w^2(\mathbf{x}-\mathbf{p})\, f(\mathbf{x})\, d\mathbf{x} \tag{12.3}$$

and a residue image $f(\mathbf{x}) - \overline{f}(\mathbf{p})$. The weighted energy (or amplitude) of this residue image, i.e.,

$$\begin{aligned} e(\mathbf{p}) = a^2(\mathbf{p}) &= \int_{\mathcal{F}} w^2(\mathbf{x}-\mathbf{p})\, \left[f(\mathbf{x}) - \overline{f}(\mathbf{p})\right]^2\, d\mathbf{x} \\ &= \int_{\mathcal{F}} w^2(\mathbf{x}-\mathbf{p})\, f^2(\mathbf{x})\, d\mathbf{x} - \overline{f}^2(\mathbf{p}), \end{aligned} \tag{12.4}$$

can be used to infer if the image is locally uniform around the position $\mathbf{p}$. The quadratic filter that maps the image $f(\mathbf{x})$, for $\mathbf{x} \in \mathcal{F}$, into the energy $e(\mathbf{p})$, or amplitude $a(\mathbf{p})$, for $\mathbf{p} \in \mathcal{F}$, can be implemented very efficiently. As shown in Figure 12.2, it consists of applying the linear filter with impulse response $w^2(-\mathbf{x})$ to both the input image $f(\mathbf{x})$ and the squared image $f^2(\mathbf{x})$, followed by simple pointwise combination of both filter outputs [Mar94, Mar95]. The result of applying this quadratic filter to an image is illustrated in Figure 12.3 for the case of a Gaussian window $w^2(\mathbf{x}) = w_g^2(\mathbf{x};\sigma)$ with standard deviation σ, where σ is equal to 2.5 times the sampling distance in the example.

3.2 ENERGY AND AMPLITUDE HISTOGRAMS

Since images usually contain input noise, the outputs $e(\mathbf{p})$ will most often not be equal to zero, not even for regions in the image that most

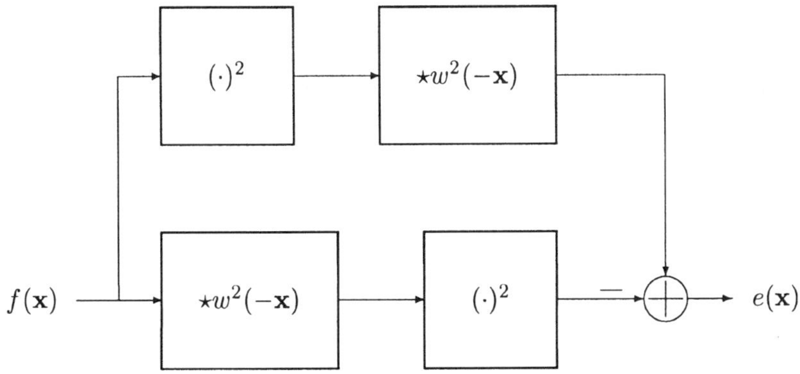

Figure 12.2. Algorithmic structure for calculating the residue energy.

human observers will consider to be uniform. A threshold value for the derived energy measure $e(\mathbf{p})$, below which image positions are classified as belonging to uniform regions, hence has to be decided. As part of this process, noise characteristics can be estimated.

As a first step towards finding an adequate threshold, the first order statistic (i.e., the histogram) of the residue energy $e(\mathbf{p})$ is analyzed. For the case of a uniform window $w(\mathbf{x})$ containing $2q+1$ pixels and an input signal $f(\mathbf{x})$ consisting of white Gaussian noise on a uniform background, the following analytic expressions were derived for the probability density functions (PDFs) of the energy e, i.e.,

$$P(e) = \frac{1}{e_o\,(q-1)!}\left(\frac{e}{e_o}\right)^{q-1} \exp\left(-\frac{e}{e_o}\right) \tag{12.5}$$

and the amplitude a, i.e.,

$$P(a) = \frac{2}{a_o\,(q-1)!}\left(\frac{a}{a_o}\right)^{2q-1} \exp\left[-\left(\frac{a}{a_o}\right)^2\right] \tag{12.6}$$

where $e_o = a_o^2$ [Mar95]. The energy is hence χ^2-distributed with $2\,q$ degrees of freedom (DOF. The average values for the energy and amplitude are

$$\mu_e = q \cdot e_o \text{ and } \mu_a = a_o \frac{\Gamma(q+0.5)}{\Gamma(q)}, \tag{12.7}$$

while the corresponding variances are

$$\sigma_e^2 = q \cdot e_o^2 \text{ and } \sigma_a^2 = q \cdot a_o^2 - \mu_a^2. \tag{12.8}$$

Figure 12.3. The left figure shows the original "terrace" image, while the right figure shows the corresponding residue amplitudes for a Gaussian window with standard deviation σ equal to 2.5 times the sampling distance.

The maximum values for the energy and amplitude PDFs occur at

$$e_m = e_o\,(q-1) \text{ and } a_m = a_o\sqrt{q-0.5}, \tag{12.9}$$

respectively. These PDFs for the residue energy and amplitude are plotted in Figure 12.4.

Because of the central limit theorem, this χ^2-distribution for the residue energy should also be valid in the limit for any sufficiently large (non-uniform) window, even if the noise is not Gaussian or white. In case of a non-uniform window or correlated noise, the number of DOF will however have to be adjusted. For example, if the input signal $f(\mathbf{x}) = \overline{f} + n(\mathbf{x})$ is the sum of a uniform background $\overline{f}$ and zero-mean additive noise $n(\mathbf{x})$ with autocorrelation function $R(\mathbf{x})$, then the mean

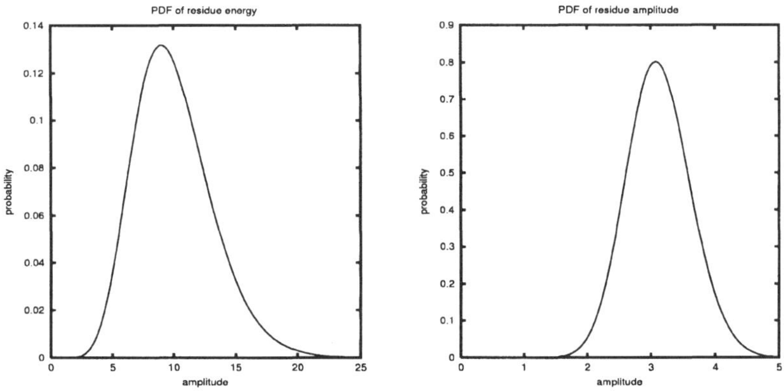

Figure 12.4. Chi-squared PDF for the energy (left) and corresponding amplitude PDF (right) in case the parameter values are $e_o = a_o = 1$ and $q = 10$.

value of the energy measure is

$$\mu_e = R(0) - \int\int d\mathbf{x}_1 \, d\mathbf{x}_2 \, w^2(\mathbf{x}_1) \, w^2(\mathbf{x}_2) \, R(\mathbf{x}_1 - \mathbf{x}_2). \tag{12.10}$$

In case of a Gaussian window $w(\mathbf{x}) = w_g(\mathbf{x}; \sigma)$ with standard deviation σ and Gaussian noise with correlation length l_c, i.e.,

$$R(\mathbf{x}) = \sigma_n^2 \exp\left(-\frac{\|\mathbf{x}\|^2}{2\, l_c^2}\right), \tag{12.11}$$

the following equations

$$\mu_e = \sigma_n^2 \frac{\sigma^2}{\sigma^2 + l_c^2} = \sigma_n^2 \frac{1}{1 + (l_c/\sigma)^2} \quad \text{and} \quad \sigma_e^2 = \frac{\mu_e^2}{F[l_c/\sigma]} \tag{12.12}$$

express how the mean value and the variance of the residue energy depend on the normalized correlation length l_c/σ (see Appendix 12.A). Equating these results to the expressions for the mean and variance in case of the χ^2-squared distribution, implies that the number of DOF

$$DOF[l_c/\sigma] = 2\, q = 2\, F[l_c/\sigma] \tag{12.13}$$

also depends on the normalized correlation length l_c/σ. This function $DOF[l_c/\sigma]$ is plotted in Figure 12.5.

Energy or amplitude histograms for an actual image can be compared against the above χ^2-distribution in order to find the PDF parameters

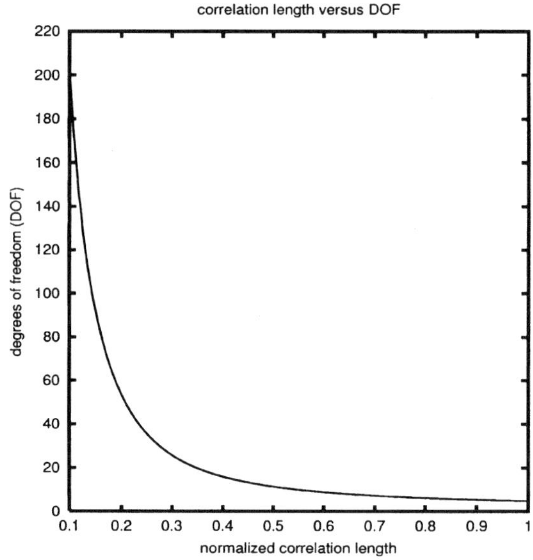

Figure 12.5. Degrees of freedom (DOF) in the chi-squared expression for the PDF of the residue energy as a function of the normalized correlation length l_c/σ in case of a Gaussian window and Gaussian noise with the autocorrelation function of Equation (12.11).

e_o (or a_o) and q that determine the average energy $\mu_e = q \cdot e_o$. There are a number of reasons why the actual energy histogram usually deviates from the χ^2-distribution. First, the χ^2-distribution constitutes only an approximate expression. Second, images contain non-uniform regions. The corresponding responses for image features such as edges will influence the shape of the histogram, especially for large energy (or amplitude) values. This implies that only the part of the histogram at low energy (or amplitude) values can be used for noise estimation and curve fitting against the χ^2-distribution. Third, the uniform regions may contain additional distortions that are not noise (such as blockiness in DCT-coded images, for instance). Such quantization effects typically distort the shape of the energy (amplitude) histogram. A significant discrepancy between the actual histogram and the χ^2-distribution is therefore often an indication of such an additional source of distortion of the uniform regions in the image. The degree of correspondence between the actual amplitude histogram $h(a)$ and the amplitude PDF $P(a)$ of equation (12.6) can for instance be expressed by the inner-product correlation

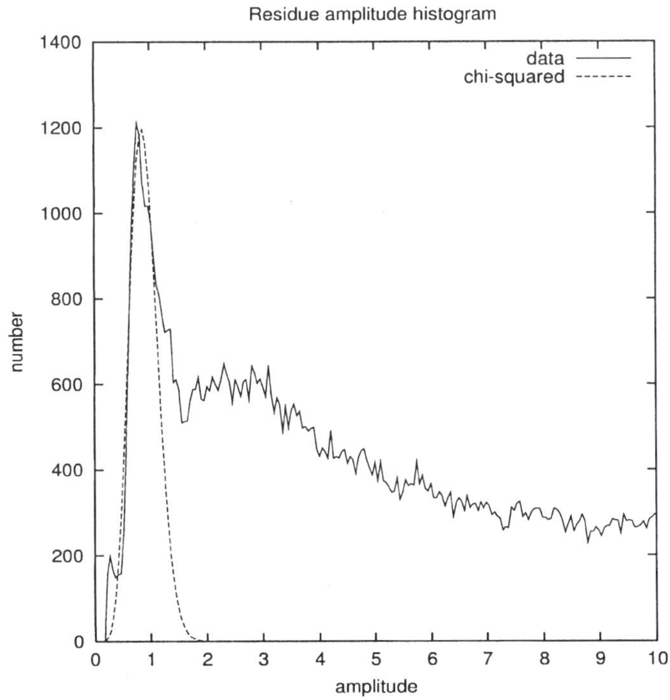

Figure 12.6. The drawn curve shows the histogram of the residue amplitude for the image in Figure 12.3. The dashed curve corresponds to optimal fitting of the χ^2-distribution ($q = 3.40$ and $a_o = 0.50$, resulting in a peak value at $a_m = 0.86$, an average energy $\mu_e = 0.86$ and a inner-product correlation coefficient of $\rho_n^2 = 0.989$).

coefficient

$$\rho_n^2 = \frac{\left[\int_{a_1}^{a_2} h(a) \cdot P(a)\, da\right]^2}{\int_{a_1}^{a_2} h^2(a)\, da \cdot \int_{a_1}^{a_2} P^2(a)\, da}, \tag{12.14}$$

where the limit values a_1 and a_2 are selected such that only the peaks of the distributions are compared. For instance, we can choose a_1 and a_2 to lie on both sides of the peak amplitude a_m and to satisfy $P(a_1) = P(a_2) = P(a_m)/2$. In Figure 12.6, we show the histogram of the residue amplitude for the image in Figure 12.3, together with the best-fitting amplitude PDF according to equation (12.6).

Especially for medical images, such as X-ray images, where Poisson noise is an important noise source, the noise level typically depends on the mean level $\overline{f}$ and can hence not be adequately described by additive white noise. In such a case, one will have to perform an analysis of the energy (or amplitude) histogram, conditional on the mean-level value $\overline{f}$. We can divide the range of mean values in a small number of bins. A histogram of the mean values can for instance be used to determine the bin boundaries such that all bins contain (approximately) the same number of pixels. We can then determine energy (or amplitude) histograms, such that only pixels for which the mean value $\overline{f}$ falls into the same bin contribute to the same histogram. The above histogram analysis can subsequently be applied separately to each histogram. This allows to check if a level-dependent additive (white) noise model provides a significantly more accurate description of the data.

3.3 ENERGY MAXIMA

The residue energy (or amplitude) image can also be used to locate important image features such as edges. For example, let us assume that the input pattern is a one-dimensional ideal edge with amplitude Δf, gradient direction $\mathbf{n}$, and located at a distance d from the center $\mathbf{p}$ of the window, i.e.,

$$f(\mathbf{x}) = f_o + \Delta f \cdot u[(\mathbf{x} - \mathbf{p}) \cdot \mathbf{n} - d], \tag{12.15}$$

where

$$u(x) = \begin{cases} 0, & \text{if } x < 0 \\ 1, & \text{if } x > 0 \end{cases} \tag{12.16}$$

is an indicator function that satisfies $u(x) = u^2(x)$. It is easily verified that the residue energy at position $\mathbf{p} \in \mathcal{F}$ is then of the form

$$e(\mathbf{p}) = (\Delta f)^2 \left[\beta(d) - \beta^2(d)\right], \tag{12.17}$$

where

$$\beta(d) = \int w^2(\mathbf{x})\, u(\mathbf{x} \cdot \mathbf{n} - d)\, d\mathbf{x}. \tag{12.18}$$

If the window $w(\mathbf{x})$ is circularly symmetric, then the residue energy $e(\mathbf{p})$ depends only on the distance d, and not on the edge gradient direction $\mathbf{n}$. For the important special case of the above-mentioned Gaussian window $w(\mathbf{x}) = w_g(\mathbf{x}; \sigma)$ with standard deviation σ, the function $\beta(d)$ can be expressed as

$$\beta(d) = \frac{1}{2}\left[1 + \operatorname{erf}\left(\frac{d}{\sigma}\right)\right]. \tag{12.19}$$

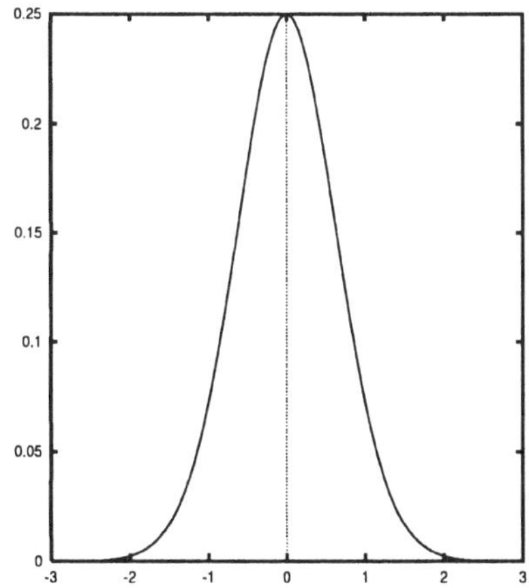

Figure 12.7. Residue energy $e(\mathbf{p})$ as a function of the normalized edge distance d/σ for a Gaussian window with standard deviation σ.

The residue energy $e(\mathbf{p})$ as a function of the normalized edge distance d/σ is hence a unimodal function that reaches its maximum value at the position of the edge (i.e., when $d = 0$), as shown in Figure 12.7.

For other symmetric or asymmetric features, such as lines and bars, the energy will also reach a local maximum at the symmetry or asymmetry point of the feature. Hence, these positions where the energy is maximum are interesting candidates for a more in-depth analysis of the underlying image pattern. If the underlying image features are one-dimensional, then these maxima of the energy will form one-dimensional ridges in two dimensions. These ridges can be detected by means of non-maximum suppression.

First, we have to determine the local orientation in the residue energy image. One of the simplest ways to accomplish this is to apply derivative filters such as the Sobel filters

$$S_x = \begin{bmatrix} -1 & 0 & 1 \\ -2 & 0 & 2 \\ -1 & 0 & 1 \end{bmatrix} \quad \text{and} \quad S_y = \begin{bmatrix} -1 & -2 & -1 \\ 0 & 0 & 0 \\ 1 & 2 & 1 \end{bmatrix} \tag{12.20}$$

to the energy image $e(\mathbf{p})$ to obtain $e_x(\mathbf{p}) = S_x \star e(\mathbf{p})$ and $e_y(\mathbf{p}) = S_y \star e(\mathbf{p})$, and to define the local orientation as

$$\phi(\mathbf{p}) = \arctan\left(\frac{e_y(\mathbf{p})}{e_x(\mathbf{p})}\right). \tag{12.21}$$

Figure 12.8. The left image shows the ridge chains in the residue energy image of Figure 12.2 with a length exceeding 10 pixels. The right image shows the "uniform" pixels that arise by excluding all pixels in a radius of $3\,\sigma$ from the ridge points in the left image.

Second, for each position $\mathbf{p}$ in the energy image, we consider the pair of positions $(\mathbf{p}_{li}, \mathbf{p}_{ri})$ in a 3x3 neighborhood of $\mathbf{p}$ that lie closest to this local orientation. If $e(\mathbf{p})$ is larger than $e(\mathbf{p}_{li})$ and $e(\mathbf{p}_{ri})$, then position $\mathbf{p}$ is a local maximum. Local maxima at neighboring positions can be linked into chains, which allows to define the chain length and average chain energy. For example, the ridge chains in the residue energy image of Figure 12.2 with a length exceeding 10 pixels are displayed in the left part of Figure 12.8.

We can define "uniform" points as positions in the image that lie more than a window length away from a ridge point. In case of a Gaussian window with standard deviation σ, we can for instance select a distance equal to $3\,\sigma$. The resulting "uniform points" for the example image are shown in the right part of Figure 12.8. In Figure 12.9, we show a similar residue amplitude histogram as in Figure 12.6, but now only the "uni-

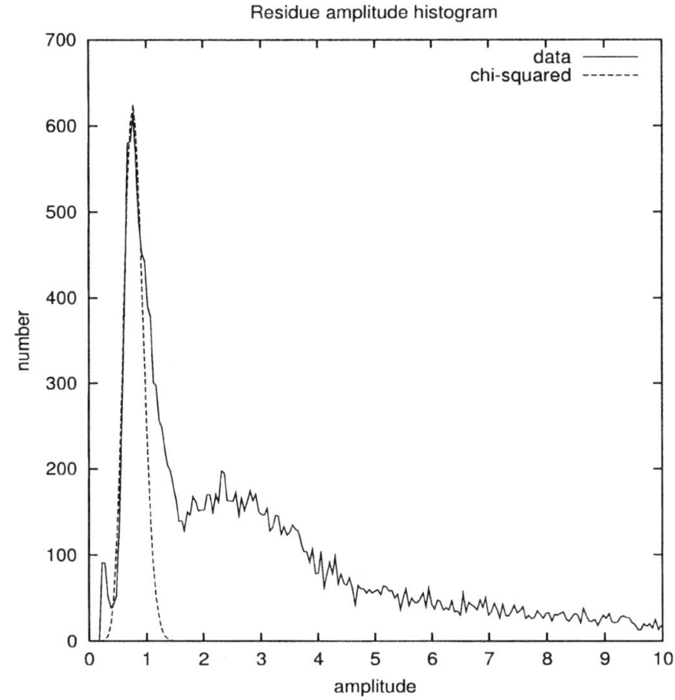

Figure 12.9. The drawn curve shows the histogram of the residue amplitude for the "uniform" points in the image in Figure 12.3. The dashed curve corresponds to optimal fitting of the χ^2-distribution ($q = 5.68$ and $a_o = 0.34$, resulting in a peak value at $a_m = 0.76$, an average energy $\mu_e = 0.64$ and a inner-product correlation coefficient of $\rho_n^2 = 0.988$).

form" points contribute to this histogram. Note that this restriction to the uniform points reduces the number of high-amplitude contributions to the histogram. Usually, this has only limited influence on the noise-estimation process, which is based on detection of the low-amplitude peak position in the histogram.

4. EDGE PARAMETER ESTIMATION

In order to decide if the (ridge) positions that have been pre-selected on the basis of the residue energy criterium correspond to features of interest, a more in depth analysis of the image is carried out at those selected positions. The analysis that we describe next is specifically aimed at finding and analyzing edges. It consists of three steps:

1 The image is mapped to its local derivatives by means of the **Hermite transformation** [Mar90c, Mar90b]. The derivative-of-Gaussian filters in the Hermite transformation have been proposed as model for simple-cell responses in the visual cortex [You85], and hence are a feasible choice for the spatial filtering performed in the early visual system.

2 The derivatives are analyzed in order to decide if the image is locally one-dimensional (1-D), and to find the optimum direction. The two-dimensional (2-D) derivatives are mapped to 1-D derivatives along this optimum direction [Mar97]. This could be seen as our interpretation of some of the nonlinear processing that is being performed in the visual cortex.

3 The 1-D derivatives are used to decide if the image is an edge, and to perform the (nonlinear) estimation of the edge parameters. This mechanism has also been proposed as a model for edge and bar localization in human perception [GF97].

4.1 TWO-DIMENSIONAL HERMITE TRANSFORMATION

A set of filters is applied to the image $f(\mathbf{x})$ in order to find local derivatives (also called Hermite coefficients) $f_{n-m,m}(\mathbf{p})$, for $n = 0, 1, 2, \ldots$ and $m = 0, \ldots, n$, at selected positions $\mathbf{p} \in \mathcal{F}$. The filters that are used have polar-separable Fourier transformations of the form

$$\hat{D}_{n-m,m}(\omega, \phi) = \hat{D}_n(\omega) \cdot \alpha_{n-m,m}(\phi), \tag{12.22}$$

for $m = 0, \ldots, n$, where

$$\hat{D}_n(\omega) = \frac{1}{\sqrt{2^n\, n!}} (-j\omega\sigma)^n \exp\left[-(\omega\sigma)^2/4\right] \tag{12.23}$$

is the 1-D Fourier transformation of the n-th order Gaussian derivative, for $n = 0, 1, 2, \ldots$, and (ω, ϕ) are polar coordinates in the Fourier plane. Several alternative choices, the pros and cons of which are discussed in [Mar97], can be made for the angular functions $\alpha_{n-m,m}(\phi)$. For the current discussion, the so-called polar angular functions

$$\alpha_{n-m,m}(\phi) = \sqrt{B_{n-m,m}} \cdot \begin{cases} \sqrt{2}\cos(n-2m)\phi, & 0 \leq m < \frac{n}{2}. \\ 1, & m = \frac{n}{2} \\ \sqrt{2}\sin(n-2m)\phi, & \frac{n}{2} < m \leq n \end{cases} \tag{12.24}$$

with

$$B_{n-m,m} = \frac{n!}{2^n\,(n-m)!\,m!}, \tag{12.25}$$

are most appropriate.

In the remainder, we will no longer explicitly denote the dependency of the derivatives on the position $\mathbf{p}$. It is understood that the estimation process that we discuss is to be repeated at all selected positions $\mathbf{p}$ (i.e., at the ridge positions of the residue energy).

4.2 ONE-DIMENSIONAL HERMITE TRANSFORMATION

The local derivatives $\{f_{n-m,m}; m = 0, \ldots, n\}$ of the same order n together describe how the directional derivative of order n varies as a function of the orientation. For instance, for the above-mentioned polar angular functions, it can be proven that

$$f_n(\phi) = \sum_{m=0}^{n} f_{n-m,m} \cdot \alpha_{n-m,m}(\phi) \tag{12.26}$$

is the n-th order derivative along the axis that makes an angle ϕ with the horizontal axis [DM97a, Dij97, Mar97]. If we are interested in 1-D patterns, such as edges, then $|f_n(\phi)|$ is maximum if the angle ϕ agrees with the orientation of the 1-D pattern. This property is used to find this orientation from the Hermite coefficients by maximizing the energy measure

$$E_N^{1D}(\phi) = \sum_{n=1}^{N} f_n^2(\phi), \tag{12.27}$$

where most often $N = 1$, if we are only interested in asymmetric patterns such as edges, or $N = 2$, if both asymmetric and symmetric patterns (edges and lines) are to be detected. Comparing this 1D energy measure to the overall energy measure

$$E_N = \sum_{n=1}^{N} \sum_{m=0}^{n} f_{n-m,m}^2 \tag{12.28}$$

can furthermore reveal if the pattern is indeed locally one-dimensional along direction ϕ. If $E_N^{1D}(\phi) \approx E_N$, then the analysis of the differential structure can be reduced to an analysis of the 1-D derivatives $f_n(\phi)$.

4.3 COMPOSITE EDGES

The Hermite coefficients at the origin for a 1D pattern $f(x)$ are given by

$$f_n(\sigma) = \int_{-\infty}^{+\infty} f(x)\, d_n(-x, \sigma)\, dx, \tag{12.29}$$

for $n = 0, 1, \ldots,$ where the analysis filter

$$d_n(x,\sigma) = \frac{1}{\sqrt{2^n\, n!}}\, \frac{d^n}{d(x/\sigma)^n} \left[\frac{1}{\sigma\sqrt{\pi}} \exp\left(-\frac{x^2}{\sigma^2}\right)\right], \tag{12.30}$$

is proportional to the n-th derivative of a Gaussian. These filters were depicted in Figure 2.4 for orders $n < 10$.

We use the following **composite edge model**

$$f(x) = k_0 + \frac{k_1}{\sigma}\, x + \Delta k_0 \cdot e_0(x-d) + \frac{\Delta k_1}{\sigma}\, e_1(x-d) \tag{12.31}$$

to approximate real image edges. This edge model combines the effect of an overall gradient, an **ideal edge**

$$e_0(x) = \begin{cases} -\frac{1}{2} & x < 0 \\ +\frac{1}{2} & x \geq 0 \end{cases} \tag{12.32}$$

and an **ideal crease**

$$e_1(x) = \begin{cases} -\frac{x}{2} & x < 0 \\ +\frac{x}{2} & x \geq 0 \end{cases} \tag{12.33}$$

(with $e_1'(x) = e_0(x)$). This composite edge pattern is completely specified by the parameters k_0, k_1, Δk_0, Δk_1 and d, some of which may of course be equal to zero. The value and gradient change abruptly at position $x = d$. This composite edge model, as well as simplified versions of it, has been proposed by several authors for performing edge detection in 2-D images [LM88, LMAR89, GM94, JA95, GM97]. In order to decide if this pattern can model local image structures, we must be able to estimate the edge parameters. We present a method based on the Hermite coefficients that is closely related to the moment-based methods that are typically adopted in the above-mentioned references.

We first determine the Hermite coefficients for all the subpatterns in the composite edge pattern. The gradient pattern $k_0 + (k_1/\sigma) \cdot x$ gives rise to the Hermite coefficients

$$\begin{aligned} f_0(\sigma) &= k_0 \\ f_1(\sigma) &= \frac{k_1}{\sqrt{2}} \\ f_n(\sigma) &= 0 \end{aligned} \tag{12.34}$$

for $n = 2, 3, \ldots$ The Hermite coefficients for the ideal edge are

$$\begin{aligned} f_0(\sigma) &= -\frac{\Delta k_0}{2} \operatorname{erf}\left(\frac{d}{\sigma}\right) \\ f_n(\sigma) &= \frac{\Delta k_0 \cdot \sigma}{\sqrt{2n}} d_{n-1}(-d, \sigma) \\ &= \frac{\Delta k_0}{\sqrt{\pi\, 2^n\, n!}} H_{n-1}\left(\frac{d}{\sigma}\right) \exp\left(-\frac{d^2}{\sigma^2}\right) \end{aligned} \tag{12.35}$$

for $n = 1, 2, \ldots$, while those for the ideal crease are

$$\begin{aligned} f_0(\sigma) &= \frac{\Delta k_1}{2\sqrt{2}} \exp\left(-\frac{d^2}{\sigma^2}\right) \\ f_1(\sigma) &= -\frac{\Delta k_1}{2\sqrt{2}} \operatorname{erf}\left(\frac{d}{\sigma}\right) \\ f_n(\sigma) &= \frac{\Delta k_1 \cdot \sigma}{\sqrt{4n(n-1)}} d_{n-2}(-d, \sigma) \\ &= \frac{\Delta k_1}{\sqrt{\pi\, 2^n\, n!}} H_{n-2}\left(\frac{d}{\sigma}\right) \exp\left(-\frac{d^2}{\sigma^2}\right) \end{aligned} \tag{12.36}$$

for $n = 2, 3, \ldots$, with $H_n(x)$ the Hermite polynomial of order n.

If the composite edge is blurred by a Gaussian filter with a two-dimensional impulse response $b(\mathbf{x}) = w_g^2(\mathbf{x}; b)$ with blur parameter b (and standard deviation $b/\sqrt{2}$), then the Hermite coefficients for the blurred pattern are given by

$$f_n^{blurred}(\sigma) = R_b^n \cdot f_n^{ideal}(\sigma_b), \tag{12.37}$$

where

$$\sigma_b = \sqrt{b^2 + \sigma^2} \text{ and } R_b = \frac{\sigma}{\sigma_b}, \tag{12.38}$$

as was derived by Kayargadde & Martens [KM94].

The resulting **blurred composite edge** can be expressed as

$$\hat{f}(x) = k_0 + \frac{k_1}{\sigma} x + \frac{\Delta k_0}{2} \operatorname{erf}\left(\frac{x-d}{b}\right) + \frac{\Delta k_1}{2\,\sigma} b \operatorname{ierf}\left(\frac{x-d}{b}\right), \tag{12.39}$$

where $\operatorname{erf}(x)$ is the standard error function, and

$$\operatorname{ierf}(x) = x \cdot \operatorname{erf}(x) + \frac{1}{\sqrt{\pi}} \exp(-x^2) \tag{12.40}$$

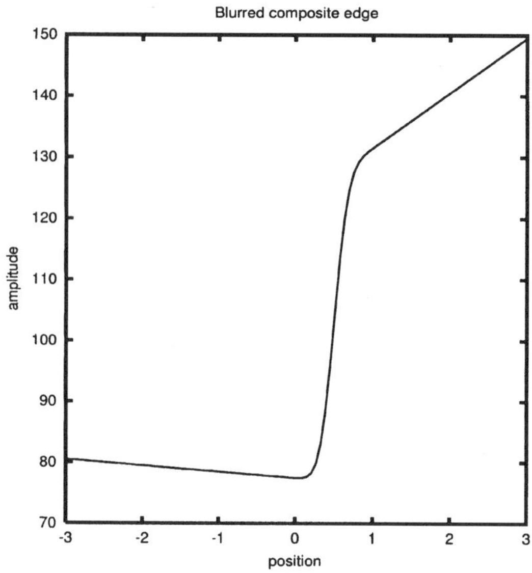

Figure 12.10. Blurred composite edge with parameter values $k_0 = 100$, $\Delta k_o = 50$, $k_1/\sigma = 4$, $\Delta k_1/\sigma = 2$, $b = 0.2$ and $d = 0.5$.

is the integrated error function. An example of such a blurred composite edge is plotted in Figure 12.10.

The Hermite coefficients for a blurred composite edge are obtained by summing al the above contributions. This results in

$$\begin{aligned}\hat{f}_0 &= k_0 + a_0 \cdot h_{-1}(D) + a_1 \cdot R_b^{-1}\, h_{-2} \\ \hat{f}_1 &= \frac{1}{\sqrt{2}}\,[k_1 + a_0 \cdot R_b + a_1 \cdot h_{-1}(D)] \\ \hat{f}_n &= \frac{1}{\sqrt{2^n n!}}\left[a_0 \cdot R_b^n\, H_{n-1}(D) + a_1 \cdot R_b^{n-1}\, H_{n-2}(D)\right] \quad (12.41)\end{aligned}$$

for $n = 2, 3, \ldots$, with $h_{-2} = \sqrt{\pi/8}$ and

$$h_{-1}(D) = -\sqrt{\frac{\pi}{4}}\,\exp(D^2)\,\mathrm{erf}(D), \quad (12.42)$$

$(h_{-1}(D) \approx -D$ for $D < 0.2$, and $h'_{-1}(D) = 2D \cdot h_{-1}(D) - 1)$.

The parameters

$$D = \frac{d}{\sigma_b} \tag{12.43}$$

$$a_0 = \frac{\Delta k_0}{\sqrt{\pi}} \exp(-D^2) \tag{12.44}$$

$$a_1 = \frac{\Delta k_1}{\sqrt{\pi}} \exp(-D^2) \tag{12.45}$$

and R_b can be estimated from given Hermite coefficients f_n of orders $n = 2, 3, 4, 5$. It is assumed that the actual coefficients f_n correspond to coefficients $\hat{f}_n$ of a blurred composite edge, so that equations (12.41) can be solved to find the parameters. This estimation process is discussed in more detail in Appendix 12.B.

It is possible that no solution for the parameters can be derived from the actual Hermite coefficients f_n, in which case the edge hypothesis is rejected. In addition, only solutions that satisfy

$$\frac{1}{1+\beta_{max}^2} \leq R_b^2 \leq \frac{1}{1+\beta_{min}^2} \tag{12.46}$$

are considered acceptable. The upper bound corresponds to a minimum possible blur of $b = \beta_{min} \cdot \sigma$, while the lower bound corresponds to a maximum allowable blur of $b = \beta_{max} \cdot \sigma$ (where β_{min} and β_{max} are selected a priori). The limit β_{min} can for instance describe the blurring that is performed in the front-end, in which case $\beta_{min} = b_{eye}/\sigma$. A nonzero value for β_{min} also helps to avoid problems that may otherwise arise with the digital approximation of the continuous derivatives used throughout this chapter. The limit β_{max} is imposed because there is a limit to the size of the blurring that can be estimated from an analysis within a Gaussian window with standard deviation equal to σ.

Similarly, an a priori condition

$$| D | \leq \alpha \tag{12.47}$$

is imposed on the distance parameter in order to guarantee that the edge is sufficiently close to the center of the analysis window to allow for a reliable estimate.

If a set of parameters is found, then they define the blurred composite edge signal $\hat{f}(x)$ of equation (12.39). We additionally require that the residue energy in the original and estimated signal are approximately equal, i.e.,

$$\begin{aligned} e &= \int_{\mathcal{F}} w_g^2(\mathbf{x};\sigma)\ \left[f(\mathbf{x}) - \overline{f}\right]^2\ d\mathbf{x} \\ &\approx \int_{-\infty}^{+\infty} \frac{1}{\sqrt{\pi}\,\sigma} \exp\left(-\frac{x^2}{\sigma^2}\right) \cdot \left(\hat{f}(x) - \hat{f}_0\right)^2\ dx \end{aligned} \qquad (12.48)$$

for the case of the 2-D Gaussian window $w(\mathbf{x}) = w_g(\mathbf{x};\sigma)$ with standard deviation σ, where $\hat{f}(x)$ and $\hat{f}_0$ denote the estimated 1D edge pattern and its average value, respectively. If this latter condition is not satisfied, then the estimates are also rejected, since the edge model cannot provide an accurate description of the actual signal.

Once a generalized edge model has been estimated from the Hermite coefficients, we must decide if we want to include the parameter estimates in an instrumental measure. A possible requirement is for instance that the generalized edge should approximate an ideal edge, since otherwise the blur estimate may not be very reliable. The generalized edge varies linearly from $k_o - \Delta k_o/2 - k_1 + \Delta k_1/2$ at $d - \sigma$ to $k_o - \Delta k_o/2$ near the edge position d, makes a discontinuous jump to $k_o + \Delta k_o/2$ at edge position d, followed by a linear variation to $k_o + \Delta k_o/2 + k_1 + \Delta k_1/2$ at position $d + \sigma$. The requirement that the linear variations at both sides of the edge are small in comparison with the step variation can for instance be mapped into a condition

$$\sqrt{k_1^2 + \left(\frac{\Delta k_1}{2}\right)^2} < \varepsilon \left|\frac{\Delta k_0}{2}\right|, \qquad (12.49)$$

where ε is again determined a priori.

4.4 SUMMARY MODEL

Three instrumental measures for image quality attributes have been implemented using the signal detection and estimation procedures described above. They are summarized in Figure 12.11.

The input image $f(\mathbf{x})$ in Figure 12.11 is the output of the front-end processing in Figure 12.1. The residue energy is calculated for all positions in the image, using a Gaussian window $w_g(\mathbf{x};\sigma)$, where the parameter σ is expressed in multiples of the sampling distance. The maxima in the residue energy are determined by non-maximum suppression and linked into maximum-energy ridges. Only ridges with a length exceeding $\ell_{min} = 10$ pixels and an average amplitude above $a_{ridge} = 5$ on a lightness scale are retained. The positions that remain after excluding all points within a radius of 3σ of a ridge point are retained as "uniform" points.

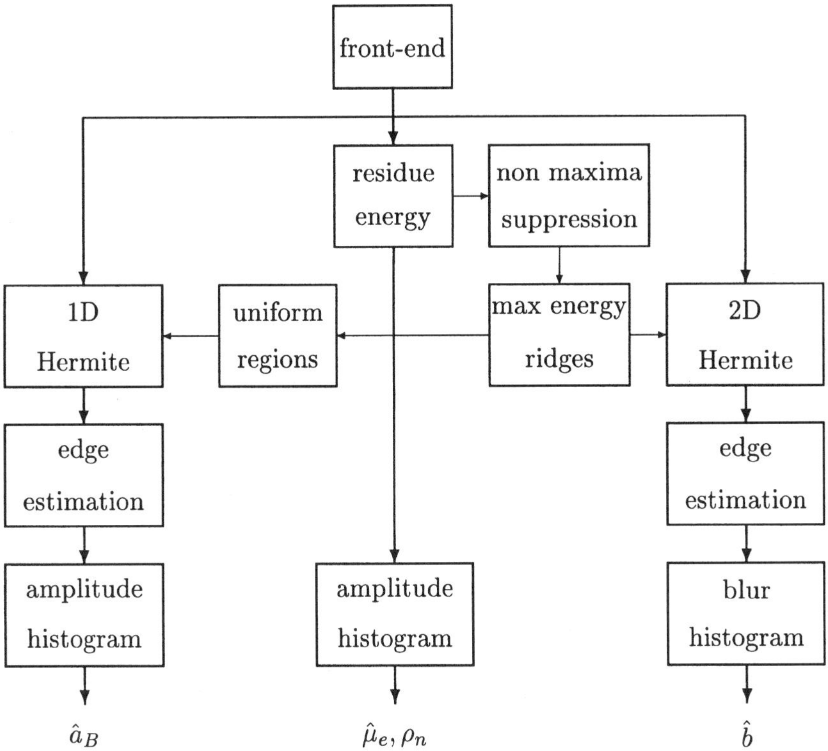

Figure 12.11. Flow chart showing how single-ended instrumental measures for blockiness, noise and blur are derived from an input image.

The histogram of the residue amplitude is determined and matched against the χ^2-distribution of equation (12.6). This histogram can be determined over the entire image, or only over the uniform regions in the image. The estimated parameters q and a_o of the χ^2-distribution are used to define the average residue energy $\mu_e = q \cdot a_o^2$. If the noise can be assumed to be white, then $\hat{\sigma}_n = \sqrt{\mu_e}$ is the estimate for the noise standard deviation[1]. The correlation ρ_n between the actual histogram and the χ^2-distribution is also retained as an indicator of whether additional sources of distortions are present in uniform regions.

A generalized edge model is fitted at the positions of the 'maximum energy ridges'. For this purpose, 2D Hermite coefficients up to order

[1]The case of colored noise will be discussed in the next section.

5 are calculated at the selected positions. Positions for which the 1d energy $E_2^{1D}(\phi)$ for the optimum angle ϕ is more than 95% of the total energy E_2 are retained. The parameters of the generalized edge parameter estimation are fixed to: $\alpha = 0.5$, $\beta_{min} = b_{eye}/\sigma$ (with $b_{eye} = 0.65$), $\beta_{max} = 1.2$, and $\varepsilon = 0.05$. The histogram of the estimated blur parameters $\hat{b}(i)$, where i ranges over all positions for which a valid estimate has been obtained, is used to derive a weighted average, i.e.,

$$\hat{b} = \frac{\sum_i \hat{b}(i)\, |f_1(i)|^2}{\sum_i |f_1(i)|^2}, \tag{12.50}$$

where $|f_1(i)|$ is the absolute value of the first-order Hermite coefficient (i.e., the first-order Gaussian derivative) at position i. This weighted average, which assigns a higher weight to more prominent edges, was also used in the original blur-estimation study [KM96a].

Especially for JPEG-coded images, an artefact called blockiness arises at lower bitrates. This artefact manifests itself as low-amplitude edges along horizontal and vertical directions in uniform regions of the image. In order to detect these artifacts, 1D Hermite coefficients up to order 3 are determined in the horizontal and vertical direction for the positions in uniform regions. A blurred ideal edge model is estimated at these positions, based on these Hermite coefficients. The parameters of the block estimation are $\alpha = 0.25$, $\beta_{min} = b_{eye}/\sigma$ and $\beta_{max} = 0.5$. The assumptions underlying the instrumental blockiness measure are that blockiness edges are sharp (i.e., contain very little blur) and have low amplitudes. Therefore, only amplitude estimates between $a_{min} = 1$ and $a_{max} = 5$ on a lightness scale are retained. In [Mee02, MM02], a large number of alternative integration rules were examined in order to arrive at an overall blockiness estimate. The rule that performed best was an unnormalized Minkowski sum of the estimated block amplitudes, i.e.,

$$\hat{a}_B = \left[\sum_i |\widehat{\Delta k}_0(i)|^p\right]^{1/p} \quad \text{for } a_{min} \leq \widehat{\Delta k}_0(i) \leq a_{max}, \tag{12.51}$$

with exponent $p = 2.8$. This blockiness measure will therefore be adopted here. We refer to [Mee02, MM02] for more details on this blockiness measure.

5. IMAGES WITH CORRELATED NOISE

In order to study the relationship between estimated noise standard deviation and perceived noisiness, the following experiment was conducted. Correlated noise was added to two of the natural scenes, Wanda and Terrace, that were also used in the experiments of chapter 5. Correlated noise samples with variance $\sigma_n^2 = 25$ and $\sigma_n^2 = 50$ were obtained by filtering white noise with variable standard deviation σ_w by means of binomial filters of order B (with $B = 0, 1, 2, 3, 4, 6, 8$). This (discrete) binomial filtering is approximately equivalent to a (continuous) Gaussian filter $w^2(\mathbf{x}; l_c)$ with parameter $l_c = \sqrt{B/2}$. In case of such a Gaussian filter, it can be shown that the relationship between the standard deviation of the unfiltered white noise and the standard deviation of filtered colored noise is given by $\sigma_n = \sigma_w/(l_c\sqrt{2\pi})$. In order to obtain a constant value for the standard deviation σ_n of the colored noise, the standard deviation σ_w of the white noise hence has to increase with the filter order B. For each value of σ_n and B, the appropriate value for σ_w was determined iteratively. The 14 images with colored noise were presented together with the original image in a numerical category scaling experiment in which the six subjects were asked to judge the noisiness of the images on an integer scale from 0 to 10. All stimuli were presented 4 times in random order. The display characteristic and the viewing distance were identical to those described in chapter 5.

The correlated noise estimation which supplies the instrumental measure for noisiness is based on equation (12.12). This equation expresses the dependence of the average residue energy μ_e on the noise standard deviation σ_n and correlation length l_c for a Gaussian window with size parameter σ. This implies that the noise parameters σ_n and l_c can be estimated, provided that the average residue energy μ_e is known for at least two window sizes σ. In order to accommodate the fairly large range of noise correlation lengths, we actually determined the residue amplitude histograms for four window sizes $\sigma = 1$, $\sigma = 1.5$, $\sigma = 2$ and $\sigma = 2.5$ times the sampling distance. Each of these histograms was fitted against the χ^2-distribution of equation (12.6) in order to obtain the optimal parameters $\hat{q}$ and $\hat{a}_o$, which in turn determined the estimated average energy $\hat{\mu}_e = \hat{q} \cdot \hat{a}_o^2$. Subsequently, minimization of the quadratic error between the model of equation (12.12) and the experimental data ($\hat{\mu}_e$ as a function of σ) was used to estimate the noise standard deviation $\hat{\sigma}_n$ and the correlation length $\hat{l}_c$.

The noisiness estimates, with corresponding 95 % uncertainty intervals, were obtained by analyzing the subjective scores of the six subjects with the maximum-likelihood estimation program XGms for discrete data (see chapter 5). If mean square error would be an adequate criterium for characterizing the noisiness, then subjective noisiness judgements should correlate well with the estimated noise standard deviation $\hat{\sigma}_n$. The graphs in the upper part of Figure 12.12 show that this is clearly not the case. We therefore investigated the hypothesis that the

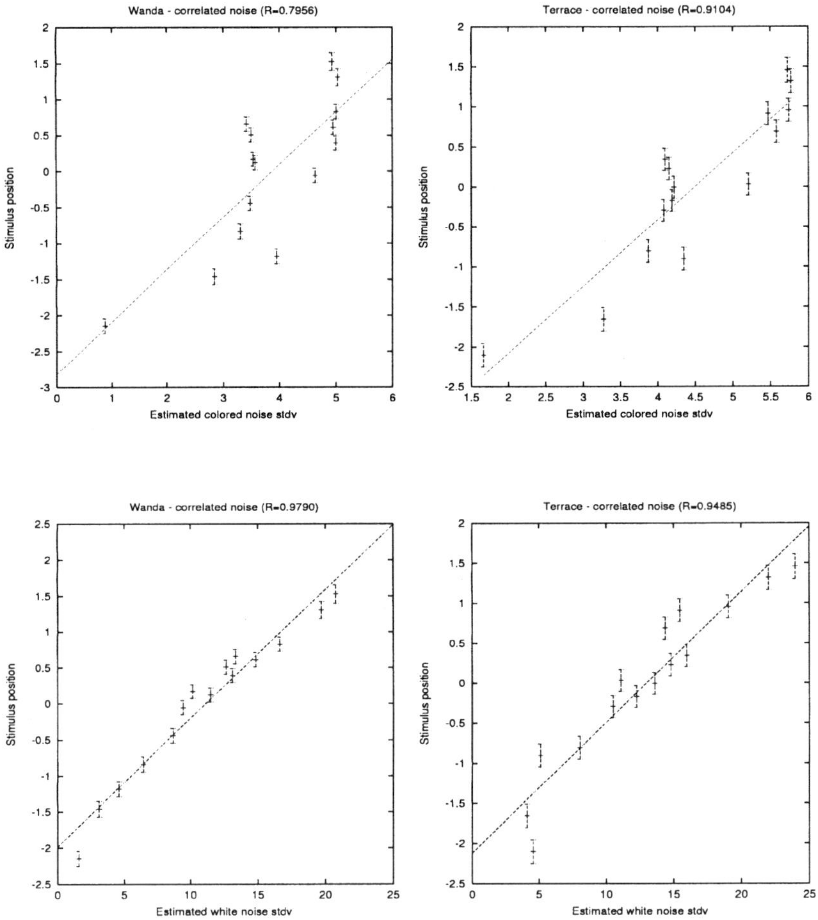

Figure 12.12. Perceived noisiness versus the instrumental measures $\hat{\sigma}_n$ (top) and $\hat{\sigma}_w$ (bottom) for images with correlated noise. The measure $\hat{\sigma}_n$ is the estimated standard deviation of the colored noise, while $\hat{\sigma}_w$ is the estimated standard deviation of the equivalent white noise (i.e., before filtering).

standard deviation of the unfiltered white noise, which can be estimated by $\hat{\sigma}_w = \hat{\sigma}_n \cdot (\hat{l}_c\sqrt{2\pi})$, is a better correlate for perceived noisiness. The relationship between the perceived noisiness and the estimated noise parameter $\hat{\sigma}_w$ is plotted in the lower part of figure 12.12. The significantly larger linear-correlation coefficients indicate an approximately linear relationship between perceived noisiness and the estimated noise parameter $\hat{\sigma}_w$. However, since the linear regression line does not pass through all uncertainty intervals, there are some effects within the experimental data that are still not explained by the estimated noise parameter $\hat{\sigma}_w$. Note for instance that the data points corresponding to the noise-free images deviate most from the linear regression line. It should therefore be possible to improve on the simple measure $\hat{\sigma}_w$.

6. IMAGES WITH NOISE AND BLUR

In chapter 5, we presented experimental data for images with noise and blur [KM96c]. A numerical category scaling procedure was used to evaluate images with respect to the attributes noise, blur and quality. All combinations of four levels of additive white Gaussian noise and four levels of spatial filtering with different binomial filter kernels resulted in a total of 16 images per scene. Three different scenes were used, and all images were judged four times by all seven subjects. Only the results for the two natural scenes, Wanda and Terrace, will be discussed here.

The data for all subjects were analyzed per attribute with the maximum-likelihood estimation program XGms. Subsequently, a linear regression analysis was performed between the experimentally obtained attribute scores and the instrumental measures for noise and blur (i.e., the estimated noise standard deviation $\hat{\sigma}_n$ and image blur parameter $\hat{\sigma}_b$). A Gaussian window with parameter $\sigma = 2.5$ times the sampling distance was used to perform all parameter estimates. The noise correlation length l_c was assumed to be much smaller than the window parameter σ, so that the estimated noise variance $\hat{\sigma}_n^2$ could be approximated by the average residue energy $\hat{\mu}_e$.

In Figure 12.13, the perceived noisiness is plotted against the estimated noise standard deviation $\hat{\sigma}_n$. The 95% uncertainty intervals on the experimental data are indicated by the bars. The linear regression model does not provide a perfect fit to the experimental data, since the regression line does not pass through all measurements. Especially for the Terrace image, the perceived noisiness for a fixed noise standard deviation varies when the image is filtered, even though the noise is added after the filtering. The estimates $\hat{\sigma}_n$ are however hardly influenced by the image filtering. This interaction between noise and blur could of course be better predicted if the perceived noisiness was allowed to depend on both the estimated noise standard deviation $\hat{\sigma}_n$ and the estimated blur $\hat{\sigma}_b$.

In Figure 12.14, the perceived blur is plotted against the instrumental blur measure $\hat{\sigma}_b$. Note that the original "Terrace" image is less blurred than the original "Wanda" image. Again, the linear regression model has a high correlation with the experimental data, but does not provide a perfect fit due to the interaction between noise and blur. A linear

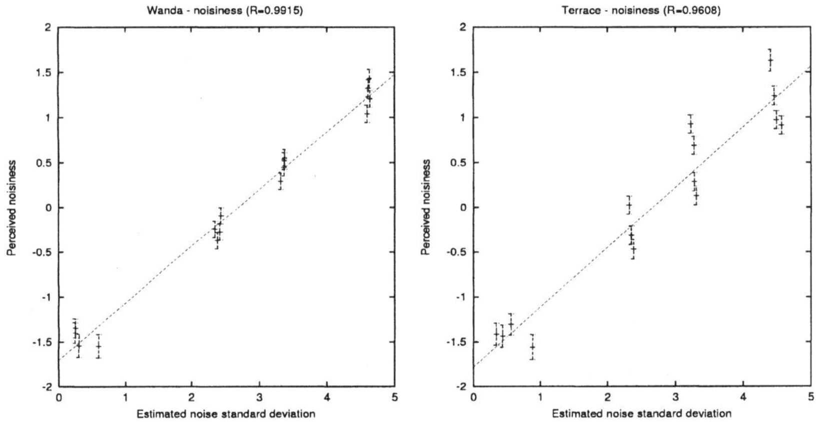

Figure 12.13. Perceived noisiness versus the instrumental measure $\hat{\sigma}_n$ for images with blur and noise.

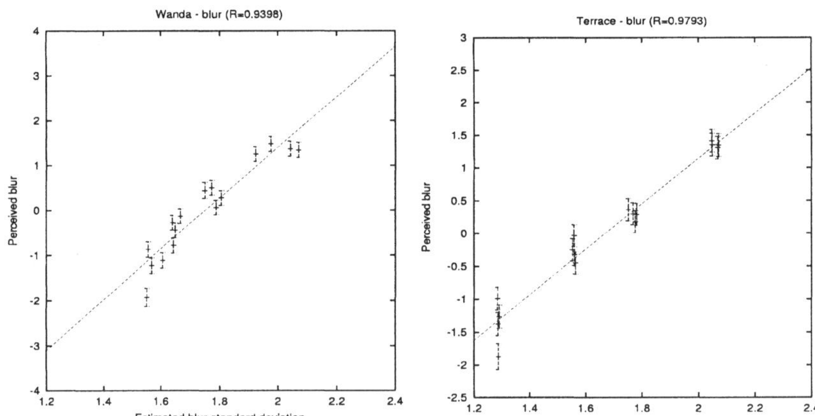

Figure 12.14. Perceived blur versus the instrumental measure $\hat{\sigma}_b$ for images with blur and noise.

model based on both predictors can not predict all the variance in the data either. This is because some of the interactions between noise and blur are clearly nonlinear. More specifically, the experimental data show that the perceived blur can either increase or decrease with the noise standard deviation, depending on whether the size of the filter kernel is small (little blur) or large (much blur), respectively. Only one of these effects, i.e., a decrease in the estimate $\hat{\sigma}_b$ with increasing noise standard

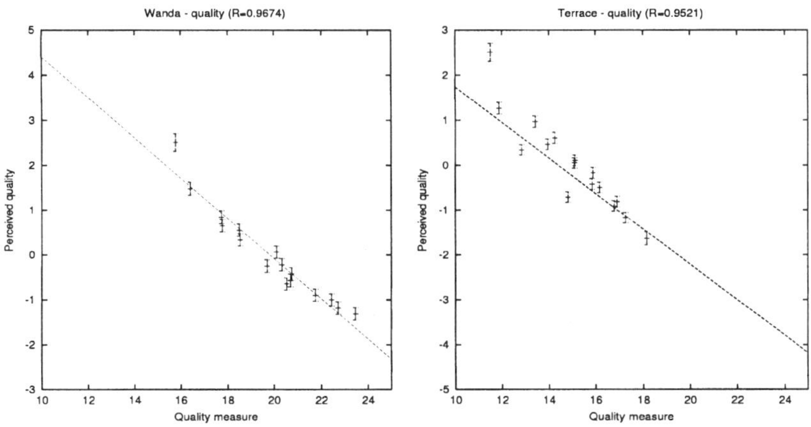

Figure 12.15. Perceived quality versus an instrumental measure that is proportional to $\tau \cdot \hat{\sigma}_b + \hat{\sigma}_n$ for images with blur and noise.

deviation, is observed in the instrumental measures. This explains why the upper points in the graphs of Figure 12.14 more or less follow the regression line. In Figure 12.15, the perceived quality is plotted versus an instrumental quality measure that is proportional to a weighted sum $\tau \cdot \hat{\sigma}_b + \hat{\sigma}_n$ of the blur and noise measures. The relative contribution τ of both instrumental measures to overall quality is of the same order for both images (Wanda: $\tau = 9.8$, Terrace: $\tau = 7.0$), so that similar results would have been obtained with a fixed value, say $\tau = 8.5$.

7. JPEG-CODED IMAGES

Predicting variations in image quality (or one of its attributes) within a scene is very likely less demanding than making predictions across scenes. Instrumental measures for image quality are however supposed to also offer consistent predictions across scenes. Unfortunately, very few experiments have been carried out in which subjects were asked to compare image quality attributes both within and across scenes. All experimental data sets presented in this book up to now for instance only compared images of the same scene. In this section, we present a noticeable exception, in which JPEG-coded images were compared both within and across scenes [Mee02].

Three scenes entitled "Roses" (34), "Museum" (86) and "Boat" (15) were used in the experiment. A public-domain software package for

Figure 12.16. Images used in the JPEG experiment: "Roses' (left), 'Museum' (middle) and 'Boat' (right).

JPEG encoding and decoding (Independent JPEG Software Group[2]) was used to code these images at quality levels 60, 40, 25 and 15. Including the original, there were hence five images per scene. A region of interest of 480 rows by 240 columns was selected from the images in order to allow simultaneous display of two images. The uncoded images are shown in Figure 12.16. The non-interlaced display was calibrated to have the gray-value-to-luminance characteristic of equation (11.11). In order to allow subjects to adequately judge the small coding differences, the viewing distance was 80 cm.

There were 15 images in the experiment (5 images times 3 scenes) that were combined pairwise. Only 93 of the 105 possible combinations were actually used in the experiment. More precisely, some of the image pairs consisting of the same scene at different coding levels were removed from the data set. The six subjects were asked to rate differences in overall quality and in the three major attributes (blockiness, blur and ringing). Attribute differences were rated on a category scale from -4 to 4. Only the experimental results for blockiness and quality will be discussed here.

The experimental data were analyzed with the discrete version of the MDS program XGms. The data were assumed to be non-metric, and a non-linear quantization characteristic based on a monotonic spline function with 2 internal knot points was used in all cases to relate the

[2] http://www.jig.org/

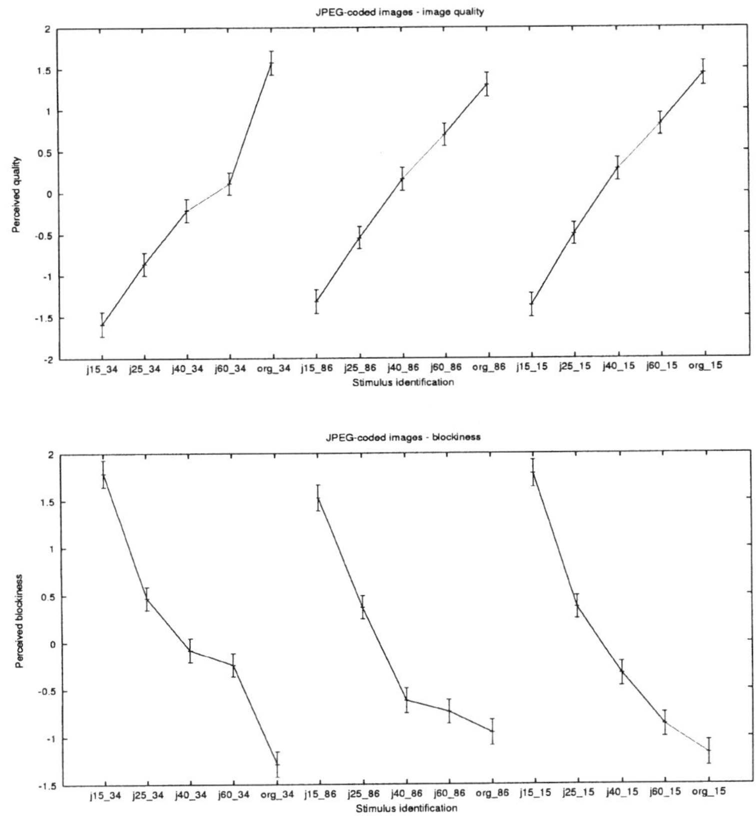

Figure 12.17. Attributes strengths derived from the experiment with JPEG-coded images. The two different attributes, image quality and blockiness, are presented in the upper and lower part, respectively.

stimulus coordinates to the scale values attributed by the subjects. The resulting one-dimensional stimulus configurations for image fidelity and blockiness are presented in Figure 12.17.

We subsequently compared the perceived blockiness with the single-ended blockiness measure presented earlier. The results are shown in the left part of Figure 12.18. Although a reasonable correlation can be obtained between the instrumental blockiness measure and the perceived blockiness, it is obvious that some effects in the experimental data remain unaccounted for by this simple measure. Since blockiness has a very high negative correlation with overall image quality, we expect that the blockiness measure correlates about equally well with image quality

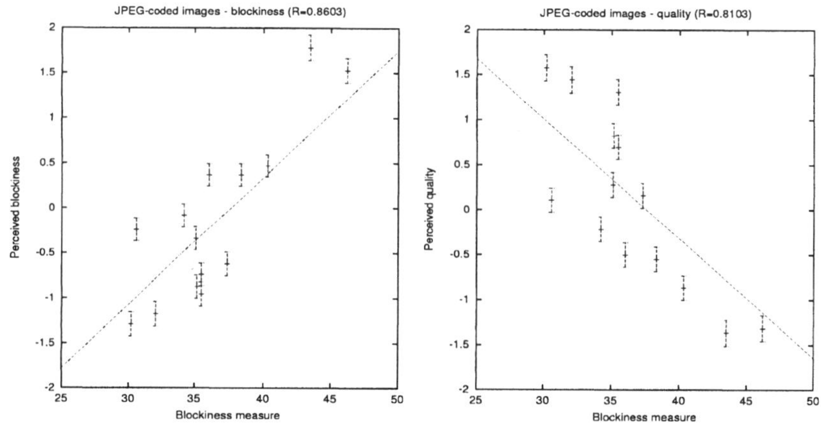

Figure 12.18. Perceived blockiness and image quality of JPEG-coded images plotted as a function of the instrumental blockiness measure.

as with blockiness. This is indeed observed in the right part of Figure 12.18.

8. CONCLUSIONS

We have demonstrated how single-ended instrumental measures for the image quality attributes noise, blur and blockiness can be constructed from detected uniform regions and edge features. Although a good correlation was obtained between the instrumental measures and subjective data for the example data sets, there is ample room for improvements. The algorithms presented here are fairly straightforward up to the point of edge (or uniform region) detection and parameter estimation. Many alternative choices for compressing these local estimates into a single statistic are however possible, some of which remain to be explored. In addition, measures for other impairments such as ringing may also be required to obtain a better quality prediction across images that have undergone different processing or coding. Last but not least, alternative implementations of the underlying computational principles of section 2 are also possible and likely to be useful.

APPENDIX 12.A: RESIDUE ENERGY

The residue energy at the origin for the signal $f(\mathbf{x}) = \overline{f} + n(\mathbf{x})$ can be expressed as

$$e = \int d\mathbf{x}\, w^2(\mathbf{x})\, n^2(\mathbf{x}) - \int d\mathbf{x}_1\, w^2(\mathbf{x}_1)\, n(\mathbf{x}_1) \cdot \int d\mathbf{x}_2\, w^2(\mathbf{x}_2)\, n(\mathbf{x}_2).$$

If the zero-mean noise signal $n(\mathbf{x})$ has autocorrelation function $R(\mathbf{x})$, then the average residue energy is obtained through

$$\mu_e = \int d\mathbf{x}\, w^2(\mathbf{x})\, R(0) - \int\int d\mathbf{x}_1\, d\mathbf{x}_2\, w^2(\mathbf{x}_1)\, w^2(\mathbf{x}_2)\, R(\mathbf{x}_1 - \mathbf{x}_2) = R(0) - I_1,$$

since the window $w^2(\mathbf{x})$ is normalized. Similarly, the second-order moment is equal to

$$\begin{aligned} E[e^2] = \sigma_e^2 + \mu_e^2 & \\ = & \int\int d\mathbf{x}_1\, d\mathbf{x}_2\, w^2(\mathbf{x}_1)\, w^2(\mathbf{x}_2)\, E[n^2(\mathbf{x}_1) \cdot n^2(\mathbf{x}_2)] \\ - & \; 2 \cdot \int\int d\mathbf{x}_1\, d\mathbf{x}_2\, d\mathbf{x}_3\, w^2(\mathbf{x}_1)\, w^2(\mathbf{x}_2)\, w^2(\mathbf{x}_3)\, E[n^2(\mathbf{x}_1) \cdot n(\mathbf{x}_2) \cdot n(\mathbf{x}_3)] \\ + & \int\int d\mathbf{x}_1\, d\mathbf{x}_2\, d\mathbf{x}_3\, d\mathbf{x}_4\, w^2(\mathbf{x}_1)\, w^2(\mathbf{x}_2)\, w^2(\mathbf{x}_3)\, w^2(\mathbf{x}_4) \\ & E[n(\mathbf{x}_1) \cdot n(\mathbf{x}_2) \cdot n(\mathbf{x}_3) \cdot n(\mathbf{x}_4)]. \end{aligned}$$

The higher-order moments can be expressed in terms of the correlation function in case of Gaussian noise (see Appendix 10.A), i.e.,

$$\begin{aligned} E[n^2(\mathbf{x}_1) \cdot n^2(\mathbf{x}_2)] &= 2 \cdot R^2(\mathbf{x}_1 - \mathbf{x}_2) + R^2(0) \\ E[n^2(\mathbf{x}_1) \cdot n(\mathbf{x}_2) \cdot n(\mathbf{x}_3)] &= 2 \cdot R(\mathbf{x}_1 - \mathbf{x}_2) \cdot R(\mathbf{x}_1 - \mathbf{x}_3) + R(0) \cdot R(\mathbf{x}_2 - \mathbf{x}_3) \\ E[n(\mathbf{x}_1) \cdot n(\mathbf{x}_2) \cdot n(\mathbf{x}_3) \cdot n(\mathbf{x}_4)] &= R(\mathbf{x}_1 - \mathbf{x}_2) \cdot R(\mathbf{x}_3 - \mathbf{x}_4) \\ &+ R(\mathbf{x}_1 - \mathbf{x}_3) \cdot R(\mathbf{x}_2 - \mathbf{x}_4) \\ &+ R(\mathbf{x}_1 - \mathbf{x}_4) \cdot R(\mathbf{x}_2 - \mathbf{x}_3), \end{aligned}$$

which leads to the following expressions for the second-order moment

$$E[e^2] = \sigma_e^2 + \mu_e^2 = R^2(0) + 2\, I_2 - 4\, I_3 - 2\, R(0)\, I_1 + 3\, I_1^2$$

and the variance

$$\sigma_e^2 = 2\, I_2 - 4\, I_3 + 2\, I_1^2,$$

where

$$\begin{aligned} I_1 &= \int\int d\mathbf{x}_1\, d\mathbf{x}_2\, w^2(\mathbf{x}_1)\, w^2(\mathbf{x}_2)\, R(\mathbf{x}_1 - \mathbf{x}_2) \\ I_2 &= \int\int d\mathbf{x}_1\, d\mathbf{x}_2\, w^2(\mathbf{x}_1)\, w^2(\mathbf{x}_2)\, R^2(\mathbf{x}_1 - \mathbf{x}_2) \\ I_3 &= \int\int\int d\mathbf{x}_1\, d\mathbf{x}_2\, d\mathbf{x}_3\, w^2(\mathbf{x}_1)\, w^2(\mathbf{x}_2)\, w^2(\mathbf{x}_3) \\ & \quad R(\mathbf{x}_1 - \mathbf{x}_2) \cdot R(\mathbf{x}_1 - \mathbf{x}_3) \end{aligned}$$

More explicit expressions can be obtained for these integrals in case the window $w(\mathbf{x}) = w_g(\mathbf{x};\sigma)$ is Gaussian and the autocorrelation function is expressed by equation (12.11), i.e.,

$$\begin{aligned} R(0) &= \sigma_n^2 \\ I_1 &= \sigma_n^2 \frac{l_c^2}{l_c^2 + \sigma^2} \\ I_2 &= \sigma_n^4 \frac{l_c^2}{l_c^2 + 2\,\sigma^2} \\ I_3 &= \sigma_n^4 \frac{4\,l_c^4}{(2\,l_c^2 + \sigma^2)\,(2\,l_c^2 + 3\,\sigma^2)}, \end{aligned}$$

which leads to the following results

$$\mu_e = \sigma_n^2 \frac{\sigma^2}{l_c^2 + \sigma^2}, \quad \sigma_e = \frac{\mu_e^2}{F[l_c/\sigma]},$$

for the mean value and the variance of the residue energy where

$$F(\ell) = \frac{(2\,\ell^2 + 1)\,(2\,\ell^2 + 3)\,(\ell^2 + 2)}{2\,\ell^2\,(2\,\ell^4 + 4\,\ell^2 + 3)}$$

is proportional to the number of degrees of freedom in the approximate χ^2-distribution (see main text).

APPENDIX 12.B: EDGE PARAMETERS

We show, for both the composite edge model and a number of special cases, how edge parameters can be estimated from Hermite coefficients.

COMPOSITE EDGE MODEL

Solving the equations (12.41) in the general case can be done as follows. First, the blur factor R_b is found by solving the equation

$$32\, r_2^4 \cdot R_b^6 + 48\, r_2^2\, \left(r_2\, r_4 - r_3^2\right) \cdot R_b^4 + 24\, \left(r_2\, r_4 - r_3^2\right)^2 \cdot R_b^2 = \\ 4\, \left(r_2\, r_4 - r_3^2\right) \cdot \left(r_3\, r_5 - r_4^2\right) - \left(r_3\, r_4 - r_2\, r_5\right)^2,$$

where $r_n = f_n \cdot \sqrt{2^n\, n!}$, for $n = 2, 3, 4, 5$. This equation is of the form

$$a^2\, z^3 + 3\, a\, b\, z^2 + 3\, b^2\, z = c,$$

with $z = R_b^2$, and has a real root equal to

$$z = \frac{-b + (b^3 + a\, c)^{1/3}}{a}$$

if $a \neq 0$, and $z = c/(3\, b^2)$ if $a = 0$. Only solutions for which

$$\frac{1}{1 + \beta_{max}^2} \leq R_b^2 \leq \frac{1}{1 + \beta_{min}^2}$$

are retained, as is discussed in the text.

Second, the distance factor D is found by solving the equation

$$4\, s_2 \cdot D^2 - 4\, s_3 \cdot D + (2\, s_2 + s_4) = 0,$$

where $s_n = r_n / R_b^n$, for $n = 2, 3, 4$. The coefficients have to satisfy the condition

$$s_3^2 \geq s_2\, (2\, s_2\; + s_4)$$

in order to guarantee real solutions to this equations. Again, an a priori condition $\mid D \mid \leq \alpha$ is imposed. Even with this condition, there may be two solutions for D, so that additional information may be required (such as higher-order derivatives) to select between these alternatives. If no such information is available, we select the solution with the smallest absolute value.

Third, the remaining amplitude factors can be expressed as

$$a_0 = s_2 \cdot D - \frac{s_3}{2}$$

and

$$a_1 = R_b\, \left[s_2 \cdot (1 - 2\, D^2) + s_3 \cdot D\right] = R_b\, \left[2\, s_2 - s_3 \cdot D + \frac{s_4}{2}\right].$$

In the case of a composite edge with known blur value R_b, we can still use the latter equations to estimate D, a_0 and a_1. For instance, $R_b = 1$ corresponds to fitting an unblurred edge model.

BLURRED STEP EDGE WITH GRADIENT

This corresponds to $a_1 = 0$ in the general treatment. The remaining parameters D, a_0 and R_b can be estimated from the coefficients up to order 4.

Solving the equations

$$\begin{aligned} f_2 &= \frac{R_b^2}{2\sqrt{2}} a_0 \cdot 2D \\ f_3 &= \frac{R_b^3}{4\sqrt{3}} a_0 \cdot (4D^2 - 2) \\ f_4 &= \frac{R_b^4}{8\sqrt{6}} a_0 \cdot (8D^3 - 12D) \end{aligned}$$

is done as follows. First, the blur factor R_b is found by solving the equation

$$16\, r_2^3 \cdot R_b^4 + 2\, r_2 \left(4\, r_2\, r_4 - 3\, r_3^2\right) \cdot R_b^2 + r_4 \left(r_2\, r_4 - r_3^2\right) = 0,$$

where $r_n = f_n \cdot \sqrt{2^n\, n!}$, for $n = 2, 3, 4$. Second, the distance factor D is found by solving the equation

$$2\, s_2 \cdot D^2 - s_3 \cdot D - s_2 = 0,$$

where $s_n = r_n / R_b^n$, for $n = 2, 3$. Third, the remaining amplitude factor can be expressed as

$$a_0 = \frac{s_2}{2D} = \frac{s_3}{4D^2 - 2}.$$

BLURRED STEP EDGE WITHOUT GRADIENT

This corresponds to $a_1 = 0$ and $k_1 = 0$ in the general treatment, which in turn implies that parameters D, R_b and a_0 can be estimated from the coefficients up to order 3. This is the case that was discussed in the original blur-estimation paper [KM94].

Solving the equations

$$\begin{aligned} f_1 &= \frac{R_b}{\sqrt{2}} a_0 \\ f_2 &= \frac{R_b^2}{2\sqrt{2}} a_0 \cdot 2D \\ f_3 &= \frac{R_b^3}{4\sqrt{3}} a_0 \cdot (4D^2 - 2) \end{aligned}$$

is done as follows. First, the blur factor R_b can be expressed as

$$R_b^2 = \frac{r_2^2 - r_1\, r_3}{2\, r_1^2},$$

where $r_n = f_n \cdot \sqrt{2^n\, n!}$, for $n = 1, 2, 3$. Second, the distance factor D is given by

$$D = \frac{s_2}{2\, s_1},$$

where $s_n = r_n / R_b^n$, for $n = 1, 2$. Third, the remaining amplitude factor is $a_0 = s_1$.

References

[AAH$^+$86] M. Annegarn, J. Arragon, G. de Haan, J. van Heuven, and J. Jackson. Hd-mac: een stap vooruit in de evolutie van de televisietechniek'. *Philips Technisch Tijdschrift*, pages 213–230, 1986.

[Agr89] A. Agresti. Categorical data analysis. *Psychological Bulletin*, 105:290–301, 1989.

[Agr90] A. Agresti. *Categorical Data Analysis.* John Wiley and Sons Inc., New York, 1990.

[AHLL88] R.C. Atkinson, R.J. Herrnstein, G. Lindzey, and R.D. Luce, editors. *Stevens' Handbook of Experimental Psychology — Perception and Motivation.* John Wiley and Sons Inc.: New York, 1988.

[AJ93] A.J. Ahumada Jr. Computational image-quality metrics: a review. In *SID 93 Digest*, pages 305–308, Society for Information Display, Santa Ana, California, 1993.

[AN94] P. Artal and R. Navarro. Monochromatic modulation transfer function of the human eye for different pupil diameters: an analytical expression. *Journal of the Optical Society of America A*, 11:246–249, 1994.

[ANS93a] ANSI. American national standard it8.7/1: Graphic technology - color transmission target for input scanner calibration. Technical report, American National Standards Institute, Reston, Virginia, 1993.

[ANS93b] ANSI. American national standard it8.7/2: Graphic technology - color reflection target for input scanner calibration. Technical report, American National Standards Institute, Reston, Virginia, 1993.

[AP81] J.A. Anderson and P.R Philips. Regression, discrimination and measurement models for ordered categorical variables. *Applied Statistics*, 30:22–31, 1981.

[BA83] P.J. Burt and E. Adelson. The laplacian pyramid as a compact image code. *IEEE Transactions on Communications*, 31:532–540, 1983.

[BA02] F.A. Baqai and J.P Allebach. Computer-aided design of clustered-dot color screens based on a human visual system model. *Proceedings of the IEEE*, 90:104–122, 2002.

[Bar87] P.G.J. Barten. The sqri method: a new method for the evaluation of visible resolution on a display. *Proceedings of the Society for Information Display*, 30:253–262, 1987.

[Bar90] P.G.J. Barten. Evaluation of subjective image quality with the square-root integral method. *Journal of the Optical Society of America A*, 7:2024–2031, 1990.

[Bel99] N. Belaïd. *Perceptual Linearization of Soft-Copy Displays for Achromatic Images.* PhD thesis, Eindhoven University of Technology, 1999.

[Ber77] J.M.F. ten Berge. Orthogonal procrustes rotation for two or more matrices. *Psychometrika*, 42:267–276, 1977.

[BHN+96] S. Bech, R. Hamberg, M. Nijenhuis, C. Teunissen, H. Looren de Jong, P. Houben, and S.K. Pramanik. The rapid perceptual image description method. In B.E. Rogowitz and T.N. Pappas, editors, *Human Vision and Electronic Imaging*, volume 2657, pages 317–328, San Jose, California, January 29 - February 1 1996. Proceedings of the SPIE.

[BK80] C.A. Burbeck and D.H. Kelly. Spatiotemporal characteristics of visual mechanisms: Excitatory-inhibitory model. *Journal of the Optical Society of America*, 70:1121–1126, 1980.

[BKT86] K.R. Boff, L. Kaufman, and J.P. Thomas, editors. *Handbook of Perception and Human Performance — Sensory Processes and Perception.* John Wiley and Sons Inc.: New York, 1986.

[BL96a] C.J. van den Branden Lambrecht. Color moving pictures quality metric. In *IEEE International Conference on Image Processing*, volume I of III, pages 885–888, Lausanne, Switzerland, September 16–19 1996.

[BL96b] C.J. van den Branden Lambrecht. *Perceptual models and architectures for video coding applications.* PhD thesis, Ecole Polytechnique Fédérale de Lausanne, 1996.

[Blo87] F.J.J. Blommaert. *Vision of details in space and time.* PhD thesis, Eindhoven University of Technology, 1987.

[BM95] H. Blume and E. Muka. Presenting medical images on monochrome displays. *Information Display*, 6:14–17, 1995.

[BM98a] N. Belaïd and J.-B. Martens. Grey scale, the crispening effect, and perceptual linearization. *Signal Processing*, 70:231–245, 1998.

[BM98b] M.J. Blacl and D.H. Marimont. Robust anisotropic diffusion. *IEEE Transactions on Image Processing*, 7:421–432, 1998.

[BNS87] D.A. Baylor, B.J. Nunn, and J.L. Schnapf. Spectral sensitivity of cones of the monkey macaca fascicularis. *Journal of Physiology*, 390:145–160, 1987.

[BOM97] N. Belaïd, I. van Overveld, and J.-B. Martens. Display fidelity: a link between psychophysics and contrast discrimination models. *Spatial Vision*, 11:205–223, 1997.

[Bos01] M.C. Boschman. Difscal: A tool for analysing difference ratings on an ordinal category scale. *Behavior Research Method, Instruments and Computers*, 33:10–20, 2001.

[Box93] P.A.M. van Boxtel. Perceptual assessment of sampled images. Technical Report 946, Institute for Perception Research, Eindhoven, the Netherlands, 1993.

[Bra89] D. Brainard. Calibration of a computer controlled color monitor. *Color Research and Application*, 14:23–34, 1989.

[BSL+02] A. Buja, D.F. Swayne, M.L. Littman, N.H. Dean, and H. Hoffman. Visualization methodology for multidimensional scaling. *Journal of Classification*, 19:7–44, 2002.

[Car82] C.R. Carlson. Sine-wave threshold contrast-sensitivity function: Dependence on display size. *RCA Revue*, 43:675–683, 1982.

[CC78] C.R. Carlson and R.W. Cohen. Visibility of displayed information: image descriptors for displays. Technical Report ONR-CR213-120-4F, Office of Naval Research, Arlington, VA, 1978.

[CC94] T.F. Cox and M.M.A. Cox. *Multidimensional Scaling.* Chapman & Hall, London, 1994.

[CF91] M.W. Cannon and S.C. Fullenkamp. A transducer model for contrast perception. *Vision Research*, 31:983–998, 1991.

[CG65] F.W. Campbell and D.C. Green. Optical and retinal factors affecting visual resolution. *Journal of Physiology*, 181:576–593, 1965.

[CG66] F.W. Campbell and R.W. Gubisch. Optical quality of the human eye. *Journal of Physiology*, 186:558–578, 1966.

[CIE71] CIE. Colorimetry. Technical Report CIE Publication no. 15, Commission Internationale de l'Eclairage, Paris, France, 1971.

[CIE79] CIE. Recommendation on uniform color spaces, color-difference equations and psychometric color terms. Technical Report Supplement no. 2 to CIE Publication no. 15, Commission Internationale de l'Eclairage, Paris, France, 1979.

[CIE83] CIE. The basis of physical photometry. Technical Report CIE Publication no. 18.2, Commission Internationale de l'Eclairage, Paris, France, 1983.

[CIE95] CIE. Industrial colour-difference evaluation. Technical Report CIE Publication no. 116, Commission Internationale de l'Eclairage, Paris, France, 1995.

[CK82] J. Cohen and W. Kappauf. Metameric color stimuli, fundamental metamers, and wyszecki's metameric blacks. *American Journal of Psychology*, 95:537–564, 1982.

[CK85] J. Cohen and W. Kappauf. Color mixture and fundamental metamers: Theory, algebra, geometry, application. *American Journal of Psychology*, 98:171–259, 1985.

[CKL66] F.W. Campbell, J. Kulikowski, and J. Levinson. The effect of the orientation on the visual resolution of gratings. *Journal of Physiology*, 187:427–436, 1966.

[CKT+99] T. Carney, S.A. Klein, C.W. Tyler, A.D. Silverstein, B. Beutter, D. Levi, A.B. Watson, A.J. Reeves, A.M. Norcia, C.-C. Chen, W. Makous, and M.E. Eckstein. The development of an image/threshold database for designing and testing human vision models. In B.E. Rogowitz and T.N. Pappas, editors, *Human Vision and Electronic Imaging IV*, volume 3644, pages 542–551, San Jose, California, January 25–28 1999. Proceedings of the SPIE.

[CN89] C.M. Cicerone and J.L. Nerger. The relative numbers of long-wavelength-sensitive to middle-wavelength-sensitive cones in the human fovea centralis. *Vision Research*, 29:115–128, 1989.

[Coh88] J.B. Cohen. Color and color mixture: Scalar and vector fundamentals. *Color Research and Application*, 13:5–39, 1988.

[Cow85] M. Cowlishaw. Fundamental requirements for picture presentation. *Proceedings of the SID*, 26:101–107, 1985.

[Dal92] S. Daly. Visible differences predictor: an algorithm for the assessment of image fidelity. In B.E. Rogowitz, editor, *Human Vision, Visual Processing, and Digital Display III*, volume 1666, pages 2–15, San Jose, California, February 10–13 1992. Proceedings of the SPIE.

[Dal93] S. Daly. Visible differences predictor: an algorithm for the assessment of image fidelity. In A.B. Watson, editor, *Digital Images and Human Vision*, pages 179–206. MIT Press, Cambridge, Massachusetts, 1993.

[Dau92] I. Daubechies. *Ten lectures on wavelets.* Society for Industrial and Applied Mathematics, 1992.

[Dij97] A.M. van Dijk. *Image representation and compression using steered Hermite transforms.* PhD thesis, Eindhoven University of Technology, 1997.

[DM96] A.M. van Dijk and J.-B. Martens. Quality assessment of compressed images: a comparison between two methods. In P. Delogne, editor, *Proceedings of the IEEE International Conference on Image Processing*, volume II of III, pages 25–28, Lausanne, Switzerland, September 16–19 1996.

[DM97a] A.M. van Dijk and J.-B. Martens. Image representation and compression with steered hermite transforms. *Signal Processing*, 56:1–16, 1997.

[DM97b] A.M. van Dijk and J.-B. Martens. Subjective quality assessment of compressed images. *Signal Processing*, 58:235–252, 1997.

[DMW95] A.M. van Dijk, J.-B. Martens, and A.B. Watson. Quality assessment of coded images using numerical category scaling. In N. Otha, H.U. Lemke, and J.C. Lehureau, editors, *Advanced Image and Video Communications and Storage Technologies*, volume 2451, pages 90–101, Amsterdam, March 20–23 1995. Proceedings of the SPIE.

[DS98] N.R. Draper and H. Smith. *Applied Regression Analysis.* John Wiley and Sons Inc., New York, 1998.

[Dub85] E. Dubois. The sampling and reconstruction of time-varying imagery with application in video systems. *IEEE Proceedings*, 73:502–521, 1985.

[EB98] M.P. Eckert and A.P. Bradley. Perceptual quality metrics applied to still image compression. *Signal Processing*, 70:177–200, 1998.

[Ell87] D.F. (Ed.) Elliott. *Handbook of Digital Signal Processing: Engineering Applications.* Academic Press, San Diego, California, 1987.

[ER92] B. Escalante-Ramírez. *Perceptually-assessed digital processing of medical images.* PhD thesis, Eindhoven University of Technology, 1992.

[ERMR95] B. Escalante-Ramírez, J.-B. Martens, and H. de Ridder. Multidimensional characterization of the perceptual quality of noise-reduced computed tomography images. *Journal of Visual Communication and Image Representation*, 6:317–334, 1995.

[ERRM94] B. Escalante-Ramírez, H. de Ridder, and J.-B. Martens. Multidimensional perceptual quality of noise-reduced computed tomography images. In H.L. Kundel, editor, *Medical Imaging 1994: Image Perception*, volume 2166, pages 119–130, Newport Beach, California, February 17–18 1994. Proceedings of the SPIE.

[FA91] W.T. Freeman and E.H. Adelson. The design and use of steerable filters. *IEEE Transactions on Pattern Analysis and Machine Intelligence*, 13:891–906, September 1991.

[Fal86] J.C. Falmagne. Psychophysical measurement and theory. In Boff et al. [BKT86], pages 1–66.

[Fau95] O. Faugeras. Stratification of three-dimensional vision: projective, affine, and metric representations. *Journal of the Optical Society of America A*, 12:465–484, 1995.

[FBR97] E.A. Fedorovskaya, F.J.J. Blommaert, and H. de Ridder. Perceptual quality of color images of natural scenes transformed in cieluv color space. In R. Eschbach and K. Braun, editors, *Recent Progress in Color Science*, pages 42–45. Springfield, 1997.

[Fib97] D.K. Fibush. Practical application of objective picture quality measurements. In *IBC97*, pages 123–135, International Broadcast Union, 1997.

[Flo97] L. Florack. *Computational Imaging and Vision.* Kluwer Academic Publishers, Dordrecht, the Netherlands, 1997.

[FRB97] E.A. Fedorovskaya, H. de Ridder, and F.J.J. Blommaert. Chroma variations and perceived quality of color images of natural scenes. *Color Research and Application*, 22:96–110, 1997.

[FRYB97] E.A. Fedorovskaya, H. de Ridder, S. Yendrikhovskij, and F.J.J. Blommaert. Multidimensional structure of colorfulness: chroma variation in color images of natural scenes. In R. Eschbach and K. Braun, editors, *Recent Progress in Color Science*, pages 93–96. Springfield, 1997.

[FS86] G. Fawcett and G. Schrack. Halftoning techniques using error correction. *Proceedings of the SID*, 27:305–308, 1986.

[GC98] P.D. Gowdy and C.M. Cicerone. The spatial arrangement of the l and m cones in the central fovea of the living human eye. *Vision Research*, 38:2575–2589, 1998.

[GCJS89] P.E. Green, F.J. Carmone Jr., and S.M. Smith. *Multidimensional Scaling, Concepts and Applications.* Allyn & Bacon, Boston, Massachusetts, 1989.

[Gei84] W. Geisler. Physical limits of acuity and hyperacuity. *Journal of the Optical Society of America*, pages 775–782, 1984.

[GF97] M.A. Georgeson and T.C.A. Freeman. Perceived location of bars and edges in one-dimensional images: Computational models and human vision. *Vision Research*, 37:127–142, 1997.

[GH73] E.M. Granger and J.C Heurtley. Visual chromaticity-modulation transfer function. *Journal of the Optical Society of America*, 63:1173–1174, 1973.

[GM94] S. Ghosal and R. Mehrotra. Detection of composite edges. *IEEE Transactions on Image Processing*, 3:14–25, January 1994.

[GM97] S. Ghosal and R. Mehrotra. A moment-based unified approach to image feature detection. *IEEE Transactions on Image Processing*, 6:781–793, June 1997.

[GM98] E.J. Giorgianni and T.E. Madden. *Digital Color Management - Encoding Solutions.* Addison-Wesley, Reading, Massachusets, 1998.

[GNF43] W.C. Granville, D. Nickerson, and C.E. Foss. Trichromatic specifications for intermediate and special colors of the munsell system. *Journal of the Optical Society of America*, 33:376–385, 1943.

[GP79] J.S. Goodman and D.E. Pearson. Multidimensional scaling of multiply-impaired television pictures. *IEEE Transactions on Systems, Man and Cybernetics*, 9:353–356, 1979.

[GRK+97] D. Gurr, D.A. Rubin, B.H. Kart, A.M. Peterson, C.R. Fuhrman, H.E. Rockette, and J.L. King. Forced choice and ordinal discrete rating assessment of image quality: A comparison. *Journal of Digital Imaging*, 10:103–107, 1997.

[GS99] K.R. Gegenfurtner and L.T. Sharpe, editors. *Color Vision: From Genes to Perception.* Cambridge University Press: Cambridge, 1999.

[Gui31] J. Guild. The colorimetric properties of the spectrum. *Philosophycial Transactions of the Royal Society of London, Series A*, 230:149–187, 1931.

[GvL83] G.H. Golub and C.F. van Loan. *Matrix Computations.* John Hopkins University Press, Baltimore, Maryland, 1983.

[GW93] R.C. Gonzalez and R.E. Woods. *Digital Image Processing.* Addison-Wesley Publishing Company, Reading, Massachusetts, 1993.

[GWA90] R. Gentile, E. Walowit, and J. Allebach. Quantization and multilevel halftoning of color images for near original image quality. In B.E. Rogowitz and J.P. Allebach, editors, *Human Vision and Electronic Imaging: Models, Methods, and Applications*, volume 1249, pages 249–260, Santa Clara, California, February 12–14 1990. Proceedings of the SPIE.

[Hal87] P. Halmos. *Undergraduate texts in Mathematics.* Springer Verlag, New York, 1987.

[Har99] J.Y. Hardeberg. *Acquisition and reproduction of colour images: colorimetric and multispectral approaches.* PhD thesis, Ecole Nationale Supérieure des Télécommunications, 1999.

[HB69] G. van der Horst and M.A Bouman. Spatiotemporal chromaticity discrimination. *Journal of the Optical Society of America*, 59:1482–1488, 1969.

[Hei94] F. van der Heijden. *Image Based Measurement Systems - Object Recognition and Parameter Estimation.* John Wiley & Sons, Chichester, 1994.

[Hop55] H.H. Hopkins. The frequency response of a defocused optical system. *Proceedings of the Royal Society of London, Series A*, 231:91–103, 1955.

[Hor84] B. Horn. Exact reproduction of colored images. *Computer Vision, Graphics and Image Processing*, pages 135–167, 1984.

[Hor86] B. Horn. *Robot Vision.* MIT Press, McGraw-Hill, Cambridge, Massachusetts, 1986.

[HS93] R.M. Haralick and L.G. Shapiro. *Computer and Robot Vision.* Addison-Wesley, Reading, Massachusetts, 1993.

[HSC+98] Tung-Chang Hseue, Yu-Chuan Shen, Po-Chi Chen, Wen-Hsing Hsu, and Yuan-Te Liu. Cross-media performance evaluation of color models for unequal luminance levels and dim surround. *Color Research and Application*, 23:169–177, 1998.

[Hun78] R. Hunt. Colour terminology. *Colour Research and Application*, 3:79–87, 1978.

[Hun88] R. Hunt. *The Reproduction of Colour in Photography, Printing and Television*. Fountain Press, Tolworth, England, 1988.

[Hun91] R. Hunt. *Measuring colour*. Ellis Horwood, Chichester, England, 1991.

[Hun98] R.W.G. Hunt. Revised colour-appearance model for related and unrelated colours. *Color Research and Application*, 16:146–165, 1998.

[Hur81] L. Hurvich. *Color Vision*. Sinauer Associates, Sunderland, Massachusetts, 1981.

[IBS93] J.K. Ijspeert, T.J.T.P. van den Berg, and H. Spekreijse. An improved mathematical description of the foveal visual point spread function with parameters for age, pupil size and pigmentation. *Vision Research*, 33:15–20, 1993.

[IR98] ITU-R. Methodology for the subjective assessment of the quality of television pictures. Technical Report Recommendation ITU-R BT.500-8, International Telecommunications Union, 1998.

[Ive15] H.E. Ives. The transformation of color-mixture equations from one system to another. *Journal of the Franklin Institute*, 16:673–701, 1915.

[JA95] K. Jensen and D. Anastassiou. Subpixel edge localization and the interpolation of still images. *IEEE Transactions on Image Processing*, 4:285–295, March 1995.

[Jai89] A. Jain. *Fundamentals of Digital Image Processing*. Information and Systems Sciences Series, Prentice-Hall, Englewood Cliffs, New Jersey, 1989.

[Jan99] T.J.W.M. Janssen. *Computational Image Quality*. PhD thesis, Eindhoven University of Technology, 1999.

[JB97] T.J.W.M. Janssen and F.J.J. Blommaert. Image quality semantics. *Journal of Imaging Science and Technology*, 41:555–560, 1997.

[JB00a] T.J.W.M. Janssen and F.J.J. Blommaert. A computational approach to image quality. *Displays*, 21:129–142, 2000.

[JB00b] T.J.W.M. Janssen and F.J.J. Blommaert. Predicting the usefulness and naturalness of color reproductions. *Journal of Imaging Science and Technology*, 44:93–104, 2000.

[JHG99] B. Jahne, H. Haußecker, and P. Geißler. *Handbook of Computer Vision and Applications*. Academic Press, San Diego, 1999.

[JN84] N. Jayant and P. Noll. *Digital Coding of Waveforms*. Signal Processing Series, Prentice-Hall, Englewood Cliffs, New Jersey, 1984.

[Jud51] D. Judd. Basic correlates of the visual stimulus. In S.S. Stevens, editor, *Handbook of Experimental Psychology*, pages 811–867. John Wiley & Sons, New York, 1951.

[Kan97] H.R. Kang. *Color technology for Electronic Imaging Devices.* SPIE Optical Engineering Press, 1997.

[Kay95] V. Kayargadde. *Feature extraction for Image Quality Prediction.* PhD thesis, Eindhoven University of Technology, 1995.

[KB96] P.K. Kaiser and R.M. Boynton. *Human Color Vision.* Optical Society of America, Washington DC, 1996.

[Kel79a] D.H. Kelly. Motion and vision i: Stabilized images of stationary gratings. *Journal of the Optical Society of America*, 69:1266–1274, 1979.

[Kel79b] D.H. Kelly. Motion and vision ii: Stabilized spatio-temporal threshold surface. *Journal of the Optical Society of America*, 69:1340–1349, 1979.

[Kel83] D.H. Kelly. Spatiotemporal variation of chromatic and achromatic contrast thresholds. *Journal of the Optical Society of America*, 73:742–750, 1983.

[KGN43] K.L. Kelly, K.S. Gibson, and D. Nickerson. Tristimulus specification of the munsell book of color from spectrophotometric measurements. *Journal of the Optical Society of America*, 33:355–376, 1943.

[KK95] S.A. Karunasekera and N.G. Kingsbury. A distortion measure for blocking artifacts in images based on human visual sensitivity. *IEEE Transactions on Image Processing*, 4:713–724, 1995.

[KM94] V. Kayargadde and J.-B. Martens. Estimation of edge parameters and image blur using polynomial transforms. *Computer Vision, Graphics and Image Processing: Graphical Models and Image processing*, 56:442–461, 1994.

[KM96a] V. Kayargadde and J.-B. Martens. Estimation of perceived image blur using edge features. *International Journal of Imaging Systems and Technology*, 7:102–109, 1996.

[KM96b] V. Kayargadde and J.-B. Martens. An objective measure for perceived noise. *Signal Processing*, 49:187–206, 1996.

[KM96c] V. Kayargadde and J.-B. Martens. Perceptual characterization of images degraded by blur and noise: experiments. *Journal of the Optical Society of America A*, 13:1166–1177, 1996.

[KM96d] V. Kayargadde and J.-B. Martens. Perceptual characterization of images degraded by blur and noise: model. *Journal of the Optical Society of America A*, 13:1178–1188, June 1996.

[Kru64a] J.B. Kruskal. Multidimensional scaling by optimizing goodness of fit to a nonmetric hypothesis. *Psychometrika*, 29:1–27, 1964.

[Kru64b] J.B. Kruskal. Nonmetric multidimensional scaling: a numerical method. *Psychometrika*, 29:115–129, 1964.

[KW78] J.B. Kruskal and M. Wish. *Multidimensional Scaling.* Sage University Paper Series 07-011 on Quantitative Applications in the Social Sciences, Sage Publications, Beverley Hills, California, 1978.

[LBL96] P. Lindh and C.J. van den Branden Lambrecht. Efficient spatio-temporal decomposition for perceptual processing of video sequences. In *IEEE International Conference on Image Processing*, volume III of III, pages 331–334, Lausanne, Switzerland, September 16–19 1996.

[LCR01] M.R. Luo, G. Cui, and B. Rigg. Further improvements to the cielab based colour difference formulae. *Color Research and Application*, 26:340–350, 2001.

[Leg83] G.E. Legge. A power law for contrast discrimination. *Vision Research*, 21:457–467, 1983.

[Lev71] H. Levitt. Transformed up-down methods in psychoacoustics. *Journal of the Acoustical Society of America*, 49:467–476, 1971.

[LH98a] M.R. Luo and R.W.G. Hunt. A chromatic adaptation transform and a colour inconstancy index. *Color Research and Application*, 23:154–158, 1998.

[LH98b] M.R. Luo and R.W.G. Hunt. The structure of the cie 1997 colour appearance model (ciecam97s). *Color Research and Application*, 23:138–146, 1998.

[LH98c] M.R. Luo and R.W.G. Hunt. Testing colour appearance models using corresponding-colour and magnitude-estimation data sets. *Color Research and Application*, 23:147–153, 1998.

[LK88] R.D. Luce and C.L. Krumhansl. Measurement, scaling, and psychophysics. In Atkinson et al. [AHLL88], pages 3–74.

[LLH00] C.J. Li, M.R. Luo, and W. Hunt. A revision of the ciecam97s model. *Color Research and Application*, 25:260–266, 2000.

[LM88] E.P. Lyvers and O.R. Mitchell. Precision edge contrast and orientation estimation. *IEEE Transactions on Pattern Analysis and Machine Intelligence*, 10:927–937, November 1988.

[LMAR89] E.P. Lyvers, O.R. Mitchell, M.L. Akey, and A.P. Reeves. Subpixel measurements using a moment-based edge operator. *IEEE Transactions on Pattern Analysis and Machine Intelligence*, 11:1293–1309, December 1989.

[LNS93] M.A. Losada, R. Navarro, and J. Santamaria. Relative contributions of optical and neural limitations to human contrast sensitivity at different luminance levels. *Vision Research*, 33:2321–2336, 1993.

[Lub93] J. Lubin. The use of psychophysical data and models in the analysis of display system performance. In A.B. Watson, editor, *Digital Images and Human Vision*, pages 162–178. MIT Press, Cambridge, Mass., 1993.

[Lub95] J. Lubin. A visual discrimination model for imaging system design and evaluation. In E. Peli, editor, *Visual Models for Target Detection and Recognition*. World Scientific Publishers, River Edge, N.J., 1995.

[LWV92] S.P. Lipshitz, R.A. Wannamaker, and J. Vanderkooy. Quantization and dither: a theoretical survey. *Journal of the Audio Engineering Society*, 40:355–375, 1992.

[Mal99] L.T. Maloney. Physics-based approaches to modeling surface color perception. In Gegenfurtner and Sharpe [GS99], pages 387–416.

[Mar72] L.E. de Marsh. Optimum telecine transfer characteristics. *Journal of the SMPTE*, 10:784–787, 1972.

[Mar82] D. Marr. *Vision*. W.H. Freeman and Company, San Francisco, 1982.

[Mar90a] J.-B. Martens. Deblurring digital images by means of polynomial transforms. *Computer Vision, Graphics and Image Processing*, pages 157–176, 1990.

[Mar90b] J.-B. Martens. The hermite transform – applications. *IEEE Transactions on Acoustics, Speech, and Signal Processing*, 38:1607–1618, 1990.

[Mar90c] J.-B. Martens. The hermite transform – theory. *IEEE Transactions on Acoustics, Speech, and Signal Processing*, 38:1595–1606, 1990.

[Mar94] J.-B. Martens. Noise reduction and enhancement by means of adaptive residue-image processing. In P. Delogne, editor, *Proceedings of the first IEEE International Conference on Image Processing*, pages 467–471, Austin, Texas, November 13–16 1994.

[Mar95] J.-B. Martens. Adaptive contrast enhancement through residue-image processing. *Signal Processing*, 44:1–18, 1995.

[Mar97] J.-B. Martens. Local orientation analysis in images by means of the hermite transform. *IEEE Transactions on Image Processing*, 6:1103–1116, August 1997.

[Mar02] J.-B. Martens. Multidimensional modeling of image quality. *Proceedings of the IEEE*, 90:133–153, 2002.

[MB01] J.-B. Martens and M. Boschman. The psychophysical measurement of image quality. In C.J. van den Branden Lambrecht, editor, *Vision Models and Applications to Image and Video Processing*, pages 71–99. Kluwer Academic Publishers, 2001.

[McC80] P. McCullagh. Regresssion models for ordinal data. *Journal of the Royal Statistical Society*, 42:109–142, 1980.

[Mee73] A. van Meeteren. *Visual Aspects of Image Intensification.* PhD thesis, University of Utrecht, 1973.

[Mee02] L. Meesters. *Predicted and Perceived Quality of Bit-Reduced Gray-Scale Still Images.* PhD thesis, Eindhoven University of Technology, 2002.

[MK96] J.-B. Martens and V. Kayargadde. Image quality prediction in a multidimensional space. In P. Delogne, editor, *Proceedings of the IEEE International Conference on Image Processing*, volume I of III, pages 877–880, Lausanne, Switzerland, September 16–19 1996.

[MM89] J.-B. Martens and G.M.M. Majoor. The perceptual relevance of scale-space image coding. *Signal Processing*, 17:353–364, 1989.

[MM98] J.-B. Martens and L. Meesters. Image dissimilarity. *Signal Processing*, 70:155–176, 1998.

[MM01] J.-B. Martens and L. Meesters. Single-ended intrumental measurement of image quality. In C.J. van den Branden Lambrecht, editor, *Vision Models and Applications to Image and Video Processing*, pages 101–127. Kluwer Academic Publishers, 2001.

[MM02] L. Meesters and J.-B. Martens. A sinle-ended blockiness measure for jpeg-coded images. *Signal Processing*, 82:369–387, 2002.

[MN75] H. Marmolin and S. Nyberg. Multidimensional scaling of subjective image quality. Technical Report FOA Report C 30039-H9, Swedish National Defence Research Institute, Stockholm, 1975.

[MNP77] F.W. Mounts, A.N. Netravali, and B. Prasada. Design of quantizers for real-time hadamard-transform coding of pictures. *Bell System Technical Journal*, 56:21–48, 1977.

[Mul85] K.T. Mullen. The contrast sensitivity of human colour vision to red-green and blue-yellow chromatic gratings. *Journal of Physiology*, 359:381–400, 1985.

[Mul90] K.T Mullen. The chromatic coding of space. In C. Blakemore, editor, *Vision: Coding and Efficiency*, pages 150–158. Cambridge University Press, 1990.

[MW92] D.H. Marimont and B. Wandell. Linear models of surface and illuminant spectra. *Journal of the Optical Society of America A*, 9:1905–1913, 1992.

[MW94] D.H. Marimont and B. Wandell. Matching color images: the effects of axial chromatic aberration. *Journal of the Optical Society of America A*, 11:3113–3122, 1994.

[Nad00] M. Nadenau. *Integration of human color vision models into high quality image compression.* PhD thesis, Ecole Polytechnique Fédérale de Lausanne, 2000.

[NB96] M.R.M. Nijenhuis and F.J.J. Blommaert. A perceptual error measure for sampled and interpolated complex colour images. *Displays*, 17:27–36, 1996.

[NB97a] M.R.M. Nijenhuis and F.J.J. Blommaert. Perceptual-error measure and its application to sampled and interpolated single-edged images. *Journal of the Optical Society of America A*, 14:2111–2127, 1997.

[NB97b] M.R.M. Nijenhuis and F.J.J. Blommaert. Perceptual error measure for sampled and interpolated images. *Journal of Imaging Science and Technology*, 41:249–258, 1997.

[Nes68] F.L. van Nes. *Experimental Studies in Spatiotemporal Contrast Transfer by the Human Eye.* PhD thesis, University of Utrecht, 1968.

[NH88] A. Netravali and B. Haskell. *Digital Pictures: Representation and Compression.* Plenum Press, New York, 1988.

[NHT+97] M. Nijenhuis, R. Hamberg, C. Teunissen, S. Bech, H. Looren de Jong, P. Houben, and S.K. Pramanik. Sharpness, sharpness related attributes, and their physical correlates. In *Very High Resolution and Quality Imaging II*, volume 3025, pages 173–184, San Jose, California, February 10–11 1997. Proceedings of the SPIE.

[Nij93] M.R.M. Nijenhuis. *Sampling and interpolation of static images: a perceptual view.* PhD thesis, Eindhoven University of Technology, 1993.

[NNJ43] S.M. Newhall, D. Nickerson, and D.B. Judd. Final report of the o.s.a. subcommittee on the spacing of the munsell colors. *Journal of the Optical Society of America*, 33:385–418, 1943.

[Nov69] S.B. Novick. The effects of viewing conditions on required tone characteristics in colour photography. *British kinematography sound and television*, 7:268–275, 1969.

[OC90] J. Oackley and M. Cunningham. A function space model for digital image sampling and its application in image reconstruction. *Computer Vision, Graphics and Image Processing*, 49:171–197, 1990.

[OCL90] J. Oackley, M. Cunningham, and G. Little. A fourier-domain formula for the least-squares projection of a function onto a repetitive basis in n-dimensional space. *IEEE Transactions on Acoustics, Speech and Signal Processing*, 38:114–120, 1990.

[Ove96] I.M.C.J. van Overveld. Contrast, noise, and blur affect performance and appreciation of digital radiographs. *Jpurnal of Digital Imaging*, 8:168–179, 1996.

[PC89] D. Post and C. Calhoun. An evaluation of methods for producing desired colors on crt monitors. *Color Research and Application*, 14:172–186, 1989.

[Pel90] E. Peli. Contrast in complex images. *Journal of the Optical Society of America A*, 7:2032–2040, 1990.

[Piz81] S.M. Pizer. Intensity mappings to linearize display devices. *Computer Graphics and Image Processing*, 17:262–268, 1981.

[PM93] W.B. Pennebaker and J.L. Mitchell. *JPEG still image compression standard.* Van Nostrand Reinhold, New York, 1993.

[Poy93] C. Poynton. Gamma and its disguises: the nonlinear mappings of intensity in perception, crts, film, and video. *SMPTE Journal*, 102:1099–1108, 1993.

[Pra91] W. Pratt. *Digital Image Processing.* John Wiley & Sons, New York, 1991.

[Pub91] Lexicon Publications. *The New Lexicon Webster's Dictionary of the English Language.* Lexicon Publications, Inc., New York, 1991.

[PV00] K.N. Plataniotis and A.N. Venetsanopoulos. *Color Image Processing and Applications.* Springer Verlag, Berlin, 2000.

[PW86] A. Parducci and D.H. Wedell. The category effect with rating scales: Number of categories, number of stimuli, and method of presentation. *Journal of Experimental Psychology: Human Perception and Performance*, 12:496–516, 1986.

[PW93] A.B. Poirson and B.A. Wandell. Appearance of colored patterns: Pattern-color separability. *Journal of the Optical Society of America A*, 10:2458–2470, 1993.

[PW96] A.B. Poirson and B.A. Wandell. Pattern-color separable pathways predict sensitivity to simple colored patterns. *Vision Research*, 36:515–526, 1996.

[PYGR91] E. Peli, J. Yang, R. Goldstein, and A. Reeves. Effect of luminance on suprathreshold contrast perception. *Journal of the Optical Society of America A*, 8:1352–1358, 1991.

[RA88] T.S. Rao and G.R. Arce. Halftone patterns for arbitrary screen periodicities. *Journal of the Optical Society of America A*, 5:1502–1511, 1988.

[RAA90] T.S. Rao, G.R. Arce, and J. Allebach. Analysis of ordered dither for arbitrary sampling lattices and screen periodicities. *IEEE Transactions on Acoustics, Speech and Signal Processing*, pages 1981–2000, 1990.

[Ram77a] J.O. Ramsay. Maximum likelihood estimation in multidimensional scaling. *Psychometrika*, 42:241–266, 1977.

[Ram77b] J.O. Ramsay. Monotonic weighted power transformations to additivity. *Psychometrika*, 42:83–109, 1977.

[Ram78] J.O. Ramsay. Confidence regions for multidimensional scaling analysis. *Psychometrika*, 43:145–160, 1978.

[Ram80] J.O. Ramsay. The joint analysis of direct ratings, pairwise preferences, and dissimilarities. *Psychometrika*, 45:149–165, 1980.

[RB81] J.A.J. Roufs and F.J.J. Blommaert. Temporal impulse and step responses of the human eye obtained psychophysically by means of a drift-correcting perturbation technique. *Vision Research*, 21:1203–1221, 1981.

[RBF97] H. de Ridder, F.J.J. Blommaert, and E.A. Fedorovskaya. Naturalness and image quality: chroma and hue variation in color images of natural scenes. In R. Eschbach and K. Braun, editors, *Recent Progress in Color Science*, pages 170–176. Springfield, 1997.

[RBR91] J.A.J. Roufs, F.F.J. Blommaert, and H. de Ridder. Brightness-luminance relations: future developments in the light of the past. In *CIE Devision 1: Vision and Colour. Minutes of the 7th Meeting*, volume annex 1, pages 1–14, Melbourne, Australia, July 2–11 1991.

[Rid92] H. de Ridder. Minkowski metrics as a combination rule for digital image coding impairments. In B.E. Rogowitz, editor, *Human Vision, Visual Processing and Digital Display III*, volume 1666, pages 16–26, San Jose, California, February 10–13 1992. Proceedings of the SPIE.

[Rid96a] H. de Ridder. Current issues and new techniques in visual quality assessment. In P. Delogne, editor, *Proceedings of the IEEE International Conference on Image Processing*, volume I of III, pages 869–872, Lausanne, Switzerland, September 16–19 1996.

[Rid96b] H. de Ridder. Naturalness and image quality: saturation and lightness variation in color images of natural scenes. *Journal of Imaging Science and Technology*, 40:170–176, 1996.

[Rid97] H. de Ridder. Naturalness and image quality: influence of chroma variation at various lightness levels. In R. Eschbach and K. Braun, editors, *Recent Progress in Color Science*, pages 101–103. Springfield, 1997.

[RKT94] J.A.J. Roufs, V.J.F. Koselka, and A.A.A.M. van Tongeren. Global brightness contrast and the effect on perceptual image quality. In B.E. Rogowitz and J.P. Allebach, editors, *Human Vision, Visual Processing, and Digital Display V*, volume 2179, pages 80–89, San Jose, California, February 8–10 1994. Proceedings of the SPIE.

[RKT97] J.A.J. Roufs, V.J.F. Koselka, and A.A.A.M. van Tongeren. Global brightness contrast and the effet on perceptual image quality. In R. Eschbach and K. Braun, editors, *Recent Progress in Color Science*, pages 233–240. Springfield, 1997.

[RM90] H. de Ridder and G.M.M. Majoor. Numerical category scaling: an efficient method for assessing digital image coding impairments. In B.E. Rogowitz and J.P. Allebach, editors, *Human Vision and Electronic Imaging: Models, Methods and Applications*, volume 1249, pages 65–77, Santa Clara, California, February 12–14 1990. Proceedings of the SPIE.

[Rob66] J. Robson. Spatial and temporal contrast-sensitivity functions of the visual system. *Journal of the Optical Society of America*, 56:1141–1142, 1966.

[Rou92] J.A.J. Roufs. Perceptual image quality: concept and measurement. *Philips Journal of Research*, 47:35–62, 1992.

[RtBS84] J.O. Ramsay, J. ten Berge, and G.P.H. Styan. Matrix correlation. *Psychometrika*, 49:403–423, 1984.

[SACM97] M. Stokes, M. Anderson, S. Chandrasekar, and R. Motta. A standard default color space for the internet srgb. Technical report, International Color Consortium, http://www.color.org, 1997.

[SB55] W.S. Stiles and J.M. Burch. Interim report to the cie zurich, 1955, on the national physical laboratory's investigation of colour matching (1955) with an appendix by stiles, w.s. and burch, j.m. *Optica Acta*, 2:168–181, 1955.

[SB59] W.S. Stiles and J.M. Burch. Npl colour-matching investigation: final report (1958). *Optica Acta*, 6:1–26, 1959.

[SCB88] M.C. Stone, W.B. Cowan, and J.C. Beatty. Color gamut mapping and the printing of digital color images. *ACM Transactions on Graphics*, 7:249–292, 1988.

[SCB98] D.F. Swayne, D. Cook, and A. Buja. Xgobi: Interactive dynamic data visualization in the x window system. *Journal of Computational and Graphical Statistics*, 7:113–130, 1998.

[Sch69] I.J. Schoenberg. Cardinal interpolation and spline functions. *Journal of Approximation Theory*, 2:167–206, 1969.

[Sch70] Sch onemann, P.H. and Carroll, R.M. Fitting one matrix to another under choice of a central dilation and a rigid motion. *Psychometrika*, 35:245–255, 1970.

[SE90] C.P. Sandbank (Ed.). *Digital Television*. John Wiley & Sons, Chichester, 1990.

[SF96] D.A. Silverstein and J.E. Farell. The relationship between image fidelity and image quality. In *IEEE International Conference on Image Processing*, volume I of III, pages 881–884, Lausanne, Switzerland, September 16–19 1996.

[She62] R.N. Shepard. The analysis of proximities: Multidimensional scaling with an unknown distance function. *Psychometrika*, 27:219–246, 1962.

[She74] R.N. Shepard. Representation of structure in similarity data: Problems and prospects. *Psychometrika*, 39:373–424, 1974.

[SO91] A. Stuart and J.K. Ord. *Kendall's Advanced Theory of Statistics: Volume 2 - Classical Inference and Relationship*. Edward Arnold, London, 5 edition, 1991.

[SP72] V.C. Smith and J. Pokorny. Spectral sensitivity of color-blind observers and the cone photopigments. *Vision Research*, 12:2059–2071, 1972.

[SP75] V.C. Smith and J. Pokorny. Spectral sensitivity of the foveal cone photopigments between 400 and 500 nm. *Vision Research*, 15:161–171, 1975.

[Spe59] N.I. Speranskaya. Determination of spectrum color co-ordinates for twenty-seven normal observers. *Optics and Spectroscopy*, 7:424–428, 1959.

[SS99] A. Stockman and L.T. Sharpe. Cone spectral sensitivities and color matching. In Gegenfurtner and Sharpe [GS99], pages 53–87.

[ST85] W. Schreiber and D. Troxel. Transformation between continuous and discrete representations of images: a perceptual approach. *IEEE Transactions on Pattern Analysis and Machine Intelligence*, 7:178–186, 1985.

[Sta90] P. Stalmeier. *On the measurement of large color differences.* PhD thesis, University of Nijmegen, 1990.

[Ste57] S.S. Stevens. On the psychophysical law. *Psychological Review*, 54:153–181, 1957.

[Sze59] Szegö. *Orthogonal Polynomials.* American Mathematical Society, 1959.

[Teu96] C. Teunissen. The validity of ccir quality indicators along a graphical scale. *SMPTE Journal*, 105:144–149, 1996.

[TH94] P. Teo and D. Heeger. Perceptual image distortion. In *IEEE International Conference on Image Processing*, pages 982–984, Austin, Texas, November 1994.

[THBL99] Xin Tong, D. Heeger, and C. van den Branden Lambrecht. Video quality evaluation using st-cielab. In B.E. Rogowitz and T.N Pappas, editors, *Human Vision and Electronic Imaging IV*, volume 3644, pages 185–196, San Jose, California, January 25–28 1999. Proceedings of the SPIE.

[THM97] C. Teunissen, R. Hamberg, and J.-B. Martens. The optimal glass transmission coefficient as a function of the level of diffuse ambient illumination. *Displays*, 17:94–99, 1997.

[Tho00] K. Thomsen. A euclidean color space in high agreement with the cie94 color difference formula. *Color Research and Application*, 25:64–65, 2000.

[Thu27] L.L. Thurstone. A law of comparative judgement. *Psychological Review*, 34:273–286, 1927.

[Tor58] W.S. Torgerson. *Theory and Methods of Scaling.* John Wiley and Sons Inc.: New York, 1958.

[Tre68] H.L. van Trees. *Detection, Estimation an Modulation Theory.* John Wiley and Sons Inc., New York, 1968.

[TW96] C. Teunissen and J.H.M.D. Westerink. A multidimensional evaluation of the perceptual quality of television sets. *SMPTE Journal*, 105:31–38, 1996.

[UAE91] M. Unser, A. Aldroubi, and M. Eden. Fast b-spline transforms for continuous image representation and interpolation. *IEEE Transactions on Pattern Analysis and Machine Intelligence*, 13:277–285, 1991.

[UAE92] M. Unser, A. Aldroubi, and M. Eden. Polynomial spline signal approximations: filter design and asymptotic equivalence with shannon's sampling theorem. *IEEE Transactions on Information Theory*, 38:95–103, 1992.

[UAE93] M. Unser, A. Aldroubi, and M. Eden. B-spline signal processing. *IEEE Transactions on Signal Processing*, 41:821–848, 1993.

[Uli88] R. Ulichney. Dithering with blue noise. *Proceedings of the IEEE*, pages 56–79, 1988.

[Vos78] J.J. Vos. Colorimetric and photometric properties of a 2 degree fundamental observer. *Color Research and Applications*, 3:125–125, 1978.

[VW71] J. Vos and P.L. Walraven. On the derivations of the foveal receptor primaries. *Vision Research*, 11:799–818, 1971.

[Wag75] W.A. Wagenaar. Stevens versus fechner: A plea for dismissal of the case. *Acta Psychologica*, 39:225–235, 1975.

[WAJF86] A.B. Watson, A.J. Ahumada Jr., and J.E. Farrell. Window of visibility: a psychophysical theory of fidelity in time-sampled visual motion displays. *Journal of the Optical Society of America A*, 3:300–307, 1986.

[Wan87] B. Wandell. The synthesis and analysis of color images. *IEEE Transactions on Pattern Analysis and Machine Intelligence*, 9:2–13, 1987.

[Wan95] B.A. Wandell. *Foundations of Vision.* Sinauer Associates, Inc., Sunderland, Massachusetts, 1995.

[WB01] Z. Wang and A.C. Bovik. A universal image quality index. *IEEE Signal Processing Letters*, 9:81–84, 2001.

[WBMN94] D.R. Williams, D.H. Brainard, M.J. McMahon, and R. Navarro. Double-pass and interferometric measures of the optical quality of the eye. *Journal of the Optical Society of America A*, 11:3123–3135, 1994.

[Wea61] R.A. Weale. Notes on the photometric significance of the human crystalline lens. *Vision Research*, 1:183–191, 1961.

[Wes91] J.H.D.M. Westerink. *Perceived sharpness in static and moving images.* PhD thesis, Eindhoven University of Technology, 1991.

[Whi92] P. Whittle. Brightness, discriminability and the crispening effect. *Vision Research*, 32:1493–1507, 1992.

[Wic98] T.D. Wickens. Tutorial on modeling ordered categorical response data. *Annual Review of Psychology*, 48:537–558, 1998.

[Wil80] H.R. Wilson. A transducer function for threshold and suprathreshold vision. *Biological Cybernetics*, 38:171–178, 1980.

[WM83] R.J. Watt and M.J. Morgan. The recognition and representation of edge blur: evidence for spatial primitives in human vision. *Vision Research*, 23:1465–1477, 1983.

[WP92] H. Widdel and D.L. Post. *Color in Electronic Displays.* Plenum Press, New York, 1992.

[Wri28] W.D. Wright. A re-determination of the trichromatic coefficients of the spectral colours. *Transactions of the Optical Society*, 30:141–164, 1928.

[WS82] G. Wyszecki and W. Stiles. *Color Science: Concepts and Methods, Quantitative Data and Formulas.* John Wiley & Sons, New York, 1982.

[XI97] ACR/NEMA Working Group XI. Gray-scale display function standard. Technical report, NEMA Standards Publication, 1997.

[YBR98a] S.N. Yendrikhovskij, F.J.J. Blommaert, and H. de Ridder. Optimizing color reproduction of natural scenes. In *Final Program and Proceedings of IS&T/SID, the 6th Color Imaging Conference: Color Science, Systems, and Applications*, pages 140–145, Scottsdale, Arizona, November 17–20 1998.

[YBR98b] S.N. Yendrikhovskij, F.J.J. Blommaert, and H. de Ridder. Perceptually optimal colour reproduction. In B.E. Rogowitz and T.N. Pappas, editors, *Human Vision and Electronic Imaging III*, volume 3299, pages 274–281, San Jose, California, January 26–29 1998. Proceedings of the SPIE.

[YBR98c] S.N. Yendrikhovskij, F.J.J. Blommaert, and H. de Ridder. Towards perceptually optimal colour reproduction of natural scenes. In *Proceedings of the Conference on Colour Imaging in Multimedia, CIM-98*, pages 237–245, Derby, UK, March 26–27 1998.

[YBR99a] S.N. Yendrikhovskij, F.J.J. Blommaert, and H. de Ridder. Color reproduction and the naturalness constraint. *Color Research and Application*, 24:52–67, 1999.

[YBR99b] S.N. Yendrikhovskij, F.J.J. Blommaert, and H. de Ridder. Representation of memory prototype for an object color. *Color Research and Application*, 24:393–410, 1999.

[Yen98] S.N. Yendrikhovskij. *Color reproduction and the naturalness constraint.* PhD thesis, Eindhoven University of Technology, 1998.

[YH87] F.W. Young and R.M. Hamer. *Multidimensional Scaling: History, Theory and Applications.* Erlbaum, New York, 1987.

[You84] F.W. Young. Scaling. *Annual Review of Psychology*, 35:55–81, 1984.

[You85] R. Young. The gaussian derivative theory of spatial vision: Analysis of cortical cell receptive field line-weighting profiles. Technical Report Report GMR-4920, General Motors Research Labs, 1985.

[You88] N. Young. *Cambridge Mathematical Textbooks.* Cambridge University Press, Cambridge, Massachusetts, 1988.

[YRFB97] S.N. Yendrikhovskij, H. de Ridder, E.A. Fedorovskaya, and F.J.J. Blommaert. Colourfulness judgments of natural scenes. *Acta Psychologica*, 97:79–94, 1997.

[YW98] M. Yuen and H.R. Wu. A survey of hybrid mc/dpcm/dct video coding distortions. *Signal Processing*, 70:247–278, 1998.

[ZH89] C. Zetzsche and G. Hauske. Multiple channel model for the prediction of subjective image quality. In B.E. Rogowitz, editor, *Human Vision, Visual Processing, and Digital Display*, volume 1077, pages 209–216, Los Angeles, California, January 18–20 1989. Proceedings of the SPIE.

[ZW98] X. Zhang and B.A. Wandell. Color image fidelity metrics evaluated using image distortion maps. *Signal Processing*, 70:201–214, 1998.

Index